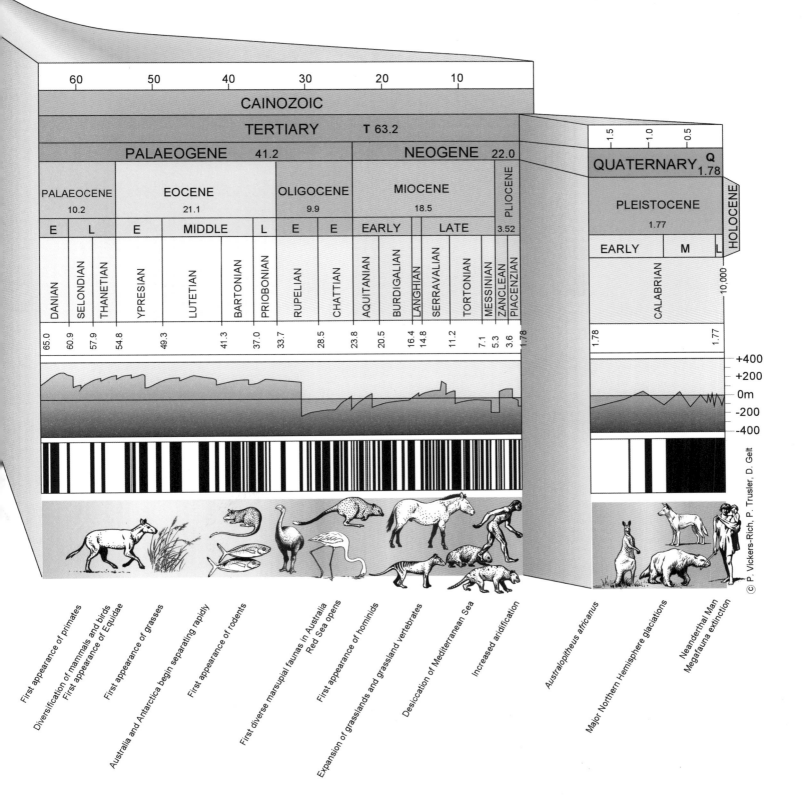

60	50	40	30	20	10

CAINOZOIC

TERTIARY T 63.2

| PALAEOGENE 41.2 | NEOGENE 22.0 |

| PALAEOCENE 10.2 | EOCENE 21.1 | OLIGOCENE 9.9 | MIOCENE 18.5 | PLIOCENE |

| E | L | E | MIDDLE | L | E | E | EARLY | LATE | 3.52 |

| DANIAN | SELONDIAN | THANETIAN | YPRESIAN | LUTETIAN | BARTONIAN | PRIOBONIAN | RUPELIAN | CHATTIAN | AQUITANIAN | BURDIGALIAN | LANGHIAN | SERRAVALIAN | TORTONIAN | MESSINIAN | ZANCLEAN | PIACENZIAN |

| 65.0 | 60.9 | 57.9 | 54.8 | 49.3 | 41.3 | 37.0 | 33.7 | 28.5 | 23.8 | 20.5 | 16.4 | 14.8 | 11.2 | 7.1 | 5.3 | 3.6 | 1.78 |

| -1.5 | -1.0 | -0.5 |

QUATERNARY Q 1.78

PLEISTOCENE 1.77

HOLOCENE

| EARLY | M | L |

| CALABRIAN |

| 1.78 | 1.77 | 10,000 |

+400
+200
0m
-200
-400

© P. Vickers-Rich, P. Trusler, D. Gelt

First appearance of primates
Diversification of mammals and birds
First appearance of Equidae
First appearance of grasses
Australia and Antarctica begin separating rapidly
First appearance of rodents
First diverse marsupial faunas in Australia
Red Sea opens
First appearance of hominids
Expansion of grasslands and grassland vertebrates
Desiccation of Mediterranean Sea
Increased aridification

Australopitheus africanus
Major Northern Hemisphere glaciations
Neanderthal Man
Megafauna extinction

WILDLIFE
OF
GONDWANA

LIFE OF THE PAST
James O. Farlow, Editor

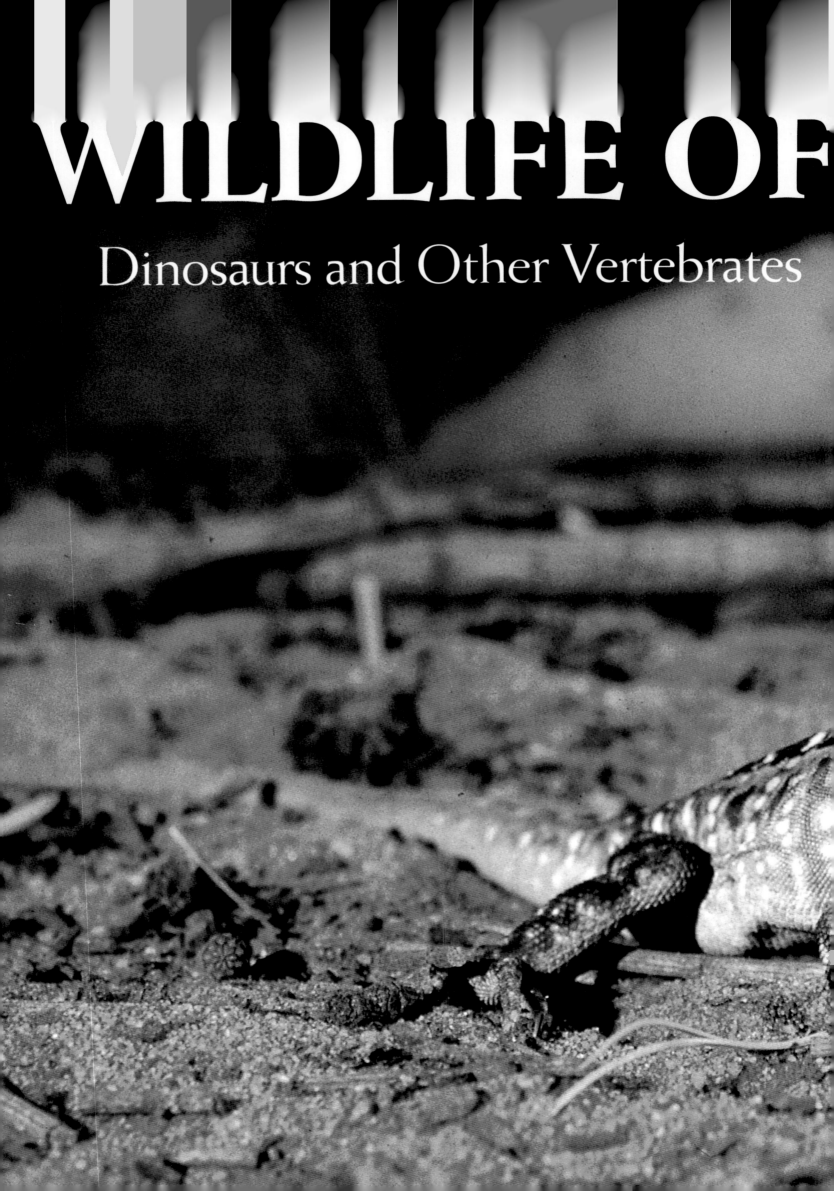

WILDLIFE OF

Dinosaurs and Other Vertebrates

GONDWANA

from the Ancient Supercontinent

PATRICIA VICKERS-RICH AND
THOMAS HEWITT RICH

PRINCIPAL PHOTOGRAPHY BY FRANCESCO COFFA AND STEVEN MORTON
RECONSTRUCTIONS BY PETER TRUSLER

INDIANA UNIVERSITY PRESS
BLOOMINGTON · INDIANAPOLIS

1 *Sulphur-crested Cockatoos, Cacatua galerita.
One of a great variety of parrot species that
characterize the Australian fauna of today. (F. Coffa)*

*TITLE PAGE:
Painted Dragon, Amphibolurus pictus, in the
Mallee, Sunset Country, north-western Victoria.
(F. Coffa)*

*OPPOSITE:
The fish Leptolepis talbragarensis and the plant
Agathis jurassica from the Jurassic Talbragar Fish
Beds, New South Wales. (F. Coffa, courtesy of the
Australian Museum)*

*OVERLEAF:
Pygmy Possum, Cercartetus lepidus, the smallest
of all the possums, having a total length of about 110
to 140 millimetres and weighing between 6 and 9
grams. A nocturnal marsupial, it occupies a wide range
of tree-covered environments, from the Victorian
Mallee to the wet sclerophyll forests of Tasmania. It
feeds primarily on insects. (F. Coffa)*

*CONTENTS PAGES 8-9:
Anterolateral view of the skull and lower jaws of the
dipnoan Chirodipterus australis from the Early
Devonian Gogo Formation of Paddy's Springs, Gogo
Station, Western Australia. (F. Coffa, courtesy of the
Australian National University and K. Campbell)*

*CONTENTS PAGES 10-11:
Red Kangaroo, Macropus rufus, at Lake Mungo,
western New South Wales. A kangaroo of the arid
plains of Australia, the Red is well suited to its
environment: it has a relatively low oxygen
metabolism, development of young in the pouch,
and the ability to retain a fertilized egg in diapause
as well as to limit water loss in urine, faeces and
through evaporation. (F. Coffa)*

This book is a publication of

Indiana University Press
601 North Morton Street
Bloomington, Indiana
47404-3797 USA

www.indiana.edu/~iupress

Telephone orders 800-842-6796
Fax orders 812-855-7931
Orders by e-mail iuporder@indiana.edu

Library of Congress Cataloging-in-Publication Data

Rich, Pat Vickers.
    Wildlife of Gondwana : dinosaurs and other vertebrates from the ancient superconti-
   nent / Patricia Vickers-Rich and Thomas Hewitt Rich ; principal photography by Francesco
   Coffa and Steven Morton ; reconstructions by Peter Trusler.
      p.   cm. — (Life of the past)
   Includes bibliographical references and index.
   ISBN 0-253-33643-0 (cl : alk. paper)
   1. Vertebrates, Fossil—Southern Hemisphere. 2. Paleontology—Southern Hemisphere.
3. Gondwana (Geology) I. Rich, Thomas H. V. II. Title. III. Series.
QE841.R5   1999
566—dc21   99-36298

1 2 3 4 5 04 03 02 01 00 99

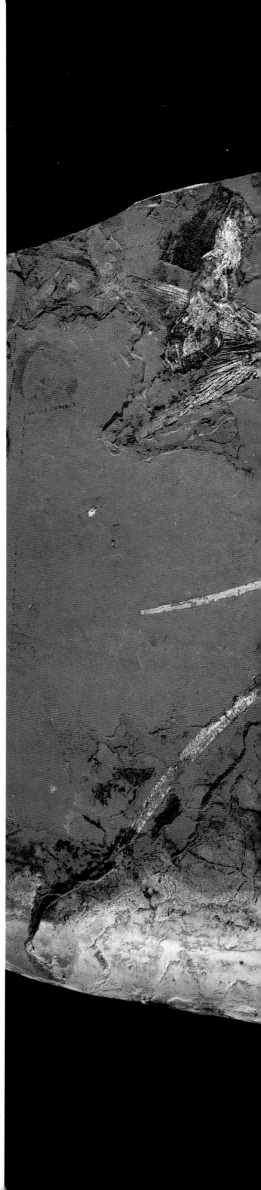

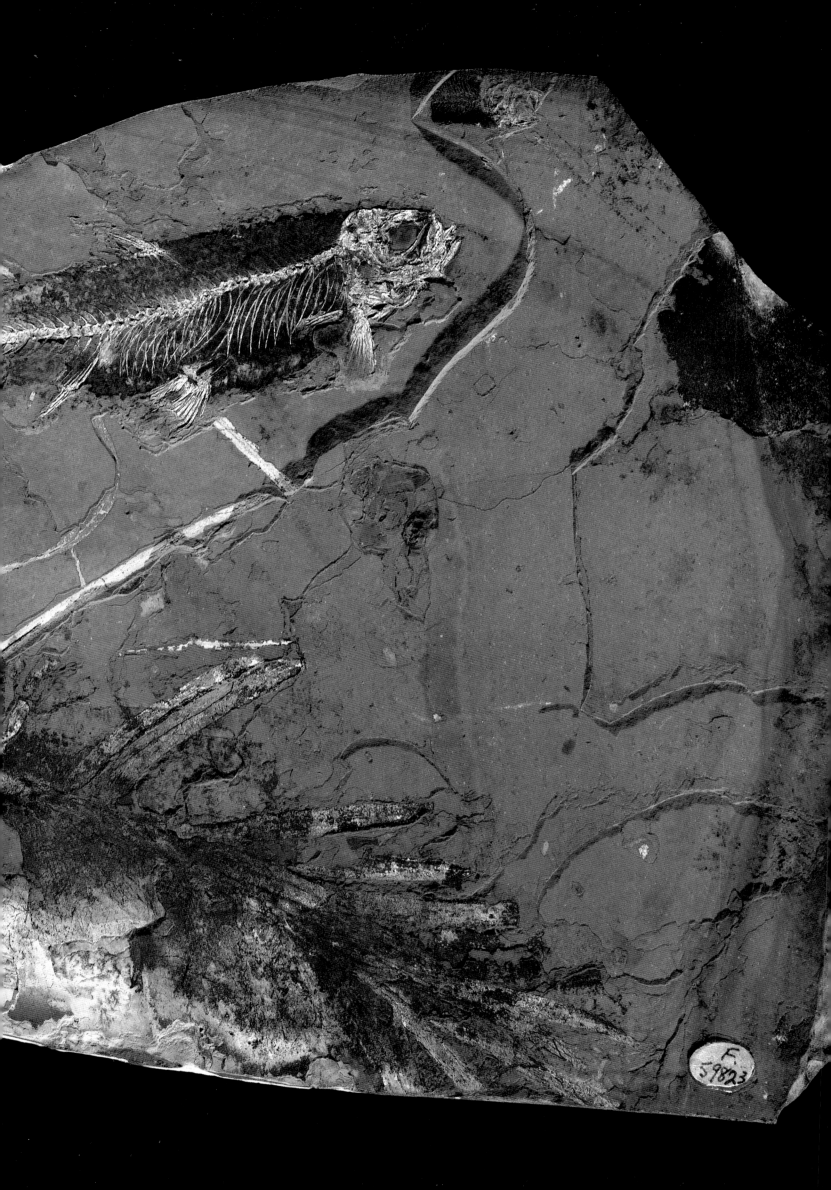

F.
59823

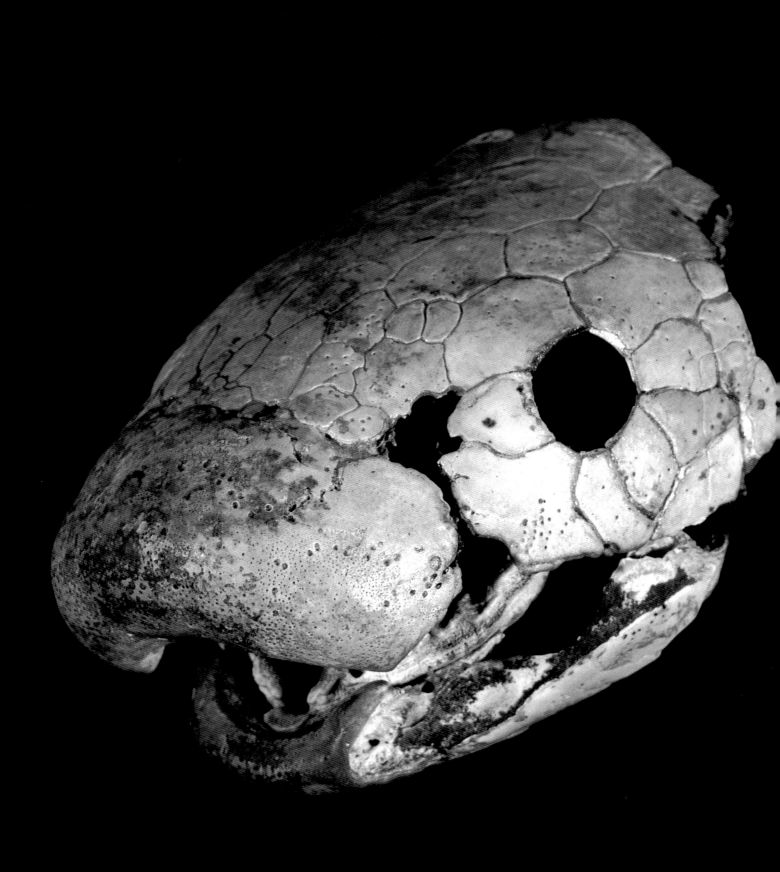

CONTENTS

INTRODUCTION   13

INTRODUCTION TO THE REVISED EDITION   14

PART ONE
GONDWANA IN PERSPECTIVE   15

Birth Of The Gondwana Concept — The Rocky Framework Of Gondwana — The Way The Earth Works — *Measuring Geological Time* — 500 Million Years Of Continents Adrift — The Origins And Pathways Of Life — Life In The Cryptozoic — Life Emerges: A Product Of Chance — The Path To Sexuality And Many Cells — Towards A New Beginning — *Putting Time In Perspective* — Life In The Phanerozoic — Beginnings Of A Backbone — The First Vertebrates — An Invertebrate Ancestry — The Vertebrate Pedigree: Bone And The Neural Crest — Conquest Of The Waters: A Palaeozoic Beginning — Invasion Of The Land — A Time Of Crisis: The Palaeozoic–Mesozoic Boundary — Diversity In A Warm World — From Greenhouse To Icehouse: The Making Of A Modern World — Development Of A Unique Australian Fauna — *Australia's Modern Flora*

PART TWO
THE SEARCH FOR BEGINNINGS   47

Digging Up Bones In The Antipodes — French And British Coastal Surveys — Exploration Of The Interior — "Foreign Experts"— Mineral Wealth: Catalyst For Independence — The Earliest Museums — *"The Raid On The Museum"* — Universitites: Beginnings Of Home-grown Training — Scientific Societies: A Forum For Local Debate — Pioneers In Australian Vertebrate Palaeon-tology — An Independent Australian Vertebrate Palaeontology — Do Fossils Lie? Bias In The Fossil Record — Topography And Vegetation — Environments Of Deposition — Shape And Quality Of Bones — The Biology And Behaviour Of Extinct Vertebrates — Philosophies And Techniques Of Palaeontologists

PART THREE
THE FOSSIL VERTEBRATES OF AUSTRALIA   61

Chapter 1
AND THEN THERE WERE BONES   62
THE ORDOVICIAN TO SILURIAN PERIODS, FROM 490 TO 410 MILLION YEARS AGO
Mineralized Tissues And Biotic Diversity — Gondwana's Oldest Vertebrates — Life Without Jaws

Chapter 2
AN AGE OF FISHES   68
THE DEVONIAN PERIOD, FROM 410 TO 354 MILLION YEARS AGO
Jaws And Paired Fins — *Australia's Oldest Jawed Fishes* — An Explosion Of Variety: The Early Devonian Taemas–Wee Jasper And Buchan Faunas — Placoderms, The Great Experimenters — The Flexible Arthrodires — The Armoured Antiarchs — The Petalichthyids, Haunters Of The Ocean Bottom — Other Placoderm Types — Osteichthyes: A Hint Of Things To Come — Ray-fins, Ancestors Of Modern Fishes — Lungfish, A Group Apart? — Chondrichthyans, Fish Without Bones — *Jawless Survivors: The Resilient Agnaths* — An Exquisite Trove: Middle And Late Devonian Faunas Of Australia — The Gogo Reef Faunas — Ancestors Of The Tetrapods, Crossopterygians Or Dipnoans? — The Mount Howitt And Canowindra Faunas — Microscopic Remains Of Devonian Vertebrates — Global Faunal Similarities In The Devonian — Changes To Come

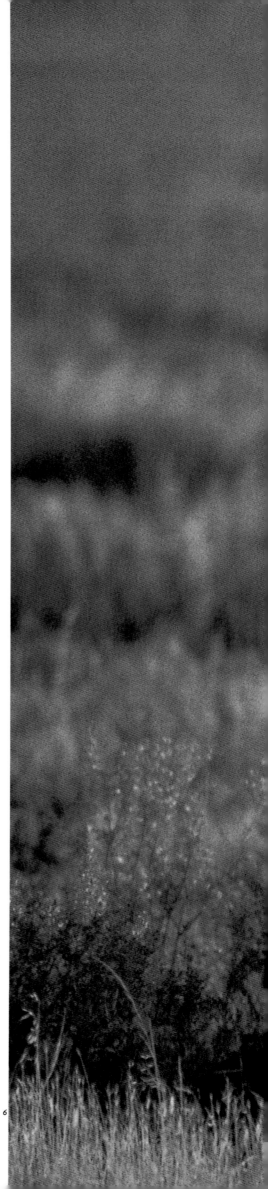

Chapter 3
ATTACK ON THE LAND **99**
THE DEVONIAN TO TRIASSIC PERIODS, FROM 410 TO 205 MILLION YEARS AGO
On Becoming Terrestrial — The Coming Of The Amphibians — The Labyrinthodonts — The Silverband Trackway Of The Early Devonian — The Late Devonian Genoa River Prints — Australia's Oldest Terrestrial Bone — A Rare Permian Find — *Extinctions In The Late Permian* — *Permo–Carboniferous Fishes In Australia* — The Triassic Heyday Of The Labyrinthodonts — Lifestyles Of Australian Triassic Labyrinthodonts — Australia's Elusive Triassic Reptiles — A Global Perspective Of Triassic Tetrapods — Major Extinctions At The Close Of The Triassic

Chapter 4
DINOSAURS AND POLAR NIGHTS **123**
THE EARLY JURASSIC TO LATE CRETACEOUS PERIODS, FROM 205 TO 65 MILLION YEARS AGO
Australia's Oldest Dinosaur — Jurassic Footprints, But Precious Few Skeletons — Late Jurassic And Early Cretaceous Vertebrates — The Talbragar And Koonwarra Lakes — Floodplain Faunas Of The Great Southern Rift Valley — An Unusual Environment For Dinosaurs — Opalized Denizens Of Australia's Great Inland Sea — Rare Remains Of Flying Vertebrates — *Muttaburrasaurus* And Other Northern Australian Dinosaurs — Late Cretaceous Vertebrates — The Winton Fauna — The Demise Of The Dinosaurs

Chapter 5
AN ARK TO THE TROPICS **149**
THE TERTIARY PERIOD TO THE PLEISTOCENE EPOCH OF THE QUATERNARY PERIOD,
FROM 65 MILLION TO 10, 000 YEARS AGO
Continents On The Move — Change In Australia's Centre — The Age Of Mammals In Gondwana — Marsupials Of The Americas And Australia — The Earliest Australian Mammalian Faunas, At Tingamarra And Geilston Bay — An Explosion Of Variety In The Middle Tertiary — Lakes And Streams Of Central Australia: The Lake Eyre And Tarkarooloo Basins — The Northern Limestone Country: Riversleigh And Bullock Creek — Other Middle Cainozoic Vertebrates With The Mammals — Alcoota, A Faunal Switch-point In The Late Miocene — Increasing Aridity In Post–Miocene Australia, From 5.2 Million Years Ago To The Present — Unpredictability In The Last 500,000 Years — Extinction Of The Quaternary Megafauna — Australian Quaternary Caves — Lake Callabonna — Dwarfing And Low Diversity On Islands — The Missing Predators — The Developing Psychrosphere, And Vertebrates At Sea — Marine Birds Of The Cainozoic — Towards Our Modern World

Chapter 6
THE LIVING VERTEBRATES OF AUSTRALIA **213**
THE QUATERNARY PERIOD, FROM 1.78 MILLION YEARS AGO TO THE PRESENT
The Uniqueness Of The Australian Biota — The Effects Of Drifting Continents And Changing Climates — Australia's Modern Biogeographic Provinces — Heat And Water In Australia Today — And What Of Man's Dalliance?

PART FOUR
GONDWANAN FAUNAS IN GLOBAL CONTEXT **227**

The Palaeozoic Record — The Mesozoic Record — The Cainozoic Record

AFTERWORD **253**
ACKNOWLEDGEMENTS **269**
GLOSSARY **271**
SYSTEMATIC, GEOGRAPHIC AND GEOLOGIC INDEX **277**
BIBLIOGRAPHY **285**
INDEX **297**

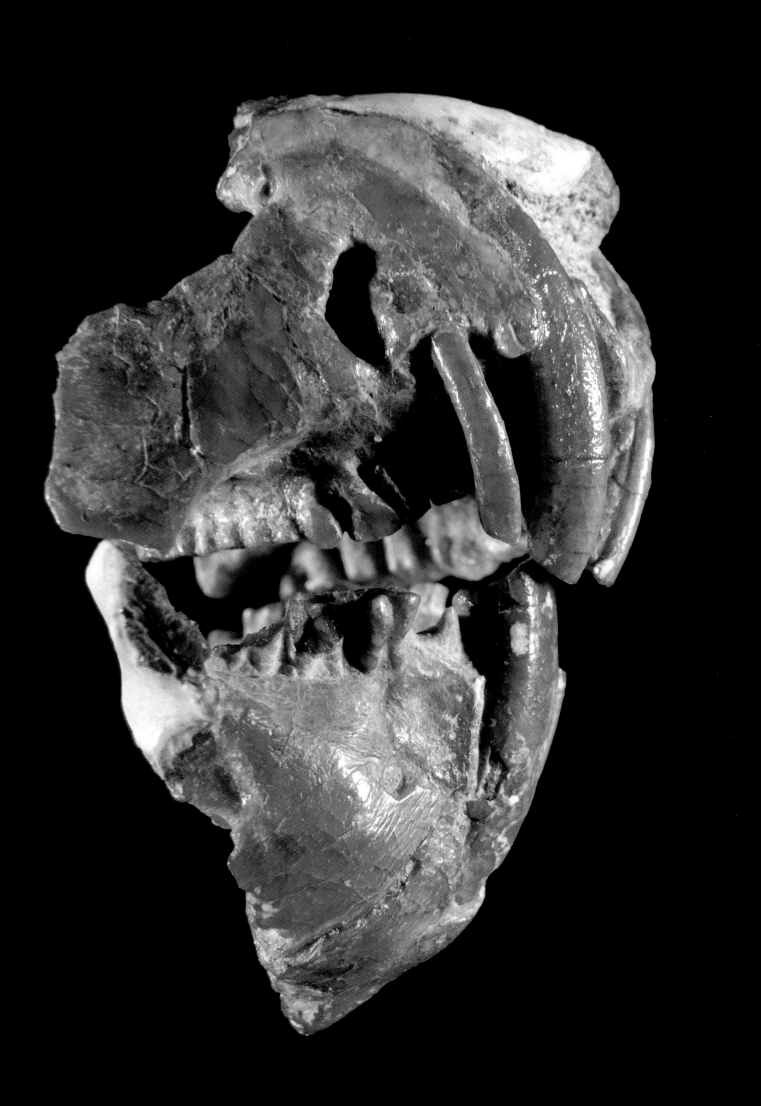

INTRODUCTION

Alice of Wonderland fame made a keen observation during her travels that things are not always as they seem. Smiles can exist without the whole Cheshire Cat. Alice's keen insight is also applicable when one steps back through the "looking glass" of time to view the world as it has been. The view beheld is often fragmental, just like the Cheshire Cat. We must interpret the past from bits and pieces — from the rocks and their magnetic properties, and from fossils and the like left behind, oft times destroyed during the passage of millions, even billions of years on our restless, dynamic planet Earth.

Fifty years ago, most geologists thought that the Earth's continents were immoveable across the face of the planet. There were some dissidents — Alfred Wegener, Alexander du Toit — but they were in the minority. As the magnetic properties and age of the Earth's rocks were better understood, however, more fragments of the past were pieced together. Both by extensive data gathering and by the birth of new theories, a completely novel interpretation of Earth history emerged. The outer few tens of kilometres of our planet appear to have been involved in a magnificent waltz, carrying with it continents and old ocean basins. New ocean basins formed as continents and older ocean basins split asunder. Volcanoes and earthquakes are but manifestations of this restlessness, and their global distribution has been neatly explained in the *Theory Of Plate Tectonics* proposed in the early 1960s.

Gondwana, the great southern landmass, was imagined because of the similarity of living organisms, rock sequences and fossils on now far distant continents. The very rocks that were to provide the name for this supercontinent were originally noted by geologists working for the Indian Geological Survey in 1872, when they described the repetitive sequences of sediments and volcanics bearing abundant plant fossils of late Palaeozoic to middle Mesozoic age in the Narbada Valley of Peninsular India. At the time of such descriptions, the idea of continents moving about had not been mooted, and certainly no mechanism to explain how such movement could take place was abroad. Alfred Wegener published *The Origin of Continents and Oceans* in 1912, and there suggested that several other now southern continents had similar sequences and fossils to those of India. Thus, the idea of a once great southern continent encompassing Africa, Antarctica, India, South America and Australia came into being. It has been only during the past three decades, however, that the details of how that supercontinent was born, when it existed and how it died have fully come to light.

Gondwana, or Gondwanaland as it was formerly known, has been a physical entity at two times in the past — once during the early to middle Palaeozoic, and again in the middle Mesozoic. Ironically, during the time that the Gondwanan sequence of rocks with its massive Jurassic dolerites were solidifying and Permian *Glossopteris* floras was being deposited, Gondwana was *not* physically isolated from the rest of the world. During this late Palaeozoic to middle Mesozoic time, the southern and northern continents were locked together forming a massive continent, Pangaea, which stretched from the North to the South poles. The regionalization of faunas on this megacontinent was purely the result of climatic variations, the Gondwanan biota inhabiting the southern temperate to polar areas.

Today the six major continents of the globe, Eurasia being considered as one, are separated by vast seas. At times in the past the lands were not nearly so fragmented, the modern arrangement being a product of only the past 100 million years or so. Before that, the relentless motion of the Earth's crust sometimes joined, sometimes separated, many partners; twice, such a union occurred for the great southern landmass of Gondwana. This book will follow the history of the vertebrate animals that inhabited this southern supercontinent, with special emphasis on Australia, in an attempt to understand how the evolutionary paths of these animals have been shaped both by the mobile Earth beneath their feet and by changing climate, itself related somehow to the forces that move the great crustal plates.

Patricia Vickers-Rich
January 1993

7 Partial skull and lower jaws of Groeberia cf. minopriori from the Early Oligocene Divisadero Largo Formation of Argentina. Relationships of this unique mammal are the centre of scientific debate. x 10 (F. Coffa, courtesy of the Museo De La Plata and R. Pascual)

INTRODUCTION TO THE REVISED EDITION

Since the publication of the original edition of *Wildlife of Gondwana* in 1993 there has been an acceleration of discoveries on the Gondwana continents—from Antarctica to Africa to Australia and New Zealand. The flamboyantly-crested *Cryolophosaurus* from the windblown slopes of Antarctica discovered by William Hammer; the tiny, cold-adapted hypsilophodonts from southern Australia; the embryos "en egg" of the mighty sauropod dinosaurs found in northern Patagonia—these are only a few of a vast array of new forms coming to light in the last decade. And the stories that many of these new finds are telling indicate that Gondwana was certainly the nursery of novelty—a place where innovation did take place and move out to occupy other, more northerly places. Many groups (perhaps placental mammals, ceratopsian, and ornithomimosaurian dinosaurs) have very old records on Gondwana, and, thus, it is now not so certain that such groups originated on the northern continents. In some cases, too, Gondwana served as a sanctuary for some groups (the allosaurid dinosaurs, the labyrinthodont amphibians) that became extinct elsewhere much earlier.

We have no doubt that with the increased activity of amateur and professional palaeontologists and with the founding and nurturing of more museums on the Gondwana continents, even more exciting discoveries await us in the near future. And discoveries on these continents are undoubtedly going to fundamentally change prevailing views on the course of vertebrate evolution geographically. This new edition sets the stage for such a future age of discovery on these fascinating southern lands.

We have taken the opportunity of this Indiana University Press edition to correct a number of typographical mistakes in the original edition, to revise and update the Systematic, Geographic and Geologic Index and the Bibliography, to correct or replace some of the original illustrations, and to add a new Afterword, which includes many new photographs.

Two brief additions to the original text could not be incorporated for reasons of space. At the end of page 45: "A new find in the Eocene of Queensland suggests that primitive placentals (condylarths?) may have been present in Australia in the early Cainozoic, but more material is need to firmly document this." On page 132, column 2, end of paragraph 5: "Recent discoveries of neocerotopsians (horned dinosaurs) and ornithomimosaurs (ostrich-mimic dinosaurs) further reinforce the idea that new major groups will continue to turn up as the sample size of Early Cretaceous faunas in Australia increases."

Patricia Vickers-Rich
June 1999

GONDWANA
IN
PERSPECTIVE

BIRTH OF THE GONDWANA CONCEPT

In ancient times many Gond kingdoms lay to the south of the Narbada Valley of Peninsular India, south of the majestic Himalayas. The Gonds lived their lives and died in this land, their kingdoms prospered and crumbled, only to be hidden by succeeding realms, all without ever knowing the mysteries locked in the rocks beneath their feet. Not until 1872 when H.B. Medicott and W.T. Blandford published their *Manual of the Geology of India* did these very rocks that had felt the sandals of the ancients and the hooves of their goats reveal a most complex and unexpected history of India. And that history involved not only India, but many other landmasses as well.

Geologists Medicott and Blandford, and later Eduard Suess, noticed a distinctive and repetitive sequence of rocks in the Narbada Valley, in the Jabalpur, Nágpur and Chatisgarh regions. They examined, described and thoroughly studied these rocks and found that the "marine older and middle Mesozoic, and probably the upper Palaeozoic formations . . . are represented in the Peninsula of India by a great system of beds, chiefly composed of sandstones and shales, which appear . . . to have been entirely deposited in fresh water, and probably by rivers. Remains of animals are very rare in these rocks, and the few which have hitherto been found belong chiefly to the lower vertebrate classes of reptiles, amphibians, and fishes. Plant remains are more common, and evidence of several successive floras has been detected. The subdivisions of this great plant-bearing series have been described under a number of local names . . . but the Geological Survey [of India] has now adopted the term Gondwána for the whole series" — a name meaning "land of the Gonds", the ancient caretakers of these rocks and sediments.

8 Antarctic Beech forest was observed in the mid-nineteenth century in Tasmania, New Zealand, South America and islands in the Southern Ocean, and its disjunct distribution could only be explained by a previous connection of the now widely separated southern lands. The first serious consideration by Earth scientists of the concept of a southern supercontinent — Gondwana — resulted from this observation of extant flora. (Jim Frazier)

9 The Glossopteris Flora that characterizes the Permian Coal Measures of all the southern continents and India demonstrates the previous union of these lands. The first palaeobotanical investigation of a Glossopteris flora was carried out in that part of India where the Gond people lived. When a name was sought for the southern supercontinent, "Gondwanaland" — and later "Gondwana" — was chosen. x 2.0 (Jim Frazier)

9

THE ROCKY FRAMEWORK OF GONDWANA

The Gondwanan rocks range from the Carboniferous to the Jurassic, that is, about 354 to 141 million years in age, and form a long and nearly continuous sequence. The sequence — which is known well on five now disjunct landmasses (India, Africa, South America, Australia and Antarctica) — begins, at the bottom, with chaotic sediments called tillites. The particles in these sediments are of many different sizes, from clay up to boulder-sized, and they are randomly distributed throughout the rocks, certainly showing no repetitive layering or grading in size in any direction. Many of the boulders show parallel scratch marks on their surfaces where they must have rubbed against each other or some other abrasive surface. These tillites lay over much older rocks, and their zone of contact is a smoothed, sometimes highly polished and almost always thoroughly scratched surface thought to have been made as gigantic continental glaciers moved across the landscape more than 290 million years ago. As the glaciers moved, a cargo of loose boulders and rocks of many sizes was incorporated into their bases. Acting as a giant rasp, this cargo cut furrows and grooves into the underlying land surface, providing a record of the passage of these great ice sheets long after they disappeared. When these glaciers melted, the boulders and rocks held within them were dropped on that same basement surface, leaving behind the chaos of tillites, a second piece of evidence of the presence of massive ice accumulations.

The Seaham Formation in the Hunter Valley of New South Wales rests on one of these striated floors, and tillites of the Late Carboniferous and Early Permian are abundant in many parts of Australia including such areas as Bacchus Marsh just west of Melbourne. Similarly, far away from Australia, the Buckeye Tillite of Antarctica, the Dwyka Tillite in South Africa, the Itarare Series and the Tupe Tillite of Argentina and the Talchir Tillite of India are examples of material deposited during the same glacial events in the Late Carboniferous or Early Permian on many of the Gondwanan supercontinental fragments. Gondwana at this time lay far south, and many of its pieces were locked in the frozen grip of ice sheets of truly massive proportions. When the direction of flow of these ice sheets on all the Gondwanan fragments is plotted, based on the deeply incised striations on the basement rocks and the distribution of the tillites, it shows that many fragments of the great supercontinent were united, with a South Pole located near the centre of the continental mass.

Interbedded with the tillites, and very much a part of the Gondwanan sequence, are a number of important coal seams. The Newcastle Coal Measures and the Illawarra Coal Measures are examples in Australia, as are the coals in the Beacon Supergroup in Antarctica, the Ecca Coal Measures in South Africa, the Rio Bonito Beds of South America, and the Raniganj and the Barakar Coal Measures of India. All are from the Late Carboniferous and Early Permian, about 315 to 270 million years in age. Their deposition is cyclic, with the terrestrial coal seams alternating with sands and shales. The cyclicity appears to be tied to the growth and melting of the massive continental glaciers that typified the times, the sea rising with the melt and the land emerging during the freeze which tied up more and more sea water in the ice masses. The coals deposited in these ancient ice age swamps fuel our factories and kitchen lights today, especially in eastern Australia, as well as the steam engines of South Africa's vast railway system. They are very much a part of our everyday life, and although of seemingly endless extent, they are finite — a part of the Gondwanan heritage of more than 300 million years ago.

Some of the most beautiful and distinctive fossils entombed in these Gondwanan coal measures are the leaves of an ancient seed-fern, *Glossopteris*, and a variety of other related plants. The leaves were abundant, and the trunks of the trees that bore them are marked with growth rings, reflecting the effects of strong seasonality. They are known in the Narbada Valley of Peninsular India and in similar looking sands and shales of South America, South Africa and Madagascar, Australia and even Antarctica. As *Glossopteris* seeds were reasonably large, wind dispersal to such far distant parts of the world seems unlikely. *Glossopteris* fossils also seem to occur in areas where the lycopsid flora, made up of ancient plants related to the diminutive living clubmosses, so typical of places like North America

10

and Europe, are not found. Trees of the lycopsid flora lack much in the way of growth rings; their habitat was evidently more equable, perhaps almost unchanging from day to day. An interpretation of this evidence is that the lycopsid flora probably represents a more tropical assemblage of plants, in marked contrast to the *Glossopteris* assemblage which is a flora of more temperate, seasonal climes. *Glossopteris* and a variety of seed-ferns and ferns from the Gondwanan sequence thus bind many now far distant places with a common thread, at a time more than 250 million years in the past.

Plants are not the only fossils that tie the disparate parts of Gondwana together. So, too, do numerous vertebrate fossils. Small aquatic vertebrates called mesosaurs have been recovered from Early Permian rocks on both sides of the present Atlantic Ocean — from the Dwyka shales of South Africa and the Irati Shales of South America. They are little reptiles, with a long flattened tail, paddle-like feet and a long snout graced with a collection of elongate teeth thought to have been useful in straining tiny crustaceans and other invertebrates from the water. In order to explain the apparently disjointed distribution of the mesosaurs, Africa and South America are thought to have once been closely apposed, as part of Gondwana.

Lystrosaurus is another vertebrate that occurs in many of the Gondwanan sequences, in Early Triassic rocks (about 251–241 million years old) of southern Africa, India, southern China and Antarctica, as well as in the non-Gondwanan USSR. This animal was a mammal-like reptile, with a distinctive skull that possessed two impressive tusks. It was truly terrestrial, and its distribution seems to demand a continental connection to explain its dispersal. A similar dicynodont reptile that is thought to be a close relative of *Lystrosaurus* is known from the Arcadia Formation, also of Early Triassic age, in eastern Australia.

Above the coals containing *Glossopteris* the younger part of the Gondwanan sequence is mainly non-marine sediments — red beds, wind-blown sands, riverine sands and silts — but in some places such as eastern Africa, northern India and South America these sediments

10 Glacial dropstone emplaced when an iceberg melted and dropped its cargo of rocks and sediments. These sediments at South Durras in New South Wales are of Permian age and signal the major glaciation that was affecting much of Gondwana at this time. The large boulder is about 2 metres across. (R. Cas)

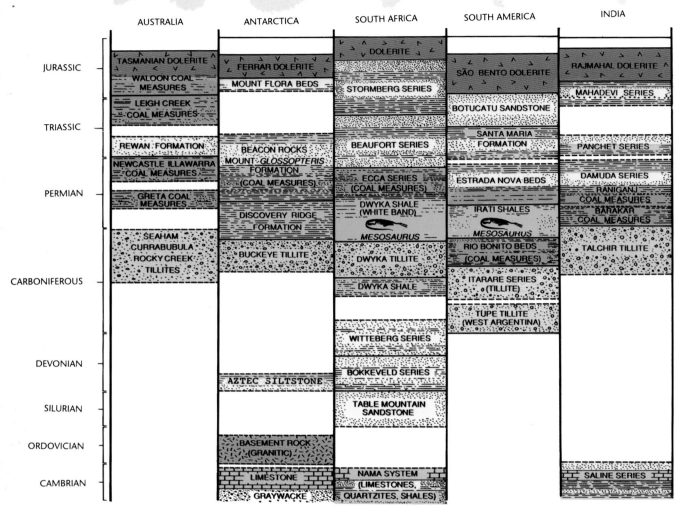

	AUSTRALIA	ANTARCTICA	SOUTH AFRICA	SOUTH AMERICA	INDIA
JURASSIC	TASMANIAN DOLERITE	FERRAR DOLERITE	DOLERITE	SÃO BENTO DOLERITE	RAJMAHAL DOLERITE
	WALOON COAL MEASURES	MOUNT FLORA BEDS	STORMBERG SERIES		MAHADEVI SERIES
	LEIGH CREEK			BOTUCATU SANDSTONE	
TRIASSIC	COAL MEASURES			SANTA MARIA FORMATION	
	REWAN FORMATION	BEACON ROCKS MOUNT *GLOSSOPTERIS* FORMATION (COAL MEASURES)	BEAUFORT SERIES		PANCHET SERIES
	NEWCASTLE ILLAWARRA COAL MEASURES		ECCA SERIES (COAL MEASURES)	ESTRADA NOVA BEDS	DAMUDA SERIES
PERMIAN	GRETA COAL MEASURES		DWYKA SHALE (WHITE BAND)	IRATI SHALES	RANIGANJ COAL MEASURES
		DISCOVERY RIDGE FORMATION	*MESOSAURUS*	*MESOSAUHUS*	BARAKAR COAL MEASURES
	SEAHAM CURRABUBULA ROCKY CREEK TILLITES	BUCKEYE TILLITE	DWYKA TILLITE	RIO BONITO BEDS (COAL MEASURES)	TALCHIR TILLITE
CARBONIFEROUS			DWYKA SHALE	ITARARE SERIES (TILLITE)	
				TUPE TILLITE (WEST ARGENTINA)	
			WITTEBERG SERIES		
DEVONIAN			BOKKEVELD SERIES		
		AZTEC SILTSTONE			
SILURIAN			TABLE MOUNTAIN SANDSTONE		
ORDOVICIAN		BASEMENT ROCK (GRANITIC)			
CAMBRIAN		LIMESTONE	NAMA SYSTEM (LIMESTONES, QUARTZITES, SHALES)		SALINE SERIES
		GRAYWACKE			

The Gondwanan sequence as known on several continental fragments that once made up the great southern supercontinent. The most convincing correlation is found where glacial debris, or tillite, lies under terrestrial sediments that are rich in plant and animal fossils. These sediments and volcanic rocks and their biota seem to have been deposited when all of the Gondwanan continents were close to one another and located far south of the Equator. Gondwana during the late Palaeozoic to Early Jurassic was isolated by climate, not by physical severence, for this southern landmass was actually connected with the northern continents during much of this time, forming an even larger continent, Pangaea. (Fig. modified from Scientific American illustration.)

meet and interdigitate with marine sands and clays. This junction allows geologists to correlate the Gondwanan sediments and events with those in far distant places. The marine organisms that inhabited the sea between Gondwana and other continents have also thus left their skeletons associated with now distant land areas. They serve as a time line, allowing us to link faunas on land which look very different and have had distinct evolutionary histories.

And finally, the Gondwanan sequence is capped by massive volcanic outpourings. In some places volcanic flows covered large tracts of the ground surface, while in others massive bodies of hot rock were emplaced and slowly cooled underground. The Ferrar Dolerite that forms impressive ice-covered cliffs along the Transantarctic Mountains is but one example of the manifestations of this violent time that marked the end of the Gondwanan sequence; so, too, are the Sao Bento dolerites of South America and the Rajmahal Dolerite in Peninsular India, the dolerites in southern Africa including those in the Drakensberg of Lesotho, and in Australia the Tasmanian dolerites such as those that make up part of the Mt Wellington sill that towers over Hobart harbour.

Above the Jurassic dolerites, the Gondwanan sequence ceases and

11-15 A variety of skulls of the mammal-like reptile Lystrosaurus *from the Gondwanan continents. Most species were sheep-sized. (Courtesy of the Natural History Museum Of Los Angeles County)*

thereafter the geological records of India, Africa, South America and Australia–Antarctica become unique in themselves. These lands no longer seem to share a common heritage. And there is a reason. The supercontinent of Gondwana began to sunder.

The Gondwanan sequence is a most compelling reason for reuniting now separate continents at a time in the past. There are other reasons, too, like the "jigsaw fits" of such continental margins as those of eastern South America and western Africa, and of southern Australia and the opposing coast of Antarctica. The history of mountain building with its resulting twisting and faulting of the rocks in distinct belts can be explained for southern Africa, South America, eastern Australia and parts of Antarctica if these continental pieces are all shoved back into close juxtaposition. The mountains form a continuous chain across these reunited continents. Ancient mountain chains now chopped off abruptly at the south-eastern edge of South America match up with another sundered chain at the west coast of Africa, and the Transantarctic Mountains fit into this sinuous chain as well.

Perhaps the most striking argument, however, which convinces most people that a Gondwana did once exist has been developed fully only within the last 25–30 years. It explains not only how continents have moved, and how Gondwana could be reassembled, but also why oceans are where they are, why volcanoes erupt and earthquakes shake only in certain places, and why parts of the Earth are hotter than others. It is a general theory of which the explanation of Gondwana is only a part, the *Theory Of Plate Tectonics.*

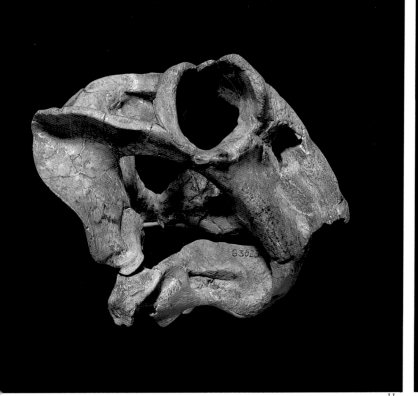

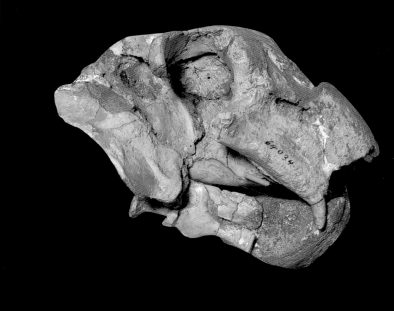

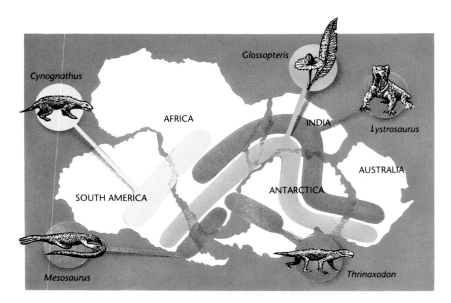

Some of the palaeontological evidence used to reconstruct Gondwana. Distributions of the primitive freshwater reptile *Mesosaurus*, the mammal-like reptiles *Cynognathus* and *Lystrosaurus*, and *Thrinaxodon* as well as many plant groups, such as *Glossopteris*, all favour a rejoining of several now separate continents during the Permian and Triassic, and even into the Jurassic. (Modified from Hamblin, 1985)

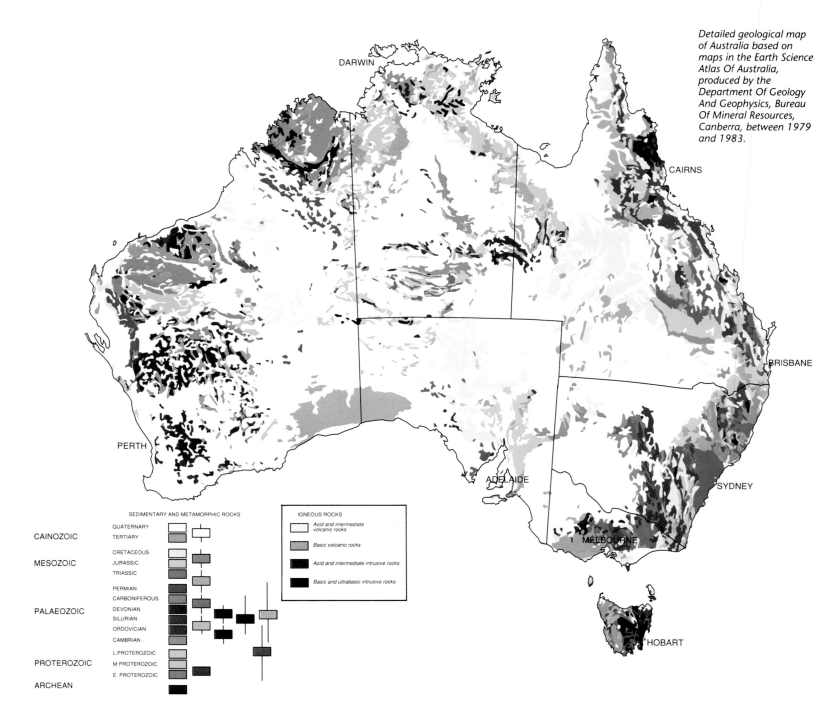

Detailed geological map of Australia based on maps in the Earth Science Atlas Of Australia, produced by the Department Of Geology And Geophysics, Bureau Of Mineral Resources, Canberra, between 1979 and 1983.

SEDIMENTARY AND METAMORPHIC ROCKS

IGNEOUS ROCKS

Acid and intermediate volcanic rocks

Basic volcanic rocks

Acid and intermediate intrusive rocks

Basic and ultrabasic intrusive rocks

CAINOZOIC
QUATERNARY
TERTIARY

MESOZOIC
CRETACEOUS
JURASSIC
TRIASSIC

PALAEOZOIC
PERMIAN
CARBONIFEROUS
DEVONIAN
SILURIAN
ORDOVICIAN
CAMBRIAN

PROTEROZOIC
L.PROTEROZOIC
M.PROTEROZOIC
E.PROTEROZOIC

ARCHEAN

THE WAY THE EARTH WORKS

Just because continents are where they are today does not mean they will be there always. The Earth is not unchanging. Earthquakes can impressively alter the landscape in a few seconds and have a devastating effect on populations. Floods, too, can cause mighty rivers to dramatically change course in a very short period of time. So, change is not an unexpected part of nature. But in past times these "small scale" observed changes were not thought to be applicable to the bigger picture.

The idea that continents have always remained where they are today has been a topic of debate ever since Medicott and Blandford first described the Gondwanan sequence back in 1872. Indeed, centuries before, Francis Bacon had pointed out the amazing jigsaw fit of eastern South America and West Africa. Serious consideration that continents might have moved in the past was not forthcoming, however, until 1912 when Alfred Wegener published *The Origin of Continents and Oceans*. Over the next two decades geologists argued the issue intensely. Then in 1937 Alexander L. du Toit, a South African geologist, published an enlightening book entitled *Our Wandering Continents* which brought together, in one place, a summary of the evidence in favour of continental drift, of continents wandering about in times past. Du Toit's book presented the picture of a world with a very different arrangement of continents than that of 1937. He further pointed out the merits and problems of this controversial idea of wandering continents, and challenged the geological world to consider it seriously as a better hypothesis about the workings of the Earth than the stabilist one then in vogue. He had no ready answers concerning what might be the driving force behind the drift of the continents, but that explanation was not long away — the evidence for it being provided by the magnetic properties of the Earth and its surface rocks.

The Earth has a magnetic field, a fact clearly evident when we use a pocket compass. It has this field because of the properties of the materials deep within this planet. At the centre of the Earth is a solid metallic core, the Inner Core. But about 1200 kilometres from the centre, metals become liquid and temperatures reach those of the Sun's surface, about 5800° Celsius. Pressures exceed a million atmospheres. This part of the Earth, the molten, principally iron, Outer Core, churns around the solid Inner Core, propelled by the rotation of the Earth. The intense heat of this moving Outer Core acts like a dynamo to generate an electric current which in turn produces the Earth's magnetic field. Every decade this field weakens by a small percentage, and on average about every million years it reverses its polarity, that is, the magnetic North Pole becomes the magnetic South and the magnetic South Pole becomes the North. The geographic poles, of course, don't change, but the direction of the magnetic force lines change 180°, just as an electric current would change if the positive and negative electrodes were suddenly reversed. This variable behaviour in the Earth's magnetic field has had its effect on surface rocks, and provides the most convincing evidence for continental movement.

Many volcanic rocks include particles of iron and nickel bound up in a number of different minerals such as hematite and magnetite (both oxides of iron, Fe_2O_3 and Fe_3O_4 respectively). Before these rocks harden from a molten melt, the magnetic particles align themselves like tiny magnets parallel to the prevailing magnetic field. As the rock cools, a temperature (the Curie Point) is reached at which the particles can move about in response to the Earth's magnetic field, and with further cooling the orientation of these particulate magnets becomes "frozen" into the crystalline structure of the rock. For any specific time in the past, when a rock has cooled beyond the Curie Point, the prevailing magnetic field of that time is recorded forever — unless, of course, the rock is remelted again.

The tiny metallic particles in these rocks not only aligned themselves parallel to the prevailing field but they also dipped or tilted towards the North or South pole, depending upon which hemisphere they were in at the time, thus in effect behaving like miniature compass needles. If the rocks can be precisely dated, by techniques that utilize what we know about radioactive decay (for example, by methods which measure ratios of unstable radioactive isotopes of elements that decay into stable "daughter" products such as the Potassium–Argon method, or examine fission tracks left in crystals by particles emitted when a radioactive isotope decays), then the direction of the North and South poles and the sense of the prevailing magnetic field can be determined for each continent from the orientations of these natural compasses, for many times in the past.

When these palaeomagnetic measurements were begun in the 1950s, scientists quickly realized that the magnetic poles had not always been where they are today. By plotting where the "palaeopoles" were relative to each continent over the past few hundred million years, a polar wander path for the individual continents was produced. These paths were all different, indicating a number of poles in different positions. Because there is no reason to suspect that there was ever more than one South Pole and one North Pole, it was clear that the continents must have moved in order to explain the apparent multiplicity of palaeopoles. If it had been the magnetic pole, and not the continents, that had wandered, the polar wandering curves of all the continents would have been the same. Since they weren't, so the conclusion: continents had moved with respect to one another at times in the past. Wegener and du Toit were correct.

Contemporaneous with the discoveries of polar wandering was a growing awareness about the topography of the world's ocean floors. Maurice Ewing, Bruce Heezen and Marie Tharp at the Lamont–Doherty Geological Laboratories of Columbia University, New York, had amassed sonar data gathered by ships criss-crossing the Atlantic Ocean. Most maps prior to their work depicted the ocean floors as flat, nearly featureless plains, except in areas of deep trenches and volcanic islands. Ewing, Heezen and Tharp discovered that this was far from true. An enormous underwater mountain chain, of which the Mid-Atlantic Ridge was a part, split many of the world's ocean basins. This ridge sometimes surfaces as volcanically active islands, for example, Iceland. Further oceanographic exploration demonstrated that this mountain chain connected to others and encircled the entire Earth, extending for over 75,000 kilometres and including the ridge that lies mid-way between Australia and Antarctica. This continuous sequence of mountain chains was found to have a number of interesting characteristics. It was a hot area compared to the ocean floor on either side of it, and seismologists were able to ascertain that it was also an area of shallow earthquakes. For the most part the entire structure is 1000–1200 kilometres wide, with a steep-sided valley being characteristic of the central region. When later explored by submersibles the central area was found to be volcanically active, with abundant pillow lavas, indicating that most of its rocks had been erupted from underwater volcanoes or fissures.

With the information on ocean floor topography, the palaeomagnetic story that had begun with wandering poles continued to become even more indicative that continents had moved. Fred Vine and D.H. Matthews noticed a strange pattern in the Earth's magnetic character in the ocean floor across the Atlantic Ocean basin, utilizing information gathered by towing magnetometers behind ships and aeroplanes back and forth across the ocean basin. They plotted this magnetic information directly onto topographic maps and found that there were magnetic stripes which were mirror images of one another on either side of the Mid-Atlantic Ridge, separating Europe and Africa on the east from North America and South America on the west. From these observations they proposed that the stripes, which they called "Zebra Stripes", represented periods of reversed and normal magnetic polarity that had been recorded in the basalt rocks when they erupted and then cooled through the Curie Point. The stripes reflected the changing polarity of the Earth at many times in the past.

The reason that the stripe pattern was bilaterally symmetrical with respect to the Mid-Atlantic Ridge was that new molten material was constantly being added along the ridge and, as this material cooled and moved away from the ridge and solidified, it "froze" the prevailing magnetic alignment into its structure. It was then pushed away from the ridge where new material continued to arise and then "freeze" the magnetic signature of the time. So, the "Zebra Stripes" were parallel to this ridge and bilaterally symmetrical because material was moving away from the ridge in two directions at the

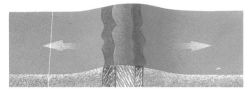

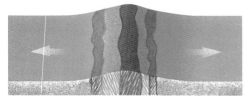

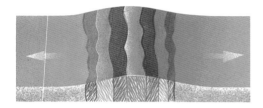

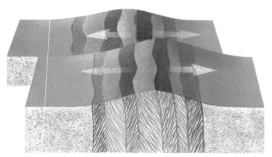

Magnetic patterns, or "Zebra Stripes", preserved as mirror images on either side of mid-ocean ridges are registers of the Earth's changing magnetic field through time. As new molten basalts are emplaced along these ridges, and harden, the magnetic materials in them record the prevailing magnetic field of the time. These tiny natural magnets are frozen into the rocks once they solidify, and because they are essentially on a conveyor-belt that is continually moving, they will give an ongoing record of the behaviour of the Earth's magnetic field as long as the ocean basin continues to expand. That record will be emplaced on either side of the ridge. The sequence of magnetic reversals recorded on the ocean floor basalts will be the same pattern found in stacked basalt flows on land, and the similarity in pattern between these two different places can be used to correlate them with each other.

Theory Of Plate Tectonics. This new theory married the ideas of continental drift and sea floor spreading and gave a model of the Earth that most geologists use today, and will continue to use until something better is proposed. It suggests that the outer few kilometres of the Earth (up to 100 kilometres in continental areas) is divided into thin, cold, rigid plates that will fracture if stressed too much rather than bend or flow. Geologists call these lithospheric plates, and they include the Earth's Crust plus part of the uppermost Mantle. These plates meet each other along one of three kinds of boundary — *ridges, trenches* (or *subduction zones*) and *transform faults*. The plates are relatively stable — volcanic activity and earthquakes are for the most part confined to their edges, along one of the three types of boundary.

The Mid-Atlantic Ridge is an example of a ridge boundary. It is a hot area where volcanic material is being added and vulcanism and earthquakes occur (such as in Iceland), but where the earthquakes are always shallow and the faulting is generally tensional (that is, the rocks are being stretched, pulled apart). The valleys that form atop the mid-ocean ridges — downdropped valleys like the Red Sea and the East African Rift Valley on land — are tensional features.

The New Hebrides and Japan trenches are examples of the trench type of boundary. Here earthquakes begin shallow and deepen away from the trench. Vulcanism occurs some distance away from the trench, an example of which produced Mt Fuji to the west of the Japan Trench. This vulcanism seems related to the heating up of the lithospheric plate as it plunges into the Japan Trench and underneath the Asian continent. Once heated to a certain point, the rock melts and rises to the surface, resulting in volcanic eruption. The rocks near the trench itself are quite cold, as they form the oldest part of the ocean basin which has decidedly cooled after a fiery origin on the central ocean ridge. And, of course, this type of boundary is characterized by a topographic low, a deep trench.

The third type of boundary, the transform fault, appears as great offsets of ridges and trenches in the ocean basins, such as the Kangaroo Fracture Zone that offsets the mid-ocean ridge south of Adelaide. Some vulcanism and sometimes submarine volcanic islands can be associated with these fractures, and they are defined by shallow earthquake activity that records lateral movement between the massive lithospheric plates. Transform faults, however, are a most unusual kind of fault. The movement recorded by analysis of the earthquake data is usually in just the opposite direction as would appear to be the case. This characteristic is due to the fact that transform faults have a two-fold history: they are initially fractures that offset the ridges but then further movement follows, brought about by the continuous addition of new material along the ridges.

same time. Since the magnetic field of the Earth was spontaneously reversing polarity at intervals ranging from every 10,000 to 1 million years, this pattern was repeated over and over again. As the ocean basins were further explored by a number of oceanographic expeditions that drilled deep into the sediments and ultimately into the volcanic basement that underlies it all, it was found that those volcanic rocks nearest the ridges are youngest and those furthest away, near the continental edges, are the oldest. So, it seems that new molten rock is added at the ridges, "entombing" the magnetic field signature of the time, and moves away from the ridges as more material is generated — a theory that was named *Sea Floor Spreading.*

A further fragment of information was to prove crucial in finally piecing together the whole story of Gondwana. Seismology provided details of the global distribution of oceanic earthquakes and the differing depths at which they occurred. While the ridges and the great faults that displace them exhibit only shallow earthquakes, the trenches have a more complex association of shallow, intermediate and deep earthquakes. In this latter case, the earthquakes are shallower directly beneath the topographic trenches and gradually increase up to about 600 kilometres in depth the further away they occur from the actual trenches.

All of this information on ocean floor topography, heat flow, distribution of earthquakes, volcanic activity, magnetic properties of the sea floor, polar wander paths and ages of many of the features — most of which was gathered during or after the 1950s — led to one final theory that related this great variety of geological phenomena, and ultimately explained the origin and evolution of Gondwana, the

None of the Earth's ocean floors are older than Jurassic. The youngest parts of ocean basins are along oceanic ridges, with age increasing symmetrically in either direction away from the crests of those ridges. Older ocean floor has been recycled down subduction zones. (Modified from Hamblin, 1989)

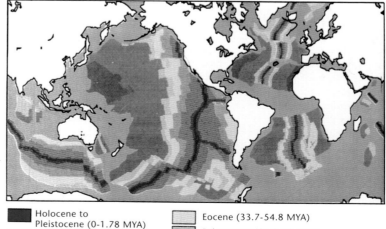

■ Holocene to Pleistocene (0-1.78 MYA)	▨ Eocene (33.7-54.8 MYA)
▨ Pliocene (1.78-5.32 MYA)	▨ Palaeocene (54.8-65 MYA)
▨ Miocene (5.32-23.8 MYA)	▨ Cretaceous (65-141 MYA)
▨ Oligocene (23.8-33.7 MYA)	▨ Jurassic (141-205 MYA)

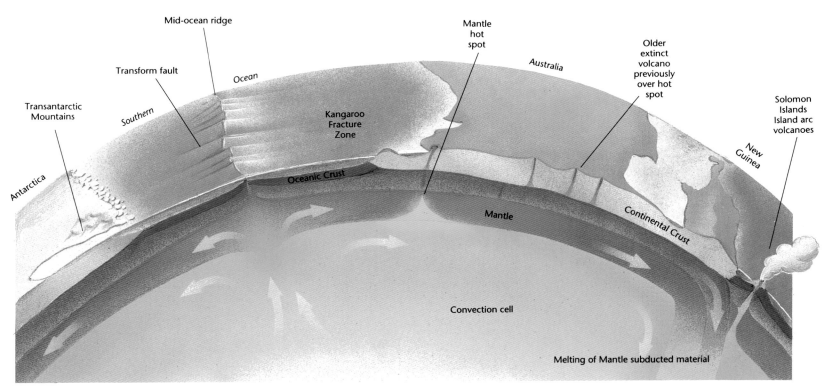

A cross-section showing major divisions in the outer few hundred kilometres of the Earth's Crust and Mantle according to plate tectonic theory. Ridges are places where lava rises from the deeper parts of the Mantle towards the surface.

Once the molten rock reaches the surface at these ridges, it causes volcanic activity and expansion, indicated by a zone of shallow earthquakes. The mid-oceanic ridge between Antarctica and Australia is such a feature.

Trenches or subduction zones are places where one crustal plate (lithospheric plate) dives under another, for example, the Tonga Trench and the Japan Trench, and where earthquakes signal the interaction of two

converging plates. Deep topographic trenches, earthquakes that range from shallow to deep and volcanic activity located some distance from the topographic trench all characterize this type of tectonic boundary.

Transform faults are areas where the lithospheric plates slide past one another laterally, such as those offsetting the mid-ocean ridge between Australia and Antarctica. It is the movement of these

crustal plates that effects the movement of continents, which are really just light-weight passengers on the backs of the great lithospheric plates. (Drawings modified from those of L. Sykes and P. Wyllie)

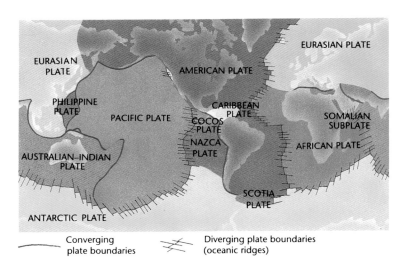

| Converging plate boundaries | Diverging plate boundaries (oceanic ridges) |

Major lithospheric plates that make up the Earth's outer shell. It is the movement of these rigid plates relative to one another that is responsible for the "drift of continents" over at least the last 600 million years of Earth history, and probably for a much longer period. The boundaries between them are marked by ridges (heavy lines), trenches or subduction zones (light lines) and transform faults, which displace both trenches and ridges. (Modified after Hamblin, 1975)

The movement along all of these boundaries, bringing about the constant reshuffling of the lithospheric plates, seems to be the result of convection in the underlying Mantle of the Earth (a part called the asthenosphere), which behaves like a viscous liquid. Convection is a way of transferring heat from the interior of the Earth to its surface: the heat is carried by the movement of the hot Mantle rising, cooling, and then the cold Mantle sinking again. This circulation, a convection cell, is much like that in a boiling pot of tomato soup: heated on the bottom first, the hot soup rises and as it cools it sinks back to the bottom of the pot. The convection cells of the Mantle move at rates of 1–10 centimetres per year, and so, too, do the overlying lithospheric plates, and ultimately the continents.

The continents are just parts of the great lithospheric plates that make up the Earth's surface. Australia, for example, lies on the Australian–Indian Plate, separated on the west and south from the African and Antarctic plates by a ridge, on the east from the Pacific Plate by a series of trenches and transform faults, and on the north

and west from the Eurasian Plate also by a series of trenches and transform faults.

It is the restless movement of the plates that results in changes of continental positions. This restless movement is the reason that Gondwana once existed as an isolated free-standing entity and remained that way for 350 million years until the Middle Carboniferous, some 320 million years ago. It is also the reason for Gondwana's ultimate death, by first merging with the northern lands of Baltica and Laurentia, and then fragmenting and dispersing to the far corners of the globe. This final break-up began more than 208 million years ago in the Triassic, each piece eventually to be wed to another player on the Earth's majestic chessboard. It is of interest that the Gondwanan rock sequences and the very distinct Gondwanan biota were for the most part younger than the time of geographic isolation of the southern continents. Isolation of the flora and fauna during the late Palaeozoic and much of the Mesozoic was not imposed by geography, but by climate.

500 MILLION YEARS OF CONTINENTS ADRIFT

By using two types of information it is possible to estimate where continents have been in the past relative to their present positions. For the Mesozoic and Cainozoic eras reconstructions of continental positions are based mainly on information from the sea floor — the magnetic bands, or "Zebra Stripe" anomalies, that reflect the production of new oceanic floor. These bands can be dated and identified, and the "tape recorder" run backwards to close ocean basins, thus refitting continents together in an ancient jigsaw. This method, however, will work only for the last 200 million years or so because no oceanic Crust is older than Jurassic age. Where the rocks are older than this, the repositioning of the continents must depend on plotting of the polar wander paths obtained from each of the continents relative to the poles and to each other.

From those two types of information, a general picture of continental rearrangement has emerged. At the beginning of the Phanerozoic, about 570 million years ago, the world looked very different than it does now. In the oldest period of this eon, the Cambrian Period, one very large continent, Gondwana, existed, straddling the Equator. What happened before this time is not very certain because of extensive alteration of older rocks, although some attempts at reconstruction have been made. But, from the point of view of vertebrate history, such attempts are of no consequence, for vertebrates first appear in Late Cambrian times.

Included in Cambrian Gondwana were Africa, South America, Australia and Antarctica, and around its margins were a number of islands composed of bits of Europe and Asia. A smaller continent, Laurentia, also lay along the Equator as did a number of smaller, isolated fragments such as Siberia and parts of Kazachstan — strung out like beads on a string. Only Baltica, which included some of the present Baltic states and Poland, and the fringes of Gondwana at this time lay in the temperate latitudes to the south. Fragments of China and South-East Asia lay to the north of the Equator. Opinions differ on just how far apart these fragments were from one another, some palaeontologists suggesting that interchange between many was easily possible.

The continents at this time were constructed of different jigsaw pieces than those of today. Some were in very different places than they are now. Antarctica was nowhere near the South Pole; instead it lay near the Equator, as did Australia. The general climates on these equatorially located continents, as indicated by the distribution of reefs and salt deposits (evaporites), were dominated by warmth and humidity — they were truly tropical.

During the Ordovician and Silurian periods, from 510 to 409 million years ago, due to readjustment of the lithospheric plates, the continents moved into radically new positions, mainly away from the Equator. Gondwana remained a free-standing entity, by far the largest landmass on Earth. It began to shift southwards, while Laurentia and such smaller fragments as Siberia and Kazachstan remained Equator-bound. Some continents, like Baltica, raced toward the Equator from a near polar position, much as India and Australia would later in time.

By Siluro–Devonian times, 439 to 363 million years ago, much of Gondwana lay over or close to the South Pole. Australia formed a northern peninsula, and its north-eastern border lay in the tropics, close to parts of China. Antarctica, however, was now close to the pole, virtually the same position having been maintained until the present day. But Gondwana did not remain an isolated supercontinent. At least 360 million years ago Pangaea began to coalesce, with direct connections being established between the Gondwanan landmass and many northern continental fragments — North America, Siberia, Europe, and parts of Asia.

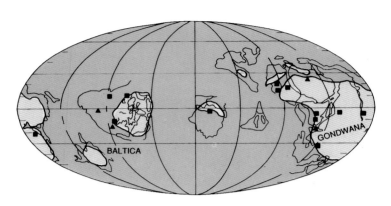

LATE CAMBRIAN 490 MY

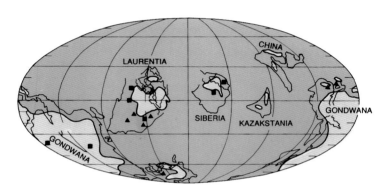

MIDDLE ORDOVICIAN 465 MY

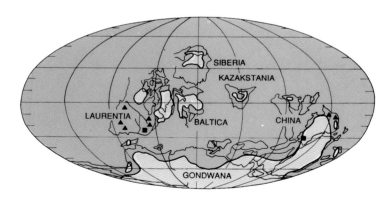

MIDDLE SILURIAN 420 MY

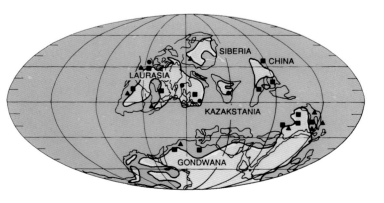

EARLY DEVONIAN 390 MY

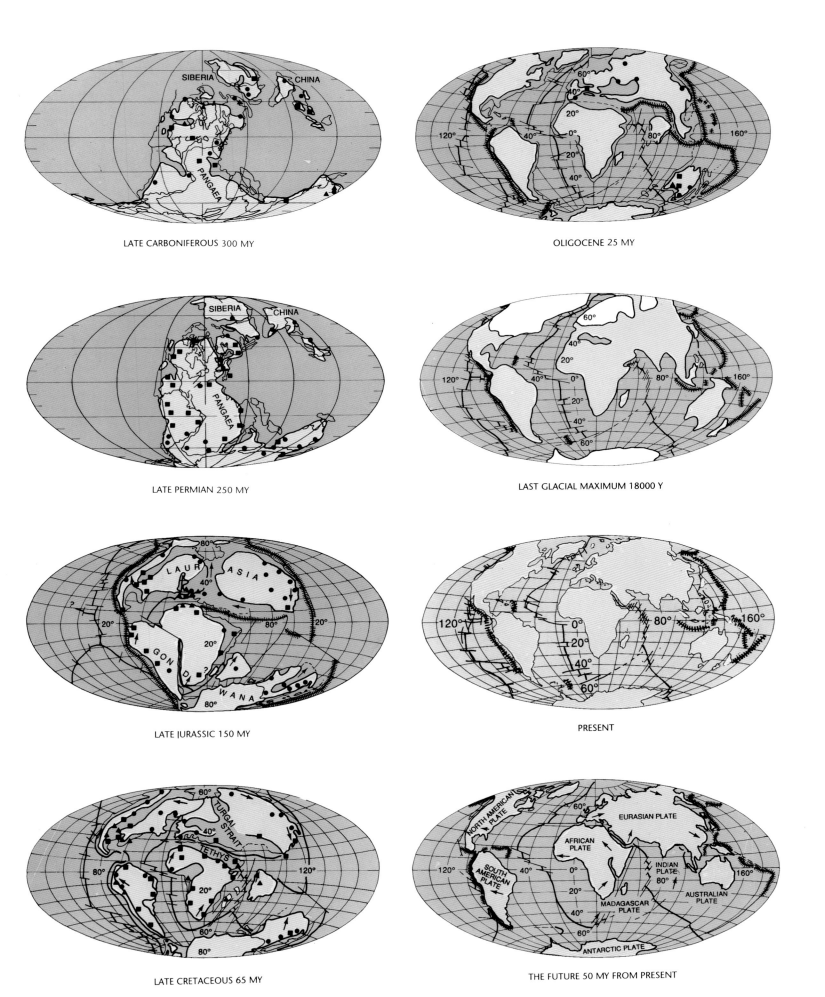

LATE CARBONIFEROUS 300 MY

OLIGOCENE 25 MY

LATE PERMIAN 250 MY

LAST GLACIAL MAXIMUM 18000 Y

LATE JURASSIC 150 MY

PRESENT

LATE CRETACEOUS 65 MY

THE FUTURE 50 MY FROM PRESENT

25

All along the "southern margins" of Gondwana, from Colombia to Papua New Guinea, was an active mountain chain, the Gondwanides, long ago recognized by Alex du Toit when he named the Samfrau Geosyncline, which included rock sequences in South America, Antarctica, the Cape region of South Africa and parts of eastern Australia. These mountains apparently formed along a convergent boundary, where two lithospheric plates were pushing into one another, the kind of boundary that today is reflected by the Andes. The northern margin of Gondwana, from the early Palaeozoic onwards, resembled the kind of quiet margin that characterizes the western Atlantic Ocean today along the eastern seaboards of North America and South America. No mountain belts were present, only stable continental shelves. From the middle Palaeozoic, small continental plates began to break away, drift north and embed themselves in what is now Eurasia. Evidence of these collisions can be found in such places as Turkey, Iran, Tibet and Malaysia.

Sometime during the Carboniferous, which began about 363 million years ago, Australia shifted dramatically southward, and much of the old Gondwanan landmass lay in close proximity to the South Pole. By the Middle Carboniferous, about 325 million years ago, the western parts of this continent were in broad contact with the northern landmasses, forming the largest continent of the entire Phanerozoic: Pangaea. During this nuptial period, the world was locked in a severe ice age which affected a major part of this impressive continental mass. This glaciation left the tillites on Gondwana and led to the low diversity of the *Glossopteris* floras and associated faunas. It was a glaciation that brought about dramatically lowered sea levels, as more and more sea water was bound up in continental ice sheets, and following this came unprecedented extinctions of nearshore marine organisms. From that time until the later fragmentation of the continents forming Gondwana, the great southern land-mass was directly connected to the northern continents. Yet the fauna and flora of Gondwana were distinct from those of Laurasia (North America, Europe and most of Asia), a factor which seems to be related not to the physical isolation of Gondwana but to its climatic isolation in the temperate and polar latitudes of the Southern Hemisphere.

During Mesozoic times Pangaea began to break apart, beginning in the Triassic sometime after 251 million years ago. By the Late Jurassic, North America and Eurasia had broken from Africa leaving Gondwana for a short time isolated once again. But then Gondwana began to lose its integrity. In the Early Cretaceous, about 140 million years ago, Africa broke with Antarctica, and the South Atlantic began to grow, as indicated by the oldest of the magnetic anomalies in that ocean basin. The rumblings of this separation had begun as early as 220 million years ago with the great outpouring of the Karoo basalts of South Africa. The creation of the Paraná basalts and coastal ophiolites of southern South America about 120 million years ago signalled the beginning of the separation of that continent from Africa. This separation was well under way by 95 million years ago. Contemporaneously, India broke from Africa, Antarctica and Australia, beginning its long trek north. In the Late Cretaceous, New Zealand broke from Antarctica and Australia with the opening of the Tasman Sea, and finally, sometime after this, Australia broke from Antarctica.

The oldest sea floor dated between Australia and Antarctica was laid down about 96 million years ago. At first the separation of the two continents was slow, about 5 millimetres per year, but about 50 million years ago the sea floor spreading dramatically speeded up to about 27 millimetres a year. But the first rumblings of the divorce suit which was to spell the end of Gondwana came even before the formation of actual sea floor, with a massive outpouring of volcanic debris that fell into a deepening rift valley between the two continents, much like the great rift valley of East Africa today. In one area where Gondwana was breaking asunder, southern Africa, the heat and tremendous pressures generated deep within the associated volcanic Kimberlite pipes yielded diamonds, a precious gift to the future and a lasting reminder of the violent past that gave birth to our modern geography.

16 *The Gondwanan seed-fern Glossopteris sp. from Permo-Carboniferous sediments between Walleramany and Mudgee, New South Wales. x 1.0 (F. Coffa, courtesy of the Museum Of Victoria)*

In order to construct a sequential history of life or of geological events, a way must be found to date individual events. Over the past two centuries a number of methods have been devised to do just that. Some of these methods give only relative dates, ordering when one event occurred "relative" to another (before or after). Other methods give an absolute date, or a date measured in years or millions of years.

Relative dates can be determined by examining the sequence of rock layers. The oldest rocks lie below the younger ones unless, of course, tectonic activity has overturned the sequence. So, if there is no evidence of faulting or bending of the rocks, then the *Law Of Superposition* applies and the oldest rocks lie below.

Fossils can also be used to determine the relative ages of the sediments in which they occur. Trilobites, marine arthropods with many body segments and a myriad of legs, always occur in rocks older than those containing dinosaurs. And, dinosaurs occur in sediments older than those with kangaroos. Sequences based on tens of thousands of different kinds of fossil organisms have been established around the world, which allow refined dating, particularly of marine rocks, to be carried out using the first and last occurrences of certain fossil groups. Such dating techniques must take into account that not all organisms lived in the same kinds of environments, thus absence of a particular species can be governed by environment as well as by time. Nevertheless, once environment is accounted for, fossils can allow detailed relative dating. The best fossils for this purpose are index fossils — species with a wide geographic and ecologic range, a short life span (for the species, not for the individual) and a characteristic shape so that they can be easily identified and not confused with any other species.

Fossils, in fact, are the primary basis for the definition of the *Geologic Time Scale*, which divides Earth History into a series of large time slices — eons, eras, periods and epochs.

The Phanerozoic Eon, during which vertebrates appeared and prospered, is the time of "Evident Life", whereas the Cryptozoic before it — comprising the Hadean Eon, the Archean Eon and the Proterozoic Eon, and making up the greatest part of Earth History — is the time of "Hidden Life". During most of the Cryptozoic the only organisms were single-celled microbes, and not until the very end of this time did multicellular forms appear — examples being the beautiful Ediacara jellyfish and worms best known from the Flinders Ranges in South Australia.

The Phanerozoic Eon is subdivided into three eras: the Palaeozoic, or era of "Ancient Life"; the Mesozoic, or era of "Middle Life" (informally called the "Age Of Reptiles"); and the Cainozoic, or era of "Modern Life" (also called the "Age Of Mammals"). These names and the numerous period and epoch subdivisions within them that will be used throughout this book were originally defined in continental Europe and England, and many were based on collections of invertebrates that occurred in the

mainly marine sedimentary sequences there. For example, the Cambrian Period and the Silurian Period were based on fossil assemblages found in Wales in the early nineteenth century, and were named after the ancient tribes that once inhabited the countryside in which the fossiliferous rocks occurred. Later, these names were applied to rocks of similar age in Australia and other parts of the world, though, of course, the dating nowadays involves both fossils and radioactive dating techniques.

Absolute dating (physical time dating) of rocks was not possible until well after the beginning of the twentieth century. Many absolute dating techniques are used and they all rely on the principle that radioactive processes proceed at a constant rate under the range of temperature, pressure and chemical conditions typical of the Earth's surface. One of the best known of these methods is the Carbon 14 technique, often used to date charcoal and plant material found in sites not older than about 35,000 years. Another radiometric method, the Potassium–Argon technique, is valuable for rocks ranging from 1 million years to as much as 4000 million years old. Both of these techniques rely on the fundamental principle that after a known period of time, called the "half-life", the amount of original radioactive material remaining is reduced by one-half because of radioactive decay. And, this decay continues halving the radioactive material at the same rate, until the amount remaining is infinitesimally small, essentially undetectable due to background radiation from space.

Carbon 14 has a half-life of about 5700 years. The "14" in Carbon 14 refers to the total number of particles making up the nucleus of the carbon atom. Carbon 14 is only one of the three forms, or isotopes, of carbon, the other two being Carbon 12 and Carbon 13. While Carbon 14 is radioactive, Carbon 12 and Carbon 13 are not. Thus, while the quantity of Carbon 14 declines with decay, the amount of Carbon 12 and Carbon 13 remains unchanged forever, unless the rock is unduly heated or chemically altered.

Carbon 14 is produced when cosmic rays from space bombard nitrogen in the atmosphere. This isotope of carbon is constantly produced and forms a part of the air we and other organisms breathe, and is incorporated into the plants we eat. The naturally occurring ratio of Carbon 14 to Carbon 12 and Carbon 13 in the atmosphere is known. That ratio is the same in an animal or plant as long as it is alive. Once it dies, however, no new carbon is incorporated, and the Carbon 14 begins to decay. So, by measuring the ratio of Carbon 14 to Carbon 12 and Carbon 13 in fossil material, the length of time that Carbon 14 has been decaying can be determined and, thus, we can also pinpoint the age of the fossil.

Still another radioactive technique is that of fission track dating, which relies on the spontaneous splitting, or fission, of an atom of Uranium 238. The large and heavy nucleus of uranium spontaneously splits into two pieces

17

17 *Zircon crystal with fission tracks etched out in it. Fission tracks in the crystal are produced when decaying radioactive atoms emit particles that damage the crystal structure. By acid etching the crystal, the tracks are highlighted and their density is directly related to the length of time decay has been occurring. By counting the number of tracks in the crystal, its age can be determined. (I. Duddy)*

that fly apart from each other at high speed. These fragments move through the enclosing material and leave a trail of damage behind them. So, the number of trails left in a crystal where Uranium 238 has decayed is a measure of the age of that structure. The trails left are too small to be seen unless they have been chemically etched and then the material containing the tracks has been finely sliced and viewed under a high-powered microscope. These etched holes are the "fission tracks". By knowing the amount of original uranium in the sample and the half-life of the Uranium 238, then counting the number of tracks, the age of the enclosing rock can be estimated. The minerals most useful for this sort of study are apatite, zircon and some volcanic glasses such as obsidian, and the technique has been quite successfully applied to such sedimentary sequences as the Otway and Strzelecki groups of Cretaceous age in southern Australia. The zircons dated in these rocks were derived from volcanic eruptions occurring as Australia began to part company with Antarctica: the fission track dates pin down the time of those eruptions by identifying when the radioactive clocks started ticking in the zircon crystals as they cooled from their molten beginnings.

Such radioactive techniques have allowed dates in years to be applied to sequences that previously had only relative dates. Thus, the boundary between the Palaeozoic and the Mesozoic can be pinpointed at 251 million years before the present. These dates may vary somewhat from one publication to the next, depending upon the half-life value that has been used and the philosophy of the researcher providing the data. But, despite a slight variance, the ages are reasonably similar, and can be used to relate rock sequences all over the world. The classic sequences in Europe have thus been directly tied to those in Australia and other fragments of the ancient Gondwanan supercontinent.

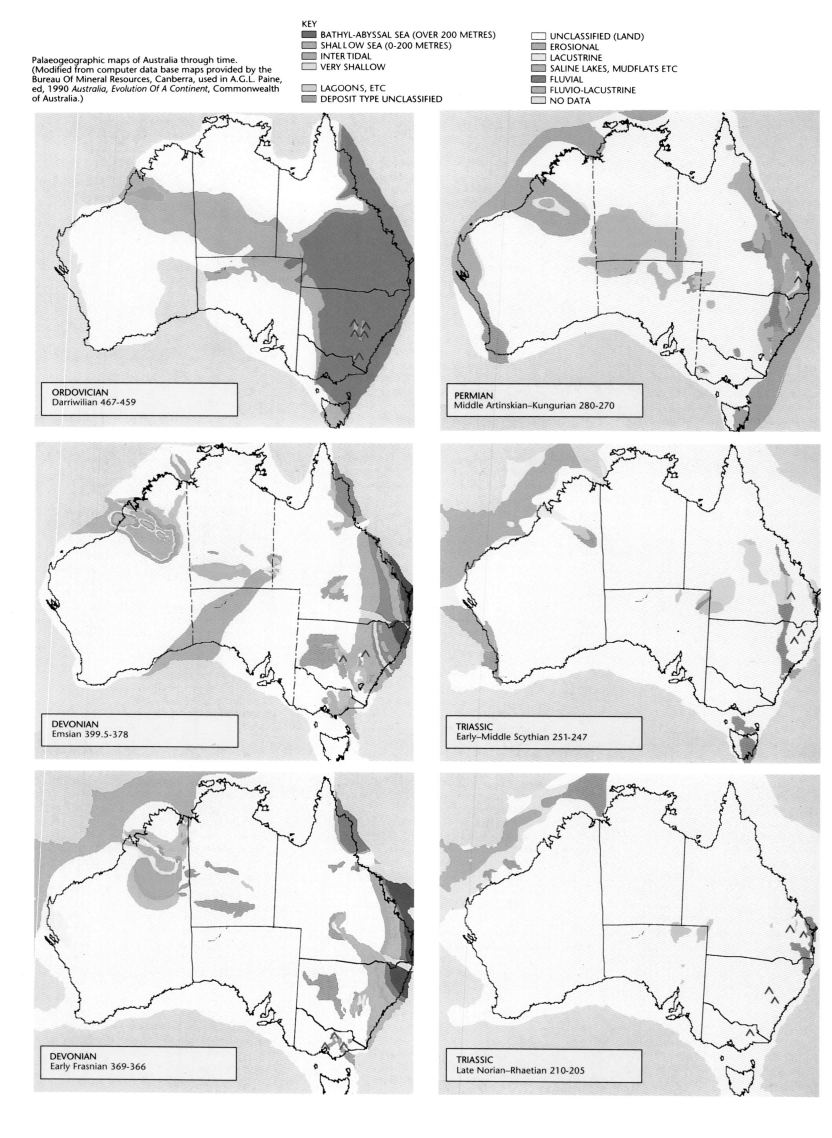

Palaeogeographic maps of Australia through time.
(Modified from computer data base maps provided by the
Bureau Of Mineral Resources, Canberra, used in A.G.L. Paine,
ed, 1990 *Australia, Evolution Of A Continent*, Commonwealth
of Australia.)

KEY

BATHYL-ABYSSAL SEA (OVER 200 METRES)
SHALLOW SEA (0-200 METRES)
INTER TIDAL
VERY SHALLOW

LAGOONS, ETC
DEPOSIT TYPE UNCLASSIFIED

UNCLASSIFIED (LAND)
EROSIONAL
LACUSTRINE
SALINE LAKES, MUDFLATS ETC
FLUVIAL
FLUVIO-LACUSTRINE
NO DATA

ORDOVICIAN
Darriwilian 467-459

PERMIAN
Middle Artinskian–Kungurian 280-270

DEVONIAN
Emsian 399.5-378

TRIASSIC
Early–Middle Scythian 251-247

DEVONIAN
Early Frasnian 369-366

TRIASSIC
Late Norian–Rhaetian 210-205

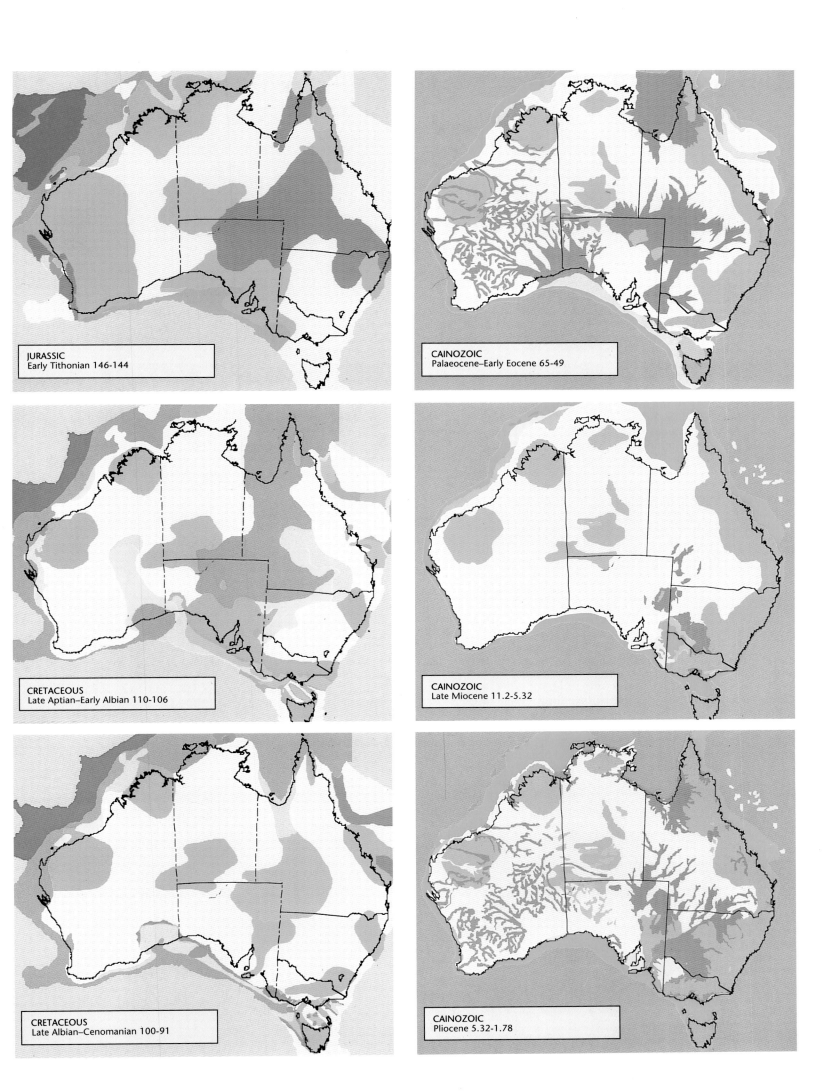

JURASSIC
Early Tithonian 146-144

CAINOZOIC
Palaeocene–Early Eocene 65-49

CRETACEOUS
Late Aptian–Early Albian 110-106

CAINOZOIC
Late Miocene 11.2-5.32

CRETACEOUS
Late Albian–Cenomanian 100-91

CAINOZOIC
Pliocene 5.32-1.78

THE ORIGINS AND PATHWAYS OF LIFE

The Earth is a solid planetary body, an anomaly in space — most of which is empty of matter. What's more, Earth happens to be "just right" in a number of ways, which is why life exists on it at all: it is just the right distance from a star of just the right brightness and temperature to allow it to gain and retain an atmosphere; it generates its own radiation-shielding magnetic field; and its surface temperatures are conducive to water remaining liquid, not vapour as on Venus or ice as on Mars. Life has originated and prospered under these conditions, and although such conditions may not be unique in the Universe, they are not common. So far, we know of life in only this one place.

Earth condensed out of a nebula of compressed gases, starstuff, that already had a long history, perhaps as much as 15 billion years. The Universe, according to one popular theory — the Big Bang Hypothesis — began some 18 to 20 billion years ago with intense heat and rapid expansion. The Earth and other planets in the Solar System apparently began to condense about 4600 million years ago from the dense edges of gas rings flung out by the Sun as it contracted and increased its rotation speed or as the result of a near encounter with a more massive passing star. Astrophysicists still debate their theories about how the Universe came to be, but in many respects their theories share common threads: expansion, contraction, condensation, consolidation.

The outermost planets — the Jovian Planets, including Jupiter and those beyond — are composed of elements in proportions similar to the rest of the Universe. The inner four — the Terrestrial Planets, including Earth — are different. They are small and have weaker gravitational fields than the Jovian Planets, and because they are closer to the Sun they are warmer, encouraging the lighter, more volatile elements to escape into space. After their initial condensation, these planets continued to accumulate additional material, sweeping up space debris over the next 4 billion years or so, thus adding to their original bulk.

LIFE IN THE CRYPTOZOIC

Initially, the Earth heated up, but then cooled. Its Core, Mantle and Crust — its internal structure —began to differentiate some 4300 million years ago. Its surface at this time was fiery with volcanic activity. Like that of its near companion, the Moon, the Earth's surface was also pummelled by meteors until about 3900 million years ago. During these tempestuous times, life first arose and left its sketchy, but nonetheless real, signature in the rocks on this third planet from the Sun.

Some 4200 to 4300 million years ago, when the oldest known rocks on Earth were forming (during the Hadean Eon, the oldest subdivision of the Cryptozoic) — based on dates derived from tiny zircon crystals that formed a part of a sedimentary rock from Mount Narrayer in Western Australia — the Earth would have been a bleak place. It was a hellish time, just as its name suggests, with volcanoes blasting lava fountains in the air, fumeroles steaming, and meteors crashing into the Crust. There was no stable surface on which soil could form, no water, no atmosphere. In short, there was no place that life as we know it could have begun. But, due to the gases and water produced by volcanic eruption, and to a lesser extent by comets, the oceans and atmosphere began to form.

By 3800 million years ago Crust was present, and on it a variety of different kinds of sediments were being deposited, slightly metamorphosed (metamorphosis is a process that subjects rocks to heat and/or pressure and changes them from their original structure). Sediments and volcanic rocks of this age are associated with layered, iron-rich rocks in south-western Greenland. Their occurrence defines the earliest part of the Archean Eon. Bands of silica-rich chert and carbonate-rich dolomite also occur in this sequence, giving rare insights into the state of the Earth in this ancient time.

The Earth by now had both atmosphere and hydrosphere, most likely derived from volcanic exhilations. The atmosphere had significant amounts of carbon dioxide and water vapor, which is indicated by the kinds of sediments present and by the weathering that had occurred. It is also evident that the iron in the dolomites and the banded-iron rocks had been precipitated under low oxygen conditions. Oxygen was clearly not yet an abundant element in the atmosphere. The kinds of sedimentary and volcanic rocks in the Greenland sequence, furthermore, imply that they were deposited in a shallow marine basin near emergent land. The chemical composition of the volcanic rocks was like that of rocks which are erupted in ocean basins today: so, an archipelago in an ocean basin seems the best model to explain these 3800-million-year-old rocks; rocks rich in iron and magnesium —mafic rocks.

Greenland is not the only place that rocks of this age and with much the same nature occur. They are also known from northeastern India, Enderby Land in Antarctica, South Africa and Zimbabwe, and even the north-western corner of the Yilgarn block in Western Australia. Today in the Pilbara region of Australia is a stacking of late Archean rocks that range in age from 3600 to about 2800 million years old, the most complete sequence of sediments, volcanics and granites of this great age anywhere in the world. Places like the Pilbara are windows allowing us to gaze out on our most primordial past, and they tell us that about 3500 million years ago the continents were just beginning to build.

It is in Archean sediments that the first evidence of life appears in the rock record, somewhere between 2800 and 3500 million years old, perhaps as much as 3800 million years. The latter part of this eon is also a revolutionary time for geological processes, a time of change from a primarily oceanic regime of volcanic rocks dominated by magnesium and iron to those dominated by lightweight silica and aluminium, the seeds that formed continents and continental shelves typical of the modern world. In fact, as much as 60 per cent of today's continental masses may have been produced by 2600 million years ago, at the beginning of the Proterozoic Eon.

During the Proterozoic the mechanisms that cause plate tectonics probably turned on, as a consequence of the convective cooling of the Earth's interior, thus providing this planet with its modern geodynamic systems. It was also a time that produced the physical environments — the continents and continental shelves — which today are home for almost all life.

Plate tectonics may have made one more vital contribution to the modern world; it offered a mechanism for recycling carbon dioxide (CO_2) in a fashion that prevented massive build-up of this gas in the atmosphere, which would have led to a runaway greenhouse effect and alarmingly elevated temperatures such as exist on Venus today. Much of the carbon dioxide incorporated into the structure of living organisms eventually settles to the bottom of the ocean basins — either as calcium carbonate bound up in skeletons or shells, or as organic matter constituting the remains of the soft parts of animals and plants (which becomes kerogen, coal or oil in sedimentary rocks). This material can eventually follow a tectonic plate down a subduction zone near an oceanic trench and be incorporated into the Crust and upper Mantle where, upon heating, it is often mobilized as molten rock rising into volcanoes and erupted into the air or water to start the cycle all over again. Or, alternatively, sediments can be uplifted without cycling down a subduction zone, and be subjected to chemical weathering where soil gases and acids act on the kerogen in the presence of oxygen to produce carbon dioxide. The biological recycling of carbonates is more complex, but is well summarized by Robert Berner and Antonio Lasaga (1989): "To sum up this part of the [carbonate] cycle, each molecule of atmospheric carbon dioxide produces a molecule of carbonic acid in the soil. The carbonic acid molecule dissolves carbonate minerals to produce two bicarbonate ions. One bicarbonate ion is transformed by marine organisms into calcium carbonate and buried on the sea floor, eventually to become sedimentary rock; the other is transformed into carbon dioxide. In this way all the atmospheric carbon dioxide taken up during carbonate weathering is ultimately returned to the atmosphere." In addition, recycling brought about by plate tectonics occurs on Earth but not on Venus, and probably is equally critical in determining the fates of each planet's atmosphere. Mars once may have experienced plate tectonic activity, but does so no longer. The atmosphere of Mars is formed almost exclusively of carbon dioxide.

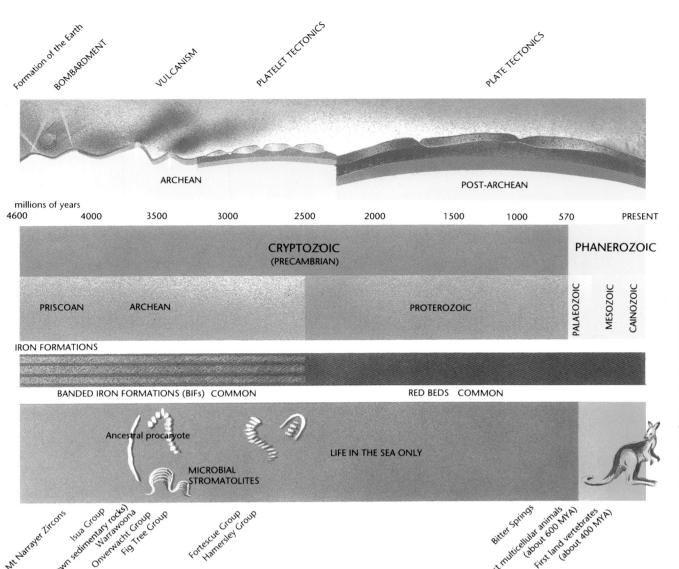

millions of years

4600 | 4000 | 3500 | 3000 | 2500 | 2000 | 1500 | 1000 | 570 | PRESENT

ARCHEAN · POST-ARCHEAN

CRYPTOZOIC (PRECAMBRIAN) · PHANEROZOIC

PRISCOAN · ARCHEAN · PROTEROZOIC · PALAEOZOIC · MESOZOIC · CAINOZOIC

IRON FORMATIONS

BANDED IRON FORMATIONS (BIFs) COMMON · RED BEDS COMMON

Ancestral procaryote · MICROBIAL STROMATOLITES · LIFE IN THE SEA ONLY

Mt Narrayer Zircons · Isua Group (oldest known sedimentary rocks) · Warrawoona Group · Onverwacht Group · Fig Tree Group · Fortescue Group · Hamersley Group · Bitter Springs · First multicellular animals (about 600 MYA) · First land vertebrates (about 400 MYA)

Major events in Earth history since the origin of the planet. The Archean and the Hadean represents almost half of the Earth's history from its formation until about 2500 million years ago (2.5 Ga), when plate tectonics began. Evidently, before that time the dynamic system of the Earth's interior was not active in the way it is now and has been since 2.5 Ga ago. During the Archean, meteoric/comet impact on the Earth was the main surface process in action. Some of the oldest known sedimentary rocks, the 3800 million year old Isua Group in Greenland, made up of water-deposited material, reflects a climatological regime not terribly different from that at present. The oldest remains of living systems occurs in rocks of around or slightly older than 3500 million years. (Figure modified from Cloud, 1988)

LIFE EMERGES — A PRODUCT OF CHANCE

Life was around for some time before it diversified broadly. The oldest life dates from the early Archean, perhaps back as far as 3800 million years based on material from Western Australia. From about 3500 to 2800 million years ago the record of life is definite, being recorded by the widespread occurrence of stromatolites. These generally calcareous structures, laminated and three-dimensional, occur in sedimentary rocks. They seem to represent deposits left behind by microbial communities that formed mats, which in turn trapped sedimentary particles producing impressive mound-like structures. Microbes included a variety of forms — bacteria, blue-green algae — and sometimes the cells of these microscopic beings were trapped in the multitudinous layers of the stromatolites. Modern counterparts of such structures exist today, in places like Shark Bay in Western Australia and the Persian Gulf, inhabiting nearshore hypersaline environments, generally quite restricted in geographic extent. In the Cryptozoic, these same stromatolite-forming organisms were quite widespread, and the reason they were so successful yet are so restricted now seems related to lack of direct competition and total lack of predation in Cryptozoic times.

Both stromatolites and microscopic remains of cells are known throughout the Cryptozoic. The rocks of the Fig Tree Group of South Africa, about 3200 million years in age, contain some of the oldest remains of single cells, those without distinct nuclei, called procaryotes. At least two kinds of fossils can be recognized: tiny rod-shaped forms that are quite similar to some species of living bacteria, and spherical structures reminiscent of blue-green algae. These cells are small, most no longer than 20 microns (20/1000ths of a millimetre). The Warrawoona Group in Western Australia has likewise produced filamentous and colonial fossil micro-organisms that may be as old as 3800 million years. In the age range between 2200 and 1800 million years old many locales throughout the world, including some in Western Australia, have produced a variety of single-celled forms. Even younger cherts (silica-rich sedimentary rocks) from Bitter Springs in the Macdonnell Ranges of the Northern Territory have yielded 900-million-

year-old single-celled organisms, beautifully preserved, of about the same size as the older ones, and also apparently lacking a nucleus.

Besides the presence of structures that can be identified as the fossilized remains of once living organisms, there are chemical signals in the rock sequences of about this age range which foreshadow the presence of life at the time the sediments were being deposited. In the Onverwacht Group of South Africa there is a distinct change in the ratio of one isotope of carbon (Carbon 13) relative to another (Carbon 12). In the lower parts of the Onverwacht sediments the ratio of Carbon 13 to Carbon 12 is very high, but in the upper part of the formation there is a real reduction in this ratio. Carbon 13 and Carbon 12 both naturally occur in the atmosphere, but when green plants metabolize and use carbon dioxide they selectively concentrate Carbon 12 in the carbohydrates that they produce in the photosynthetic process. This process seems to have been occurring when the upper Onverwacht Group was being deposited, and the ancient algae have left their indelible mark.

During this Cryptozoic time of procaryotes and stromatolites, of changing carbon isotope ratios, the oceans and atmosphere had distinctly different compositions than those of today. One element conspicuously absent was free oxygen. The environment was reducing, not oxidizing. Humans and most life forms that prosper in today's world would have been poisoned by such a hideous environment. It was plants, using carbon dioxide and the Sun's light energy to produce their own food, that gave the world oxygen as a byproduct of their metabolism. Not until about 2000 million years ago did oxygen become a noticeable component of the world's environment, and this chemical change brought about a major revolution.

Because of the low oxygen level prior to 2500 million years ago the sedimentary record is neatly sliced into two parts, with a slight gradational boundary between. The "time before oxygen" is characterized by the presence of BIFs (Banded Iron Formations), long sedimentary sequences of iron-rich and iron-poor laminae sandwiched with other bands of the silica-rich chert. After oxygen becomes abundant, sedimentary red beds become widespread. The two

31

18

biological processes were at work. Photosynthesis seems to have been an integral part of these processes.

Glaciation or some other environmentally cyclic event could have controlled the rhythmic upwelling of iron-rich waters. The iron was utilized by microbes that photosynthesized, a process which produced oxygen as a byproduct, which in turn favoured the precipitation of insoluble ferric hydroxide and its conversion to hematite. Other hypotheses exist, but a thorough explanation of just how BIFs formed is not at hand. Despite our poor understanding of just how they came to be, we do know that their presence indicates low oxygen levels in the environment. Red beds indicate the presence of atmospheric oxygen, and they replaced the BIFs by 1900 million years ago, when they began forming for the first time, almost simultaneously in many places around the globe.

Evidence of life existing on planet Earth for nearly 3800 million years is frozen in the Archean and Proterozoic rocks. Although many details need to be filled in, there is some agreement amongst scientists on how life may have originated.

Life most likely came about through the self-assembly of small organic molecules into larger, more complex structures. One hypothesis uses the surface of clays or crystals to form the template upon which the more complex structures formed. Short chains of amino acids or nucleotides could have been linked into more complex proteins and nucleic acids, the building blocks of DNA. The short chains of amino acids and nucleotides have been experimentally constructed under laboratory conditions since the 1950s, beginning with the famous experiments carried out by Stanley Miller and Harold Urey at the University Of Chicago. Their work simulated Archean Earth conditions by using an electric spark representing lightning (or solar ultraviolet radiation or some other form of energy) and reacting it with what is thought to be a primitive atmosphere of hydrogen, methane, ammonia and water vapour. Their experiment produced amino acids, and later versions of it produced some of the bases involved in the structure of the DNA molecule itself.

There is no doubt that the Earth had an ample supply of the basic organic compounds needed to build the more complex structures of life. Processes known to encourage more complex linking include dehydration and freezing. But at the moment, the steps from more complex proteins and nucleic acids to full-blown, self-replicating DNA have not been made. Keep in mind, however, that humans have been working with this process for only 40 years. Nature had more than 1000 million years of experimentation before life arose on Earth. It is not wise to too quickly dismiss as failure human efforts along these lines!

Life was well established towards the end of the Proterozoic. It and plate tectonics most likely played an important role in ensuring moderate temperatures on Earth and retention of an atmosphere conducive to life's continuation. Only one critical substance was missing that was to provide the building blocks for the modern biota — and the longest winter the world has ever known was soon to remedy that.

kinds of sediment do not co-exist for long, and BIFs are not being formed now.

BIFs are important not only for understanding past atmospheres but also because they are now vital to the survival of modern industrial nations —they are the major sources of iron ore in the world. The Hamersley Basin of Western Australia and the neighbouring Nabberu Basin, together with three other areas in Brazil, Canada and the Transvaal of South Africa, hold 92 per cent of the world's minable iron reserves. This iron was all deposited as BIFs in Archean times, when oxygen was essentially missing in the environment. But how did these massive and incredibly cyclic accumulations form? There are many ideas and no complete agreement amongst geologists, but one widely supported hypothesis relates them to the rhythm of seasonal algal blooms which would have produced a temporary abundance of oxygen in the environment, coupled with upwelling of iron-rich waters from deep anoxic basins. The deep iron-rich water may have come into contact with high oxygen concentrations on continental shelves in areas like those along coastal Chile today, leading to deposition of the ferric oxides (the red bands). Just as in the uppermost part of the Onverwacht Formation in the earlier part of the Archean, light carbon is in abundance in these layers, suggesting that

---

18 Banded iron formation (BIF) from Precambrian rocks of Western Australia. BIFs are our major source worldwide of metallic iron, necessary for the industrial society in which we live. They are an integral part of the older Proterozoic sequences, greater in age than 2 billion years. More than 90 per cent of the world's iron reserves are BIFs that occur in the Hamersley Basin of Western Australia and in Brazil, eastern Canada, the Transvaal of South Africa and the Ukraine — more than a hundred trillion (10^{14}) tonnes are known from these areas. The Nabberu Basin, neighbouring the Hamersley, has known reserves of 10^{13} tonnes of iron ore! x 2.0 (S. Morton)

19 Single cells from the silica-rich cherts of Bitter Springs, near Ellery Creek in the Macdonnell Ranges, Northern Territory. These cells, about 900 million years old, are the right size (about 10/1000ths of a millimetre, or 10 microns) for procaryotes, cells that lack a nucleus. The nucleus-like structures in the cells in most cases appear to be pyrite crystals. (J. Warren and I. Stewart)

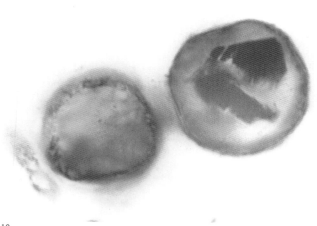

19

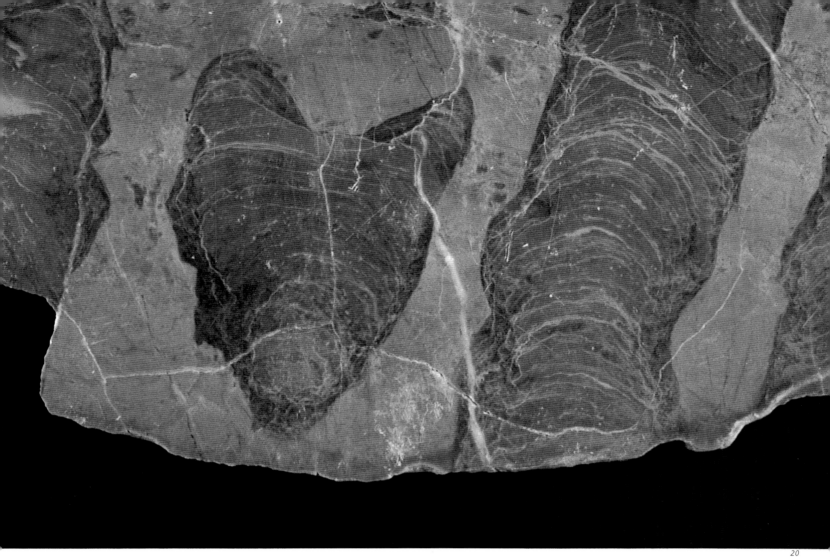

THE PATH TO SEXUALITY AND MANY CELLS

The Earth had known glaciations several times during its history. The ice sheets and low temperatures that gripped the planet from about 1000 million to about 600 million years ago, however, lasted the longest of any ice age ever. Not only did the ice persist for a long period, but it blanketed most continental masses, even those near or on the Equator. And it was not confined to small alpine glaciers, but occurred in full-blown, thick continental ice sheets. In Australia, this glaciation is well recorded in the Umberatana Group exposed in the Flinders Ranges of South Australia. Tillites, striated surfaces and cyclic lake deposits called varvites are all there. So, too, are drop-stones, rocks that have been carried seaward in icebergs and unceremoniously dropped into fine-grained sediments lining the ocean bed beneath where the icebergs melted. Indeed, ice sheets gripped much of Australia from about 900 million years ago to about 670 million years ago, imposing frigid desolation across two-thirds of the present-day continent from Tasmania to the dead heart of Central Australia and spreading to the north-west as far as the Indian Ocean.

Many theories explaining the causes of ice ages have been offered. Several factors seem to be involved, and not all are responsible for every ice age. Reduced solar radiation may be involved, and greater reflectivity of the growing accumulations of ice and snow further reduce retained heat. For ice and snow to build up, moisture must be available. Variations of the Earth's orbit around the Sun (that is, how close the Earth is to the Sun at any one time), variations in the inclination of the Earth's axis to the Sun, volcanic activity, positions of the continental masses with respect to the rotational axis, and even the effects of Apollo objects such as meteors and comets, are all implicated as triggers of ice ages. But, for the most part, glaciation has affected only the poleward parts of continental masses, and only occasionally throughout geological history. How the ice ages at the end of the Proterozoic came to happen, affecting so many continents thought to lie near the Equator, is a dilemma which is far from resolved. Undoubtedly, it was a most unusual time in the history of the planet.

This late Proterozoic glaciation had a profound effect on life. Extinctions were rampant, dramatically reducing diversity. Something else occurred during this time, however, that had positive effects: the first clumping of cell-like structures in an arrangement different than just a linear chain or flat mat. The first truly multicellular animals

20 Fossil stromatolite columns, sliced and polished, from the Precambrian of the Macdonnell Ranges, Northern Territory. These ancient algal structures are about 900 million years old. x 2.4 (S. Morton)

21 Living stromatolites in Shark Bay, Western Australia, are very similar to those known from Precambrian rocks around the world. Today the algae and bacteria that deposit these structures are much more limited in geographic extent, perhaps because of predation pressure. The circular, massive growth form of these stromatolites reflects the energetic nature of their environment when the tide is in. (B. Bolton)

arose out of this long winter. Some of the first evidence of that multicellular life is recorded in rocks about 670 million years of age in the Flinders Ranges of South Australia — the Ediacara Fauna, a biota of segmented worms, jellyfish, sea pens and even the oldest known spiny-skinned animals (the echinoderms), close relatives of the vertebrates. Ice may have been one trigger to the development of multicellularity, but so, too, may have been oxygen. Oxygen levels in the atmosphere had been on the rise since at least 2000 million years ago. Multicellular animals definitely have intolerance of oxygen levels below 10 per cent of present atmospheric concentrations. There is evidence for considerable oxygen build-up in the atmosphere in the 400 million years before the Proterozoic–Phanerozoic boundary, which may be implicated in the development of multicellularity and may also be somehow interrelated with the glacial events. Much is yet to be learned about this revolutionary time in the history of the biosphere.

Sometime before this 400-million-year period single-celled organisms had developed a nucleus, which concentrated material critical for recording and transmitting genetic information from one generation to the next. This development may have occurred as far back as 1400 million years ago, because at about that time the average size of cells increased dramatically. Cells without a nucleus, procaryotic cells, are small, generally less than 20 microns in diameter. At about 1400 millions years ago cells of about 60 microns in diameter appear in the record, more the size of living nucleated cells (eucaryotic cells). It is not until much later that a nucleus is clearly visible in a fossilized cell; nevertheless, this size increase at 1400 million years is very suggestive of the evolutionary jump to nucleus-bearing — a jump made possible by the development of a nuclear membrane to isolate the DNA, perhaps by the union of two procaryotes.

Sometime later sexual reproduction began, giving organisms the ability to vastly increase variation from one generation to the next. No longer was an offspring a carbon copy of a single parent's own DNA, but a mixture of two parents, combined with the possibility of innovation by mutation and recombination of existing DNA material. By the time of the Ediacaran animals, all this and multicellularity had come to pass; it had taken 85 per cent of the known history of the Earth for this development to happen.

Then, the Earth began to warm.

TOWARDS A NEW BEGINNING
The grip of the glaciation began to ease at the end of Proterozoic time. Its last effect, its parting shot, however, may have been the mobilization of frigid, phosphate-rich water, generally confined to the depth of ocean basins. One theory proposes that the intense and lengthy glaciation of the late Proterozoic cooled the ocean's surface waters so much that they became more dense than the deep waters of the ocean basins. The surface waters sank under their own weight, and in doing so forced to the surface the phosphate-enriched bottom waters. Suddenly an element that had been in short supply, even rare, became abundant. Those organisms that had the ability to utilize such a newcomer may, in fact, have deposited the first skeletons. But not all palaeontologists would accept this idea. Perhaps skeletons are the result of more sophisticated modulation of cellular calcium and phosphate ions or simply the evolution of more efficient predators. Regardless of the cause, so began the Eon Of Abundant Or Evident

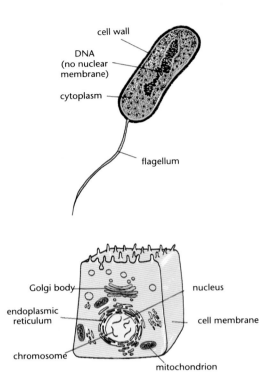

Procaryotic cells have a simple organisation. They are single cells, each surrounded by a sturdy cell wall. There are no substructures within the cell and the DNA is in direct contact with the cytoplasm — that is there is no distinct nucleus. Bacteria and blue-green algae are typical procaryotes. (Modified from Raven and Johnson, 1986)

Eucaryotic cells are complex in that they contain many organelles and encase their DNA within a double membrane. A series of membranes, the endoplasmic reticulum, separates the different parts of the cell into compartments which each have different functions. (Modified from Raven and Johnson, 1986)

Life, the Phanerozoic — "abundant" and "evident" because so many of the animals and plants that lived for the next 590 million years or so left behind hard skeletons that were easily fossilized. Their forerunners had not been so graciously endowed.

LIFE IN THE PHANEROZOIC
At the beginning of the Phanerozoic the Earth moved from an icehouse to a greenhouse regime. However, throughout this eon climates oscillated between the two extremes. During this time, too, plate tectonics reshuffled the continents several times. Such climatic and geographic dynamism had a profound effect on the course of evolution.

BEGINNINGS OF A BACKBONE
Once organisms were able to produce hard tissues, to deposit skeletons and shells of various kinds, the fossil record literally explodes with variety. Skeletons have many advantages for their bearers. Onto them muscles can be attached, with a resulting greater efficiency of energy use in locomotion and feeding. External skeletons can provide armour protective against the ever-present enemies, both predators and competitors.

Skeletons provided flexibility and opportunity, and by their very nature were necessary ingredients for the development of the vertebrates.

The history of vertebrates is not an even spacing of events. There have been periods of innovation followed by frantic expansion, with the development of a great variety of new forms produced by their evolving into previously unoccupied ecological space or by fierce competition for that space and concomitant displacement of the unfortunate "less fit" species. Then, for a time, stability may have prevailed, often followed by extinctions. And then the entire play may have begun again. Radiations of new forms also may have been triggered by innovation "forced" upon a population already present, or it may have been preceded by mass extinctions caused by some external factor such as a deadly visitor from outer space (possibly a comet or meteor producing an explosion surpassing a hundred million Hiroshima bombs) or perhaps the advance of massive continental glaciers accompanied by lowered sea levels as the ice demanded fodder from the ocean waters. Expansion into an ecological void thus produced may have been rapid, delayed or may never have occurred at all. Vertebrates, like all organisms, have had an uneven history, one full of both danger and opportunity, and also guided by chance.

22-25 A variety of multicellular animals from late Precambrian rocks in the Flinders Ranges, South Australia. (N. Pledge, South Australian Museum)

22 Dickinsonia costata, probably some sort of an annelid worm. It is about 6.5 centimetres long.

23 Tribrachidium heraldicum, about 2.5 centimetres in diameter. Its relationships are unknown.

24 Parvancorina minchami, an animal of about 2 centimetres in length, whose relationships may be with crustaceans. (N. Pledge, pers. comm.)

25 Mawsonites spriggi, a jellyfish-like animal of about 8 centimetres in diameter.

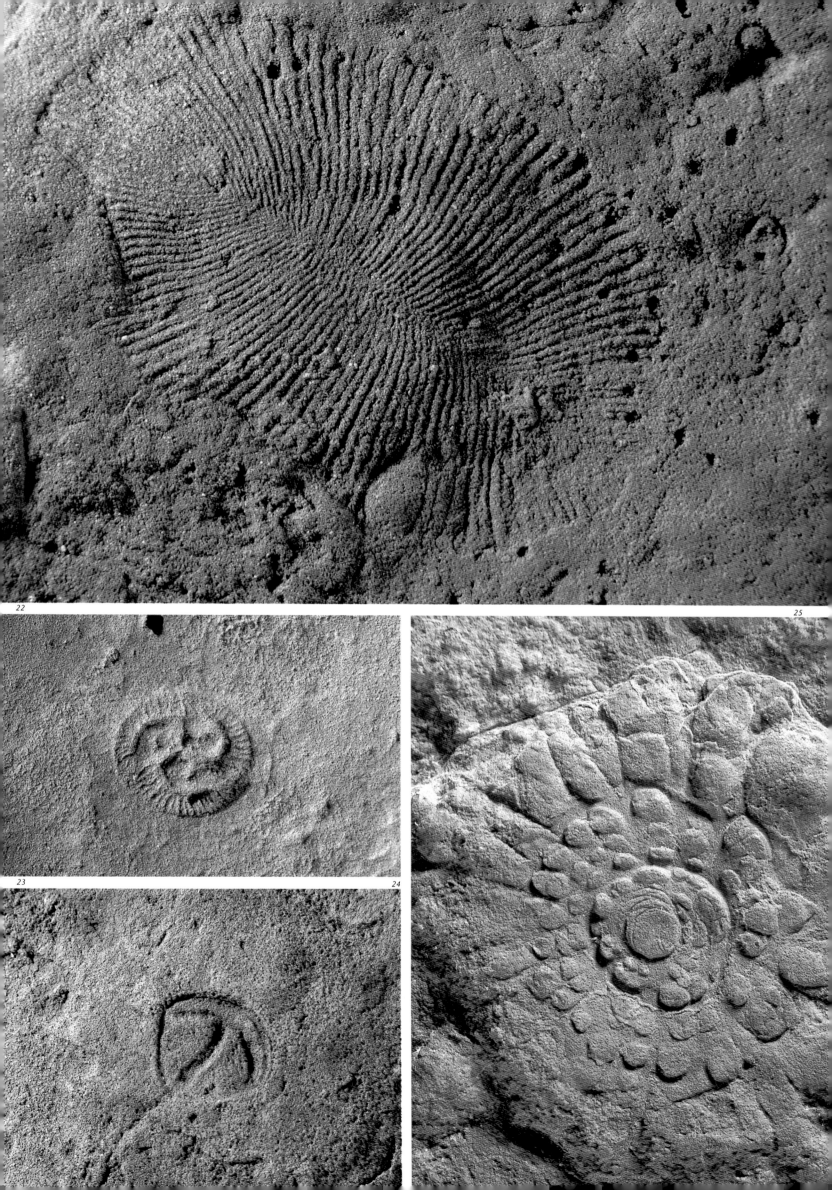

22

23

24

25

26

THE FIRST VERTEBRATES

The earliest record of vertebrates, that is, animals with backbones, is from the Cambrian Period, but they are probably not the very first vertebrates. These organisms, which left behind their mosaic of phosphatic plates and scales, lived in the shallow seas of the Late Cambrian and Early Ordovician in several parts of the world including North America, Greenland and Spitzbergen. They left only tiny scales and body armour made of apatite, the mineral constituent of bone. These remains were only a few millimetres square and less than one-tenth of a millimetre in thickness. They were scattered amongst the skeletons of trilobites, brachiopods and the microscopic conodonts, all marine forms.

26 *Trilobites, which at the beginning of the Phanerozoic, some 545 million years ago, were among the first animals to develop hard parts. They rapidly diversified and together with the brachiopods dominated the marine realms for much of the Early Palaeozoic. x 2.0 (S. Morton)*

27 *The living Lancet, Branchiostoma, also called amphioxus (meaning "sharp at both ends"), which is found in shallow marine waters today. It is an invertebrate with a notochord-like structure, and may be the closest living analogue to the ancestral vertebrate. The notochord is a dorsally situated support rod found in primitive vertebrates, which is replaced by vertebrae in more advanced forms. x 4.5 (S. Morton, courtesy of Monash University)*

Australia and Bolivia, two Gondwanan fragments, provide the first glimpses of what these early vertebrates, primitive fish, looked like whole. *Sacabambaspis* from 470-million-year-old brachiopod-bearing sediments in Bolivia and the somewhat younger *Arandaspis* from the early Middle Ordovician of Central Australia are the first partial or nearly complete vertebrates known. Both are of simple design, having no fins other than a tail fin, and no moveable jaws — they were essentially filter-feeding organisms.

These early records give little hint at vertebrate ancestry; those clues come primarily from the anatomy, embryology and biochemistry of living invertebrates.

AN INVERTEBRATE ANCESTRY

Vertebrates belong to a group called Chordata, which includes animals with a single nerve chord situated on the dorsal or top side of the major internal organs — it is a hollow, fluid-filled tube, not a solid one as in many other animal groups. At some stage in their development all chordates have gill slits in the throat region; even humans possess these in their embryonic life. And, as their name implies, at some stage in development they possess a notochord, a structure that in the more primitive chordates is the main support for the body and lies just above the gut. The notochord is a significant support structure in primitive fish such as *Arandaspis*, but wanes in importance as the calcified or truly bony vertebrae develop in the course of evolution. In higher vertebrates this support chord is surrounded and eventually replaced by the vertebral column. Reduced away to almost nothing, it is the jelly-like substance that occupies the hollow area of bony fish vertebrae.

Extrapolating back from the most primitive vertebrates known, some estimates can be made about what the ancestral vertebrate looked like. It would have been a small animal, probably just a few centimetres long. It would most likely have had a fusiform, fish-like body that was segmented longitudinally into many parts — basically muscle segments associated with propulsion. This ancient vertebrate would have had a notochord to which its segmented muscles attached. At the end of its body would have been a tail fin, perhaps the only fin, whose main purpose would have been for propulsion. Head and body would have been closely integrated; there would have been no distinct neck. The dorsal, hollow nerve chord would have lain above the notochord, and would have been connected and controlled by a small brain situated in the front of the head. There were probably a number of paired organs in the head region associated with sight, smell and balance and, in addition, there would have been a system of lateral line organs, canals probably containing fluid of some sort, forming a branching array over the surface of the body. This system was sensitive to changes in water pressure produced by movement, and may also have registered electrical currents. Finally, this early vertebrate would have had a mouth and a set of gill slits through which water passed. These slits were probably associated with thin, lamellar gills that allowed for gas exchange, enabling this "fish" to obtain the needed oxygen from the water. Feeding was most likely an engulfing affair, for no true jaws were present — prey was simply surrounded and dispatched whole. The development of active swimming habits provided opportunity for diversification, as well as for a less dangerous lifestyle — a variety of benthic arthropods were a major threat to sedentary bottom dwellers. About the only carnivorous forms in the water column during the early history of vertebrates were the nautiloids, distant relatives of the *Nautilus* of today. New feeding opportunities and the ability to escape were the main gifts of the newly acquired vertebrate pedigree.

One tiny marine invertebrate today that has many of the characteristics expected in an ancestral vertebrate is the lancelet (*Branchiostoma*). It lives, often half buried, in the sands of tropical and warm temperate marine waters around the world. Lancelets can swim with fish-like undulations, utilizing their segmented muscles which are attached to a well-developed notochord. They feed by drawing water into the mouth and sifting food out using their "gill" slits. These slits do not open directly to the outside as they do in fishes, but into a chamber that surrounds the throat region. This chamber in turn opens to the outside through a pore some distance to the rear. Perhaps this arrangement is a specialization with lancelets, or perhaps it hints at what early chordates, the ancestors to vertebrates, were indeed like. *Branchiostoma* lacks eyes and a brain

associated with its dorsal hollow nerve chord (the chord in front simply tapers to a point), both conditions which may be due to the sedentary bottom-dwelling habit of this genus. Despite some specializations that have undoubtedly occurred in this form, simply because of the passage of time, it is still the closest living approximation to the primitive chordate that could have given rise to animals with a true backbone. A chordate, very like *Branchiostoma*, is known to have existed in the Middle Cambrian of North America — a form called *Pikaia* from the Burgess Shale of British Columbia, Canada.

There are a few other living "invertebrate chordates" besides the lancelet, such as acorn worms and sea squirts that will shoot water at you as you wander along a beach, but they do not resemble the primitive vertebrate so strikingly as does the lancelet. Nevertheless, all of these animals share at some stage in their lives those three characteristics of chordates: notochord, dorsal hollow nerve chord, and gill slits. No other group in the world of living things has this same combination of attributes.

How then are these animals related to others? The evidence of relationships is provided mainly by embryological and biochemical studies. Although it may seem unlikely, the group most closely related to the chordates, and so to the vertebrates, includes starfish and sea urchins, the Echinodermata. The echinoderms share with the chordates a similar method of embryological development of the mesoderm, or the body layer which eventually differentiates into muscles, blood vessels and the coelom (the body cavity which houses vital organs, such as the liver and spleen). In echinoderms and chordates, the mesoderm first appears as pouches growing outward from the walls of the embryonic gut, and the body cavity develops as these pouches close off enclosing space within. This method of development is very different from that in worms and clams, where the mesoderm begins as a solid mass of cells that bud off from the posterior end of the embryo and in which the body cavity develops as slits within this tissue.

A further link between chordates and echinoderms is the construction of the larval forms of some of the invertebrate chordates, which are strikingly similar to those of the starfish. Yet another bond is the similarity of phosphorus compounds instrumental in energy release during muscular activity. Vertebrates and some echinoderms utilize the same phosphate compound, creatine, whilst almost all other non-vertebrate groups use argenine. There are even some fossil echinoderm groups, such as the carpoids, that may form links with the chordates when the detailed morphologies of each are compared, but further work is needed before general agreement amongst palae-ontologists and biologists is reached.

THE VERTEBRATE PEDIGREE — BONE AND THE NEURAL CREST

The transition from echinoderms and lancelets to true vertebrates can be tied to the development of three kinds of embryonic tissue: the muscular hypomere, the ectodermal placodes and the neural crest. The mature organs derived from these three different embryonic tissues were responsible for the "birth" of active, mobile animals

capable of a predatory lifestyle from the passive, filter-feeding invertebrate chordates. Active predation involved the development of decidedly more efficient patterns of locomotion, a major re-organization of the pharynx, and elaborations of the circulatory system as well as the digestive, nervous and special sense organ systems.

The muscular hypomere, an extension of mesodermal tissue beneath the segmented muscles along the body axis, contributes to the muscles of the heart and those involved in more efficient transport of food and water in the vertebrate body. This elaboration allowed development of a higher metabolic rate that is typical of vertebrates when compared to the more primitive chordates and many other invertebrate groups.

Structures derived from the ectodermal placodes give rise to skin, and also to the brain and special sensory structures like eyes and the lateral line systems present in early vertebrates. These latter two are especially important features in the evolution of a vertebrate from an invertebrate chordate ancestor.

The neural crest cells are found only in vertebrates. They are derived from the edges of the forming nerve chord and are crucial contributors to much of the skeleton (both dermal bone formed in the skin and endochondral bone formed from a cartilage precursor), the muscles of the jaws and gill arch skeleton, many parts of the nervous system associated with sight, hearing and so forth, and even pigment and brain cells. From the neural crest came bones and teeth, unique features to vertebrates and vital to both the success and the widespread fossilization of this group.

When did bone first appear? In the oldest known fossil vertebrates, bone (dermal bone) is present only as a superficial body covering. A complex tissue, it has characteristics of both the bones and the teeth of living vertebrates. It is made up of a basal layer of stratified bone in which no holes or lacunae (which are sites in living vertebrates of the bone-producing cells) occur. This acellular bone developed around openings that resemble the pulp cavities in modern teeth. In fact, the whole structure of this ancient bone resembles the structure of a

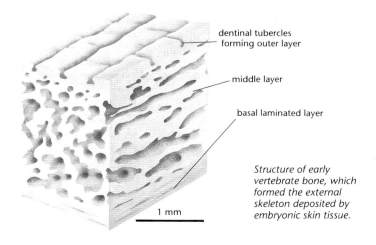

dentinal tubercles forming outer layer

middle layer

basal laminated layer

1 mm

Structure of early vertebrate bone, which formed the external skeleton deposited by embryonic skin tissue.

tooth: the acellular bone is very like the enamel and dentine that surrounds a tooth's pulp cavity, which itself has a good blood and nerve supply.

Such superficial bone definitely pre-dated bone of the internal skeleton. The ability to ossify tissues within the body, thus, seems to have evolved later than the ability to produce an exoskeleton. Cartilage most likely preceded bone as an internal support.

Calcified cartilage, that is, cartilage with calcium carbonate deposited within its structure, is first known in Ordovician vertebrates. But, although such calcified cartilage was an effective support for the vertebrate body, bone had many superior features. Bone is composed of the mineral apatite — $Ca_5 (F, Cl, OH)(PO_4)_3$. The inclusion of the elements fluorine and phosphorous in bone makes it much stronger and harder than calcified cartilage. A smaller amount of bone is needed for the same structural effect.

Just how bone came to be is a topic of great debate, certainly not yet resolved. Some palaeontologists have argued that bone first developed not as protection but as a phosphate reserve. Phosphate is needed for storage and transfer of energy, such as that used to activate a muscle. A lack of phosphate can be a limiting factor in the success of a species. Superficial bone of some early fishes shows changes that may correspond to the seasonal deposition and resorption of phosphate; in this case the bone may have been a "phosphate bank", where this element was kept in reserve until needed during times of stress. Once present, no matter what the initial reason for formation, bones could have taken on other functions such as muscle attachment, body support and protective covering. Another possible trigger for the development of bone might be related to its ability to insulate. As some biologists have suggested, it may have insulated the electrosensory organs of many early vertebrates (those associated with the lateral line system).

Perhaps, in fact, both ideas are to a certain extent correct. Maybe bone was originally deposited as a calcium phosphate reserve. Then this material may have preferentially crystallized near electrosensory organs, providing an added service to the vertebrate. Once deposition around certain features was possible, natural selection may have favoured its deposition in other places. Certainly primitive fishes with some sort of external protection, a suit of armour, would have had many advantages over their naked counterparts — and even a slight difference in individuals within one population would have ensured the armoured ones left more offspring in the next generation, a basic principle that in time led to the armoured group succeeding over the other.

Whatever the trigger for the development of bone, it did develop by the Late Cambrian, and to its possessors it gave distinct advantage. A rich tapestry of finned and scaled vertebrates evolved over the next 50 million years. By the beginning of Devonian times, some 400 million years ago, many vertebrate groups, all fish, had developed the capacity to ossify not only the dermal or "skin-derived" skeleton, but the internal skeleton as well. If formed as in modern vertebrates, the bones of the internal skeleton were preceded by cartilage precursors. Cartilage is deposited by much the same kind of tissue that forms bone, but instead of collagen fibres being laid down, a matrix of sulphated polysaccharides are deposited and they produce a firm gel through which is spread a spiderweb network of connective tissue fibres. This structure is then gradually replaced by bone. First the cartilage is invaded by blood vessels, and the cells that previously had been laying down cartilage switch function and begin breaking it down. Then the bone-forming cells, the osteoblasts, switch on and begin depositing bone. By having such precursor supports made of

28 A skull, in anterolateral view, of the lungfish Chirodipterus australis from the Late Devonian Gogo Formation at Paddy's Valley, Gogo Station, Western Australia. Lungfish develop, through their history, very distinctive teeth in both the upper and lower jaws — not separate isolated teeth as in most other fishes, but fanlike crushing plates that are used to process a variety of foods, mainly plant remains. Two such tooth plates occur in the upper jaw that are opposed by a single plate in each of the rami of the lower jaw. They provide a broad surface area over which food material can be pulverized. (J. Frazier, courtesy of the Western Australian Museum and K. McNamara)

cartilage, growth from juvenile to adult is allowed — the final ossification does not occur until the adult size is reached, with the more central parts of bones ossifying first, followed by the ends.

During the first 100 million years of vertebrate history, from the Late Cambrian to the Late Silurian, a myriad of jawless fishes prospered. Today, jawless fishes are novelties — only the lamprey and the hagfish live on. The demise of the early fishes came in the latest Silurian when another major evolutionary novelty arose.

CONQUEST OF THE WATERS — A PALAEOZOIC BEGINNING
The development of bone and the consolidation of vertebrae in the internal skeleton had provided protection for the vital nerve chord and an anchor for locomotor muscles. The next major innovation in vertebrate history was the development of jaws.

Jaws must have first developed sometime during the Ordovician, however the first jawed fishes don't appear in the fossil record until the Early Silurian, 100 million years later. Jaws allowed the development of the first truly predatory vertebrates, and also enabled herbivorous forms to cope with a greatly expanded variety of foods. Almost simultaneously with the appearance of jaws, a variety of fins — pectoral, pelvic, dorsal — developed, allowing greater manoeuvrability and finer control of movement. All in all, a much more versatile vertebrate arose in the years of the Silurian, one capable of utilizing a great deal more of the ecological space in the watery environments of the Earth.

Where did jaws come from? Primitive sharks may provide part of the answer and embryology some as well. The jaws and the gill supports in jawed fishes, especially primitive sharks, are similar in many ways — and it may be that jaws evolved from the specialization of several of the most anterior gill arches. Both the gill bars and the jaws have a hinge between top and bottom elements, with the bottom elements bending forward. The muscles that move the jaws are quite similar in structure, and in their attachment on the jaw, to muscles that operate the gill arches which open and close the gills. Both the jaws and the gill arches form from the same embryonic tissue, the neural crest, whereas much of the remainder of the vertebrate skeleton does not (it forms from the mesoderm not the ectoderm). Through time the core of the upper jaw may have formed from the top half of the most anterior gill arch. This core became associated with the braincase, sometimes only lying beside it loosely attached with connective tissue and muscles, sometimes completely fused with it. The lower jaw may, similarly, have developed from the bottom half of this most anterior gill arch. Later, other bones that formed in the skin became associated with these two core components. The hyomandibular bone, the top half of the second gill arch from the front, early in the history of jawed vertebrates formed a support for the upper jaw, often being involved in its direct attachment to the braincase.

All this sounds plausible, but another theory, recently proposed, suggests that the jaws had no precursor in the gill arches at all, but instead developed as totally new structures associated with the mouth. Which is right? Further study of fossil material is the most likely way to resolve this conflict. Palaeontologists certainly do not agree on which theory is correct at the moment.

With jaws came two other major new developments, paired fins and a braincase very similar to that of modern fishes. The combination of these characteristics endowed their possessors with greater feeding opportunities as well as greater manoeuvrability, better coordination and better protection for vital organs. In the later Silurian, some 420 million years ago, several groups of fish appeared, nearly at the same time, that had all of these attributes. One group included the precursors of modern fishes, the bony fishes (the Osteichthyes) and the sharks (the Chondrichthyes). Another was a collection of archaic fishes that would not last out the Palaeozoic Era, the spiny "sharks" (the acanthodians) and the generally heavily armoured placoderms. The acanthodians possessed a number of characteristics that suggest they are related to the modern bony fishes, but the placoderms were a group unique to themselves. For a while the placoderms were fabulously successful, only to succumb to the late Palaeozoic extinctions as the whole world cooled.

The great wealth of fishes that had developed by the middle of Palaeozoic times invaded and prospered in much of the environment previously tended by invertebrates or by nothing at all. Herbivores

with many different lifestyles and carnivores of all sizes, up to several metres, developed in seas, lakes and rivers. Scavenging and filter feeding offered other opportunities. But vertebrate life was restricted to the water until the early part of the Devonian, when it developed the ability to move on land.

INVASION OF THE LAND
Amphibians were the first vertebrates to emerge on the land, a land already inhabited by plants and a variety of invertebrates. Today, several amphibians are basically land dwellers, but many must still return to the water to mate and lay their eggs, and many have an aquatic larval stage. Amphibians of today are frogs (Anura), salamanders (Urodela) and a small group which includes the legless caecilians (Apoda). Modern amphibians, however, are only a tiny remnant of a once much more diverse group, which included the Labyrinthodontia, named for their teeth with bizarre, complexly folded dentine.

29

The labyrinthodonts first appeared in the Early Devonian of Australia, and by the latter part of that period they were found in other parts of the world. Their numbers began to dwindle globally in the Late Triassic, their last stand being in Australia where they persisted until the Early Cretaceous. Another Palaeozoic group, the lepospondyls, is a heterogeneous assemblage of early amphibians that may have been derived several times from different labyrinthodont groups. Just how these more ancient amphibian groups are related to the modern assemblages is not yet fully resolved. It does seem, however, that the labyrinthodonts themselves came from one specific group of fish, the rhipidistian crossopterygians, which share a number of features with early labyrinthodonts, including the labyrinthine infolding of dentine in their teeth. The terrestrial habit may have developed as a survival strategy for rhipidistians living in areas frequented by drought, or it may have been sparked by the lack of competition for food resources on land or even by freedom from predation in that environment.

These earliest vertebrate land dwellers, the labyrinthodonts, were somewhat crocodile-like in appearance, but much less streamlined. Most had large, often flattened skulls with complex ornamentation. They were of moderate size, averaging about a metre in total length. These animals had to solve many problems as they moved onto land, the most significant being how to support the body out of water, how to feed out of water and obtain water and air in a non-aqueous environment, how to retain water and not dehydrate, and how to reproduce. Support was organized by developing strong connections between the vertebral column and the shoulder and pelvic girdles, and those in turn with the limbs. With land life necessitating mobility between the head and the body, independent motion of the head developed which led to the loss of some bones and fusion of others to strengthen the skull and protect the braincase. Lungs and structures associated with intake and expulsion of air developed. Heavy scales aided in retention of water. The lateral line system, useful in a watery

29 *Anterior part of a skull and lower jaw of a small labyrinthodont amphibian from the Early Triassic Arcadia Formation, Rewan, Queensland. x 3.8 (A. Warren)*

environment to sense pressure changes and thus motion, was no longer effective and thus was lost by most labyrinthodonts. In its place, organs associated with sight and hearing must have been enhanced. The eardrum (tympanum) and a lightweight stapes connecting it to the inner ear developed, probably capable of registering low frequency but not high frequency sound. For reproduction, labyrinthodonts most likely returned to the water to lay eggs. Larval labyrinthodonts with external gills are known, and suggest that these primitive amphibians were not totally terrestrial.

Whilst labyrinthodonts arose during greenhouse times, they diversified when icehouse conditions were once again gripping the Earth in the late Palaeozoic. At this time, too, true terrestriality was attained in the form of reptiles, which first appeared in the Carboniferous. Reptiles had the added advantage of being able to reproduce in completely terrestrial environments. They were the first vertebrates to lay an amniote egg, a new kind of incubator, which had a hard shell and a number of embryonic membranes that provided nutrition, waste collection, dehydration protection and mechanical protection for the developing baby inside. This egg, first known in the Permian, could survive in a terrestrial environment and give rise to a fully functional juvenile when hatching occurred. There was no need to involve water in any way in the reproductive cycle.

Labyrinthodonts and reptiles co-existed and prospered throughout the icehouse times and into the early Mesozoic when greenhouse conditions returned. Triassic labyrinthodont faunas are rich, especially in Australia, but at the end of this period, about 213 million years ago, massive extinctions affected large numbers of terrestrial vertebrates as well as a great variety of both marine and non-marine organisms. The labyrinthodonts were sorely affected, and only a few taxa survived into the Jurassic and Early Cretaceous, mainly in outposts of Gondwana. Just what caused these extinctions is not clear. There was a major floral change at this time from the seed-fern dominated *Dicroidium* flora to one dominated by gymnosperms. Elements of the *Dicroidium* flora show, in the shape of their spike-like leaves and their thickened cuticle, adaptations to aridity. Many other hints, such as the abundance of evaporite (salt) deposits on a world-wide scale, also point to severe aridity during this period. Perhaps it was this climatic stringency that led to the labyrinthodonts' demise in most parts of the world. Indeed, the labyrinthodonts survived only in China, perhaps Russia, and in Australia, and interestingly the latter did not reflect such a severe climate.

A TIME OF CRISIS — THE PALAEOZOIC–MESOZOIC BOUNDARY
The Palaeozoic–Mesozoic transition was a time of immense crisis for terrestrial and marine organisms alike. It was a time of transition from icehouse to greenhouse conditions. Early in the Carboniferous, cyclicity gripped the world. This cyclicity, which manifested itself in the great coal seams of the northern continents, was but a reflection of the waxing and waning of the great glaciers to the south. Seas grew and shrank, flooded the lands and retreated from them, leaving behind a stacking of marine and non-marine sediments, cyclothems, that repeated themselves 100 times or more in the northern continents during the Carboniferous. This was also a time of coalescence of the great megacontinent Pangaea, which had a decided effect on world climate. In sum, it meant greater climatic zonation with concomitant isolation of populations of organisms into latitudinally controlled provinces — greater diversity for a while on a world scale. Near the poles diversity was curtailed, as is clear from the floras which are dominated by only a few hardy genera such as *Glossopteris*.

Perhaps it was the effects of cyclicity and glaciation that terminated in disaster at the end of the Permian, which was also the end of the Palaeozoic Era. The line between the Palaeozoic and Mesozoic eras is defined by massive extinctions and impressive floral and faunal changes. Indeed, it may well have been the greatest mass extinction in all of Earth history — certainly it was the most impressive to strike the flora and fauna since the assault of the land got underway in Early Devonian times. Not just one, but several pulses of extinction occurred at this time, just as was the case in the Late Ordovician and the Late Devonian. Extinction hit worst at forms with large body size. New forms evolved from the smaller survivors. Extinctions affected all life, not just the terrestrial biota but the aquatic as well, and the heaviest losses on a world scale were sustained at low latitudes, near the tropics.

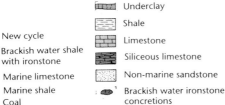

Cyclothems in the Carboniferous and Permian of North America reflect the rise and fall of the sea level, controlled by the waxing and waning of the Gondwanan glaciers. The idealized cyclothem is compared with an actual section from the Late Carboniferous of Kansas, the mid-continent of North America. A cycle begins with terrestrially derived sandstones and grades upward into swamp-deposited coals and finally into marine shales and limestones, which are usually overlain by an erosion surface from a time when the sea retreated and terrestrial conditions returned. Then the cycle begins again. (Modified from Cloud, 1988)

- ▨ Coal
- ▨ Underclay
- ▨ Shale
- ▨ Limestone
- ▨ Siliceous limestone
- ▨ Non-marine sandstone
- ▨ Brackish water ironstone concretions

New cycle
Brackish water shale with ironstone
Marine limestone
Marine shale
Coal
Underclay
Non-marine shale
Non-marine limestone
Non-marine sandstone

Previous cycle

CYCLOTHEM FROM MID-NORTH AMERICAN CONTINENT

IDEALIZED CYCLOTHEM

The cause of such a massive dying out of species is elusive. Even though the great glaciers of Gondwana were on the wane, cool climates still gripped the Earth, even into Late Permian times, as glacial dropstones transported by floating icebergs have been reported from such places as Siberia (which lay near the Permian North Pole) and south-eastern Australia (which lay near the Permian South Pole). It may have been the first time since the late Proterozoic that both poles were icy at the same time, which meant drastically lower sea levels that essentially cancelled out habitable areas for marine life on the continental shelves.

With cold waters spreading equatorward from both poles, and very narrow continental shelves, ecospace for most marine organisms would have been very limited or for many non-existent. Temperature conditions on land surely must have had their biological effect. Still, even though the details of this extinction are not clear, it is quite apparent that the faunas which survived into the earliest part of the Triassic, both on land and in the oceans, were species poor. A large percentage of the species that did survive this debacle had a broad distribution across the globe — they were evidently generalists that could cope with a variety of environmental conditions and severe ecological stress. The change to warmer conditions in the Mesozoic brought a decline in the importance of amphibians amongst land dwellers and gave impetus to the rise of reptiles. In the rivers, streams and seas the bony fishes continued to diversify.

DIVERSITY IN A WARM WORLD

Dinosaurs, along with other large reptiles, ruled the Earth, at least the terrestrial environments, for much of the Mesozoic, when greenhouse conditions had returned. Mammals were present, but always low in diversity in the shadow of the mighty and successful saurians.

At these times, the world was bathed with warmth and equability (predictability) on a scale that it had seldom, perhaps never, encountered in the past — ideal conditions for the reptiles. Indeed although there were occasional stresses causing other extinctions, notably at the end of the Triassic, the dinosaurs and other reptilian groups arose and prospered for more than 140 million years. Success was the domain, too, of the flowering plants, the angiosperms, which had only just arisen. Palaeoclimatic evidence for the Mesozoic — oxygen-isotope palaeotemperature determinations, and distribution of carbonate sediments, salt deposits, coals and hydrocarbons — all point to a warm world beginning, initially arid but gradually becoming humid, until in the Jurassic and for much of the Cretaceous it was unseasonally wet and mild. It was a time of higher concentrations of atmospheric carbon dioxide, as much as two to 10 times present levels. The carbon dioxide levels would have increased plant growth, and atmospheric moisture and cloud cover increased, resulting in a sluggish, humid greenhouse climate. The cause of these conditions may have been increased vulcanism (which produces carbon dioxide) forced by plate tectonic activity ripping Pangaea apart, a major factor dominating the Mesozoic world. The result of this climate and the biology of the time is of economic importance, for there was a build-up of coal in the extensive swamp forests and a hydrocarbon accumulation in the thick, organic-rich muds characteristic of the oxygen-starved marine basins, which is today the source of much of the world's oil including that of Libya, the Persian Gulf and the Gulf Of Mexico.

Conditions changed dramatically and apparently rapidly about 65 million years ago, around the Cretaceous–Tertiary boundary, bringing about the demise of the dinosaurs and giving mammals the opportunity they needed. That event has caused one of the greatest debates amongst geologists and palaeontologists this century.

Some geologists have suggested that increased volcanic activity during the Cretaceous Period led to a nuclear winter with so much particulate matter in the atmosphere that incoming radiation from the Sun was effectively blocked, causing temperatures to cool to the extent that much of the food-chain collapsed and a multitude of species became extinct. Another school of thought on the Late Cretaceous extinctions marshals an impressive array of evidence to suggest that an extraterrestrial visitor, a comet or an asteriod, was the culprit that changed the climate, doing so dramatically and over a short period of time. It would first have heated the Earth to unthinkable temperatures from a biological point of view, with the whole planet being swept by firestorm and that followed by a time of intense acid rain. And after all that, because of the immense amount of particulate matter thrown into the atmosphere, the temperatures cooled. Evidence supporting this theory includes a concentration of iridium, a rare element common in meteorites, present in clays at the Cretaceous–Tertiary boundary, around 65 million years of age, found in many parts of the world including Italy, New Zealand and western North America. The iridium would have been part of the particulate matter thrown into the air by an asteroid impact. Such impacts are also notable in producing a certain type of strain in small crystals, shocked quartz, near the impact which has been collected in several places astride the Cretaceous–Tertiary boundary. Recently, even the ground-zero site for the asteroid-comet impact seems to have been found in the Caribbean Sea, in the form of a 300-kilometre diameter circular depression near the Yucatan Peninsula. All around this area the iridium-rich boundary clay is excessively thick, and the telltale shocked quartz is abundant in sediments on nearby Haiti, Cuba and in Texas.

Be it volcanoes or meteors mattered not to the dinosaurs. Whatever cause or causes brought about their demise, whatever brought about the cooling of the benign climate of the late Mesozoic, the fact remains that the vast inland seas which had inundated most continents were drained, hastened by the break-up of Gondwana and even the northern continents — and leading to a very different world in the succeeding era, the Cainozoic.

FROM GREENHOUSE TO ICEHOUSE—THE MAKING OF A MODERN WORLD

Events at the end of the Cretaceous brought about a fundamental change in plant and animal communities, transition from archaic to modern. Mammals and birds replaced the reptiles in dominance, and those reptiles that did survive were diminutive in comparison to many of their Mesozoic precursors. Gone, too, were such forms as the calcareous nannoplankton that had formed the massive chalks of the Cretaceous, as seen today in the Dover Cliffs and the Kansas cornfields. The coiled ammonites and big marine reptiles that had plied shallow continental seas were replaced by a variety of bony fishes, mainly the advanced teleosts, and eventually whales, dolphins and seals, as well as a variety of new protozoan forms belonging to the foraminifera. Even the reefs changed their character. Together with corals, the rudists (unusual coralline-shaped clams) had dominated the late Mesozoic tropical reefs. In the early Cainozoic the rudists disappeared and new varieties of the scleractinian corals, with their six-fold symmetry, took over.

In the plant kingdom those with flowers, the angiosperms, were the ultimate victors, spreading their colour and perfume across the land and occupying a great variety of niches perhaps never inhabited before. Together with a variety of birds and insects, they "challenged" each other to a co-evolutionary contest which culminated in such complexities as the elaborate orchids. These cunning plants mimic their pollinators and entice them inside with impersonation and scent, only to ensnare them in their mating ritual. Pollen is thus transported from one plant to another, fertilization ensues and species perpetuation is ensured for awhile.

Those species surviving the Cretaceous crisis had before them vast opportunities, however the nature of the opportunities changed rather dramatically over the next 65 million years or so and those species capable of flexibility were the winners in the continuing battle for ecospace. Their lot was not an easy one, for, unlike the Mesozoic, conditions in the Cainozoic were anything but equable, and certainly not at all stable. Gradual cooling overall, but in fits and starts, characterized the Cainozoic. Then, at the end of the era, came the greatest test of all — massive glaciations that swept equatorward on the northern continents, lowering temperatures, lowering sea levels, and imposing a variety of new stresses on an already vulnerable biota. Predictably, extinctions and readjustments took place, and from those conditions emerged the modern world we know. From that, we ourselves emerged and began to meddle with much of which we had no understanding at all.

During the Cainozoic the continents were on the move, dramatically rearranging the chessboard upon which the biological pawns moved. India and Australia made the most dramatic shifts, but all other continents were affected by the spreading along oceanic ridges and the disappearance of lithospheric plates. The majestic Himalayas of Asia (where Permian marine limestones crown Mt Everest at over 8800 metres), the Alps of Europe and the Rocky Mountains of North America were all the progeny of this relentless growth and death of ocean basins, and ultimately the heat from within the Earth. Yet another progeny of such oceanic restlessness was New Guinea; before Cainozoic times it had lain, for the most part, below a mild, tropical sea. Pacific plate movement jacked its limestones into the clouds by Pliocene times, with rocks bearing marine mammals reaching 3000 metres in a few million years.

Continental movement and climatic changes in the Cainozoic teamed to yield great changes in the biota of Gondwana. The more homogeneous faunas of the Mesozoic were sundered. Australia, Antarctica and South America maintained contact with each other well into the Cainozoic, with New Zealand breaking away before the beginning of the era. Although there were physical connections, mainly insular, between South America and Australia–Antarctica, changing climate seems to have played a crucial role in restricting the movement of a variety of forms.

Africa and India had both drifted away from central Gondwana before the Cainozoic. India moved northward toward Asia, and the eventual collision of the two landmasses, beginning in the Eocene, was responsible for uplift of the Himalayas. This time frame is supported by fossil vertebrate faunas of India which clearly reflect interchange with the Asian landmass by this period. Africa's ties during the Cainozoic are primarily with the northern continents, with many pulses of immigration from the north.

The great Tethys Sea that separated Africa from Eurasia waxed and waned with the cooling climates of the Cainozoic, continuing to narrow during the 65 million years of that era. Sixteen million years ago Africa and Saudi Arabia met Asia. Then about 6.7 million years ago the convergence of Europe and north-west Africa, combined with the lowering of the sea level by an increasing build-up of ice at the poles, led to the Mediterranean, a *de novo* sea that came after the Tethys, being completely isolated from all other seas. The result, probably in less than 1000 years, was a dry and desolate basin separating northern and southern continents, which sometimes descended to as much as 5000 metres below sea level — a picture reconstructed from the salt deposits that underlie the present Mediterranean. (The salt deposits were discovered in cores recovered by the Deep Sea Drilling Project in 1970.) From 6.7 to about

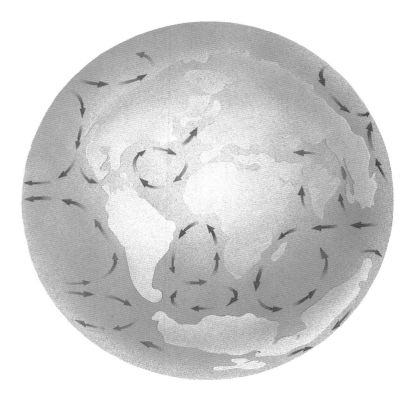

Continental positions and oceanic circulation at the end of the Eocene. Large areas of continental crust were covered with shallow, epeiric seas. Ocean circulation was distinctly different from the modern patterns because of the lack of the Circum-Antarctic Current. Warm climates prevailed rather further north and south than at present. Because of world climate and oceanic circulation patterns, a warm Pacific current bathed the south-eastern coast of Australia. The Bering Land Bridge connected Eurasia and North America, allowing a decided mixing of the faunas of these two areas. (Modified from Stanley, 1986)

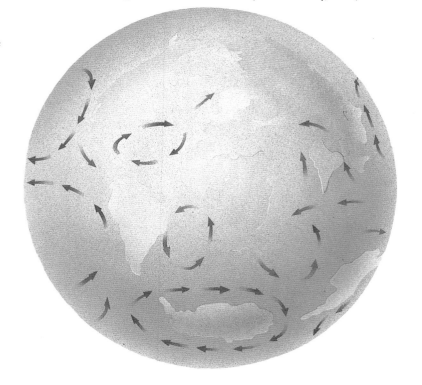

Major changes in climatic regimes and oceanic circulation occurred when Australia broke away from Antarctica, by the Early Oligocene, opening up the Southern Ocean between the two continents and allowing the development of the Circum-Antarctic Current and the psychrosphere. This development was the real beginning of the chilling of the Earth's climates that ultimately led to the glacial climate in which we live today. The warm Pacific current that had bathed south-eastern Australia in the Late Eocene was deflected further northward by the Circum-Antarctic Current. (Modified from Stanley, 1986)

5.2 million years ago the salt desert persisted, with occasional intermissions, until a briny waterfall refilled the basin through the Straits Of Gibraltar.

Africa and Eurasia exchanged fauna throughout the Cainozoic, as a result of intermittent tectonic activity, and many elements of the northern fauna remain today as relics of the Pleistocene in Africa. The repeated connections between these two large landmasses precluded Africa from developing an entirely independent fauna on the scale of the distinctive faunas in South America and Australia.

Both Australia and South America fragmented from Gondwana during the Cainozoic, though tenuous ties persisted until the Late Eocene. From either Africa or North America, South America received rodents in the Early Oligocene and primates by the Late Oligocene. Otherwise, the terrestrial mammals of South America evolved in isolation, with the possible exception of bats, from the beginning of the Cainozoic until the Late Pliocene. Then, a direct land connection between the two Americas developed, which wreaked havoc amongst many of South America's endemic mammals, both marsupial and placental. Very little reverse immigration occurred, with the American opossum, ground sloths, armadillos, hystricomorph rodents and possibly some bats being some of the few South Americans that invaded the northern continents.

Australia remained isolated through much of the Cainozoic, initially being partially separated from Antarctica and South America by climatic constraints, and finally in the Late Eocene and Early Oligocene by real physical severence. Australia then began its northward trek in earnest, and in doing so experienced profound isolation until relatively recently. Such isolation was the director of development of the modern terrestrial fauna of Australia, giving rise to its strangeness relative to that of the rest of the world. The Australian fauna is dominated by marsupials and bird groups like parrots and pigeons, and deficient in animal groups that elsewhere dominate, such as placental mammals and iguanid lizards. Evidently, some groups either originated in or immigrated to Australia before its definitive separation from the Pangaea consortium and made their own evolutionary way regardless of the activities overseas. This evolution proceeded for tens of millions of years before Australia began to "dock" with Asia and the first wave of immigrants, including the bats, made their foray into this antipodean continent. They were followed by others, such as the varanid lizards and rodents. A few hardy Australian forms such as the cuscus and some species of *Eucalyptus* moved northward, but only a few.

The invaders from the north seem to have been, just as they were when the North American–South American linkage was established in the late Tertiary, in some way competitively superior to their southern cousins. The northerners' entrance spelled disaster for many elements in the native faunas of the southern continents; for example, the demise of most endemic marsupial groups and primitive placentals in South America. The final insult to the Australian Gondwanans came with the arrival of man and his entourage of companions such as the Dingo and later the domestic cat. That, combined with the unpredictability of the climate from one year to the next, brought about enormous changes to the biota of the continent. Indeed, with each of these remaining pieces of the Gondwanan puzzle — Australia, New Zealand, South America and Antarctica — now entirely on its own, the faunas of each were on distinctly separate paths.

During the Cainozoic the climates of most continents of the globe, including most of the Gondwanan continents, experienced a gradual cooling from beginning to end, which culminated in the ice age that still grips our Earth. The pattern of cooling was not entirely a smooth and gradual process; there were times of warming and then resurgence of the cooling trend — for example, a warm phase in the Early Miocene was followed by an oscillating climate that gradually cooled thereafter.

However, the temperatures of both Australia and New Zealand for this period, particularly from the early to the middle Tertiary, are anomolous. Studies of the ratios of the isotopes of Oxygen 18 to Oxygen 16 in the shells of nearshore marine invertebrates suggest that the temperatures of Australia and New Zealand climbed from an early Tertiary low to a middle Tertiary high and then began to fall in concert with those of the rest of the world. The "out of step" nature of these antipodean temperatures, first thought extremely strange,

30

seems neatly tied with Australia's northerly drift. With the movement from the South Pole toward the Equator, the antipodean landmasses were moving into warmer and warmer climes and thus the overall global fall in temperature was not reflected. By the Miocene, Australia had neared its present position, and the rate of its northerly journey slowed. Its temperatures now began to decline in parallel with those typical of the rest of the Earth. The situation would have been mirrored by India in the Late Cretaceous and early Tertiary, when it made its pilgrimage north across the expanse of the great Tethys Sea, which closed as India and Asia collided.

When humans first arrived in Australia more than 40,000 years ago and in South America more than 10,000 years ago, they encountered animals entirely new to them. Many of the species were reminiscent of the plant-eaters and the hunters of their native lands to the north, but there were major differences, notably the dominance of marsupials rather than placentals. A record of the fauna exists in the legends passed down by these early people as well as in their rock carvings and paintings.

The wildlife was profoundly affected by the presence of the invading *Homo sapiens*. Over the next few thousand years, by man's use of fire to clear vast areas, by his concerted hunting of larger species and by his introduction of a variety of placental mammals, especially in the last few hundred years, he altered the course of evolution of the biota on the remaining fragments of Gondwana.

30 Wildfire is very much a part of most modern Australian environments, but prior to the Quaternary Period it was not nearly so prevalent, if present at all. The aridification of the continent over the last 500,000 years has been critical to the frequent natural burning of vast areas of the continent. (Courtesy of B. Poole and the Country Fire Authority, Victoria)

DEVELOPMENT OF A UNIQUE AUSTRALIAN FAUNA

31

Australia, like all the other pieces that once formed Gondwana, has not always been a distinct continental entity. Over much of Earth history, Australia simply wasn't — it didn't exist. At times the land that is now Australia formed a peninsula jutting out from a larger landmass, at times it was simply part of the coastline of the supercontinent Gondwana or the megacontinent Pangaea. At times it was mainly a terrestrial environment, much of it above sea level, yet at other times it was a series of islands criss-crossed by shallow seas in which a variety of fishes and later big marine reptiles lazed away their days.

Only during the last 50 million years or so, when the Australian continent broke away from Antarctica and the Southern Ocean widened, did this continent assume its distinctive shape and unique biota. Prior to this break-up it had developed a somewhat unusual fauna simply because of its latitudinal position and its peninsular nature at one end of Gondwana. But, after its divorce from Antarctica, Australia behaved as a "Noah's Ark", carrying with it an entourage of vertebrates that evolved in nearly complete isolation for the next few tens of millions of years. It was also a "Viking Funeral Ship" in that it carried with it a cargo of fossils of fauna and flora that had lived and died in far distant latitudes.

During this epic voyage, vertebrates on the newly distinct Australian continent evolved in response to the changing conditions of this land, which at first lay in cool temperate latitudes but quickly moved into temperate and finally tropical latitudes. Humidity on a continent-wide scale became restricted as Antarctica began to build up a major ice sheet due to the Circum-Antarctic Current which was established in the Oligocene. Even before this, the refrigeration of Antarctica gave rise to the psychrosphere, a cool bottom layer of the ocean created as the icy water near the polar continent sank due to its density into the ocean depths, dropping temperatures there 4–5°C. The upwelling

of nutrient-rich ocean waters caused by the displacement of watermasses and the cooling of ocean temperatures had an effect on marine organisms, often triggering massive faunal change, favouring those few species that could adapt to lower temperatures. Floral and faunal evidence from terrestrial environments suggests that these oceanic events had repercussions on the land as well. The greatest effects of the psychrosphere development on biota occurred in the Late Eocene and Early Oligocene, about 39 to 29 million years ago.

After the Early Oligocene, related to the establishment of the psychrosphere and the changing arrangements of continents and oceanic circulation, Australia and many other continents of the globe experienced increasing aridity, magnified in Australia because of its northerly movement. A factor, too, in this aridification was the lowered sea levels during this period, a result of a quickening of the pace of ice build-up at the poles. The inland climates of all continental masses became drier, dictated by their increased "continentality". The culmination of all this change was a switch from forested environments in the continental interiors to savannahs and grasslands. Subtropical and tropical forests, which previously had a wide distribution on most continents, were restricted to low latitudes. Central Australia lost its southern beech forests as well as the arboreal and lacustrine faunas that had been an integral part of these forest ecosystems.

In Australia, too, a further aridification occurred during the last stages of the Cainozoic, when the climatic zones contracted towards the Equator and the annual period of maximum rainfall switched from the summer to the winter in the southern half of the continent. As a consequence, the eucalypts and savannah-frequenting vertebrates became restricted to the periphery of the continent, or worse, became extinct.

The vertebrates that make up the unique Australian fauna today have thus had a variety of challenges over the past 50 to 60 million years when Australia has existed as a distinct entity. The fauna began with Gondwanan and Pangaean inheritances and was then isolated by the tectonic movement of the great lithospheric plates in the early Tertiary. Conditions did not remain at all stable from that time to the present. Instead, as Australia moved northward immense climatic changes gripped the continent, directing the course of evolution and laying down the rules of survival. Australia began as a temperate continent in the Eocene, moved into the Horse Latitudes where deserts abide and then on into the Tropics. It is now on a collision course with tropical Asia, a marriage which in 40 to 50 million years will blend the biotas of the two great continents.

To the Australian biota all of this Cainozoic variation has meant expansion of diversity at the beginning, followed by oscillations in number of species over the past tens of millions of years, and finally a diversity curtailment. What percentage of the vertebrate fauna of Gondwana's last continental outpost will remain in 50 million years is purely in the realm of speculation, but there is no doubt that humans will play a critical role in their survival or demise.

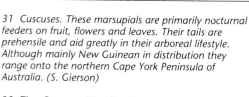

31 Cuscuses. These marsupials are primarily nocturnal feeders on fruit, flowers and leaves. Their tails are prehensile and aid greatly in their arboreal lifestyle. Although mainly New Guinean in distribution they range onto the northern Cape York Peninsula of Australia. (S. Gierson)

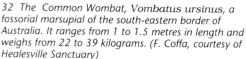

32 The Common Wombat, Vombatus ursinus, a fossorial marsupial of the south-eastern border of Australia. It ranges from 1 to 1.5 metres in length and weighs from 22 to 39 kilograms. (F. Coffa, courtesy of Healesville Sanctuary)

32

Australia's vertebrate fauna today is unique in many ways. It is composed of a number of unusual forms when compared to the faunas of other continents. Marsupials, which include kangaroos, possums, the Koala and a host of others, are a dominant element in the fauna. The only other continent which has ever had a diversity of marsupials approaching that of Australia is South America, in times prior to the Late Pliocene, but today only a few marsupial groups survive there, their demise due mainly to the invasion late in the Cainozoic of northern placentals. Parrots and pigeons, emus and cassowaries, pelodryadid and mybatrachid frogs, all are either unique or much more diverse in Australia than anywhere else in the world. And, diversity is not measured just in terms of counting the number of species, but also at the higher levels of taxonomic organization; more families of both parrots and pigeons occur in Australia than on any other continent, including South America and Asia where greater expanses of tropical environs might be thought to favour more diversity.

Concomitantly, many groups that are dominant elsewhere in the world, such as the placental mammals (those that lack pouches and hold the young within the mother as an embryo attached by a placental membrane for many months), are rare in Australia. Of the placentals, only the bats and rats arrived without the help of man, the former possibly by 55 million years ago and the latter perhaps only 5—7 million years ago. Likewise, varanid lizards have only a middle Tertiary to Recent history in Australia.

Australia has provided a refuge for a variety of animals that became extinct long ago elsewhere in the world. The echidna and platypus may be examples of this, perhaps being remnants of once more widely distributed primitive mammals. The clam *Trigonia*, found living in Australian nearshore waters by the French explorer Baudin in the early nineteenth century, had become extinct during the Eocene of Europe. The fossil record, too, shows that Australia has had a long history as a refugium. The crocodile-like labyrinthodont amphibians as well as the carnivorous dinosaur *Allosaurus* survived millions of years beyond their time elsewhere in the world, living on well into the Cretaceous Period of southern Australia.

The Australian fauna of today lacks certain functional elements that most other continents possess. There are no lions and cheetahs, no wolves and coyotes, in the modern fauna. Perhaps in the most recent past a medium-sized kangaroo, *Propleopus*, may have filled the cheetah role, being a carnivorous saltator, but marsupials never seemed to move into the open country as fast-pursuit carnivores in the way that placentals did on other continents. The perhaps recently extinct Tasmanian Tiger (*Thylacinus*) appears to have been a short-pursuit carnivore, more suited in its limb proportions for hunting in forested areas rather than grasslands.

And finally, the Australian biota is characterized by the excessive species diversification of a few genera. Marsupials are but one example of this diversity pattern, parrots another, and *Eucalyptus* with its hundreds of species certainly a third. By contrast, North America, Eurasia and to some extent Africa tend to have less species diversification but a great many genera. The greater genetic potential of the fauna on these continents, coupled with heightened competition for a greater variety of habitats, resulted in species becoming highly aggressive. When they came in contact with their southern counterparts they were in general competitively superior.

Australia today is recognized together with New Zealand and a part of the southwest Pacific as a unique biogeographic province. Its vertebrates have been used to define the limits of this province, which to the north and west fall somewhere in the area between Australia and South-East Asia. Drawing a line that separates the Australasian and Oriental biogeographic realms, however, is not simple, as the position of that line varies depending upon which group of animals or plants the biogeographer chooses to use. A. R. Wallace defined his original line in 1860 based mainly on the parrot faunas. It passes between the islands of Lombok and Bali, and is accentuated by the arid nature of the western end of Lombok and the wet tropicality of eastern Bali. Later, Thomas Henry Huxley extended and altered Wallace's Line, and its position nearly coincided with the edge of the Asiatic continental shelf. To the west lay a truly Oriental fauna. Lydekker's Line, based on a number of different vertebrate groups,

34

35

similarly corresponded with the edge of the continental shelf of West Irian and Australia. To the east lay a typical Australasian fauna. Yet another division, Weber's Line, demarks a place west of which more than 50 per cent of the bird species are of Oriental origin and east of which more than 50 per cent are of Australopapuan origin.

Why draw a line between these two areas of converging lithospheric plates at all? George Gaylord Simpson — revered evolutionary thinker, biologist and palaeontologist — thought it a useless exercise to draw lines in this area. Simpson proposed another way of coping with the biogeographic disjunction attendant to the area between the Australasian and Oriental realms: "The simplest way to cheat in the game is just to keep both Huxley's and Lydekker's lines as regional boundaries and to call the intervening area another region, not part of either the Orient or the Australian Regions." Simpson named this intervening area "Wallacea", after Alfred Russel Wallace, who as well as defining the original line was the co-proposer with Charles Darwin of the Theory Of Organic Evolution By Natural Selection. Wallacea is essentially a transition zone where two great faunas are mixing and have been mixing over the past 30 million years or so. It is one of the major battlegrounds where the remnants of the once great Gondwanan terrestrial fauna are fighting for their lives.

A FOREST IN BORNEO, WITH CHARACTERISTIC MAMMALIA

33

33-35 Oriental fauna and Australopapuan fauna, which today and for the last several million years have come into direct contact in the area known as Wallacea. (From Wallace, 1876)

PART TWO

THE SEARCH FOR BEGINNINGS

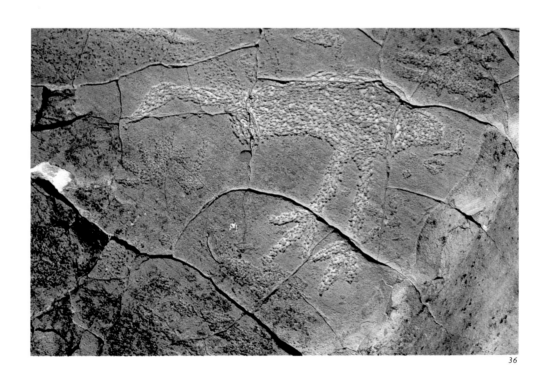

DIGGING UP BONES IN THE ANTIPODES

Long before Europeans reported the discovery of the wonderfully strange fossil vertebrates of Australia and New Zealand, Aboriginal traditions had woven a complex fabric of the past — of giant animals with both benign and malevolent countenances. Perhaps such stories were based on an acquaintance with prehistoric bones found jutting from a river bank or hillside, or lying exposed on the desolate surface of a brine-encrusted lake. But, the possibility is real that such tales of prehistoric leviathans come from tribal memory passed down from generation to generation from the time that the animals lived.

Native peoples of eastern Australia were quite fearful of the bunyip, sometimes described as a monstrous animal inhabiting deep waterholes and roaming around the billabongs at night. When confronted with the remains of some of the now extinct Australian marsupials, Aborigines would oft-times confirm them as this bunyip. The *Mihirung Paringmal* of western Victorian Aboriginal traditions may be the now extinct giant birds, the Dromornithidae, their common name given in honour of their Aboriginal contact. Some of the vivid descriptions and the understanding of such creatures by Aboriginal peoples eventually led to the discovery of rich vertebrate fossil fields, such as those at Lake Callabonna in South Australia in the late nineteenth century, which fascinated the scientific communities of England and continental Europe of the time.

38

37

Only within the past century and a half, and for the most part within the last 30 years, has a detailed understanding of the ancient vertebrates of Australia been developed. Much of the early descriptive work was accomplished by the prolific Sir Richard Owen, the English comparative anatomist who took a keen interest in antipodean vertebrate fossils. His work was only a beginning, albeit important, and there is still much to accomplish in the future. It is of no small interest to briefly consider, from the beginning to the present, just how ideas have changed with regard to Australia's ancient vertebrate history. Richness of collections and both political climate and religious convictions have played their part in influencing interpretations of this history.

39

FRENCH AND BRITISH COASTAL SURVEYS
In the early part of the ninteenth century a number of European expeditions, which were essentially to explore Australia's coasts, yielded no fossil vertebrate remains. They did, however, locate and allow the collection of invertebrate and plant fossils, as well as the living remnants of some vertebrate groups now extinct. The British Matthew Flinders Expedition of 1801-05 and the French Nicholas Baudin Expedition of 1800-04 were two such enterprises, and they were as different from one another as day and night. The Baudin Expedition, splendidly outfitted with state of the art equipment and top quality scientists, returned fossils and modern natural history

material to France to be examined in detail by the cream of that country's scientific community. Included in the Baudin collections were living specimens of the now extinct dwarf King Island Emu (*Dromaius ater*), kept alive in a Parisian zoo for some time after their arrival. J.C. Bailly, a mineralogist attached to the Baudin expedition, reported fossil ferns in shales near Parramatta, and fossil plants were collected from Tasmania. These latter were examined in Paris in 1814 by Leopold von Buch, who thought they came from a time period now called the Carboniferous.

Also included in these early *Terra Australis* collections was the clam *Trigonia*, brought up in a dredge haul off King Island in Bass Strait. Jean Baptiste Lamarck, infamous for his evolutionary theories but also a highly respected and influential invertebrate zoologist of this period, was struck by the resemblance of the living *Trigonia* with forms known only as fossils in Europe. The concept that Australia was somehow a haven, a *refuge*, for organisms that could no longer survive elsewhere had its origins in these early Lamarck observations. Australia seemingly was a land of living fossils, and this conclusion was to be further reinforced as exploration extended to the interior of the continent later in the century, mainly with the discovery of a great variety of fossil vertebrates in New South Wales. The unfortunate Flinders Expedition, in contrast to the Baudin enterprise, ended tragically in shipwreck, with the loss of most specimens, excepting a few that Robert Brown, the shipboard botanist, had surreptitiously taken ashore with him in Sydney when he departed the expedition. Although not once mentioned in Brown's catalogues, fossil invertebrates and plants that had been in his charge somehow made their way back to Europe and England. Brown himself returned to England with three cases of "minerals", part of his possessions that passed through Customs in 1805. Among these possessions, surely, were the fossils from Australian coal deposits that the Reverend William Buckland at Oxford studied and concluded were comparable to those in the Carboniferous of England. James Sowerby, likewise, described invertebrates which he must have acquired from Brown. Brown's name, in fact, is acknowledged many times in the new species then proposed by palaeontologists in both England and Europe, one example being *Glossopteris browniana*, a fossil seed-fern described by Adolphe Brongniart, which was based on fossils from New South Wales passed on to him by William Buckland.

EXPLORATION OF THE INTERIOR

Coastal surveys continued, and visitors to Australia returned to Europe with further collections of fossils. Plants and invertebrates were recovered, but no vertebrate fossils of note were found. However, this situation changed dramatically with the inland explorations carried out by Major (later Sir) Thomas Livingstone Mitchell, Surveyor-General of New South Wales from 1828 until his death in 1855. T.L. Mitchell mapped in detail, and procured many bones from, the Wellington Valley caves of New South Wales. He first visited the caves on 26 June 1830 with a local colonist, George Ranken. Ranken had previously found bone fragments in the area and had taken them to Sydney for shipment to Professor Robert Jameson of the University Of Edinburgh.

Ranken's discovery of fossil bones was announced in the *Sydney Gazette* of 25 May 1830 in an anonymous letter (signed L.) by the Reverend Dr John Dunmore Lang. Lang left Sydney on 14 August 1830 with Ranken's specimens, his own *Sydney Gazette* letter and a short manuscript by Mitchell on the Wellington caves. By early 1831 all of the specimens were in the hands of Robert Jameson, and two notes were published in the *Edinburgh New Philosophical Journal*, both credited to Lang. In the subsequent volume of this journal, however, one of the notes was correctly attributed to Mitchell.

Mitchell revisited the caves on 3 July 1830 and collected further specimens, which were apparently sent to the Geological Society in London with a letter dated 14 October 1830 and read at the society's meeting of 13 April 1831.

Various specimens collected by Ranken and possibly Mitchell were examined by William Clift, Conservator of the Hunterian Museum (College Of Surgeons), who identified dasyurids, wombats and kangaroos. Joseph Barclay Pentland who was at the time living in Paris, also examined material sent to him from the Wellington caves, as did Baron Cuvier, and independent information came from Peter Cunningham, author of the 1827 book *Two Years in New South Wales*. William Buckland considered that some of the Wellington bones represented either rhinoceros or hippopotamus.

Mitchell records that the initial exploration of Australia's first vertebrate fossils at the Wellington caves was somewhat dangerous, for as he noted in his diary:

> "The pit [Breccia Cave] had been first entered only a short time before I examined it, by Mr. Ranken, to whose assistance in the researches, I am much indebted. He went down, by means of a rope, to one landing place, and then fixing the rope to what seemed a projecting portion of rock, he let himself down another stage, where he discovered, on the fragment giving way, that the rope had been fastened to a very large bone, and thus these fossils were discovered."
> Mitchell 1838

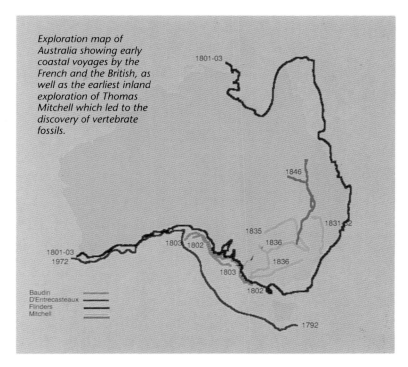

Exploration map of Australia showing early coastal voyages by the French and the British, as well as the earliest inland exploration of Thomas Mitchell which led to the discovery of vertebrate fossils.

The bone to which Ranken tied his support rope was the "lower end, mutilated, and encrusted with the red stalagmite of the cave..." of a femur that was later identified by Sir Richard Owen as belonging to a large bird, entirely new to science and the world (probably from the family Dromornithidae). It was figured in Mitchell's *Three Expeditions into the Interior of Eastern Australia*, but later disappeared, perhaps during the bombing of the Royal College Of Surgeons in London during World War II.

Although much of the publication about the Wellington caves discoveries was in English, great interest was generated in the European scientific community, notably in France and Germany. As a result, Mitchell's discoveries of fossil bones were followed by many years of active European and Australian collecting. Most of the newly found material was sent from the shores of *Terra Australis* for description and study by *"foreign experts"*, as the needed expertise and comparative collections of fossils simply did not exist in Australia. Only during the latter part of the nineteenth century did indigenous workers begin to seriously study their own fossils, even though

36 Petroglyph at Giles Creek, Mootwingee, New South Wales, depicting a large ground bird. Whether this rock carving was from the imagination or is a realistic record of a living animal is uncertain. (E. Gill)

37 Aboriginal art in caves in various parts of Australia, particularly the north, frequently depict animals that are now extinct. This cave art in the Quinkan Gallery of Cape York Peninsula, Queensland, was discovered by Percy Trezise. The large bird figure may be one of the extinct dromornithids, perhaps Genyornis. (X. Dennett)

38 The French explorer Nicolas Baudin, who at the beginning of the nineteenth century led one of the first scientific expeditions to Australia. Baudin's ship the Géographe, was splendidly outfitted, and returned to France with an array of new and exotic specimens, both fossil and recent. Examination of this material by Jean Baptiste Lamarck led to the idea that Australia was a haven for species that could no longer exist anywhere else in the world — species that had survived long beyond their time, and were thus "living fossils". (Courtesy of the Museum d'Histoire Naturelle, Paris)

39 The skin of a dwarf emu, probably from King Island, collected alive and taken to Paris, where it died and was prepared for the collections of the Museum d'Histoire Naturelle. x 0.125 (T. Rich)

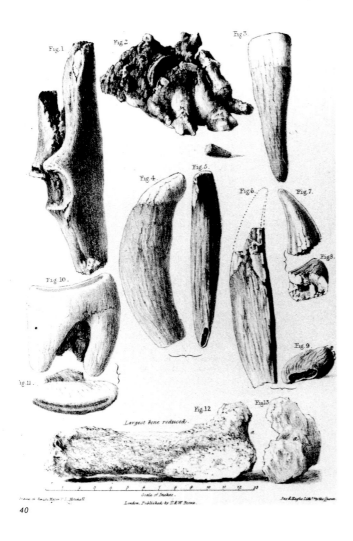

several residents such as Leichhardt and Stutchbury had previously strongly suggested that Australian material should remain in Australia.

"FOREIGN EXPERTS"

Sir Richard Owen, the renowned British comparative anatomist, described many of the new fossil vertebrates that kept pouring in from Australia and New Zealand. These fossils came to light as inland exploration and settlement expanded and as government sponsored surveys began to appear. One of Owen's earliest papers on Australian fossils, an appendix to Mitchell's volumes on his three expeditions into the interior of Australia published in 1838, described the gigantic marsupials, *Nototherium* and *Diprotodon*. Although Mitchell's and Ranken's material from the Wellington caves was examined and reported on by Cuvier and Pentland, and Darwin was, likewise, aware of it as a result of his visit to Australia in 1836, Owen was the one who undertook the tedious, yet exciting job of description and study of these antipodean jewels. He, like the French before him, suggested that it was necessary to search Britain's Mesozoic rocks "to find specimens analogous to Australia's recent marsupial fossil forms".

Owen, through his prodigious efforts, over the next 40 years made Australian and New Zealand vertebrate palaeontology his personal domain. He was supported by many resident Australians who sent him material. Friedrich Wilhelm Ludwig Leichhardt provided and helped describe bones that he found or procured from southern Queensland in 1844. The Reverend William B. Clarke and Samuel Stutchbury recovered bones in their northern surveys on the Darling Downs of Queensland as well as closer to home near Sydney.

40 Plate from T.L. Mitchell (1838) illustrating the first Australian vertebrate fossils reported in a scientific publication, in this case by the British comparative anatomist Sir Richard Owen. (Courtesy of E.B. Joyce)

Frederick McCoy and Gerard Krefft sent Owen specimens that came to their attention as the represenatives of the National Museum Of Victoria (now the Museum Of Victoria) and the Australian Museum in Sydney respectively. Often such specimens came from local pastoralists when digging wells or surveying property.

Although much of Owen's Australian work centred on fossil marsupials, he also took a keen interest in other vertebrates, and his bibliography on the subject of Australian fossil vertebrates is impressively long. None of the other foreign experts, including Darwin's "bulldog", T.H. Huxley, P. Gervais, Hochstetter and R. Lydekker, among others, published as prolifically on the Australian vertebrate record as Owen. Surprisingly, only a few mistakes crept into his work, one being his description of an elephant (supposedly a mastodont) from Australia. Many people questioned the authenticity of the elephant record in Australia, and it has been suggested that the specimen probably was brought to this continent as a trade item by some traveller or sea captain. After a well reasoned challenge by Hugh Falconer, Owen quietly abandoned his claim.

These early collections of fossil vertebrates from Australia, especially the rich discoveries at the Wellington caves, may have had some effect on scientific thought of the time. It has been suggested by Kathleen Dugan (a science historian) that the Wellington fossils were used as evidence supporting Darwin's ideas about evolution via natural selection. According to Dugan the *Law Of Succession*, generally attributed to Darwin, was formulated in part based on the types of fossils that occurred in the Wellington collections. David Ride has pointed out that Owen, in his own writings, claimed to have formulated this law. Whatever its origin, this law suggested that fossil animals in any particular geographic area are *succeeded* by other animals closely related to them, no matter what the environmental conditions. This suggestion gave no support to the creationists of the time, who believed certain animals were created to be perfectly suited to their enviroment. Following this argument, if such environmental conditons were present then certain predictable kinds of animals should occur, and thus, one would expect the same kinds of species in the tropical regions of Australia and Africa. As has been pointed out, Mitchell noted the effect that the Wellington discoveries had on certain creationists:

> "I understand Buckland's nose is put completely out of joint by the bones from Australia, their not being those of lions and hyenas is, I find, a fact which is considered in England to entirely upset his theory. And I have now heard from the best authority that the fact of their fossil bones belonging to animals similar to those now existing has worked a great change in all their learned speculation on such subjects at home."
>
> from Ranken 1916

So, even early in the history of the discovery of vertebrates on this continent, the antipodean fossils had some degree of international significance. In part, they caused Buckland to revise his hypothesis supporting a universal flood, and, in fact, eventually to abandon it altogether.

The work of foreign experts such as Sir Richard Owen was of vital importance during this period. Owen is also significant for his encouragement of colonial scientists. Indeed, the first native born Australian to study vertebrate fossils systematically was the naturalist Edmund Charles Hobson, who had studied under Owen. Born in Parramatta, New South Wales, Hobson on returning to Australia settled in Hobart where he established a medical practice in 1838 and also became a founding member of the Tasmanian Society. In 1840, for reasons of health, he moved to the Port Phillip district. An early death at the age of 34 put an untimely end to a highly active mind that had delved into many areas of science including geology and vertebrate palaeontology. Hobson's last paper, published posthumously, dealt with a marsupial fossil, a *Diprotodon* jaw. Hobson's wife Margaret (née Adamson) was, likewise, a keen naturalist and sketched specimens referred to by her husband in his papers. Her sketches were used by the accomplished lithographer, Thomas Ham, to produce the final plates for Hobson's papers. Until much later, she is the only woman who is recorded to have had any association with the field of vertebrate palaeontology in this country.

MINERAL WEALTH — CATALYST FOR INDEPENDENCE

Throughout the nineteenth century and even into the twentieth, vertebrate palaeontology in Australia for the most part provided only the grist, the data base, for the European intellectual mills. However, the local scientific community was becoming steadily more able to stand on it own feet, and most importantly, surer of its own ability. Gradually Australians were recognised as an integral part of the world's palaeontological community, and became involved in international as well as domestic projects. Local institutions began to train students and publish scientific results within Australia. Such activities were fostered by the Americans and the British in a series of co-operative enterprises over the past half century, and without such interaction Australia's palaeontological independence would have been much longer in coming.

Part of the increase in work by Australians must surely be related to the increasing population and its rapid concentration in growing urban centres such as Sydney, Melbourne and Hobart. Also important was the increasing wealth that accumulated in these communities, most notably that generated by the discovery and mining of gold and coal in several parts of Australia. Excavation of these valuable resources in itself led to the discovery of fossils. As well, economic necessity brought about the establishment of the first State geological survey whose job it was to document the rock record in some of the Victorian gold-bearing regions, information which was also useful for scientists looking for fossils. Coincident with this, the settling of the interior led to the discovery of bones on a number of sheep and cattle runs when wells were excavated or when drovers reported fossils they'd spotted as they travelled the countryside trailing their slow-moving herds.

With penetration of the interior and the mining boom of the mid-nineteenth century, several members of the European scientific community took up temporary or even permanent residence in Australia. Two men from this period stand out in the move towards an independent palaeontological community: Ralph Tate in Adelaide and Robert Etheridge Jr, first of Melbourne and later of Sydney. Both were determined advocates of a robust Australian science, and both firmly believed in co-operation as equals with foreign experts.

The discovery of gold, mainly in Australia's south-east, provided a major economic incentive that affected the course of palaeontology. The first of Australia's geological surveys, the Victorian Colonial Survey, was established in 1852, with A.R.C. Selwyn as its director. Selwyn's instructions were to document the geology of the gold-bearing regions, probably in the hope that such effort would allow prediction of further gold-bearing strata. Selwyn was well prepared for his task. He was trained by some of the most prominent geologists in Britain, geologists who were defining the basics of the international Geological Time Scale. When Selwyn assumed his duties as director, the most authoritative summary of the geology of Australia was Jukes' *A Sketch of the Physical Structure of Australia, So Far As It Is At Present Known* published in 1850 by T. & W. Boone in London. The smallness of the text (95 pages) reflects just how general was the knowledge of geology at this time and how restricted it was to the margins of the continent. Selwyn and his geologists added much to the knowledge of south-eastern Australian geology by preparing well over 60 detailed maps before political controversy and shortage of funds brought about the survey's dissolution. Before it was disbanded, however, Selwyn's survey was able to publish its own scientific results, and it had also trained a number of geologists who were important in setting up other surveys and institutions within Australia. Eventually the Victorian survey was re-established, and similar institutions were in place in all other States by 1890. These surveys fostered the collection and storage of fossils, some of which were vertebrates.

THE EARLIEST MUSEUMS

Fossils had been retained in both public and private Australian collections even before the State surveys. These specimens, together with the State survey collections, served as an ever-growing resource which allowed direct comparison of newly found material in the latter part of the nineteenth century.

Since the early coastal surveys of Australia by the French and British, and before them by the Dutch, considerable British and European interest in Australian natural history specimens had led to a lively trade in Australian oddments, both for scientific and commercial reasons. Respectable collections of Australian specimens accumulated in London, as well as in several European museums, at first through the efforts of such men as Sir Joseph Banks and Robert Brown. As a result overseas museums, especially the British Museum (Natural History), came to own many of the type specimens of the newly discovered "natural curiosities" from the antipodes.

This pattern began to change, however, with the arrival in Sydney of Alexander Macleay in 1826. Macleay was sent as Colonial Secretary to the government of New South Wales, a position he held for the next decade. From Europe he brought with him a fine library and a superb insect collection, amongst the best known anywhere at the time. Interestingly, Macleay's own son and his nephew, both with the name of William, were to add to Alexander's collections and perpetuate the family's support of science in Australia for years after his death.

William Julian Macleay, the nephew, expanded the family's private museum in their Elizabeth Bay residence (after both Alexander and William Sharp Macleay had died), in part by importing specimens from abroad. His wish was to possess a truly international collection, not just a local one. Macleay went so far as to hire a curator, George Masters, out of his own funds. He, like his uncle before him, continued the tradition of the Macleay's house serving as an intellectual hub in the community by giving "whisky parties" (actually scientific gatherings around the drink!) for staff from the University Of Sydney, as well as interested personages — explorers, doctors, visiting scientists, amongst others. This tradition undoubtedly led to his bequeathing his museum to the University Of Sydney, along with £6000 to pay for ongoing curation.

At the time Alexander Macleay stepped ashore in Sydney town, a museum of sorts already existed there. It consisted of a small room in the Colonial Secretary's Office, initially set up by the colony's first scientific society, the Philosophical Society Of Australasia. In the beginning, each of the seven members of the society paid £5 to have the collection organized and for the purchase of books. Major Goulburn provided the single room that constituted "the Museum". Australian specimens were put on display, but, perhaps more importantly, such a museum allowed the establishment of official contacts with foreign museums, which in turn encouraged the exchange of specimens between this and overseas institutions. The flow of scientific material was now in both directions, not just from Australia to the colonial powers.

To this infant museum Alexander Macleay added his own collections and, together with his enthusiasm, the foundations of the Australian Museum were built. Naturally, this first museum was mainly a storehouse for antipodean specimens. Because of that, it encouraged more and more residents to retain material in Australia as well as to be more aware of their native fauna and flora. It was quite amazing that such a museum was able to exist at this time. The Sydney colony was "a convict settlement ... racked with dissention between free immigrants and emancipists, businessmen and farmers, army and government, colony and colonial office ... an environment conducive [only] to ... activities [such as] those directed to individual survival and aggrandisement" (Strahan 1979). In this milieu Australian science, including vertebrate palaeontology, had its beginnings.

When the Philosophical Society first set up its museum in the Colonial Secretary's Office, Sydney was still small, with a population approaching 20,000. Yet, by 1837, this tiny museum had public hours each Tuesday and Friday from 11 a.m. to 4 p.m. and contained exhibits of native flora and fauna including more than 300 species of birds. Also present in the collections were ethnological and geological specimens, and by 1832 the government was providing £200 annually — certainly a beginning.

The Australian Museum, as it is now known, was Australia's first museum but soon others followed. The National Museum Of Victoria, now the Museum Of Victoria, was set up in Melbourne in 1854. Two thousand pounds were put aside by the Victorian government for the fledgling National Museum, and by March 1854 Captain Andrew Clarke, who was instrumental in the initial realization of this museum, ensured that two rooms were designated especially for it above his offices at the old Assay Office on Latrobe Street, just west of what was to become a more permanent site for this institution. The first staff appointment was made on 1 April 1854, one William Blandowski,

41

42

43

whose personality eventually led him into considerable conflict with the council that oversaw the museum and with the newly appointed director Frederick McCoy. Blandowski energetically mounted a number of expeditions, which impressively expanded the museum's holdings, and engaged the help of such people as Gerard Krefft, who was later to assume a curatorship in the Australian Museum. Blandowski finally resigned, and McCoy took over the reins of power. McCoy in turn determinedly built up the museum collections and expanded the original, somewhat limited facilities. He orchestrated and oversaw the removal of the museum from the government Assay Office to a new site at the University Of Melbourne, a move which was described vividly in a poem published by the Melbourne Punch:

THE RAID ON THE MUSEUM

There was a little man,
And he had a little plan,
The public of their specimens to rob, rob, rob,
So he got a horse and dray,
And he carted them away,
And chuckled with enjoyment of the job, job, job.

Blandowski's pickled 'possums
And Mueller's leaves and blossoms,
Bugs, butterflies, and beetles stuck on pins, pins, pins,
Light and heavy, great and small,
He abstracted one and all
May we never have to answer for such sins, sins, sins.

There were six foot kangaroos,
Native bears and cockatoos
That would make a taxidermist jump for joy, joy, joy.

And if you want to know
Who took them you should go
And should seek information from McCoy, Coy, Coy.

When one's living far away,
Up the country I dare say,
It's very nice to have such things at hand, hand, hand,
Yet it don't become professors,
When they become possessors,
Of property by methods contraband, band, band.

Pescott 1954

The collections remained at the University Of Melbourne until 1899, when shortly after McCoy's death they were moved again to the present site of the Museum Of Victoria on Russell Street.

Other museums followed the lead of Sydney and Melbourne: the Queensland Museum was established in Brisbane in 1855, the South Australian Museum in Adelaide in 1856, and then the Tasmanian Museum (Hobart), the Queen Victoria Museum (Launceston), the Western Australian Museum (Perth), and finally the Northern Territory Museum (Darwin and Alice Springs). In addition to these governmentally sponsored institutions, several private museums, such as the Kyancutta Museum in South Australia, were set up by interested collectors. Many of these private collections were later incorporated into State and federal museums, either by direct donation or by purchase.

Although not without political intrigue and funding difficulties, museums did continue to expand into the twentieth century. But they were no longer simply places for storage of Australia's heritage, its natural wonders and antiquities. They also became centres for public education and research, venues for scientific societies and gatherings, and a conscience for the country regarding the treatment of its natural resources. The museums, too, became the source of personnel and funding for locally based expeditions, one of the earliest being a trip to the northern part of Australia and New Guinea in 1875, organized and financed by the Macleays.

UNIVERSITIES — BEGINNINGS OF HOME-GROWN TRAINING
Universities, in general, developed slightly later than the State surveys and museums, but they played a critical role in establishing independent science in Australia. Frederick McCoy, a major force in the founding of the National Museum Of Victoria, was also the Professor Of Natural Science at the University Of Melbourne from 1854 to 1899. He was particularly interested in palaeontology and worked occasionally on vertebrate fossils. During the 45 years of his active scientific life, he ruled over palaeontology and geology with an autocratic air. Because of this, he attracted few students. T.S. Hall was an exception, having an ability to get along with McCoy despite disagreeing with him on a number of issues. Hall erroneously had assumed that he would be offered the Chair Of Geology at Melbourne when McCoy was gone, but instead it went to John Walter Gregory, a potent import from Scotland.

J.W. Gregory, a remarkable "international" geologist, succeeded McCoy in December 1900 as Professor Of Geology, carrying on the involvement of the University Of Melbourne in palaeontology. Gregory, who began his career as a palaeontological assistant in the British Museum, was interested in popularizing geology and teaching it as a practical subject, the antithesis of McCoy. A competent organizer, Gregory led a group of Melbourne University students on camelback into the Lake Eyre Basin during the summer of 1901-02. On that expedition he discovered a remarkable vertebrate fossil field, mainly along Cooper Creek. The account of this early vertebrate

palaeontological expedition appears in his *Dead Heart Of Australia*, published in London in 1906. This book was really a compilation of a number of local newspaper reports included in the Melbourne *Age* and avidly read by a public hungry for exploration news. It was not until half a century later that the full potential of this area was recognized when Ruben A. Stirton followed in Gregory's footsteps. Stirton not only relocated many of Gregory's Pleistocene locales, but he himself discovered the first concentrations of Tertiary terrestrial vertebrates ever found on this continent.

Gregory was not only associated with the university system but from 1901 simultaneously served as Director of the Geological Survey in the Mines Department Of Victoria. Gregory, furthermore, became involved in the Victorian Chamber Of Mines and was on the council of the Australian Institute Of Mining Engineers (now the Australasian Institute Of Mining And Metallurgy), as well as an office-holder in the Royal Society Of Victoria. Even more than this, he became involved in presenting extension courses for the general public and was decidedly interested in both primary and secondary education. This activity was, in part, a reflection of how talented people in the small scientific community within Australia often carried a multiplicity of responsibilities. There is, however, no doubt that Gregory was an extremely energetic man. "By repute he [Gregory] could at any one time nurse his infant on his knee, correct the proofs of one book with his left hand while writing another with his right, and dominate a polemical discussion on any topic" (George 1975). Gregory, despite his boundless reserves, finally resigned his post at the university in June of 1904 in despair because he was unable to secure enough funding from the Victorian government to operate the University Of Melbourne Geology Department. Ironically, in September of the same year, the same month that he took on the Chair Of Geology at the University Of Glasgow, the funds became available, but unfortunately it was too late. Gregory continued his expeditionary work in many parts of the world until he drowned on a trip to the Amazon Basin in 1932. Appropriately, he had written a poem on the fly-leaf of the first of his Peruvian notebooks just before his death:

> I wander'd till I died.
> Roam on! The light we sought is shining still.
> Dost thou ask proof? Our tree yet crowns the hill.
> Our Scholar travels yet the loved hill-side.
>
> Gregory 1932

Another early vertebrate palaeontologist in Australia was Alexander M. Thompson, Professor Of Geology at the University Of Sydney from 1866. He was the first professional to spend a considerable time in the caves in the Wellington area. Unfortunately, his obsession with this area most likely hastened his early death through respiratory problems in the 1870s, shortly after an expedition to the caves with Gerard Krefft.

Although not a vertebrate palaeontologist *per se*, Ralph Tate fostered palaeontological training and especially "home-grown"

thinking about palaeontology in Australia more than any other university personage in the nineteenth century. Tate arrived from England in 1874, immediately taking up the foundation Chair in Geology at the newly established University Of Adelaide. He taught palaeontology with an infectious enthusiasm and an open mind, enticing students and volunteers alike into the field by packing a keg of beer as part of the field gear! His research was thorough, abundantly published and of high quality, and his production of good students was unrivalled at the time. Besides his charisma, Tate also held the view that the Australian record should be considered as separate and independent from that elsewhere and "had little but scorn for those he thought believed all the rules of geology were written in Europe ... 'Sir F. McCoy appears to object to any Australian deposits being called Eocene unless the fossil species are identical

---

41 Alexander Macleay, Colonial Secretary of New South Wales from 1826 to 1837, whose library and natural history collections served as a nucleus for scientific enquiry in the early days of European settlement in Australia. Engraving from a portrait by Sir Thomas Lawrence in 1825. (Courtesy of the Historic Houses Trust and Elizabeth Bay House collection)

42 Sir Richard Owen, in the prime of his career. An outstanding comparative anatomist of the nineteenth century, Owen became the authority on the fossil vertebrates of Australia and New Zealand. (Courtesy of the British Museum, Natural History, London)

43 T.S. Hall, Frederick McCoy's student at the University Of Melbourne, was the first locally-trained palaeontologist in Australia. His specialty was invertebrates, but he found time to work on fossil whales as well. (Courtesy of the Museum Of Victoria)

44 Robert Etheridge Jr, an important figure in palaeontology. His early work in Australia was with the Selwyn survey in Victoria, and he finished his career at the Australian Museum in Sydney where he served as Director for many years. (Courtesy of A. Ritchie and the Australian Museum)

45 J.W. Gregory, who as a young man was responsible for the discovery of Pleistocene fossil vertebrates along Cooper Creek in the Lake Eyre Sub-basin, an area that was later to produce the first concentrations of Tertiary terrestrial vertebrates from the Australian continent. (Courtesy of Mrs Christopher Gregory)

46 Charles de Vis, late nineteenth and early twentieth century palaeontologist at the Queensland Museum. (Courtesy of R. Molnar and the Queensland Museum)

with those occurring in the London Clay, Paris Basin, and other European Eocenes, peculiar Australian species being open to grave suspicion' " (Vallance 1978). Tate is remembered as one of the first *Australian* palaeontologists, who was not always looking over his shoulder for advice from abroad or trying to shove the Australian record into a European mould.

As the universities became established, they provided permanent, indigenous positions for professional palaeontologists. The first Australian-educated students began to appear. T.S. Hall, who was trained by F. McCoy and worked with the Victorian Geological Survey under Selwyn, was the first of many. Although locally trained vertebrate palaeontologists did not appear until the twentieth century in Australia, much palaeontology was being carried out locally. For the sub-discipline of vertebrate palaeontology, Australian maturity was deferred, mainly due to the lack of a trained professional base in this country and the absence of local financial support for field work and study. It was generous support from agencies like the National Science Foundation in the United States and several overseas museums that enabled work to proceed, though it was fitful.

SCIENTIFIC SOCIETIES – A FORUM FOR LOCAL DEBATE
An important ingredient allowing an increase in home-based scientific endeavour in Australia was the setting up of local scientific societies. Patterned after groups such as the Royal Society of London, these societies gave the educated gentry a chance to exchange information at meetings and, more importantly, gave local scientists a rapid, home-based source of publication. One of the earliest of the societal publications was the *Tasmanian Journal Of Natural Science, Agriculture, Statistics &c* generated by the Tasmanian Society For The Advancement Of Natural Science, or The Tasmanian Society as it came to be known. This society was founded in 1839 and flourished under the leadership of Sir John Franklin. The *Tasmanian Journal*, first published in 1841, circumvented the necessity for scientific ideas to make the long sea journey back and forth to England before officially appearing in print.

PIONEERS IN AUSTRALIAN VERTEBRATE PALAEONTOLOGY
During the transition period from dependence to independence, a number of scientists, collectors and interested individuals substantially influenced the course of vertebrate palaeontology in Australia. One such person was the Reverend William B. Clarke, who arrived in Sydney in 1839. Clarke had migrated with his family to Australia because of his own ill health, and assumed the position of Rector of Willoughby, North Sydney. He had a background of study at Cambridge University, where he had been very much influenced by Adam Sedgwick, the Professor Of Geology, with whom he maintained a close friendship throughout his life. Clarke was an avid fossil collector and a keen geologist. He took every opportunity to show visiting scientists the rocks of the Sydney area and corresponded actively with many well known geologists overseas, such as Roderick Murchison as well as Adam Sedgwick in England, and J.D. Dana in the United States. He certainly did not work in an isolated atmosphere, despite the tyranny of distance from centres of geological thought. Clarke managed to do more than simply collect specimens and serve as a geological tour guide; he also published a number of articles in the local newspapers and in both Australian and overseas journals, especially concerning the geology of the Sydney environs. He drew together many of his geological and palaeontological ideas and the results of several survey trips in *The Sedimentary Formations Of New South Wales*, which appeared in several editions between 1867 and 1878. Despite this effort on his part, however, most of the vertebrate fossil material that he collected or that was sent to him was forwarded to scientists abroad for final study. Evidently, and quite unfortunately, he did not feel at ease in describing and considering the significance of this type of material.

Another early cleric with interests in Australian vertebrate fossils was Julian Tenison-Woods. A Roman Catholic priest-geologist, he led a varied and somewhat controversial life that took him from England to Australia, from Europe to China, and back to Australia, where he died in 1889. Tenison-Woods spent much of his time in the area around Penola, to the south-east of Adelaide, where he collected from the middle to late Cainozoic deposits, and he wrote a number of scientific papers based on material that he had acquired. This material included large bird bones from native wells around Penola, which were some of the first dromornithids reported. He maintained an active correspondence with geological enthusiasts as well as other palaeontologists and geologists in Australia, such as William Macleay and the formidable Sir Charles Lyell, one of the most prominent British geologists of the time. He was also an active member, even president, of several scientific societies, such as the Linnean Society Of New South Wales (1879-80), and a member of the Board Of Trustees of the Australian Museum (1880).

Tenison-Woods used what few chances he had for intellectual pursuits within his parish, for example, stopping for a while on such properties as that of Samuel Pratt-Winter at Murndel near Hamilton, Victoria. Murndel was ideal for Tenison-Woods as it combined an abundance of Miocene fossils (marine) with a superb library as well as an educated and travelled land-owner. Even though Pratt-Winter was not a Roman Catholic, he would travel to Tenison-Woods' presbytery in Penola, when going to Mt Gambier, and often would bring books for the priest to read in his absence from Murndel. Tenison-Woods certainly did not confine his explorations to southeastern Australia, but made a number of excursions to several parts of the continent, for both scientific and religious reasons. He also visited Malaysia and China. Despite the fact that Tenison-Woods described and named many of the fossils that he collected, he, like Clarke, often forwarded specimens to Melbourne or to London to the "experts" for a final decision.

A significant figure in South Australian palaeontology was H.Y.L. Brown, a Nova Scotian who received his training initially at the Royal School Of Mines in London. He worked for a time in both Canada and New Zealand and trained under Selwyn in the Victorian survey. Brown became the Government Geologist of South Australia in 1882 and held that job for 30 years, using all manner of transport to cover much of South Australia and the Northern Territory, which was then a part of South Australia, in his geological surveys. He was a man of catholic tastes, and his reports and maps detail not only the geology but also water resources, local environmental conditions, distribution of fauna and flora and the ethnology of each area he visited. Brown collected remains of *Diprotodon*, an extinct giant marsupial, and the extinct giant goanna, *Megalania*, as well as a number of other bones in the area north-east of Lake Eyre. He noted that the native peoples of this area thought that these bones belonged to the *cadimurka*, a large fish that lived in the bottom of the local waterholes. These *cadimurka* had never been seen alive by anyone, however.

Simultaneous with Brown's work in Central Australia was the first major Australian expedition to collect vertebrate fossils since Mitchell's work in the Wellington Valley of New South Wales. This expedition was to Lake Callabonna in north-eastern South Australia and was mounted by the South Australian Museum. E.C. Stirling and A.H.C. Zietz, based at the museum, are perhaps most remembered in vertebrate palaeontology for their direction of the excavations at Lake Callabonna.

Fossil bones were originally discovered at the lake by an Aboriginal stockman, Jackie Nolan, who reported them to the Callabonna Station owner, F.B. Raglass. Two days later Raglass visited the site. A few days after that, the station cook also visited the site, and, knowing a reward had been posted for the recovery of the feet of *Diprotodon*, took the bones to Adelaide to claim the money for himself. Because of the confusion concerning just who should receive the reward, no one ever did! Nevertheless, as a result of this discovery, the South Australian Museum dispatched H. Hurst to investigate the discovery in January 1893. After four months of field work, a considerable amount of material was returned via "buck-board" buggy by Hurst.

After evaluation of the Hurst work, Stirling and Zietz decided to themselves visit Lake Callabonna in August of 1893, and Hurst resigned his appointment upon their arrival. Despite appalling field conditions including bogged camels, difficulty in acquiring feed and firewood, rabbit plagues, illness and high temperatures, a major part of the world's largest collection of *Diprotodon* skeletons was recovered and subsequently transported to the South Australian Museum in Adelaide as a result of this late nineteenth century expedition.

One of the first excavations of its kind, where whole animals were recovered in numbers, Lake Callabonna gave Australia's small population, and that of the world, a glimpse of complete skeletons of such animals as the giant marsupial *Diprotodon* and the massive bird *Genyornis*. In other localities known up to that time, skeletons were disarticulated because of the jumbling that occurred during sedimentation and preservation in caves, stream channels and even swamp accumulations.

Stirling and Zietz were not simply field collectors, but also studied and published what they had found, producing a series of large format, well illustrated monographs on a variety of the Callabonna vertebrates. Sir Richard Owen, who had toiled so long and hard on understanding *Diprotodon*, would have envied such work, or would perhaps have done the work himself, had he the chance. Callabonna held the secrets that Owen so avidly sought, but ironically he died in 1892, the very year that bones were discovered, and he went to his grave not knowing what the feet of his treasured *Diprotodon* looked like.

H.Y.L. Brown prepared a report on the Callabonna area, dated 27 June 1893, in which he astutely recognized the significance of the site: "In view of the importance of preserving these relics of a bygone age for the future scientific exploration I would recommend that the whole area of the lake be reserved for that purpose, and to prevent the indiscriminate digging up and removal of portions of the specimens." This recommendation was implemented on 30 November 1901 by the South Australian government. Lake Callabonna still remains the only Quaternary site in Australia where an abundance of articulated specimens of extinct vertebrates can be consistently recovered. It is important that this site continues to be preserved intact for just this reason.

In the late nineteenth century two scientists with vertebrate palaeontological interests stand out as independent workers who did not automatically seek foreign expert opinion to give credence to their own ideas. They were Johann Ludwig Gerard Krefft of the Australian Museum in Sydney and Robert Etheridge Jr, whose final posting was at that museum. Both men strongly believed in their own ability to make reliable decisions without outside confirmation. Etheridge was an imaginative and careful scientist with an excellent record of well defended scientific argument. Together with Ralph Tate, he supported the idea that the Australian stratigraphic sequences might not be a direct reflection of those in the Northern Hemisphere. Krefft's record is not so exemplary, and his fiery determination to stick to his own opinions may have been more of a personal nature than a firm defence of an independent Australian scientific community.

Krefft is often remembered for his public disagreement with Sir Richard Owen over the feeding habits of the fossil marsupial *Thylacoleo*. Krefft had sent Owen material for study for some time, but he rather unfortunately stopped doing so when Owen interpreted the jaws and teeth of *Thylacoleo*, the marsupial lion, as belonging to a carnivorous animal. Owen was right, but Krefft adamantly and rather unscientifically disagreed, claiming that the beast was a plant-eater, something like a giant rat kangaroo. Krefft remained, throughout his life, an outspoken defender of his own ideas. He is best known for his studies on fossil mammals, but he worked on a number of other fossil vertebrates as well.

Robert Etheridge Jr came to Australia to be a part of Selwyn's geological survey group. Although he returned to England after the survey collapsed, he was drawn back to Australia by an abiding interest in this new country, and he was an important ingredient in establishing the science of vertebrate palaeontology on this continent. He served as a palaeontologist at the Australian Museum as well as with the Geological Survey of New South Wales, and was eventually appointed as Director of the Australian Museum in 1887, where he remained until 1919. During that time Etheridge worked on fossils from all parts of Australia, and had close working links with other geological survey personnel such as H.Y.L. Brown. He was "aloof, rather dour ... and shared his enthusiasms with few, though so many profited by them" (Vallance 1978). He had an incredible capacity for work, and his publication record was impressive (more than 400 papers). Even more impressive was the accuracy of his assessments in those papers — his scientific work has stood the test of time. "Etheridge's writings, like Tate's, betray a well-informed sense of historical scholarship" (Vallance 1978), and it was these two scientists who set the stage for independence in Australian palaeontology.

Still another late nineteenth century vertebrate palaeontologist

47

was Charles de Vis. He differed from Gregory, Brown, Stirling and Zietz in that he did little field collecting himself. Instead, he worked on the collections made by others, such as Brown and Gregory. De Vis's origins lay in Manchester, and he worked at many jobs before joining the Queensland Museum staff, including being a librarian in Rockhampton for some time. He often published articles in local newspapers, such as *The Brisbane Courier* using the pseudonym "Thickthorn". On assuming the curatorship at the Queensland Museum, he published profusely, naming many new species, primarily from late Tertiary and Quaternary deposits. His modern and fossil comparative collections were exceedingly small, and his communication with the remaining scientific world was hampered by distance. Because of these shortcomings and his lack of understanding of variability within species, he described much of the fossil material from Australia as new and extinct species. This view, in part, stemmed from his belief that all fossil forms must represent extinct species. Because of this misconception and insufficient resources, many of the original de Vis species have been found to be fossil remains of forms still alive. In fairness to de Vis, it is worth remembering the isolation in which he worked, the minimal funds he had available and the small comparative collections on which he relied. Besides his scientific work, de Vis made a significant contribution to the philosophy that museums should serve as educational institutions.

As well as full-time professionals, a variety of other part-time workers were important in the development of vertebrate palaeontology in Australia. Robert Broom serves as an example. Probably best known for his work on australopithecines in South Africa, his main occupation in this country was as a medical practitioner. Broom arrived in Sydney on 28 May 1892 and spent four years on the continent, living for some time in Taralga, New South Wales, where he served as the town's doctor. Still he found enough time to collect fossil vertebrates from the Wombeyan caves, despite resistance from the New South Wales government. Some of this material was deposited eventually in the Australian Museum, but the vast majority of it followed him overseas when he returned to Glasgow, perhaps partly because of the trouble he had with officialdom concerning his efforts in the caves of New South Wales and perhaps also due to a somewhat cool reception he received at times from Robert Etheridge, then in charge of the Australian Museum collections.

By the beginning of the twentieth century Australian-based vertebrate palaeontologists were collecting, describing and thinking about Australian fossils. They were no longer automatically shipping them overseas. But, there was still little funding for this science, either for

47 *The skeleton of a Diprotodon found weathering on the surface of Lake Callabonna, South Australia, during the South Australian Museum's expedition to this area in the late nineteenth century. Skeletons were far more evident on this expedition than in subsequent excursions made in the 1950s to 1980s. (From Stirling and Zietz, 1913)*

48

collection and study or for the hiring of professional vertebrate palaeontologists in permanent positions. As a result, students were not being trained in this discipline in Australia, and little expeditionary work was mounted locally.

AN INDEPENDENT AUSTRALIAN VERTEBRATE PALAEONTOLOGY

Vertebrate palaeontology in Australia has seen decided expansion during the twentieth century, especially since the 1950s. The greatest effort has been in the study of fossil fishes and mammals. Research on these groups has provided biostratigraphic information very useful in establishing the age of rock sequences, for instance, in the deformed Devonian marine and freshwater sediments of eastern Australia as well as in the flat-lying, monotonous Tertiary carbonate channel deposits that mimic the underlying Cambrian marine limestones in northern Australia.

During this period, advancement has been greatly influenced by a few energetic workers who have either discovered or developed important new fossil fields. Their efforts have led to the accumulation

48 Meeting of many of the currently practising vertebrate palaeontologists in Australia at the Conference On Vertebrate Evolution, Palaeontology And Systematics held in March 1989 at the Australian Geographic headquarters in Sydney. This gathering shows that Australian vertebrate palaeontology has greatly expanded in the past 30 years. Front row (left to right): Colin Groves, Jeanette Muirhead, Rhys Walkley, unidentified, Corrie Williams, Dietlind Knuth, Coral Gilkeson, Sue Creagh, Anne Warren, Sue Hand. Back row (left to right): Tony Thulborn, Michael Loy, Zhang Gue Rui, Gavin Young, Robert Jones, Tim Hamley, Susan Bergdolt, Bernie Cooke, Brian Mackness, Miranda Gott, Mike Durant, John Barry, Walter Boles, Neville Pledge, Julie Barry, Arthur White, Alex Ritchie, Jim Lavarack, Mary White, Sue Lavarack, Pat Rich, unidentified, Tom Rich, unidentified, John Long, Henk Godthelp, Peter Murray, Paul Willis, Michael Archer, John Scanlon. (Courtesy of Australian Geographic)

of enough information to allow meaningful summaries. Names that stand out amongst the many vertebrate palaeontologists who worked or are still working during the twentieth century are E.S. Hills (Aust.), R.A. Stirton (USA), R.H. Tedford (USA), M.O. Woodburne (USA), W.D.L. Ride (Aust.), J.A. Mahoney (Aust.), E.D. Gill (Aust.), E. Lundelius (USA), W. Turnbull (USA), M. Archer (Aust.), R. Wells (Aust.), P. Murray (Aust.), T.F. Flannery (Aust.), J. Hope (Aust.), S. Hand (Aust.), N. Pledge (Aust.), J.A. Long (Aust.), A. Ritchie (Aust.), K.S. Campbell (Aust.), S. Turner (Aust.), R. Miles (UK), G. Young (Aust.), R. Molnar (Aust.) and T. Thulborn (Aust.). All of these scientists contributed significantly in major field discoveries, prolific description of new taxa and in novel interpretation of the newly found material.

Hills' review of the entire field of Australian vertebrate palaeontology and Ride's summary of Australian palaeomammalogy were the first real attempts to draw together the rapidly accumulating information. Hills as well as Long, Young, Campbell, Ritchie and Miles all specialize in Devonian fishes, while Stirton, Tedford, Woodburne and their associates finally located the first concentrations of Tertiary mammals in Australia, in the Lake Eyre Basin where H.Y.L. Brown and J.W. Gregory had earlier trekked with camels. Gill, through his enthusiasm as a collector, and his publications, also promoted vertebrate palaeontology, especially in Victoria.

Ernest Lundelius and William Turnbull together with staff from the Western Australian Museum, especially Duncan Merrilees, explored the Pleistocene record of Western Australia as well as the unique Pliocene site of Hamilton in Victoria, one of the few radiometrically-dated vertebrate fossil localities in Australia. R.A. Stirton ("Stirt") and David Ride promoted Australian vertebrate palaeontology not only by their own field and research work, but also by training a number of students and by collaborating with scientists around the world on Australian projects. Current researchers such as Michael Archer and Michael Woodburne, who studied with Ride and Stirton respectively, have in turn supervised additional students.

The current field of vertebrate palaeontology has much expanded over that of the 1950s. The long list of professionals in this field (given in *Kadimakara — Extinct Vertebrates of Australia* edited by P.V. Rich and G.F. van Tets) is a reflection of the present level of activity. Each of these scientists has made a unique contribution, from Susan Turner who set up an economically useful microvertebrate biostratigraphy, to Timothy Flannery and Alex Ritchie who have collected and mounted impressive public displays, to David Ride and Alan Bartholomai who expanded and guided museums in acquiring vertebrate fossils. Activity is by far the highest it has ever been in this field in Australia.

Discoveries such as those in the Wellington Valley and at Lake Callabonna stand out as significant for the nineteenth century, when Australian vertebrate palaeontology was just being weaned from its Old World "parent". Several significant finds mark the twentieth century as well. They were discovered by a variety of people, mainly Americans, Australians and British.

Such localities as Gogo in Western Australia, discovered and developed by a joint venture between the British Museum (Natural History) and the Western Australian Museum, have produced some of the finest Devonian material known anywhere in the world. Likewise, the Devonian armoured fish localities in western New South Wales and south-eastern Victoria, worked by several institutions (the Australian Museum, the Museum Of Victoria, Monash and Melbourne universities, the Australian National University and the Bureau Of Mineral Resources), have provided much insight into Australia's geographic position during the Palaeozoic.

The Cretaceous terrestrial and marine sequences containing reptiles in south-western Queensland and Victoria have been explored by a number of institutions, the former by the Queensland Museum, Harvard University and the British Museum (Natural History), and the latter primarily by the Museum Of Victoria and Monash University. These explorations have provided a unique view of near-polar and polar terrestrial faunas that were developing in semi-isolation.

The Tertiary vertebrate-bearing clays and sands of the Great Artesian Basin, and the lime-rich channel and cave deposits of Riversleigh and Bullock Creek in the northern part of the continent, have allowed us to plot Australia's deepening isolation and aridification during the past 30 million years. And finally, the Quaternary cave deposits along the southern and eastern parts of Australia, investigated by a number of Australian and American workers from many institutions (the Australian National University, the University Of Tasmania, Flinders University, the Western Australian Museum, the University Of Texas, the Field Museum Of Natural History in Chicago, the University Of California at Berkeley, amongst others), have produced very large collections of vertebrate material which bear witness to the detailed climatic changes affecting the Australian continent over the past few hundred thousand years, a time during which continental glaciers in the Northern Hemisphere waxed and waned, and in which humans first invaded.

The discovery by Ruben Arthur Stirton during the 1950s of Tertiary vertebrates in the Lake Eyre Basin was a particularly momentous occasion. Not only did it bring to light a large number of totally unknown animals, but it quite clearly led to a real acceleration of activity in Australian vertebrate palaeontology.

Ruben Arthur Stirton had come to Australia in 1952, aiming to find pre-Pleistocene mammals in quantity. Driven by infectious enthusiasm, he achieved exactly what he had set out to do. From expeditions that he led came an abundance of new fossil material, but equally importantly he inspired new students. Stirton's dreams of exploration were allowed to blossom in large part due to the funding provided first by the American-based Fulbright scholarship program and later by America's National Science Foundation. In these years immediately after World War II, funding in many programs worldwide was a direct result of the United States reaching out for international influence; the exploration of Australia's interior was one of the many positive results of this "world view". The Fulbright program was especially potent for it made possible expeditions of extended reconnaissance with no guarantee of success, giving the prospectors the chance to acquaint themselves thoroughly with the special conditions in the Australian non-marine sequences and then spend lengthy periods on the ground exploring. In the 1990s such activities are not easily funded, because of the constant call for immediate practical results and political relevance.

During the 1960s work in the Australian interior became more expensive, and more and more the National Science Foundation in the United States was approached, successfully, for financial aid. It was during this time that many localities actively being worked today

were discovered by a number of joint expeditions involving such institutions as the South Australian Museum, the Bureau Of Mineral Resources, Monash University and the University Of Tasmania working with counterparts in the United States, mainly the University Of California. The richly fossiliferous Tertiary-aged Riversleigh and Bullock Creek areas in northern Australia, for example, were located or relocated, as was the Pliocene Wau district of New Guinea. The potential of Quaternary-aged caves on the Nullarbor Plain was recognized, and new Triassic sites in the Fitzroy Basin of north-western Australia were discovered. Near the end of this decade the first symposium that specifically dealt with Australian vertebrate palaeontology was held as part of a conference of the Australian And New Zealand Association For The Advancement Of Science convened in Melbourne. Americans dominated the agenda, including the first American vertebrate palaeontologist to be appointed to an academic position in Australia, Dr James Warren, who soon after taking up his place in 1961 at Monash University became the Professor of the Zoology Department there.

By the 1970s and early 1980s Australian vertebrate palaeontology had both domestic and international aspects. Australian-based vertebrate palaeontologists were quite visible, many finding permanent positions in museums and universities on this continent. They looked to their overseas colleagues as equals and co-workers. Many were immigrants; some like Michael Archer were Fulbright scholars who had remained in Australia or returned permanently to this country to pursue their research programs. There were Australians who had been trained overseas and returned, like David Ride, and others who were completely locally trained, such as Alan Bartholomai. A telling record of the growing "domesticity" of Australian vertebrate palaeontology is the membership list of the second symposium on vertebrate palaeontology of Australia in 1971, again held as a part of an ANZAAS conference. This symposium included only three speakers from the United States; the remaining 12 were Australians, who spoke on subjects as diverse as Devonian fishes and Pleistocene mammals.

Indigenous expeditions, as well as a number of joint expeditions, continued to be mounted involving such groups as the South Australian Museum, the University Of California (Berkeley and Riverside), the Bureau Of Mineral Resources, the American Museum Of Natural History, the Smithsonian Institution, the Queensland Museum, the Museum Of Victoria, the British Museum (Natural History), the Field Museum Of Natural History, the University Of Texas, the University Of New South Wales, Monash University, and the Australian Army. Specimens, especially those designated as "types" to represent newly defined species, were often studied in Australia and generally remained in Australian museums. Such a practice, initiated in the 1950s by R.A. Stirton, differed remarkably from those of the colonial days when material was collected by local residents and then promptly shipped from Australian shores for study by "foreign experts", such as Owen. No longer, either, were palaeontologists trying to shove the "unruly" Australian record into a European or North American mould.

Today, with a number of full-time and part-time positions filled by vertebrate palaeontologists in Australia, research and training of personnel locally is ensured. Although becoming increasingly restricted, funding for field work, research and publication in this discipline is available locally from both governmental sources, such as the Australian Research Council, and private industry. Scientific journals, such as *Alcheringa*, maintain international standards and encourage publication of local as well as international research. These factors, coupled with the seemingly insatiable interest of scientists and some international funding agencies, such as the National Geographic Society and Earthwatch, in the origin and evolution of Australia's unique vertebrates, will ensure a period of discovery for the remaining years of the twentieth century. Much of this work will undoubtedly be directed by Australians but it will be significantly enriched by interactions on an international level, interactions which have a strong historical basis and should be encouraged and nurtured.

The Australian vertebrate fossil record has sporadically revealed its secrets. Much has been discovered since the Wellington finds of Ranken and Mitchell, and still much more awaits discovery before the full potential of Australia in revealing the history of vertebrates on this planet Earth is realized.

DO FOSSILS LIE? BIAS IN THE FOSSIL RECORD

Only a small fraction of everything that has ever lived is fossilized. Most living things, of course, are recycled, their constituent elements moving into soil or stream, then perhaps to the sea, only to be forced down a tectonic trench deep on the ocean's floor to once again surface out of a volcanic vent into the atmosphere. After that, there is some chance the calcium and phosphate bound with carbonate and traces of many other constituents may find their way once again to the shoulder blade of a possum or the tooth of a shark. Such is the cycle of things.

Because of this recycling, the fossil record itself is decidedly biased. It does not give a balanced cross-section of life that existed at any one time in the past. Rather, it gives the viewer of past scenes only glimpses of the kaleidoscope of events that characterize the last few billion years of life on Earth—and the viewer needs to be cautious of just how to interpret these momentary glimpses. We do know that there are ways to critically judge just how complete or incomplete our vision of the past is, and those ways must be taken seriously. If not heeded, the conclusions drawn by the time sleuth may well be very open to error.

TOPOGRAPHY AND VEGETATION
One of the obvious sources of bias of the Australian fossil vertebrate record, when compared to that on other continents, is simply the lay of the land. For the most part Australia is a flat, worn down country with no majestic mountain ranges. One must go to New Zealand or New Guinea to find even a glimpse of a great mountain chain like the Andes of South America or the Rocky Mountains of North America. Mountain ranges are of extreme importance in the recovery of fossils, for in their tortured history of bending and fracturing on a massive scale they expose thousands upon thousands of metres of ancient silts and sands, conglomerates and limestones. When the sediments that make up these rocks were laid down on the Earth's surface, they simultaneously entombed millions of animals and plants which once lived in or near the rivers and oceans that deposited them. For the most part these sediments were buried, sometimes deep below the Earth's surface, only to be thrown high in the convulsions that followed collision of major crustal plates, such as India and Asia thrusting up the mighty Himalayas. As mountains rose, streams and glaciers began wearing them down. In doing so, they exposed buried fossil treasures that had been protected by the surrounding subterranean rocks for millions of years. Without the mountains, without the eroding streams, the fossils would still lie buried, not attainable by the passing fossil prospector.

Australia, from the vantage of the palaeontologist, has had far too few mountain-building cataclysms. Many of the rocks which hold important information about the evolution of vertebrates on the Australian continent still lie buried, often deep beneath the ground surface, hidden from those who could interpret their significance. Only occasionally, when a fortuitous drill rig punches into this subsurface in search of oil or water, do we recover tantalizing clues to what lies beneath our feet.

Australia does have some mountain ranges, such as the Great Divide running down the east of the continent, but they are of relatively low relief, not approaching the elevations of the Andes or the Himalayas. The country's tallest mountain is only just over 2220 metres in height! Australia's mountains lack one other important aspect, too. Most lie cloaked in vegetation. For people of today, the beauty and solitude of these mountain forests is of prime importance for our sanity, as well as our biological survival. But for allowing an understanding of the past such forests are barriers, hiding the rocks and sediments beneath them. Only with the gash inflicted by an occasional stream or river, or the occasional landslide or quarrying operation, can we view the fossil content of the rocks. What the palaeontologist needs most are mountain ranges in arid country, like Australia's Centre, and such ranges must be of just the right age— younger than 500 million years. The Macdonnell Ranges in Central Australia contain some rocks of just that sort, but they form only a small part of Australian mountains, most of which are restricted to the wetter, eastern margin of the continent. In the arid areas, erosion

and exposure are rapid, and there are vast tracts with little vegetation to hide what is being exposed. Erosion over years or even centuries concentrates the content of the entombing rocks and creates a lag on the ground surface. Such erosional remnants are the guides pointing to areas worthy of further exploration.

Because of the lack of extensive mountain ranges in our arid zones, palaeontologists have resorted to searching even the smallest exposures in these regions — playa lakes, ancient river courses, mound springs. In northern South Australia, playa lakes have been critical in producing the majority of vertebrate fossils from the Tertiary Period. The eroded banks of such lakes as Palankarinna and Namba have yielded bones of flamingoes and palaelodids, platypuses and dolphins, which point to much wetter, more predictable times 20 million years ago. The lake surfaces themselves, after long periods of drought, also expose skeletons of long extinct vertebrates that once plied the watercourses or nested along the shores of permanent lakes.

Australia has a further problem, beyond the lack of montane relief — it has spent much of its history, surprisingly, in relatively wet zones, especially over the last 200 million years, which is nearly half the time that vertebrates have existed on this continent. This situation has meant that soils and rocks near the surface have been deeply weathered by a high water-table. Bones that had managed to survive the ravages of death, decay and burial were eventually dissolved by the acidic waters circulating through the rock and sediment masses, never having the chance to be exposed by later erosion. Had Australia lain for eons in a mid-latitude position, like North America and Europe, instead of near-polar as it did during the late Palaeozoic until the early part of the Cainozoic, the continent's vertebrate story would most likely have been much richer. The red colour that so characterizes much of the land is a reminder of the country's wet and treacherous past, when mobile ground waters dissolved much of the content of near-surface sediments, leaving behind the scarlet iron oxides and the silica which forms the gibber.

ENVIRONMENTS OF DEPOSITION
Not all environments are suitable for fossilization. Bones are best preserved in slightly alkaline environments and in places where burial is quick and weathering is minimal. If a skeleton lies for weeks or months on a desert plain, attacked by scavengers, its chances of survival for a palaeontologist to study are slight. On the other hand, if an animal becomes bogged in a rain swollen river it has some chance of being quickly deposited, buried and preserved, especially if the river waters are not highly acidic.

In order to evaluate the chances of preservation of vertebrates at any time in the past, in Australia or anywhere else, palaeontologists need to know the regional picture of the time. Were there active rivers and streams depositing sediments, or was the area a desert? In Central Australia during the Pliocene and Early Pleistocene, permanent water was dwindling and environments capable of burying and preserving fossils were few. Consequently, few fossil vertebrate sites of that age are known from this area. So, when compared to the records of many other parts of the world at this time, such as western North America or sub-Andean South America, the Australian record is poor — almost certainly due largely to the lack of favourable bone-preserving environments rather than to any lack of living vertebrates on this continent. This record changes dramatically during the later parts of the Pleistocene because of the availability of cave environments for preserving vertebrate bone. Indeed, the Australian record explodes with hundreds of new species — *not* because they actually first appeared at that time, but because an environment capable of preserving bones became available.

It is not simply the availability or non-availability of environments that can bias the record. The type of depositional environments present dictates what will be preserved and in what state it will be fossilized. Lakes and marine basins are most likely to preserve skeletons intact. Current movement in these environs is generally small, and thus skeletons can lie undisturbed until they are covered by sediments. In some cases, the bottom conditions of these waterbodies can be anoxic (without oxygen), which even further enhances the possibility of perfection in preservation. Few scavengers that can disarticulate bones are able to live under such conditions, so, if the sedimentation rate is great enough, then the skeleton will remain intact.

Just the reverse is the situation with fluviatile or stream environments. They are rugged places for bone preservation, and yet they are very frequently the places bones are found. Streams, as they move towards the sea, not only batter bones and disarticulate skeletons relentlessly, but because of their continuous quickening and slowing they can deposit concentrations of bones that are carried along in their waters. Sand bars on the inside of a graceful river bend, where the water becomes lazy for awhile, are typical fossil locations. In flood, a river may breach its levees and suddenly spill out across a broad floodplain. The rushing water carries with it bones and sedimentary particles that have been sped along by the temper of the flood and and also picks up all in its path, perhaps the skeleton or skull of a decaying magpie or kangaroo that has lain for some time on the flats. As the sheet of water spreads out and shallows, it loses energy and drops its deathly collection which has been gathered over a large area and from many environments. Although the surfaces of the bones, and their fragmentation, will reflect their peregrinations, they will be concentrated and an easier mark for the palaeontologist.

SHAPE AND QUALITY OF BONES

Some bones are more robust than others and will better survive the ravages of stream transport. Such bones will bias a sample recovered from fluviatile sediments, but knowing this a palaeontologist can make sensible estimates of the living fauna that gave rise to the fossil assemblage. Most mammal bones fare better in streams than do bird bones. The reason is simple: birds have, in general, hollow bones, an advantage to a flying animal where weight retention is at a premium. So, bird bones wear out swiftly in the erosive environment of a stream, whereas the solid mammal bones are more resilient. Some mammals, like bats, are also rarely preserved in stream deposits because they, too, have hollow bones. Mammals have another advantage over birds, they have teeth. Tooth enamel is a very resistant material and will stand up to the worst battering. Some Australian sites, such as the Miocene-aged Lake Tarkarooloo locality in South Australia, preserve primarily mammalian teeth. Such sites seem to be the end result of long distance transport of vertebrate material in a stream that had essentially destroyed all of the more fragile elements.

Apart from the type of animal that provided the bone, the shape of the bones themselves can make a big difference between preservation and non-preservation in stream deposition. Bones that have a large surface and a small volume, such as a scapula (shoulder-blade), are most unlikely to be preserved. They float easily and do not drop out of the water column soon enough and will instead very likely be pulverized by collisions with many other suspended particles. Large, heavy objects such as *Diprotodon* skulls are unlikely to be moved at all from the site of death, so in riverine accumulations they are noticeably absent. The most frequently preserved bones are long ones such as the tibia (shin bone) or the femur (the thigh bone), which can be transported for a time and dropped before they are pulverized. These bones frequently take on the sense of stream motion, sometimes transverse to the current if the water is very shallow but more frequently parallel to the water flow with the densest part pointing upstream.

THE BIOLOGY AND BEHAVIOUR OF EXTINCT VERTEBRATES

Many other factors concerning the biology and behaviour of extinct vertebrates affect whether or not they will be preserved as fossils. Animals with denser populations and higher birthrates, and of medium-size (about the size of a small wallaby) with a large number of bones that are easily recognized, are more likely to be collected and later identified and reported on. Adults are more likely to be preserved than juveniles, simply because their bones are more solidly ossified and thus more resistant to destruction, especially in a stream environment. And, of course, animals that live in or near the sites of deposition are much more likely to be preserved than those that live far afield.

Flying animals, unless they frequent waterholes, are very unlikely to be fossilized. Arboreal animals and those that live out their lives far from water are also rare as fossils. Solitary animals are less frequently preserved than herding animals, not simply because of numbers but because of behaviour patterns, too. For example, panic behaviour in herding animals often leads to disaster, and if that happens near water many skeletons may be fossilized.

PHILOSOPHIES AND TECHNIQUES OF PALAEONTOLOGISTS

Such factors as how difficult it is today to reach certain geographic areas, such as Central Australia, how much finance is required to penetrate these areas, and afterwards to prospect and excavate, can be important limitations that can also bias the fossil record. Some people, too, have a special talent for prospecting, and chance can play a part as well. But some of the "lucky" ones have enhanced their luck by hard work, spending hundreds of hours going over geological reports and maps to determine the areas worthy of prospecting. The rocks must be of the right age and of the right type: looking for fossils in basalts and granites is unlikely to bring success! Weeks or months can be spent in the field searching place after place, talking to local people who know the land and its resources. In Australia new fossil fields have historically been found not only by the professional but also by an impressive number of amateur palaeontologists, who have volunteered their time and ingenuity to make expeditions work. Without this volunteer help there would never be enough funding and support to allow the level of progress in vertebrate palaeontogy which is typical of the 1990s.

Once a fossiliferous area is located, the way in which it is worked — what techniques are used, and how much effort is put into the area — can greatly colour the results. For example, the full potential of the discoveries of J.W. Gregory along Cooper Creek in South Australia in the early part of the twentieth century were not realised until further work was carried out by R.A. Stirton and his crews in the 1950s and by subsequent work by a number of institutions. Their application of quarrying and screen-washing techniques, not just surface prospecting, greatly expanded the knowledge of the small fauna such as the dasyurids, or marsupial mice, and the rails.

Preparation techniques, like collection techniques, can have a great effect on the final retrieval of fossil information. By immersing limestone blocks from middle Cainozoic localities such as Bullock Creek and Riversleigh in northern Australia and from the Late Devonian of Gogo in Western Australia in vats of 5–10 per cent acetic acid, wonderfully complete skulls and partial skeletons of many different vertebrates have been recovered. The preservation, down to the fine detail of cranial blood circulation, is exquisite, unrivalled by most sites in the world. Physical preparation techniques using needles and drills, grinders and vibratools (miniature jack hammers) cannot produce the masses of quality material that acid dissolution can. The limestone dissolves away, leaving the more acid-resistant bone behind.

Finally, such factors as the philosophies and rigour of the palaeontologists studying the fossils, how recently a thorough review of relevant material has been carried out and the funding available for post-collection preparation and study can all have dramatic effects on the final faunal list of an area and therefore on the final community interpretation given to a particular fossil assemblage. The study of some groups of animals is better funded, perhaps because there is an economic incentive such as exploration for oil or coal or because of public interest such as the enduring fascination with dinosaurs. It is a pity that certain groups are more studied than others as our view of vertebrate evolution on this continent would otherwise be far more balanced.

THE FOSSIL
VERTEBRATES
OF AUSTRALIA

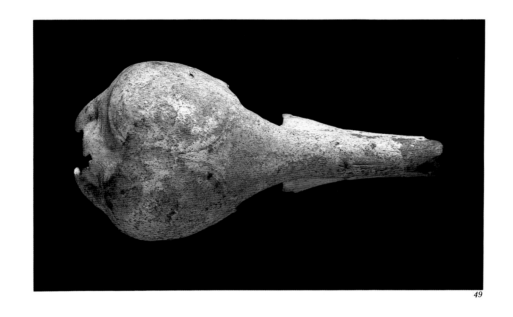

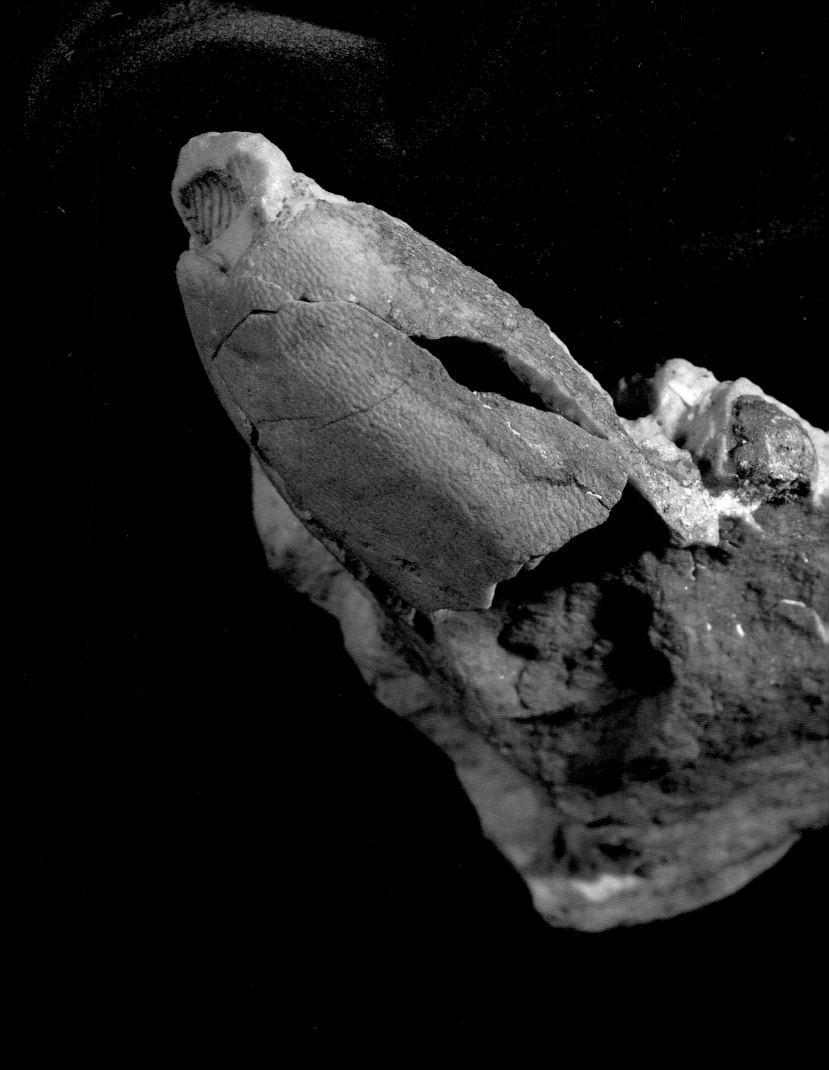

CHAPTER 1

AND THEN THERE WERE BONES

THE ORDOVICIAN TO SILURIAN PERIODS, FROM 490 TO 410 MILLION YEARS AGO

During the Ordovician the first remains of Gondwanan vertebrates appear, at localities in Australia and Bolivia. Slightly older vertebrates, of Late Cambrian age, are known from the Northern Hemisphere. For all of the Ordovician only jawless vertebrates have been found. Their feeding and locomotory options were fairly limited in that they were filter feeders with somewhat unco-ordinated movements. Only in the later part of the Silurian did backboned animals, still confined to aqueous environments, develop jaws and paired fins. Then in quick succession they explosively moved into many new lifestyles, including land-dwelling by the earliest Devonian. For much of the Ordovician and Silurian, conditions were warm. Most of the world's continents were strung like beads along a great Equatorial string. As a result, for much of the time faunas were of a cosmopolitan nature, but glacial conditions near the end of the Ordovician brought about lowered sea levels, extinctions and reduced diversity of life forms. The mingling of the cold watermasses that formed during that time with the warmer equatorial water led to enhanced deposition of phosphates in the deeper ocean basins.

49 The skull of Zaglossus cf. ramsayi, a giant echidna, in dorsal view, from Henschke's Fossil Cave at Naracoorte, South Australia. x 0.5 (S. Morton, courtesy of the South Australian Museum and N. Pledge)

50 The heavily armoured head shield of the jawless fish Arandaspis prionotolepis from the Ordovician shallow marine Stairway Sandstone Formation of Mt Watt, Northern Territory. The specimen is viewed from below and shows the mouth opening where food entered and was then strained out in the branchial basket in much the same fashion as in the invertebrate chordate Branchiostoma. This Arandaspis is the oldest vertebrate known from Australia and one of the oldest nearly complete vertebrates known from anywhere in the world. x 3.0 (F. Coffa, courtesy of the Bureau Of Mineral Resources and G. Young)

MINERALIZED TISSUES AND BIOTIC DIVERSITY

51

Vertebrates had originated by the Late Cambrian, perhaps even as early as the late Precambrian. Intriguing phosphatic fragments of Late Cambrian age strewn across several areas in Greenland, North America and Spitzbergen have been assigned to this group of animals. They occur in sediments that also entomb typical marine organisms — trilobites, brachiopods and conodonts. These small scales (named *Anatolepis*), composed of the mineral apatite (a combination of calcium, carbonate and phosphate), which is the basic ingredient of bone, are only a few square millimetres in area and about one-tenth of a millimetre in thickness. Although they have a scale-like surface ornamentation, their microstructure is not at all like somewhat later, but still primitive early fishes. There is still some question as to the exact affinity of these fossils, and more research is needed before a

reliable decision can be made on their vertebrate pedigree.

Since the fossil record only becomes robust with the development of mineralized skeletons near the end of the Precambrian, about 600 million years ago, there may have been soft-bodied ancestors or even early vertebrates that have left no record. For this reason, a continued search of rocks of this age is a worthwhile enterprise that could eventually lead to a much better understanding of the details of vertebrate origins.

By the middle part of the Ordovician, about 460 million years ago, there is no doubt about the existence of vertebrate animals. They are known on two fragments of Gondwana — Australia and South America — and in North America.

51 *An impression of the extenal armour of the jawless fish* Arandaspis prionotolepis *from the Ordovician Stairway Sandstone Formation at Mt Watt, Northern Territory. x 1.3 (F. Coffa, courtesy of the Australian Museum and A. Ritchie)*

52 *An inarticulate brachiopod. Together with trilobites, brachiopods dominated Cambrian faunas of Australia and the world. It was into this kind of fauna that the first vertebrates entered. x 2.0 (S. Morton)*

53 *Trilobites, common benthic scavengers and predators during the early part of the Palaeozoic. A few trilobites were even planktonic. The eyesight of most trilobites was highly refined, being a system based on banks of individual lenses per eye — a completely different arrangement in these invertebrate animals than was the single-lensed vertebrate eye that developed later, sometime in the Late Cambrian. x 4.0 (S. Morton)*

52

GONDWANA'S OLDEST VERTEBRATES

Arandaspis from the Middle Ordovician of Central Australia and *Sacabambaspis* from rocks of comparable age in Bolivia are the two most completely known early vertebrates. In rocks older than those yielding these two genera, all vertebrate fossils are just fragments, mainly scales, giving no idea of what the most ancient backboned animals may have looked like.

Arandaspis prionotolepis and one other less well known early vertebrate, *Porophoraspis*, come from the Stairway Sandstone Formation capping Mt Watt in the Northern Territory. *Arandaspis* is preserved only as internal and external moulds of thin bony plates from the original fish; because of weathering, the dermal skeleton has left only impressions and not original bone or even the permineralized remains of bones. Named after the Aranda, the dominant Aboriginal people of the area, this early vertebrate was not of spectacular size, reaching only 14-16 centimetres in length. Its entire head and gill (branchial) region was protected by a bony shield made up of two large plates, a deep, curved bottom (ventral) plate and a more triangular-shaped top (dorsal) plate. Separating these two major plates was a row of small, rhombus-shaped mini-plates that may have been markers of where the gill pouches lay. The exhaust from these gill pouches was most likely a common opening at the back of this scale row, a place where water taken in through the jawless mouth could exit. Behind the head some of the body scales left elongate and narrow impressions embellished with a single row of small projections that extended out behind the scale giving it a frilled appearance. Not enough of the rest of *Arandaspis* is preserved to give any idea if it had fins, although paired fins are unlikely, or what shape the body assumed, other than that it was rather deep and rounded in cross-section.

Despite the fact that *Arandaspis* is only an impression in sandstone, it left evidence of some rather interesting anatomical details. It had two laterally-placed eyes, which were located near the front of the fish and opened anterior to the row of small plates separating upper from lower bony plates. Lying between these normal eyes are two holes in the dorsal head shield plate, which appear to have housed the pineal "eyes" — outgrowths of the dorsal surface of the brain. What purpose these "third and fourth eyes" had is open to speculation, but in other vertebrates their function ranges from light sensitivity to hormone control (as in humans). In addition to the pineal eyes, grooves along the side of the head shield reflect the presence of some sort of lateral line system, which in modern fishes with similar structures is used to detect pressure changes, and thus motion, in the water. This type of structure also tells of passing competitors, mates and foes, and in some unusual forms can be modified into organs capable of detecting electric currents. Just how it functioned in *Arandaspis* is open to conjecture.

The bones of *Arandaspis* were entirely external, and apparently very thin – about one-tenth of a millimetre in thickness. They gave no hints of any fusion or resorption so they must have been deposited after the fish reached adulthood, perhaps leaving the juvenile unarmoured. There is no evidence at all of any internal bone, so the flexible notochord (the main support postulated for primitive vertebrates and vertebrate ancestors) was probably still quite important in the support of soft anatomy and in the attachment of muscles in *Arandaspis*.

The clear, shallow, tropical seas of Central Australia in the Ordovician deposited some of the oldest remains of fish in the world. They were also the home of a variety of other marine life, including shelled nautiloids and relatives of modern bivalves. The joint-legged trilobites ploughed through the bottom sediments leaving behind their traces and trails called *Cruziana*, a common fossil in some parts of the Macdonnell Ranges of the Northern Territory. Just where *Arandaspis* fits in this community is not exactly certain. Many palaeontologists have speculated that it was a bottom feeder, essentially filtering out micro-organisms as it swam along. As it had no jaws, specialized grazing was not possible so it more likely acted as a vacuum cleaner taking what might come in its path. Its movements may have rather lacked the refinement that modern fishes have, where delicate directional control is possible by adjusting pectoral or pelvic fins and using other fins to stabilize and "trim" the body as the

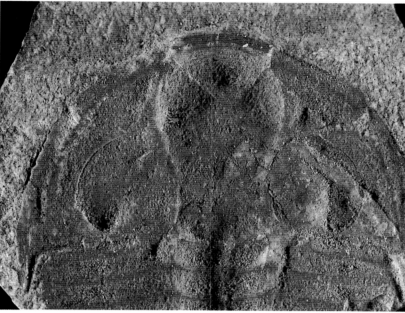

53

fish moves along. But in the Ordovician, *Arandaspis* was the best model going, and for a time it must have had many advantages over its competitors. Later, it would become outmoded when evolution produced a series of new varieties that "pushed aside" such ancient vertebrates. Nevertheless, a few of the early vertebrates did give rise to forms alive today, such as the lampreys and hagfishes which survived by moving into very specialized niches such as parasitism.

In Bolivia the Anzaldo Formation near Cochabamba in the centre of the country has produced the most complete specimens of the earliest vertebrates. As with *Arandaspis*, these fossil fishes are associated with an invertebrate fauna — trilobites and brachiopods — that suggests a nearshore marine environment. In the case of the Bolivian *Sacabambaspis janvieri*, there are at least 30 specimens and they were concentrated in a very small area, one packed on top of the other. Suggestions have been made that a storm-induced inflow of fresh water may have caused the mass mortality of these fishes and the large number of lingulid brachiopods with which they are associated.

Sacabambaspis, in contrast to *Arandaspis*, is known as a whole fish. It was about twice the size of its Australian relative, reaching about 35 centimetres in length and 8 centimetres across. *Sacabambaspis*, like *Arandaspis*, had one large dorsal and one large ventral bony plate on the head. More detail is visible on the head of this fish than is known in any other fish of comparable antiquity. It had two distinct nostrils and its eyes were protected by a circular ring of bony plates. The two pineal openings present in *Arandaspis* are also found on the Bolivian counterpart. The mouth was lined with nearly 60 rows of tiny bony plates which may have been moveable, perhaps allowing the expansion and contraction of the pharangeal cavity which in turn would make the sucking action of the "vacuum cleaner" feeding more refined. Or, perhaps there was another feeding function of these plates.

Sacabambaspis had only a caudal or tail fin and its entire body was covered with four rows of elongate, slender scales arranged in chevrons. These body scales bear the imprint of a well developed lateral line that evidently coursed along about two-thirds of its total length. Isolated scales from the Horn Creek Siltstone in Central Australia are ornamented in much the same way as those of *Sacabambaspis*, but more complete material is needed before conclusions can be drawn.

Bone is actually preserved in the head shield and scales of *Sacabambaspis*. It is acellular, quite unlike most vertebrates, and links these early Palaeozoic Gondwanans with a group much better known in the Northern Hemisphere, the heterostracans (or pteraspidomorphs).

LIFE WITHOUT JAWS

The Ordovician and parts of the Silurian were the heyday of jawless vertebrates — and although they never again reigned supreme in the vertebrate world, they have maintained a presence ever since. Two major groups of jawless fishes existed during these early times, the Pteraspidomorphi (or Diplorhina) and the Cephalaspidomorphi (or Monorhina). There is still much discussion concerning which fishes should be put in which of these groups and apparently only the Pteraspidomorphi had much of a history on the Gondwanan continents, certainly in Australia. That group included the heterostracans and the thelodonts. The Cephalaspidomorphi, with an essential restriction to the now northern continents, included the osteostracans and the anaspids.

The pteraspidomorphs are distinguished from the cephalaspidomorphs by a number of characters. A very important one is that they had paired nasal sacs located on either side of the fish's midline (called diplorhinans), whereas the cephalaspidomorphs had only a single nasal sac that was centrally located. The pteraspidomorphs are most similar in this feature to the fish-like *Arandaspis* and *Sacabambaspis*, and are thus thought to be quite primitive fish. Like *Sacabambaspis*, and probably *Arandaspis* too, the heterostracan pteraspidomorphs had a fairly solid cover over the head and throat (pharynx) area, they had a single exit for water that passed over the gills, they lacked any internal skeletal ossification, their bony external armour was acellular, and none ever developed paired fins.

The pteraspidomorphs have left a fossil record of a two-fold nature, comprising macrofossils and microfossils: while the heterostracans have left a reasonable macrofossil record, the thelodonts (which first appeared in the earliest Silurian) are for the most part known as microfossils. The thelodonts' bodies seem to have been covered by thousands of tiny, rhombic, non-overlapping scales, and it is this abundance of microscopic remains that makes them especially useful for dating rocks in Siluro-Devonian times.

Based on what is preserved in the fossil record, the jawless Agnatha, the group including both the pteraspidomorphs and cephalaspidomorphs, diversified remarkably on a world scale in the Late Silurian and Early to Middle Devonian. It was a time of great adaptive radiation for fishes of all sorts, both jawless and jawed.

The Australian record for the agnaths does not resume until the Devonian, however, and then it is essentially restricted to the thelodonts. Nevertheless, the antipodean forms seem to follow the world trends in diversity and by the middle of Devonian times are almost totally eclipsed by the impressively successful jawed placoderms, which themselves dominate only until the Devonian closes. They, in turn, are forced into extinction by the continued evolution of more advanced fishes.

The success of both the agnathans and the placoderms should not be underestimated. Agnaths were vertebrate rulers of the watery realms for nearly 100 million years and have managed to tenaciously survive through until the present. The placoderms shared agnath supremacy for 10-20 million years and remained successful, despite growing competition with sharks, "spiny-sharks" and the ancestors of modern bony fishes, for at least another 15 million years, when they rather abruptly disappear forever. One cannot lightly ignore such

54

54 The graptolite Tetragraptus (Pendeograptus) fruticosus (four-branched form) and Didymograptus sp. from the Early Ordovician (Bendigonian) deep marine sediments near Campbelltown, Victoria. These planktonic invertebrates were at this time common in the near equatorial waters of Australia and also widely distributed around the globe. Because of their cosmopolitan distribution, graptolites are important in correlating early Palaeozoic rocks in now distant parts of the world. They were advanced invertebrates that may be quite closely related to vertebrates. x 2.0 (I. Stewart)

55 The graptolite Tetragraptus approximatus from Early Ordovician deep marine sediments near Campbelltown, Victoria. x 3.1 (I. Stewart)

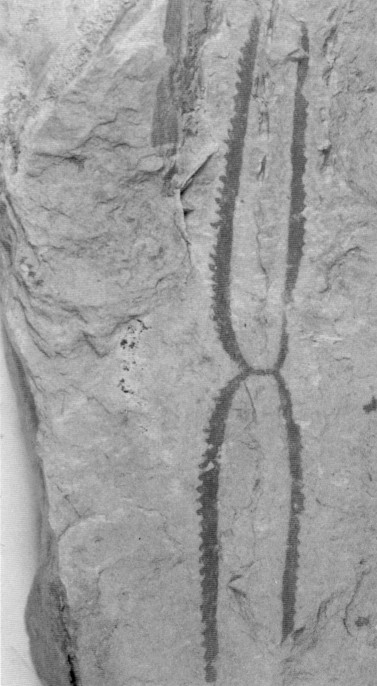

55

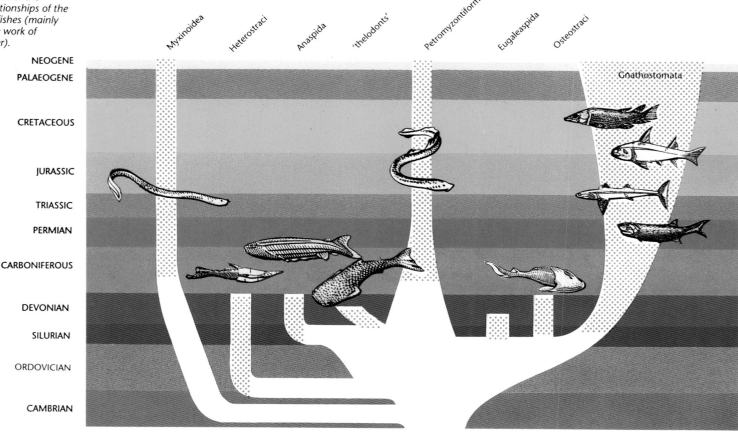

Myxinoidea   Heterostraci   Anaspida   'thelodonts'   Petromyzontiformes   Eugaleaspida   Osteostraci

Gnathostomata

NEOGENE
PALAEOGENE

CRETACEOUS

JURASSIC

TRIASSIC

PERMIAN

CARBONIFEROUS

DEVONIAN

SILURIAN

ORDOVICIAN

CAMBRIAN

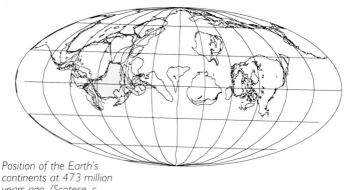

Position of the Earth's continents at 473 million years ago. (Scotese, c. 1990; "terra mobilis" program)

long-term survival and success, even if it eventually ended in near or complete extinction.

There were limitations, however, in just how far jawless forms could evolve. Agnaths never developed both paired pectoral and pelvic fins, which limited manouvrability and perhaps speed. Although a variety of body shapes developed — from fusiform to flattened, and even cylindrical (in the case of the parasitic forms such as lampreys) — feeding was essentially restricted to filtering out smaller particles from the water column, be they detritus or smaller living organisms. Definite limits were built into the structural design of these first backboned organisms, confining them to a small part of the possible niches available in the benign environment that developed in the Late Silurian and Early Devonian.

When *Arandaspis* and *Porophoraspis* plied the warm shallow seas of the Ordovician, Australia lay near the Equator and was part of Gondwana which also included Antarctica, South America and a number of other continental fragments. What is now Australia lay in the north-west sector of Gondwana and was split into two large pieces, a great island (part of present-day Central Australia) and another block just to the south. In between these two parts the continent had been flooded by the shallow, tropical Larapintine Sea.

To the South (now the east coast of Australia) lay the deeper waters of the world ocean which was the domicile of the remarkably successful graptolites, planktonic invertebrates which left their remains around the world, and in Australia notably in the black shales of Victoria. Graptolite fossils are of particular significance in dating rocks from all parts of the world because of the cosmopolitan distribution of many genera. They define contemporaneous time lines connecting such far flung areas as Nevada and Idaho in the Americas with south-eastern Australia. Graptolites for this time, then, are ideal "index fossils", because they were geographically widespread, individual species were short lived, and each is quite distinctive and easily identified, even in the field. These sorts of fossils are just what palaeontologists need to place both rocks and their included fossils into some sort of temporal sequence and thus view the evolutionary changes that have occurred through time.

Conditions were warm in the Larapintine Sea during the Ordovician, as is clearly indicated by the development of coral reefs and the high diversity of many invertebrate groups, especially the molluscs. Carbonate deposits in places like the Georgina Basin and the remains of halite (salt) crystals in sediments of the Daly River Basin also reflect warm conditions. The great Australian island in the north of the Larapintine Sea had low relief, but the country to the south was more rugged and tectonically active.

For much of the Ordovician and Silurian warm conditions persisted worldwide. While glacial conditions towards the end of the Ordovician brought about global lowering of sea levels and decreased faunal diversity, the effect was less severe in Australia than in many other places, most likely because of its equatorial position. However Australia's provincial nature seems to have lessened throughout the Ordovician, and in the later parts of that period its faunas showed great similarity to those of other parts of the world, particularly the continents strung out like beads along the equatorial necklace. As glaciation developed near the poles in the Late Ordovician, a cooler water fauna did develop that led to the definition of a distinctive faunal region called the Malfvinokaffric Realm. The cooler nutrient-rich waters that nurtured this biotic assemblage most likely caused the deposition of the phosphorites in Central Australia in Middle to Late Ordovician times as they mingled with the warmer, more tropically derived ocean waters from the east.

CHAPTER 2

AN AGE OF FISHES

THE EARLY TO LATE DEVONIAN PERIOD, FROM 410 TO 354 MILLION YEARS AGO

Australia's record of jawed fishes begins in the Late Silurian and Early Devonian. The Devonian, quite rightly named the "Age Of Fishes", is a time of expansive evolution of many fish groups, contemporaneous with the continued survival and success of some jawless forms. Australia lay much closer to the Equator than it does today, and a major reef system, the Palaeozoic great barrier reef of Gogo, developed along its present north-western margin. Minor reef-like accumulations were also present in the south-eastern part of the continent, in places like Buchan, Victoria. In the Early Devonian, Australian fish faunas were closely allied with those in other parts of Gondwana, especially Antarctica. They were quite distinct from the North American and European faunas. Some faunal similarities existed between Australia and South China, and with another continental block known as Armorica, which lay between the European and the North American landmasses and Australia. In the Late Devonian, plate movements brought southern and northern continental blocks in close proximity, and a major phase of faunal interchange took place. Many Gondwanans successfully displaced their northern counterparts, especially amongst the placoderms. Their success was to be short-lived, however, for soon the entire group was extinguished forever. The end of the Devonian, too, was the end of Eden for fishes in Australia. This antipodean continent was already on its treacherous path from the Equator to the South Pole, and the immense journey was to forever change the nature of its biota.

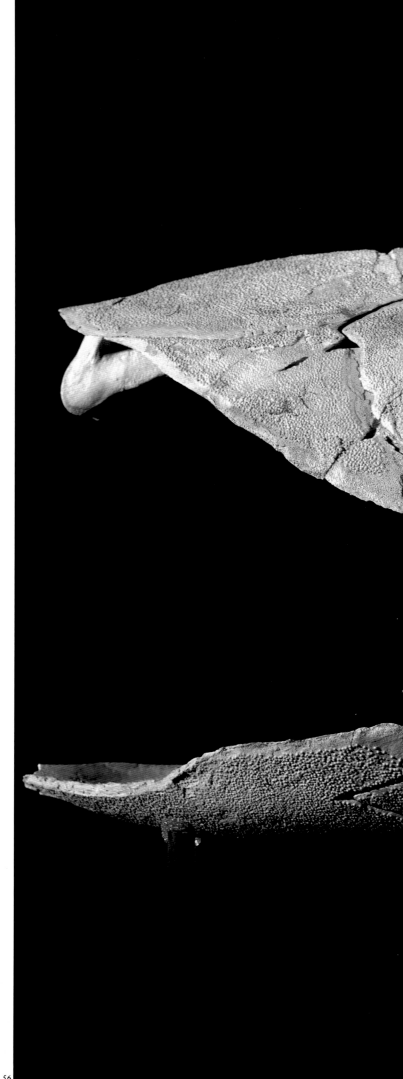

56   Side view of the skull and trunk shield of the arthrodire Eastmanosteus from the Late Devonian Gogo Station, Western Australia. x 1.3 (F. Coffa, courtesy of the Australian Museum and A. Ritchie)

56

JAWS AND PAIRED FINS

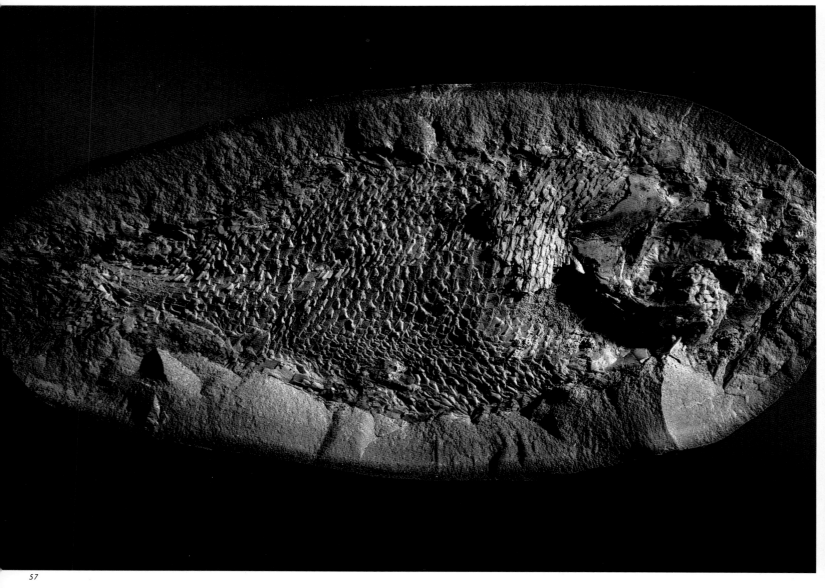

57

Australia's geographic position and the climate of the Earth in the Devonian were ideal for fostering diversity. Australia lay near the Equator, just south of it for the most part. Its position is confirmed not only by the palaeomagnetics of the rock record, but also by the abundance of salt deposits and a wide distribution of reefal limestones and oolites (concentrically banded spheres of calcium carbonate). Globally, maximum flooding of the continents occurred in Middle Devonian times, and it was then that the shallow, inland seas of tropical areas including Australia had their greatest evaporite build-up.

Invertebrate faunas at this time in Australia are amazingly diverse. Many lived on the complex reef systems, the most notable of which — a Devonian analogue of today's Great Barrier Reef — is located across Gogo Station in the Kimberleys of Western Australia. Its inhabitants included now-extinct corals (the tetracorals and the tabulates) and trilobites. But, in addition to these diverse communities of tropical marine invertebrates, the reef sheltered and entombed

57 A palaeoniscoid from the Late Devonian Gogo Formation at Paddy's Valley, Gogo Station, Western Australia. (J. Frazier, courtesy of the Western Australian Museum and K. McNamara)

58 An acanthodian fish from the late Middle Devonian Bunga Beds, near Bermagui, New South Wales. This fish group is characterized by spines supporting the major fins. x 3.3 (F. Coffa, courtesy of the Bureau Of Mineral Resources and G. Young)

some of the best preserved vertebrate remains of Devonian age known from any place on Earth.

It was into this world that the first fishes with jaws and paired fins diversified and competed with the ruling agnaths. Where these jawed forms came from and how they are interrelated is a topic of great debate, far from resolution. Several different groups seemingly appear nearly simultaneously, surely an artifact of the poor record at this time, in the Late Silurian and Early Devonian. Amongst these fishes are representatives of two modern groups, the sharks and the bony fishes, as well as two archaic groups, the spiny acanthodians and the heavily armoured placoderms. The placoderms stand quite apart from all other fish groups in possessing a number of unique features, whereas the acanthodians share enough characters with sharks and bony fishes to suggest they may have something to do with the ancestry of those groups.

All of these early fishes, the most ancient of gnathostomes, or "jawed vertebrates", had advantages over more primitive backboned forms. Their jaws provided new opportunities for feeding, including complex herbivory and, perhaps more importantly, carnivory where for the most part prey was not restricted to forms smaller than the pursuer! Most early gnathostomes also possessed paired fins with internal fin supports, not just folds of skin without any internal superstructure. Such fins allowed fine tuning of movement, which provided greater manoeuvrability. These improvements, through natural selection and survival of those most fit to leave viable offspring, led to an explosion of vertebrate life forms throughout the Devonian Period.

Although there is a rich Devonian record of fossil vertebrates in Australia, 99 per cent of which is fishes, the Silurian vertebrate record is very sparse. The record at the moment is mainly of acanthodians, the oldest gnathostomes (or jaw-bearers) known from the Australian continent. These "spiny sharks" are represented only by scales from Late Silurian rocks on the eastern part of the continent, and the known fossils indicate they appeared in Australia at about the same time as on a world scale. Acanthodian scales are distinctive, especially when compared to those of the jawless thelodonts, in not having a pulp cavity and being composed of a dentine-like material that is added in concentric layers. The superficial part of the scale is composed of dentine with the base of bone, which in primitive forms is cellular but in more advanced forms acellular.

The very oldest acanthodians in Australia come from two different places, the Mirrabooka and Silverdale formations near Canberra and the Graveyard Creek Formation of the Broken River area in northern Queensland. The scales are not from just one kind of acanthodian but from several different sorts, such as climatiids and ischnacanthids, and some scales are very similar to those of genera known from elsewhere in the world, for example, *Nostolepis* and *Gomphonchus*. Some small bony platelets recovered from Silurian sediments may even belong to placoderms, though much more effort needs to be invested in processing sediments of this age for microfossils before a realistic picture can be painted. And, the microfossil record seems to hold the best potential for success.

Acanthodians are an enigmatic group of generally small (most only reaching a few centimetres with the largest 2 metres in length), fusiform fishes with enormous eyes. They ranged from the Late Silurian to the Early Permian and look rather like modern fishes. Their bodies are covered by small, thin scales that do not overlap. Their heads are composed of a mosaic of dermal bones that vaguely resemble the patterns seen in modern bony fishes; however, when the pattern is compared in detail, homologous structures are difficult to discern. Acanthodians further differ in having a heterocercal tail, that is, the upper lobe is longer than the bottom lobe and the skeleton actually continues into the upper lobe. They have a number of solid fin spines, not only on the front of the paired fins (the pectorals in front and the pelvics behind, with sometimes up to six pairs in between) but also along the unpaired ones (one or two dorsals and an anal near the back of the fish). Vertebrae in these fishes are not formed of bone, but of cartilage, and the arches on top and below protect the nerve chord and blood vessels. No bony ribs belonging to acanthodians have ever been found, and it is uncertain if they had ribs at all.

Acanthodians possess some interesting similarities to more highly advanced fishes, notably the sharks and the true bony fishes (osteichthyans). One similarity is the arrangement of the endoskeletal bones of the jaw (those preformed in cartilage before ossification and derived, not from the superficial embryonic tissues, or ectoderm, but from deeper tissues, or mesoderm). But although the jaw parts have a similar origin, the teeth of acanthodians are another matter. Histologically the teeth are very different from those of modern fishes, for they lack enamel and apparently were not replaced but had to last for the lifetime of the fish. These teeth came in two varieties: they were either laterally compressed with many cusps on a single tooth and fused to the jaw cartilages, or they were whorl-like affairs held close to the jaw cartilages by connective tissue.

Acanthodians seem to have been active mid-water to surface feeding fish that never really developed bottom-feeding habits. Their initial success was ensured because they were not in direct competition with the bottom-living placoderms and the many kinds of jawless fishes. Throughout their history evolutionary change favoured lightening of the skeleton and making it more flexible, all associated with improvements in swimming style. The numerous spines most likely served as defensive tools, and became greatly elongated and more firmly anchored through time. It has been suggested, also, that in some acanthodians the spines may have functioned as "cutwaters".

AN EXPLOSION OF VARIETY

THE EARLY DEVONIAN TAEMAS-WEE JASPER AND BUCHAN FAUNAS

A few places in Australia have yielded a variety of fossil fishes of Early Devonian age. Amongst the most important of these are the Taemas–Wee Jasper area of New South Wales (just north of Canberra), the north-western part of New South Wales, and the Buchan area of south-eastern Victoria. These earliest of Devonian faunas are dominated by arthrodire placoderms. However, the variety of different types of fishes present (but seldom found) in these assemblages, for example, the dipnoan *Dipnorhynchus* and the palaeoniscoid *Ligulalepis*, is indicative of the explosive evolution that was occurring within the fishes at this time. In addition to the macrofossil record of skeletons and isolated head and thoracic armour, the microfossil record from these locales indicates the presence of crossopterygians, acanthodians, sharks (the elasmobranchs), a variety of placoderms and the jawless thelodonts.

PLACODERMS — *THE GREAT EXPERIMENTERS*

The earliest Australian fossil assemblages that contain more than just one or two species of vertebrates come from the Early Devonian of eastern Australia, from rocks of marine origin. From the Murrumbidgee Group of the Taemas–Wee Jasper region of New South Wales and from marine limestones of the Buchan area in Victoria come a variety of the now-extinct fish group, the placoderms, which were the first to develop true jaws. Placoderms appeared in the Early Devonian and were explosively successful throughout this period, becoming the dominant fish group both in Australia and on a world scale. But only one or two genera survived until the beginning of the Carboniferous, then the group disappeared forever.

Placoderms (whose name means "plate skin", in reference to their heavy armour formed from dermal bone) are unique fish which have left no living descendants, as far as we know, although there has been some suggestion that the ghost-fish (or chimaeras) might be related. From what group they were derived is also something of a mystery. Some of the oldest known placoderms have some resemblance to a group of jawless fish called heterostracans, in that the entire head and trunk area of both types is covered with a bony shield, and only a short, scaly tail extends behind this region. Additionally, this bony armour was built in the same way in both placoderms and heterostracans, that is, there were three distinct layers and holes within the bone for bone-producing cells. The arrangement of the armour in placoderms is, however, unique in that the head and trunk shields act as separate entities, being hinged together and moveable with respect to one another.

Just how placoderms are in turn related to other fish groups — the sharks, acanthodians and bony fishes — is uncertain. Indeed, there are about as many theories as research workers. A major stumbling block in this debate is the poor quality of the pre-Devonian record, which does not help to explain the specializations that are well developed in the placoderms and other fish groups when they appear in the Early Devonian. These specializations mask the true relationships of groups and, unless more unspecialized forms are found, make it difficult to reconstruct the family tree of early vertebrates. Perhaps the sudden appearance of the placoderms in the Early Devonian is because that was when they acquired the heavy armouring so typical of them. Perhaps they came from an unarmoured ancestral group that first developed a mosaic of small scales, more like the petalichthyids and some acanthothoracians, and

then, finally, larger plates. Perhaps, but only an older record of the group will really resolve this dilemma.

In the oldest of placoderms the braincase is heavily ossified and the upper jaw is firmly attached to both it and the bones of the cheek as a unit. Although a few placoderms have small tooth-like structures, for the most part the cutting edges of the jaws are formed of bony plates, one or two large ones in the upper jaw and a single pair in the lower. The centra of the vertebrae are generally not formed in bone but there are bony arches both above and below the cartilagenous centra, the neural and haemal arches respectively. And, to aid in the raising and lowering of the head in feeding, the front elements of this vertebral complex is generally fused into a structure called the synarcual, which has two points of articulation with the ball-like occipital condyles of the skull. Such a set up restricts motion to the vertical plane and thus ensures efficiency of forward movement.

Placoderms possessed paired fins, pectoral and pelvic. The tail fin was heterocercal, that is, the top lobe was larger than the bottom one, a design which tends to push the nose of the fish down when it swims. In concert with that design, most placoderms were flattened from top to bottom, suggesting that they were bottom dwellers that fed on the floors of oceans, lakes and streams. A few placoderms developed a more active mid-water life; one of the evolutionary trends in this group through the Devonian was to move more into that lifestyle by reducing the thickness of the bony armour and actually losing the ossification of the internal skeleton, which, clearly, reduced the weight and freed the trunk and tail for more efficient swimming.

Five or six main groups of placoderms are recognized, depending upon which palaeontologist is consulted. Of these groups, one stands out as the most successful, making up about 60 per cent of the known record — the arthrodires, also called the "euarthrodires". The arthrodires had a full complement of dermal bones in the head and trunk armour, thus there was apparently little loss or fusion of plates from the primitive forms that gave rise to earliest placoderms. As well, the arthrodires had a well-developed neck joint between the head and trunk shields. Most arthrodires were bottom or near bottom feeders, and, instead of lowering the bottom jaw, they raised the head as the mouth opened and thrust forward towards the prey. A very few moved into open-water feeding, but some that did, such as *Eastmanosteus* known from the Late Devonian of Australia, became formidable predators.

THE FLEXIBLE ARTHRODIRES

Arthrodires are one of the groups represented in the Taemas–Wee Jasper and Buchan regions. Two such forms, the broad-headed *Buchanosteus* and *Taemasosteus*, show some reduction in the extent of the dermal armour, a trend that occurs frequently within this group. This reduction seems to be an adaptation for lightening and opening up the armour, leading to greater efficiency and control in swimming as well as perhaps promoting greater speed.

The degree to which these dermal armour changes occurred has been utilized by scientists to divide up the arthrodires into a number of grades of organization, from the most primitive actinolepid arthrodires to the pachyosteomorph arthrodires, which were the first group to have a full complement of armour and no joint between the head and thoracic shield, and also the last group to have a well developed cranio-thoracic joint and a great abbreviation of the thoracic armour. The names "dolichothoracids" and "brachythoracids" have been used to describe these trends: the more primitive arthrodires, which had a long, solid trunk shield with a small opening for the pectoral fin, being the dolichothoracids; and the more advanced forms, which had a reduced thoracic shield with the opening for the pectoral fin enlarged, being the brachythoracids.

Forms similar to the Australian genera have been found in the Northern Hemisphere, for example, Morocco and Germany, and thus migration of marine forms must have been possible between Australia and these areas in the Early Devonian.

59-60 Ventral and dorsal views of the incomplete head shield of a brachythoracid arthrodire from the Early Devonian Murrumbidgee Group at Wee Jasper, New South Wales. x 0.6 (F. Coffa, courtesy of the Bureau Of Mineral Resources and G. Young)

Other arthrodires, such as *Williamsaspis, Wuttagoonaspis* and *Burrinjucosteus* from the Early Devonian, are unique to Australia and their relationships are not clear. In fact, these forms have been put within their own families, the Williamsaspidae, Wuttagoonaspidae and Burrinjucosteidae, in recognition of their uniqueness. Based on the degree of reduction of bone surface area in the trunk armour, amongst other characters, *Wuttagoonaspis* is more primitive than either *Williamsaspis* or *Burrinjucosteus*. Indeed, *Wuttagoonaspis* is considered by most researchers to have the actinolepid level of organization within arthrodires, while *Williamsaspis* and *Burrinjucosteus* have the phlyctaenaspid level. Both these latter forms have somewhat reduced armour and a well-developed ball and socket joint between the head and trunk armour, adaptations for more efficient and better co-ordinated movement and feeding.

Arthrodires came in a range of sizes from no more than 100 millimetres to greater than 6 metres, with *Buchanosteus* being one of the larger forms. Australia's largest, however, is *Westralichthyes* which reached around 2.5 metres. Many of them became formidable predators of the tropical Devonian reefs.

PACHYOSTEOMORPH LEVEL

Cranio-thoracic joint
Nuchal plate broad posteriorly
Pectoral incision
Median dorsal plate long
    or short, with keel

COCCOSTEOMORPH LEVEL

Cranio-thoracic joint
Nuchal plate broad posteriorly
Pectoral fenestra large
Median dorsal plate long,
    with keel or ridge

PHLYCTAENASPID LEVEL

Cranio-thoracic joint
Nuchal plate not broad
    posteriorly
Pectoral fenestra small
Post-pectoral wall long
Median dorsal plate long,
    without keel

ACTINOLEPID LEVEL

No cranio-thoracic joint
Nuchal plate not broad
    posteriorly
Pectoral fenestra small
Post-pectoral wall long
Median dorsal plate short
    and broad, without keel

Levels of organization of the skull and trunk armour in arthrodire placoderms. Such evolutionary changes may have occurred more than once within this group, but the main trend within any one lineage through time was to decrease the amount of armour in more advanced forms, the most advanced of which was the pachyosteomorph level. (After Moy-Thomas & Miles, 1971)

THE ARMOURED ANTIARCHS

Antiarchs, like the arthrodires, have well armoured head and trunk shields, and they also have armouring on the front (pectoral) paired fins. They look a bit like the armoured catfish of today, but, of course, are not closely related: the armoured catfish, available in many aquarium stores, are highly advanced ray-finned fish and their resemblance to the antiarchs has occurred only because of convergent evolution — a similarity of characteristics developed due to the similarity of selection pressures on a different genetic stock.

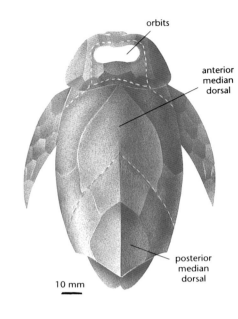

The head and trunk armour of the antiarch Remigolepis, viewed from above. (Modified from Moy-Thomas & Miles, 1971)

Bothriolepis, an antiarch placoderm well known in Devonian deposits around the world, both in fresh and marine environments. It is an important fossil in correlating rocks from many parts of the globe. (Modified from Moy-Thomas & Miles, 1971)

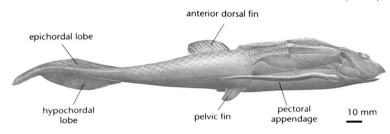

In antiarchs the head shield is always smaller than the trunk shield. Most of these fishes are relatively small, most not larger than 30 centimetres. The pectoral fin is not only heavily armoured, but also jointed in the middle. This joint allowed refined movements in both the horizontal and vertical planes, as well as rotational movements. The head shield bears articular balls and the trunk shield has sockets for articulation, just the reverse arrangement as that in arthrodires. The openings in the head shield for the eyes point upwards, suggesting that these fishes were essentially bottom dwellers.

The oldest and most primitive antiarchs, the yunnanolepidoids, come from the Late Silurian and Early Devonian of China, and are unique to that area. Slightly younger antiarchs, the sinolepidoids, occur in the Devonian of China and Australia only. Both forms have the well-developed, jointed pectoral appendages.

Two other antiarch groups are quite widespread globally during the Devonian, the asterolepidoids and the bothriolepidoids. Both

61 The long-snouted arthrodire placoderm Fallacosteus turnerae from the Late Devonian Gogo Formation on Gogo Station, Western Australia. x 1.5 (J. Long)

62 The arthrodire placoderm Latocamurus coulthardi from the Late Devonian Gogo Formation on Gogo Station, Western Australia. x 1.75 (J. Long)

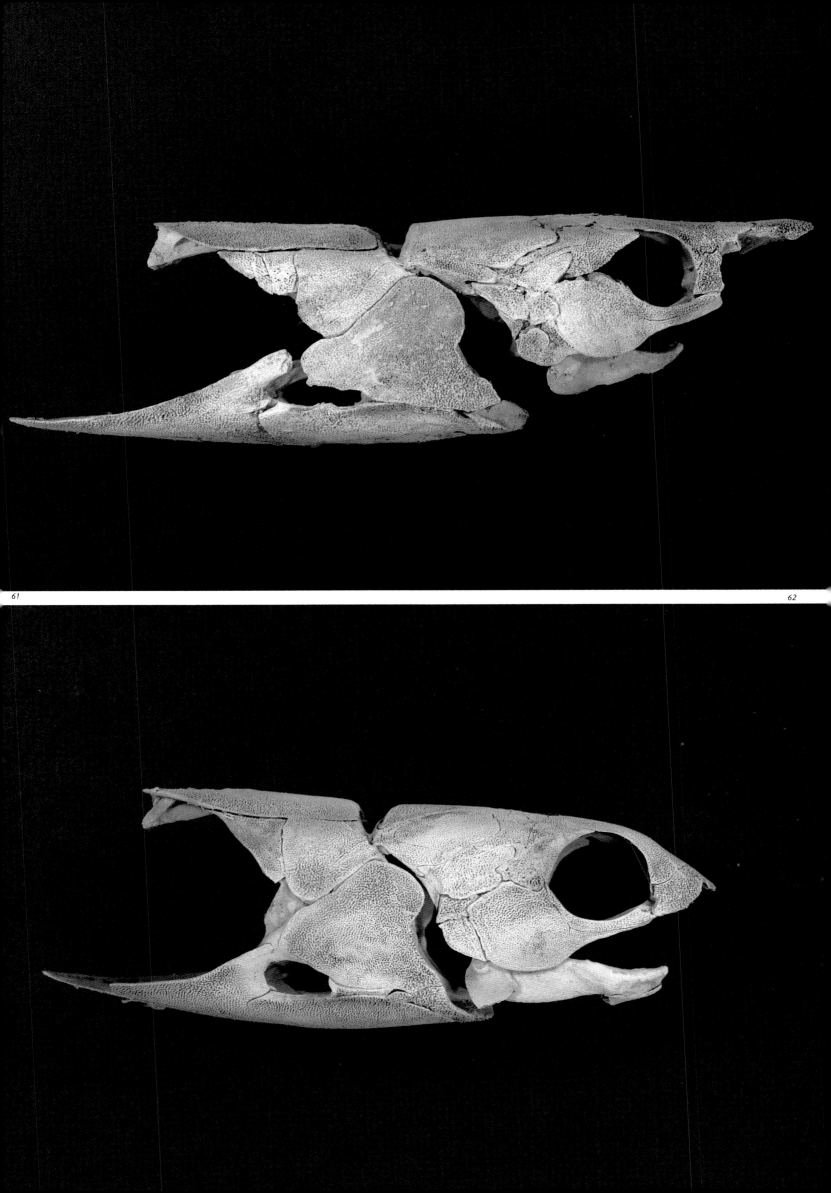

occur in Australia. Asterolepidoids possess long trunk shields, short pectoral appendages and small head shields with large openings for the eyes. Specimens from the Early Devonian Cravens Peak Beds of the Georgina Basin in south-western Queensland are amongst the oldest in the world. In bothriolepidoids the head shields are large, approximating the size of the trunk shields, and the pectoral appendages are long.

Even though the antiarchs have a presence in Australia during the Early Devonian, the tapestry of their evolution is much richer in the Middle and Late Devonian on this continent.

THE PETALICHTHYIDS, HAUNTERS OF THE OCEAN BOTTOM

Another placoderm group found in the Early Devonian of Australia, the petalichthyids, may be closely related to the arthrodires. Petalichthyids seldom exceeded half a metre in length. Quite unlike arthrodires, they were flattened, bottom-feeding fishes with a trunk armour commonly decorated with elongate spines on either side. The orbits (openings in the bony head shield for the eyes) lay on the top of the head shield. The petalichthyids are rather rare in Devonian Australia, but amongst those that do occur is *Lunaspis*, a form also known from Europe and Russia. It was a very distinctive fish with impressive spines on either side of the head and an elongate body covered with tiny, overlapping scales. In addition to this rather cosmopolitan form, several other genera are endemic to Australia including *Notopetalichthys* and *Shearsbyaspis*, both known from Early Devonian rocks in the Taemas area. Another form, *Wijdeaspis*, occurs in Early to Middle Devonian sequences in Australia and is also found in Siberia and Spitzbergen.

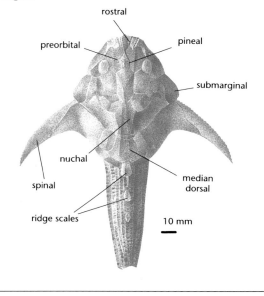

The petalichthyid placoderm Lunaspis, head and trunk shield in top view. (Modified from Moy-Thomas & Miles, 1971)

rostral
preorbital
pineal
submarginal
nuchal
spinal
median dorsal
ridge scales
10 mm

63 *Skull, trunk and left pectoral appendage armour of Bothriolepis karawaka from the Middle or Late Devonian Aztec Siltstone on Mt Ritchie in the southern Boomerang Range, Antarctica. x 1.2 (F. Coffa, courtesy of the Australian Museum and A. Ritchie)*

OTHER PLACODERM TYPES

The acanthothoracids are yet another kind of placoderm known in Early Devonian marine sediments of Australia. The group is poorly known elsewhere in the world and contains fishes with a variety of body shapes, suggesting that it may be polyphyletic, that is, derived from not one but several sources. Some acanthothoracids are rather ray-like, having expanded pectoral fins and a very flattened appearance. Their head shields are long, the trunk armour is very shortened, and their eyes and nares open upward, a characteristic of bottom-dwelling forms with the same habits as the petalichthyids.

Often the head armour in this group is not composed entirely of large interlocking plates, but instead has a few large plates that are separated by a mosaic of smaller ones. Australian forms such as *Murrindalaspis* had tall crests on the trunk armour, suggesting that they were not quite so flattened.

Some of the best preserved acanthothoracid cranial material in the world comes from Australia, in particular material of *Brindabellaspis stensioi* which shows details of the cranial nerves and vessels as well as impressions of the brain. A specimen of *Murrindalaspis* found at Taemas has even preserved a complete sclerotic capsule, the protective bones that surrounded the soft tissue of the eyeball in this ancient fish, and impressed into this bony encasement are the pathways of the arteries and veins and the surfaces for attachment of the muscles that controlled eye movement.

The Australian acanthothoracids are placed, for the most part, in either a group with unknown close affinities or in a group known nowhere else, such as *Weejasperaspis* and *Murrindalaspis* in the family Weejasperaspidae, indicating a certain amount of provincialism. These Australian forms have their closest affinities with acanthothoracids from the microcontinent Armorica located north of Gondwana.

The ptyctodontids are still another placoderm group that has some record in the New South Wales Early Devonian, a record not yet extensively published. Ptyctodontids have a superficial similarity to more modern fishes due to their great reduction in armour, which is limited to the back of the head and a short trunk shield. Their bodies were long and slim and lacked scales, and the tail was generally only a narrow whip-like affair. In fact, they bear an uncanny resemblance to the ghost-fish (chimaeroids or holocephalians).

Ptyctodonts were reasonably small fishes, generally less than 20 centimetres, and mainly, but not exclusively, marine in habit. Their teeth were massive plates which had both crushing and shearing functions, and they were reinforced with dentine. A striking feature of these fishes is their claspers, structures associated with the pelvic fin. In sharks and chimaeroids, which today have such structures, they are used in the process of internal fertilization. Although the ptyctodont claspers are not constructed in exactly the same way as those of sharks, they most likely served the same function — another example of convergent evolution in two genetically distinct groups.

Relationships of the ptyctodonts to other groups are uncertain, although some researchers have allied them with groups as diverse as placoderms and sharks. Whatever their affiliations, they were an element in the Early Devonian fish faunas of Australia, and were to become much more diverse later in the Devonian.

JAWLESS SURVIVORS—THE RESILIENT AGNATHS

The jawless fishes, which had been dominant in the early Palaeozoic, no longer ruled the waters in the Devonian. Nonetheless, they did survive and are important in dating rocks of this age and younger. They have left behind millions of microscopic scales which can be recovered in great abundance from well cores. Such scales are generally quite distinctive, albeit with some variation in form depending upon what part of the fish they came from — though once the variation is understood, individual scales can be quite diagnostic. Furthermore, many jawless fish species have very short time spans in the geologic record. So, with these attributes, thelodonts make excellent index fossils, of great value in determining the ages of rock sequences.

A variety of microscopic thelodont scales have been reported from the Early Devonian of Western Australia. *Turinia australiensis* is one bearer of such scales. In addition to the microfossil record of agnaths in the Early Devonian, a distinctive body fossil of a tiny fish with an elongate rostrum has been recovered from the latest Early Devonian or earliest Middle Devonian of the Georgina Basin of Queensland, perhaps related to the galeaspids.

65 Head shields of Bothriolepis, an antiarch, and Groenlandaspis (upper right), an arthrodire, from The Late Devonian Aztec Siltstone, Skelton Glacier, Antarctica. x 1.0. (F. Coffa, courtesy of the Australian Museum and A. Ritchie).

66 Impressions of bony plates of two types of armoured fishes, Placolepis, a phyllolepid (with fingerprint-like pattern) and Bothriolepis, an antiarch, from Late Devonian near Braidwood, New South Wales. x 1.0. (F. Coffa, courtesy of the Australian Museum and A. Ritchie).

64 Sclerotic capsule protecting the eye of Murrindalaspis wallacei, an acanthothoracid placoderm, from the Burrinjuck Dam area, south-west of Taemas Bridge in the Taemas–Wee Jasper district of New South Wales. The eye capsule was associated with parts of the head armour and pelvic supports as well as with several hundred body scales. x 2.2 (F. Coffa, courtesy of the Bureau Of Mineral Resources and G. Young)

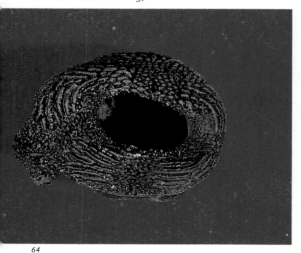

64

65

Sclerotic capsule protecting the eye of Murrindalaspis wallacei. This protective armour for the eye was evidently attached directly to the head armour and to the braincase by a cartilagenous stalk (eys) and clearly records the places of entry and exit for blood vessels (a.opt., optic artery; ov 1, ov 2, optic veins; n.f, nutritive foramen; for, foramina for arterial plexus, possibly derived from central retinal artery) and nerves (II, optic nerve; f.cil, foramina for ciliary branches of the profundus nerve) as well as the attachment areas for the six muscles controlling eye movement (m1-6; cr, crest dividing muscle insertion areas). This specimen provides the most complete anatomical detail yet known of the eye of vertebrates at an early time in their evolution, the Early Devonian (A, dorsal; B, lateral; C, mesial; D, anterior; and E, ventral views; n, notch). (Diagram modified after Long & Young, 1988)

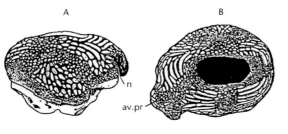

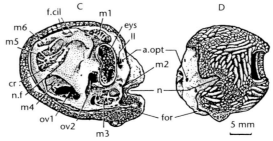

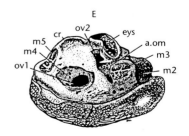

OSTEICHTHYES — *A HINT OF THINGS TO COME*

While placoderms were a diverse and important part of the marine and even freshwater communities of the Early Devonian in Australia, just as they were elsewhere in the world, they were not the only "fish in the sea". Several other groups, which would eventually assume supremacy, were also developing and diversifying.

Osteichthyes, or "bony fishes", are the dominant group of fishes today, with more than 20,000 living species. They have a well-developed internal and external bony skeleton, and a swim bladder that is important in regulating buoyancy. In some groups this structure has been modified into a set of lungs. The bony fishes are subdivided into three smaller groups: the lungfish (Dipnoi), the lobe-finned fishes (Crossopterygii) that most likely gave rise to the tetrapods, and the ray-finned fishes (Actinopterygii). It is the ray-fins, which took over from the placoderms in the Early Carboniferous, that have reigned supreme from that time until the present, dominating both fresh and salt waters.

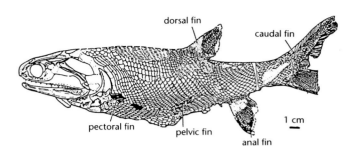

Mansfieldiscus sweeti, a palaeoniscoid from the Early Carboniferous Mansfield Group, Mansfield Basin, central Victoria. (After Long, 1988)

RAY-FINS, ANCESTORS OF MODERN FISHES

Some of the oldest actinopterygians in the world come from the Early Devonian of Australia, in the form of scales from a group called the palaeonisciforms. This primitive bony fish group, however, does not become diverse until the beginning of the Carboniferous, remaining an important group until the Triassic.

Actinopterygians are distinctive in having no internal nostrils and thus they differ from the crossopterygians, the group that eventually gave rise to tetrapods. The arrangement of the bones on the skull and the paths taken by the lateral line system are distinctive for actinopterygians and quite different from those in crossopterygians and lungfish. The scales are different, too, in that they lie beneath the skin and are periodically added to in concentric rings, on both the outside and the inside. The scales can have an outer layer that is black and glossy, called ganoine, or lack it entirely, but they never possess cosmine tissue, which has a distinctive microstructure and is found in other osteichthyans. Most importantly, the inside skeleton of the median fins is never a basal plate but always separate, slender radial bones, and the paired and median fins themselves are supported internally by a series of parallel bony fin "rays" whose motion is almost entirely controlled by muscles within the body rather than in the fin. The name of the actinopterygians, clearly, is derived from this distinctive fin structure.

Primitive actinopterygians, the chondrosteans, become important as the placoderms wane and remain so into the Triassic. The palaeonisciforms, which first appear in the Devonian, are members of this group. From the palaeonisciforms, the holostean neopterygians arose in the late Palaeozoic and diversified greatly in the Mesozoic. A more advanced neopterygian group, the teleosts, began their remarkable expansion in the late Mesozoic, the final saga in a success story which is still in progress today.

Palaeonisciform fishes are known from the Early Devonian of Australia. *Ligulalepis toombsi* scales are common in rocks of this age, for example, the Murrumbidgee Group of New South Wales. Slightly older palaeonisciform scales are known from the Late Silurian of Europe. The scales of all palaeonisciforms were thick and often rhombic in shape, with a peg and socket articulation connecting them.

Palaeonisciforms possessed very long jaws that were somewhat restricted in the types of motion they could undergo. The upper jaw, formed mainly by the maxillary, was firmly fixed to other bones of the cheek, such as the preopercular and the infraorbital. The internal bone structure of the upper jaw, the palatoquadrate, was likewise attached, albeit moveably, to the braincase in at least three places, thus making the two structures intimately involved in any motion. The elongate lower jaw, like the upper jaw, was filled with conical, sharply pointed teeth.

This arrangement of bones, muscles and ligaments allowed the mouth to be opened and the oral cavity to be expanded: first there was a contraction of the body muscles on the underside of the fish, which pulled the shoulder area backwards; this was followed by contraction of the muscles associated with the gill basket, which

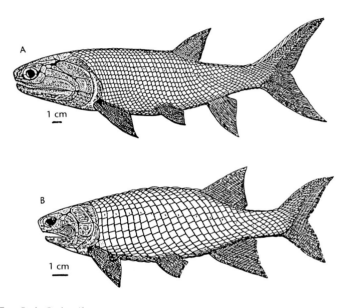

Two Early Carboniferous palaeoniscoids from the Mansfield Group, Mansfield Basin, central Victoria: A, Mansfieldiscus sweeti; and B, Novogonatodus kasantsevae. (After Long, 1988)

Restoration of the palaeoniscoid Howqualepis rostridens from the Late Devonian Avon River Group, Howqua River, at the base of Mt Howitt in central Victoria. (After Long, 1988)

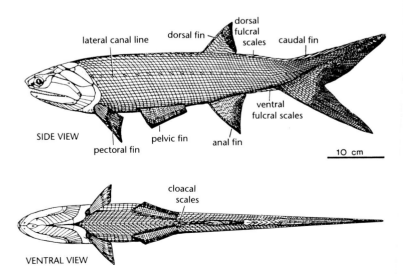

67

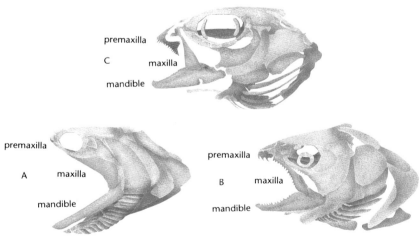

Levels of organization of the skull of actinopterygian fishes from the most primitive (A), such as occurs in palaeoniscoids, to the most advanced (C), as occurs in teleosts. In primitive forms (A) the maxilla elongates and is firmly fixed to the bones of the cheek, and the shoulder girdle is moved backwards. In a more advanced condition, for example, the holostean, (B), the premaxilla is larger and the maxilla is not fixed to the cheek but is swung forward as the mandible is depressed. In even more advanced forms (C) the maxilla is even more reduced and freed from the cheek, the premaxilla becomes the main tooth-bearing bone, and the entire skull becomes much more mobile and flexible, amongst other changes. (After Schaeffer & Rosen, 1961)

expanded the gill and mouth area; the lower jaw was then lowered by contraction of muscles in the throat area (the geniohyoids); and in some palaeonisciforms the braincase could be elevated to further widen the mouth. The movement of the jaws was much like the opening of a pair of scissors, up and down, with not too much side to side motion. Palaeonisciforms would thus have obtained their food mainly by biting and seizing it with their teeth, not by engulfing or filtering it and "pumping" it into the mouth. Larger fishes or invertebrates could have been ingested by biting, releasing the prey and swimming forward, then biting again until the whole animal was ingested. The efficiency of this bite was limited by the small size of the chamber that housed the jaw muscles (the adductor mandibulae) and the short lever arm on the lower jaw, which meant that both the force delivered on the prey and the torque developed around the jaw articulation were far less than in the more advanced ray-finned fishes; these latter developed a bony projection on the back of the jaw behind the jaw articulation (the coronoid process), thus increasing the lever arm and so the bite force. A big step in ray-fin evolution took place when the palaeonisciforms gave rise to the holosteans, sometime in the late Palaeozoic, resulting in a great variety of new fishes that dominated fresh and marine Mesozoic waters.

---

67   Fragmentary palaeoniscoid fish fossils from the Middle or Late Devonian Aztec Siltstone of Mt Crean in the Lashly Mountains, Antarctica. The Antarctic Devonian fish faunas have marked similarities to those of Australia. x 1.6 (F. Coffa, courtesy of the Bureau Of Mineral Resources and G. Young)

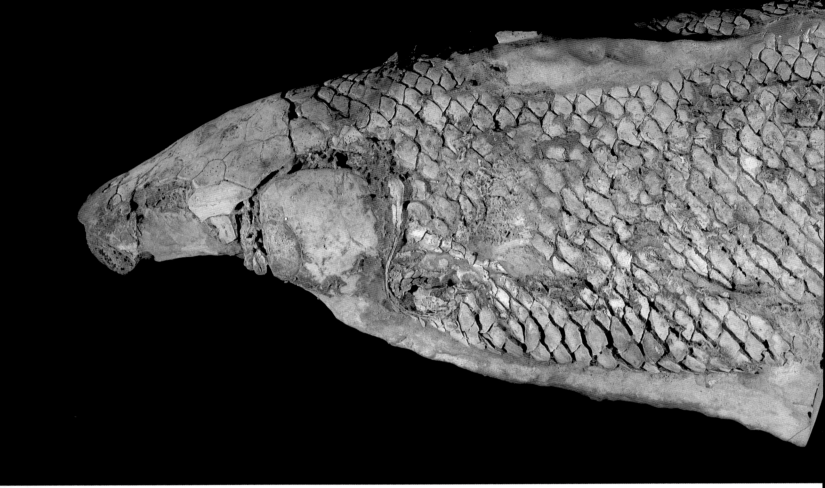

LUNGFISH – *A GROUP APART ?*

The Taemas–Wee Jasper fish fauna contains some of the oldest remains of lungfish and Australia has a rich record of this group. This evolution was most expansive in the Devonian and Carboniferous, and culminated on this continent in one of the three genera of still living forms.

Lungfish stand quite apart from other bony fishes in generally having a fusion between the upper jaw and the braincase, and in having no marginal teeth but instead fan-like tooth plates on the palate and in the lower jaw. This dentition and the structure of the skull seem tied to a powerful bite used to crush material, a duraphagous feeding style. The external bones of the skull form a mosaic composed of tiny pieces, arranged in a pattern not clearly comparable to any other fish group, not even the other "lobe-finned" fishes, the crossopterygians, which eventually gave rise to the tetrapods.

Most major events in lungfish evolution took place in Devonian times, with two patterns developing. One trend was the development

of a rasping feeding style, exemplified by the Late Devonian *Griphognathus*, well known from marine deposits of Western Australia. The other trend was the development of air-breathing, freshwater forms which unlike the marine "raspers" survived through the Devonian well into the Carboniferous and beyond. Curiously, lungfish tended to reduce the degree of ossification of the skeleton through time, so that modern lungfish are much less "bony" than their Devonian ancestors.

After the Carboniferous even the freshwater dipnoans declined dramatically, but two groups did manage to survive to the present day: the ceratodids, which once had a worldwide distribution, still inhabit a few Queensland rivers today; and the lepidosirenids continue to survive in Africa *(Protopterus)* and South America *(Lepidosiren)*. All three genera depend on atmospheric oxygen, and lepidosirenids can drown if it is not available. The African and the South American lungfish today can aestivate, that is, remain dormant during times of stress, for instance, during drought. *Protopterus* commonly burrows into the mud, sometimes for more than a year. This habit goes back at least to Permian times from which skeletons of the lungfish *Gnathorhiza* have been found preserved in burrows, still waiting after more than 200 million years for the rains to come!

Early Devonian Australian dipnoans are represented by well preserved, three-dimensional specimens. *Dipnorhynchus* is known from the Taemas–Wee Jasper and Buchan sites, and also from the somewhat older Lick Hole Limestone near Cooma in New South Wales. The palate of *Dipnorhynchus* is broad and dentine-covered, with small bumps or tuberosities. These bumps may represent an early step in the development of the tooth plates so characteristic of dipnoans. One of the oldest known lungfish, *Diabolichthyes* from the Early Devonian of China, similarly has a series of teeth on several of the palatal bones, including the premaxillary, which does not sit along the margin of the jaw in this fish. The teeth covering some of

---

68 *The Late Devonian lungfish Chirodipterus australis from the Gogo Formation on Gogo Station, Western Australia. The cut indicates where the pelvis was removed for study. x 1.0 (F. Coffa, courtesy of Australian National University and K. Campbell)*

69 *Skull of the Early Devonian lungfish Griphognathus whitei, in occlusal view, from the Gogo Formation on Gogo Station, Western Australia. At the anterior end of the skull the internal construction is visible, exposing a series of canals and tubes that in life housed nerves and blood vessels. x 0.5 (F. Coffa, courtesy of Australian National University and K. Campbell)*

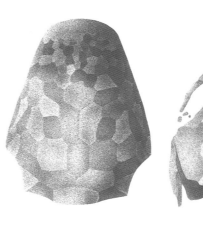
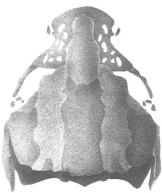

Skull in top view of the Early Devonian lungfish Dipnorhynchus (left) compared with that of the living Australian lungfish Neoceratodus (right).

Upper and lower jaw tooth plates of a late Palaeozoic lungfish Sagenodus.

The Late Devonian lungfish Dipterus (above) compared with the living Australian lungfish Neoceratodus.

the palatal bones are densely packed and form a radiating array, reminiscent of the pattern in geologically younger lungfish, but they are not yet fused to form a single tooth plate. Some bones, such as the dentary in the lower jaw, have retained teeth along the jaw margin as well. In these characters, and a number of others, this Early Devonian dipnoan is intermediate between more typical fish dentitions and the unique and specialized tooth plates of the lungfish. Despite this intermediacy, the nearest relatives of the lungfish are still not clearly identified, and there are about as many opinions on this issue as there are scientists working on these fossil fishes. Some palaeoichthyologists have even suggested that dipnoans gave rise to tetrapods, but this opinion is not widely supported.

Another Early Devonian lungfish in Australia is *Speonesydrion* from the Taemas–Wee Jasper area. It differs from *Dipnorhynchus* in having well defined tooth rows on the palate and lower jaws, but they are "teeth" that lack a pulp cavity. The structure of the teeth and the arrangement of muscles and bones in its skull suggest that *Speonesydrion* was capable of eating shelled invertebrates like brachiopods, molluscs and arthropods as well as less durable items such as annelid worms, all present as fossils in the same rocks that contain this dipnoan. Two feeding styles were probably used by *Speonesydrion*: "Larger items were cracked or crushed between the massive heels on the mandibular [lower jaw] tooth plates and the palate, and this involved direct pressure and some shearing as the mandible was pulled backwards and slightly rotated. Softer items were ground between the radial tooth rows by backward and rotational movement of the mandible, depending on the position of the food on one or both sides of the mouth. Presumably, like *Neoceratodus* [the living Australian lungfish], the animal had the capacity to extrude food from the mouth to reject hard skeletal materials and then to re-ingest the remaining soft tissues" (Campbell & Barwick, 1984).

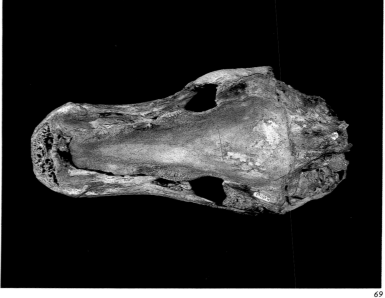

69

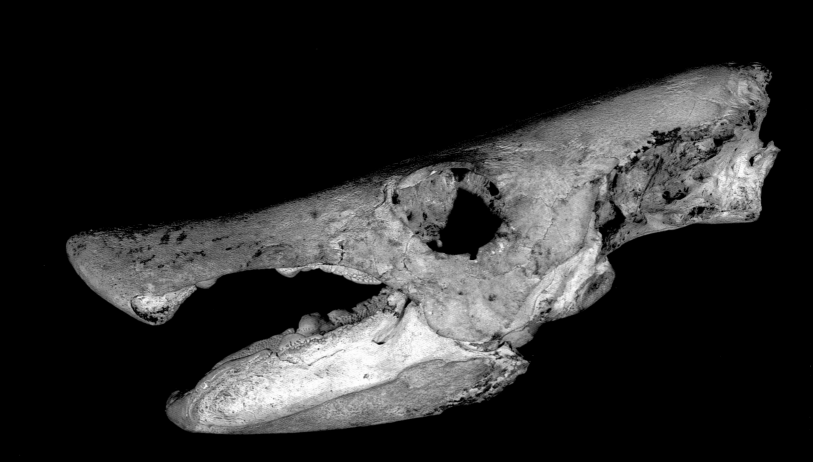

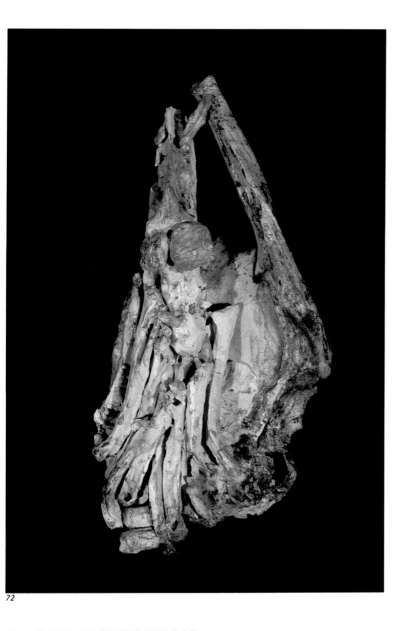

72

CHONDRICHTHYANS — *FISH WITHOUT BONES*

The chondrichthyans, or sharks and rays, are fishes to be respected and sometimes feared in today's seas. They are fishes without bone in their internal skeletons. Externally they are covered with a mosaic of small scales that have a tooth–like structure. They have several gill slits which open directly to the outside, not being covered by a bony plate as in the bony fishes. Their mouth is filled generally with hundreds of teeth, only a few of which are involved in feeding at any one time. The remaining teeth lie in wait, in line, to move into the jaw margin when a working tooth is damaged or lost. The teeth are not fused to the jaws.

Within the chondrichthyans are two groups: the more typical sharks, skates and rays (the elasmobranchs); and the chimaeras (holocephalans). They differ in many respects, but both have cartilagenous skeletons. The elasmobranchs are mainly carnivorous fish with a moveable relationship between the braincase and the palate, whereas the holocephalans have the braincase fused to the upper jaw and are essentially durophagous, that is, they feed by crushing their food, often hard-shelled invertebrates.

The oldest shark fossils in Australia come from the Early Devonian of New South Wales, scales called "*Skamolepis*". Teeth of another shark, *Mcmurdodus*, are known from the Cravens Peak Beds in western Queensland, sediments that are either latest Early Devonian or earliest Middle Devonian. Both of these genera have some similarity to the living six-gilled sharks, the hexanchids, but further study is needed to ascertain their relationships. During the later parts of the Devonian and the remaining Palaeozoic, shark fossils become increasingly abundant and more complete. Somewhat akin to the bony

fishes, the later sharks experience a number of adaptive jumps, notably marked improvements in their feeding styles and manoeuvrability which are reflected in their skull and fin structure.

The Early to Middle Devonian (Emsian to Eifelian) Cravens Peak Beds in the Georgina Basin of western Queensland contain the microscopic teeth and scales of several different sharks. Also present in these fossil faunas are the jawless thelodonts and the spiny acanthodians, heavily armoured placoderms, crossopterygians and other less well identified bony fishes. These fishes have left behind minute teeth, scales and spines, enough to allow identification of quite a diversity of forms. *Mcmurdodus* is one of the sharks represented, its name clearly flagging the fact that it is also known from the Antarctic, specifically from the Aztec Siltstone in the Beacon Supergroup of South Victoria Land. Teeth of *Mcmurdodus* closely resemble those of the living six-gilled sharks, the hexanchoids, but until the fossil material is both more abundant and better studied it is unclear as to the meaning of this similarity. Maybe it represents a completely unrelated form that convergently developed a similar-shaped tooth. Or it could mean that hexanchoids had already evolved. Perhaps, sharks had developed from the more primitive cladodontoid level to the more advanced hybodontoid.

Cladodontoids were (and are) quite primitive sharks that possessed broad-based, rather triangular fins and elongate jaws. These jaws worked in a scissor-like fashion, with the braincase firmly locked into the upper jaw via a distinct process (the postorbital process). The teeth had characteristically a large central cusp, surrounded by several smaller, generally conical cusps on either side. The xenacanth sharks in this group had rather the reverse, with two large outside cusps and a small central cusp. Xenacanths were further unusual in that they invaded fresh waters, unlike other sharks which essentially remained tied to the marine realm.

Hybodontoid sharks form a group intermediate between the primitive cladodontoids and modern sharks. The evolutionary trends involved in this transition included a freeing up of the upper jaw palatoquadrate from the braincase, thus allowing greater latitude in movement of the jaws. The bones that support the bases of the fins were reduced and consolidated, decreasing the area of contact of the fin base with the body of the fish and thereby allowing a greater range of movement of the paired fins, thus giving greater manoeuvrability. Although most of the hybodontoid sharks were marine predators, a few experimented with development of flat, solid teeth that were used to crush food such as clams and brachiopods — a lifestyle also acquired independently by many modern living sharks, and another perfect example of convergent evolution. Australia's oldest, well-preserved shark remains come from the Middle Devonian Bunga Beds of New South Wales–for example, the braincase of *Antarctilamna*.

From their origin in the Late Silurian to Early Devonian, through the remainder of the Palaeozoic and even into the Triassic, primitive sharks reigned. They looked externally much like modern sharks, except for the structure of the fins, but internally, reflected by their skeletal structure, they were clearly less advanced. The second major radiation of sharks, including the development of the formidable marine carnivores we know today, didn't begin until the Jurassic. By the Early Cretaceous, these modern sharks were abroad the Earth.

---

70 Head shield of Chirodipterus australis from the Late Devonian Gogo Formation at Paddy's Valley, Gogo Station, Western Australia. x 2.5 (J. Frazier, courtesy of the Western Australian Museum and K. McNamara)

71 Skull and lower jaws, in side view, of the Early Devonian lungfish Griphognathus whitei from the Gogo Formation on Gogo Station, Western Australia. x. 0.9 (F. Coffa, courtesy of the Australian National University and K. Campbell)

72 Branchiostegals (gill bars) of the Early Devonian lungfish Griphognathus whitei from the Gogo Formation on Gogo Station, Western Australia. x 0.7 (F. Coffa, courtesy of the Australian National University and K. Campbell)

73

Unlike the somewhat restricted Early Devonian fish fauna, those of the Middle and especially the Late Devonian are rich both in the number of localities and in the diversity of the faunas preserved. One area in particular, Gogo Station in the Kimberley region of Western Australia, stands out as a truly spectacular example. Entire fishes preserved in limestone nodules, which were deposited as a part of a great barrier reef, provide some of the best preserved specimens of Devonian fish fauna in the world. These fossils have formed the basis of many detailed anatomical studies of placoderms, lungfish and palaeonisciform ray-finned fishes. Insights have been gained into the evolutionary pattern of these groups based not only on external features but also on minute details of their internal structure which have been preserved too.

Middle and Late Devonian localities, which contain faunas dominated by fish, are scattered widely over Australia. There is a concentration in the south-eastern quadrant, in the Lachlan Fold Belt, but other areas such as central and western Queensland and the north-west of Western Australia, as well as a few places in Central Australia, have produced vertebrates of this age. Hatchery Creek and Wattagoona Station in central and western New South Wales, Bunga Beach in coastal New South Wales, Broken River in north Queensland, the Georgina Basin in western Queensland, and Tatong in Victoria have all yielded Middle Devonian macrovertebrates as well as microvertebrate associations.

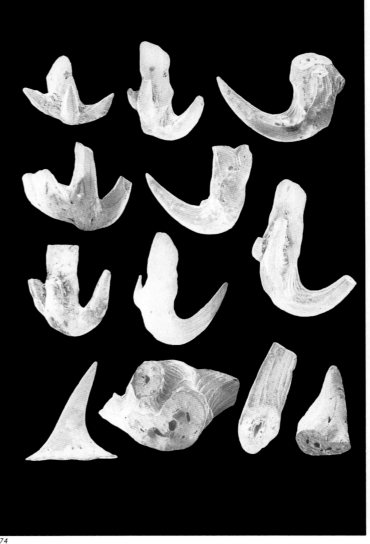

73 The left cheek and upper jaw of a crossopterygian fish *Onychodus sp.* from the Early Devonian Gogo Formation on Gogo Station, Western Australia. x 1.2 (F. Coffa, courtesy of the Bureau Of Mineral Resources and G. Young)

74 Teeth and a possible scale (lower left) of the primitive chondrichthyan *Thrinacodus (Harpagodens) ferox* from the latest Devonian or earliest Carboniferous of the Broken River embayment, north Queensland. The relationships of this group of chondrichthyans is not certain, although members of the group have some similarities to the xenacanthiforms. x 20 (University Of Queensland, Department Of Geology specimens, courtesy of Dr Susan Turner, Queensland Museum)

74

THE GOGO REEF FAUNAS

Luck seems to play a major role in discoveries, and so it was with the "finding" of one of Australia's most important palaeontological sites. In 1963 Harry Toombs from the British Museum (Natural History) spent some time in the Western Australian Museum. David Ride, palaeontologist and director of the museum, pointed out to Toombs some bone-bearing concretions from the Kimberley region of north-western Australia. Toombs, who had developed a technique using acetic acid for dissolving carbonate in concretions to remove bone, must have realized the potential of the site and took some of the concretions back to London with Ride's blessing to experiment on preparation methods. Excitement was abroad at the British Museum when Toombs found that some concretions had preserved nearly uncrushed, three-dimensional fishes with exquisitely preserved detail. A short trip back to Australia by some staff from the British Museum confirmed the richness of the site, and there followed yet other joint expeditions involving the Western Australian Museum, the British Museum (Natural History) and the Hunterian Museum in Glasgow as well as several other Australian institutions. Most recently Western Australian Museum scientists have visited the site repeatedly under the aegis of the National Geographic Society.

The concretions on Gogo Station contain fossils of vertebrates and invertebrates which inhabited an ancient barrier reef that snaked for hundreds of kilometres across a shallow, tropical sea of Late Devonian age (mainly from the Early Frasnian Epoch). Concretions found in the Gogo Formation formed as carbonate precipitated, possibly brought about by organic decomposition, and they developed before many of the very delicate bony structures, such as the gill arches, were damaged. Some concretions contain whole fishes, others only isolated bones, perhaps bits of fishes that were torn apart by carnivores. Concretions have been concentrated in the Gogo region by the high rates of erosion, in part by deflation and in part by water transport in times of flash floods and deposition in the surrounding creek beds.

The Gogo concretions are quite hard, composed mainly of calcium carbonate. They are surrounded by dark, clay-rich marine sediments. The carbonate matrix is of value in two ways. Firstly, it has protected

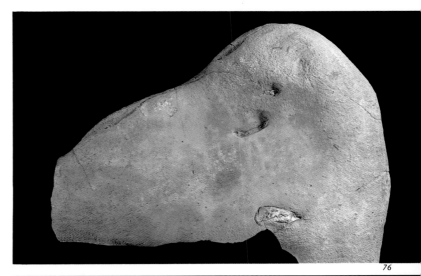

76

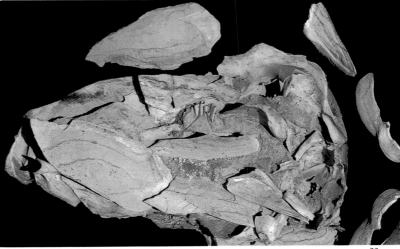

77

the fossils for long periods of time due to its hardness relative to the surrounding matrix, and thus the bones remain uncrushed. A second fortunate aspect of the carbonate is that it dissolves in weak acids, such as acetic acid, the main constituent of vinegar. Bones don't dissolve in this solution. The result of submerging such a concretion in a 10 per cent solution of acetic acid is that the carbonate gradually dissolves, leaving only the bone behind. With careful tending, using the method perfected by Harry Toombs in the 1960s, complete bony material can be recovered in a beautiful state.

The Gogo vertebrates were entirely fishes, dominantly the armoured placoderms. Coccosteomorph arthrodires were the most diverse and abundant inhabitants. Some arthrodires, like *Harrytoombsia*, are rare. While most of the fishes were of moderate size to small (most less than half a metre), others, such as the carnivore *Eastmanosteus*, reached sizes in excess of 2 metres. Other placoderms, such as the antiarch, *Bothriolepis*, are present, too. *Bothriolepis* was mainly a freshwater fish, but its presence at Gogo shows that some incursion into marine conditions was tolerable, and suggests that this fish could have migrated from one freshwater environment to the next via marine waters. This broad tolerance makes *Bothriolepis* extremely valuable to scientists relating Devonian marine and freshwater environments which generally supported distinct and exclusive faunas.

---

75 Concretions in the Gogo Formation, Gogo Station, Western Australia. Many of these limestone concretions contain fossil fishes. By carefully etching the carbonate away with dilute acetic acid and preserving the bone contained inside, three-dimensional fish specimens can be recovered. (J. Long)

76-77 Stages in the preparation of an arthrodire placoderm in a Gogo concretion, with dilute acetic acid. x 0.3 (A. Ritchie)

75

Dipnoans are well represented in the Gogo fauna, and they come in a variety of shapes reflecting a diversity of lifestyles. *Chirodipterus* was a short-snouted, tooth-plated form. *Griphognathus*, on the other hand, had a long snout and a denticulate dentition. It was probably a bottom-feeding fish that nosed about in the unconsolidated muds between high points in the reef, feeding on a variety of soft-bodied invertebrates and perhaps homing in on them using a highly sophisticated electrosensory system developed in the lateral line canals. Other dipnoans, such as the large *Holodipterus*, apparently crushed hard-shelled invertebrates. Many placoderms, including *Bullerichthys*, *Bruntonichthys* and *Kendrickichthys*, as well as some of the ptyctodontids, also assumed this durophagous lifestyle, most likely living on the reef itself and finding food amongst the cracks and crevices of that massive structure.

Besides the bottom dwellers and reef feeders, there were some Gogo fishes that apparently fed in mid-water or even on the surface. The camuropiscid arthrodires were streamlined and specialized for surface water feeding (John Long, pers. comm.), while the primitive actinopterygian palaeoniscoids *Mimia* and *Moythomasia* were probably mid-water carnivores — one specimen of *Moythomasia* had evidently eaten a conodont invertebrate just before death.

78

78 *Dorsal view of the skull and anterior part of the body of a crossopterygian fish, Koharalepis jarviki from the Middle Devonian Aztec Siltstone at Mt Crean in the Lashly Mountains, south Victorialand, Antarctica. x 0.4 (F. Coffa, courtesy of the Bureau Of Mineral Resources and G. Young)*

79 *A jaw fragment, with teeth, of a rhizodont crossopterygian, Notorhizodon mackelveyi from the Middle Devonian Aztec Siltstone at Mt Ritchie in the southern Boomerang Range, Antartica. x 0.3 (F. Coffa, courtesy of the Bureau Of Mineral Resources and G. Young)*

80 *Scales of crossopterygian fishes from the Late Devonian Hunter Siltstone of Red Cliff Mountain, north-east of Grenfell, New South Wales. x 2.5 (F. Coffa, courtesy of the Australian Museum, A. Ritchie and R. Jones)*

OVERLEAF *As the bottom-feeding dipnoan, Griphognathus whitei, probes platypus-like in the organic rich mud at a depth of about 100 metres near the Late Devonian barrier reef of Gogo, Western Australia, a curious arthrodire placoderm, Eastmanosteus calliaspis, comes to investigate the commotion. A variety of nautiloids and other cephalopods flee the scene or lie scattered about in the lime-rich muds that so beautifully entombed a great variety of life. Even at this depth some light penetrated as algae grew on the sediment-water interface. (P. Trusler)*

79

ANCESTORS OF THE TETRAPODS, CROSSOPTERYGIANS OR DIPNOANS?

The Gogo fishes, because of their preservation of fine detail, have been critical in providing answers to some controversial palaeontological questions. One fish recovered from the Gogo concretions, *Gogonasus andrewsi*, serves as an example. *Gogonasus*, discovered in 1986, is an osteolepiform crossopterygian and represents the first specimen of this group to be preserved well in three-dimensions. It is represented by several fragments and a complete skull. The skull clearly preserves the palate, and has two holes representing internal nostrils, a pre-requisite for the ancestors of the air-breathing tetrapods. Some palaeontologists had earlier suggested that these holes were not internal openings for nostrils at all, but were instead places where large tusks (the coronoid fangs) on the lower jaw fitted into the palate when the mouth of the fish was closed, thus blocking the choanae (or internal nostrils). They further suggested that lungfish, instead of crossopterygians, were indeed the group which gave rise to the tetrapods, inferring the presence of an internal nostril in the palate of Gogo dipnoans. The *Gogonasus* material from Gogo clearly shows that the tusks fit nicely into special grooves and not into what are in fact the internal nostril holes.

Further to the resolution of this argument, palaeoichthyologists Kenneth Campbell and Richard Barwick found that the position where fossil lungfish were found in the Gogo reef complex suggested that these bottom-adapted fish lived at depths of 100 to 200 metres and that their gill arches had furrows for the arterial system, typical of fishes today which use their gills to remove oxygen from water not air. In modern lungfish that do breathe air directly only the first two or three gill arches are involved in respiring. So, it seems that in the Devonian the lungfish were still confined to breathing from the water, and that they subsequently developed their ability to respire air directly, and, therefore, did so independently of the group that gave rise to the tetrapods.

Treisler 1991.

THE MOUNT HOWITT AND CANOWINDRA FAUNAS

Besides the rich Gogo reefs, other parts of Australia have also yielded a variety of Late Devonian fishes, many of them freshwater faunas. Two examples are the Mount Howitt faunas of Victoria and those from Canowindra in New South Wales.

The Mount Howitt locale has yielded perhaps the most diverse and well-preserved freshwater Late Devonian fauna in the Southern Hemisphere. This fauna occurs in the Avon River Group, a series of lake deposits of Frasnian age. Placoderms are abundant, including antiarchs in the genus *Bothriolepis*, the phyllolepid *Austrophyllolepis* and the arthrodire *Groenlandaspis*. Acanthodians are represented by the deep-bodied *Culmacanthus* and by more streamlined forms *(Howittacanthus)*. Bony fishes are present, too, for example, lungfish, as well as one of the most primitive palaeoniscoids known on a world scale *(Howqualepis)* and at least three different types of crossopterygians. Amongst these latter three was a specimen of *Marsdenichthys* which has been important in determining the relationships of fishes within the crossopterygian group. As the bone was very fragile and crumbly, study of this specimen was carried out

in a rather unique fashion. The bone itself was dissolved from the surrounding rock with a weak solution of hydrochloric acid, leaving only its imprint in the rock. Then a latex "peel" or cast was made of this impression, and it was the impression, not the real bone, that was studied. Extremely fine detail was preserved in this fashion.

Canowindra is another remarkable fish locale of similar age to Mount Howitt. Preservation of material is exquisite with impressive concentrations of placoderms packed like jackstraws on red sandstone slabs. Such antiarchs as *Bothriolepis* and *Remigolepis* are quite common, and there are rarer forms such as the crossopterygian *Canowindra* and the euarthrodire *Groenlandaspis*.

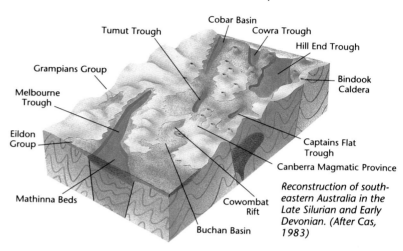

Reconstruction of southeastern Australia in the Late Silurian and Early Devonian. (After Cas, 1983)

81 A crossopterygian, Canowindra and placoderms, Remigolepis and Bothriolepis, in a large sandstone slab from the Late Devonian of Canowindra, New South Wales. x 0.2 (F. Coffa, courtesy of the Australian Museum, A. Ritchie and R. Jones)

MICROSCOPIC REMAINS OF DEVONIAN VERTEBRATES

A rich record of microfossil remains of fishes is known from Australia and is becoming increasingly important for dating rock sequences in the Palaeozoic. All major groups of fishes are represented by these microfossils.

In Devonian rocks one of the most frequently encountered groups is the thelodonts, small agnathan fishes averaging about 150 milli-metres or so in length, but some reaching up to a metre. They lived in both marine and freshwater environments, and had a body covered with minute scales. As their name suggests, their teeth look a bit nipple-like in overall shape. The structure of these teeth is very like that of mammalian teeth in that they possess a pulp cavity with dentine and enamel surrounding it. The individual scales are tiny, ranging in size from 0.1 to slightly greater than 0.3 millimetres. The shape of such scales can vary on any one fish from head to tail, but examination of rare complete fishes enables accurate species identifi-cation of individual scales. Worldwide there are several thelodont groups, but in Australia only the thelodontids are known. This group ranges from the Late Silurian to the Late Devonian and includes such genera as *Turinia* and *Nikolivia*. The Late Devonian (Early Frasnian) *Australolepis* from the Gneudna Formation of the Carnarvon Basin in Western Australia may be the youngest record for the thelodonts anywhere. In fact, the thelodonts had all but disappeared in Europe and North America at the end of the Early Devonian, while they continued to flourish along the shores of Gondwana (including Iran and Thailand in addition to Australia).

Sharks, too, have left microfossil remains. *Antarctilamna*, for example, is known in the Middle Devonian Bunga Beds of coastal south-eastern Australia, and occurs also in Antarctica and Iran, perhaps even in Bolivia. Its spines and teeth clearly mark it as a xenacanth shark, a group which successfully invaded fresh water. Another unusual shark, with characteristic horn-like teeth, was *Harpagodens ferox* (also known as *Thrinacodus ferox*), present in the latest Devonian and Early Carboniferous of Australia, as well as Thailand, China, Europe and North America.

Scales, spines and teeth from acanthodians, crossopterygians, dipnoans, actinopterygians and even placoderms are also preserved, some of which can be associated with partial or whole fishes. Many of them have very restricted time ranges, and their potential for further use as time pieces for dating rocks is great. The only limiting factor in their usefulness is the small number of people able to collect and study them. There is certainly neither lack of specimens nor potential usefulness of these fossils.

---

82 Groenlandaspid median dorsal plate and xenacanth shark tooth (arrowed) from the Middle Devonian Aztec Siltstone, south Victorialand, Antarctica.x 1.6 (F. Coffa, courtesy of the Bureau Of Mineral Resources and G. Young)

82

GLOBAL FAUNAL SIMILARITIES IN THE DEVONIAN

83

During the Early Devonian, Gondwana (which included South America, Africa, Australia and India of the major landmasses, as well as smaller continental blocks that were closely allied, Shan-Thai and South China) was separated from Euramerica. Kazachstan and Siberia were, likewise, separate blocks, though located in the Northern Hemisphere, and yet one other microcontinent may have existed, Armorica. Armorica comprised some of present-day Europe, including parts each of France, Germany, Bohemia and Spain, and probably lay in the ocean between Gondwana and Euramerica. In the Early Devonian, however, the faunas of this microcontinent seem to have had far greater affinity with those of Gondwana.

Several distinctive faunal provinces were in place during the Early Devonian: the jawless cephalaspids defined the Euramerican Province, the amphiaspids were in the Siberian Province, the jawless galeaspids and yunnanolepid placoderms were in the South China Province, and the wuttagoonaspid and phyllolepid placoderms were in the East Gondwana Province. The wuttagoonaspid-phyllolepid province is most characterized by *Wuttagoonaspis fletcheri*, a unique arthrodire with a long head shield that has small openings (orbits) for the eyes and a tall crest along the top of its trunk armour. It had a distinctive meandering ridge ornamentation on the bony plates making up its head and trunk shield, vaguely reminiscent of phyllolepids.

In the most recent palaeogeographic reconstructions, the existence of Armorica seems to be reflected by a number of genera or closely related genera that occur in common there and in Gondwana, yet are lacking in the Euramerican province to the north during the Early Devonian. The dipnoan *Speonesydrion* lived in south-eastern Australia, but had a close relative in Germany (found in the Hunsrückschiefer sediments). Likewise, the acanthodian *Machaeracanthus*, known from many parts of Gondwana including Australia and Antarctica, has also been found in the Rhineland of Germany in Early Devonian marine rocks. And, within the placoderms, *Buchanosteus*, which is known from eastern Australia, China and the Middle East, has a near relative in Brittany. Another placoderm, a petalichthyid *Lunaspis*, best known from the Rheinische Schiefergeberge of Germany, is also present in Gondwanan Australia and South China. The closely related *Wijdeaspis* in now not only known in Australia, but also in the Elburz Mountains of Iran. Microfossil assemblages of several osteichthyan groups, acanthodians and sharks likewise show a marked similarity of faunas between Spain and Gondwana in Early Devonian times.

So, it seems that in the Early Devonian a major faunal province was in place that included Gondwana and parts of present-day Europe. Only in the Late Devonian did this province come in close contact with the rest of Euramerica, as lithospheric plate movements brought the supercontinents and their passenger faunas into close apposition. It was at that time that major faunal interchange came about, with such placoderms as *Bothriolepis*, *Phyllolepis*, *Groenlandaspis* and *Remigolepis* first appearing in the European record, no doubt immigrants from the south. As the "southerners" arrived on the Euramerican scene, they had a marked impact on the endemic fauna, one example being the extinction of the bottom-dwelling psammosteids, jawless heterostracans that were evidently outcompeted by the immigrant *Phyllolepis*.

83 Head shield of the placoderm Wuttagoonaspis fletcheri from the Middle Devonian Mulga Downs Group, Mt Jack Station, near Wilcannia, New South Wales. Typical of the placoderms, Wuttagoonaspis had a heavily armoured head and thorax, but probably a relatively flexible and unprotected posterior portion. x 2 (F. Coffa, courtesy of the Australian Museum and A. Ritchie)

THE CHANGES TO COME

The high diversity of fish and invertebrate faunas of the Devonian in Australia, the presence of great expanses of carbonate deposition in the south-eastern Lachlan Fold Belt as well as in the basins of north Queensland and the Canning Basin of Western Australia, and the occurrence of oolitic limestones in the Wee Jasper area of New South Wales point to continual warm and tropical conditions for the Australian continent. But all this was to change in the Carboniferous, when Australia shifted rapidly and rather dramatically to the south. This movement coupled with global events brought a marked cooling of temperatures, placing stress on the plant and animal inhabitants. Worldwide, it was a time of massive extinctions, impressive coal deposition and an irreversible series of changes over the next hundred million years. Fishes maintained their dominance of world vertebrate faunas, but it was a different group that presided over the future than had ruled in the Devonian. The versatile ray-fins, the actinopterygians, were to displace the placoderms and acanthodians of times past. As well, a new group appeared on the scene that was to dominate terrestrial environments until late in the Triassic, the labyrinthodont amphibians to whom the crossopterygians "gave birth".

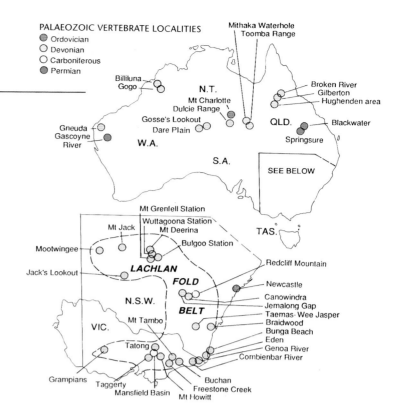

Several possible family trees for major fish groups. Palaeontologists are still not in agreement about the relationships.

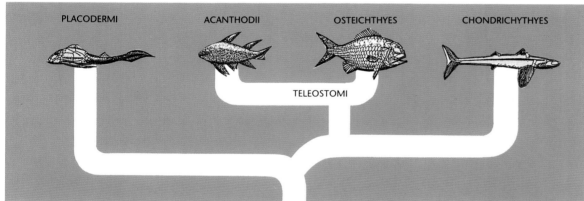

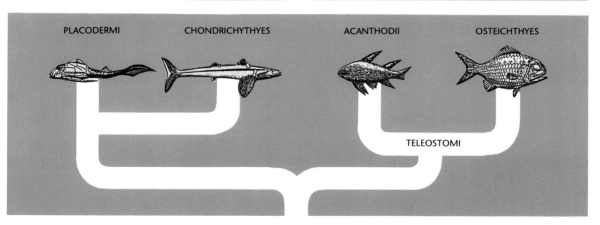

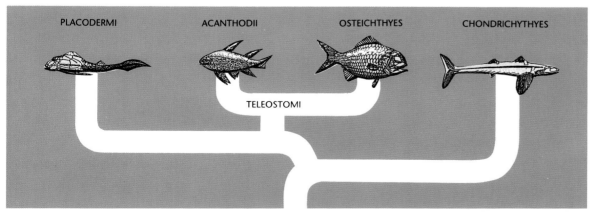

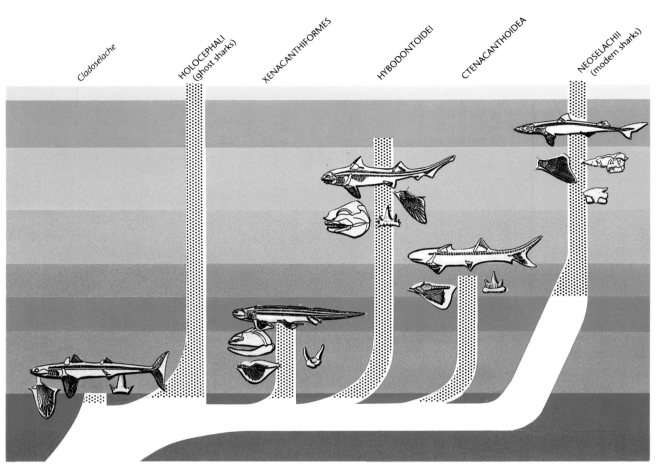

Cladoselache   HOLOCEPHALI (ghost sharks)   XENACANTHIFORMES   HYBODONTOIDEI   CTENACANTHOIDEA   NEOSELACHII (modern sharks)

Hypothesis of the interrelationships of modern and fossil sharks. (Mainly after Schaeffer & Williams, 1977; Maisey, 1984; and Long, in press)

NEOGENE

PALAEOGENE

CRETACEOUS

JURASSIC

TRIASSIC

PERMIAN

CARBONIFEROUS

DEVONIAN

SILURIAN

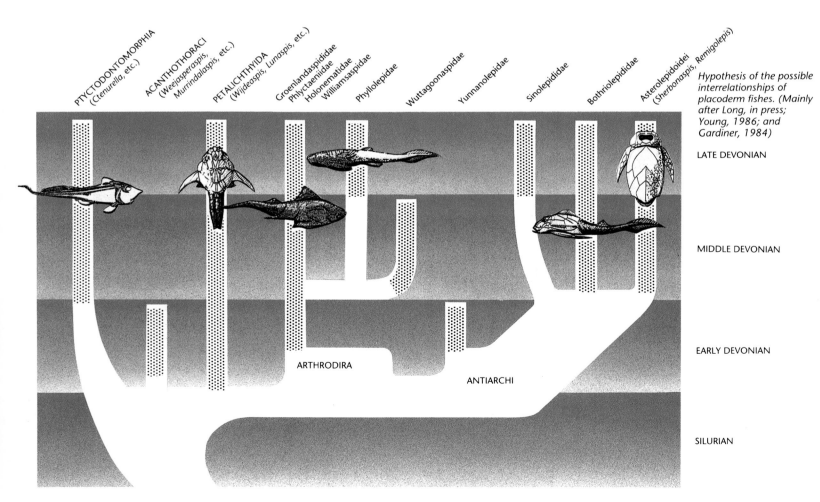

PTYCTODONTOMORPHIA (*Ctenurella*, etc.)   ACANTHOTHORACI (*Weejasperaspis, Murrindalaspis*, etc.)   PETALICHTHYIDA (*Wijdeaspis, Lunaspis*, etc.)   Groenlandaspididae   Phlyctaeniidae   Holonematidae   Williamsaspidae   Phyllolepidae   Wuttagoonaspidae   Yunnanolepidae   Sinolepididae   Bothriolepididae   Asterolepidoidei (*Sherbonaspis, Remigolepis*)

Hypothesis of the possible interrelationships of placoderm fishes. (Mainly after Long, in press; Young, 1986; and Gardiner, 1984)

LATE DEVONIAN

MIDDLE DEVONIAN

ARTHRODIRA

ANTIARCHI

EARLY DEVONIAN

SILURIAN

96

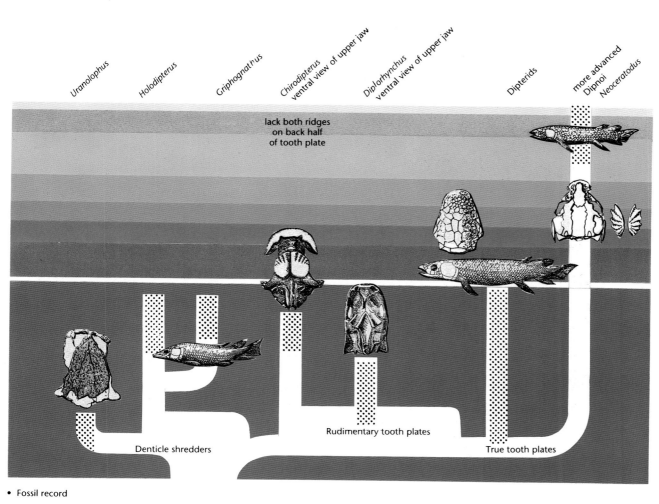

Hypothesis of the interrelationships of a variety of lungfish with several different tooth morphologies. (Modified from Long, 1988, and information in Campbell & Barwick, 1987)

Uranolophus

Holodipterus

Griphognathus

Chirodipterus
ventral view of upper jaw

Diplorhynchus
ventral view of upper jaw

Dipterids

more advanced
Dipnoi
Neoceratodus

NEOGENE

PALAEOGENE

CRETACEOUS

JURASSIC

TRIASSIC

PERMIAN

CARBONIFEROUS

lack both ridges
on back half
of tooth plate

DEVONIAN

Rudimentary tooth plates

Denticle shredders

True tooth plates

• Fossil record

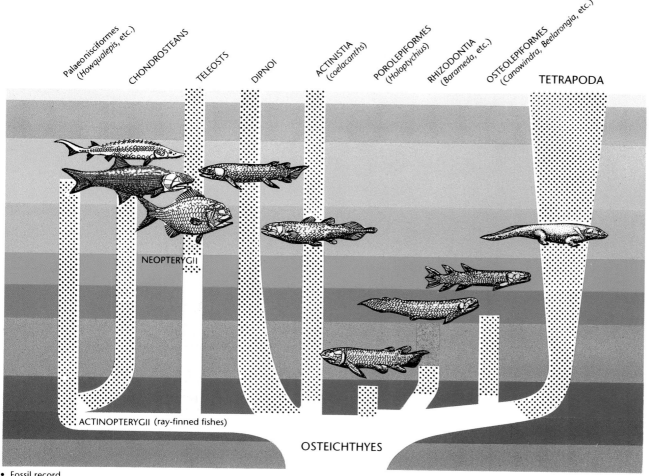

Hypothesis of the interrelationships of the bony fish, Osteichthyes. (Modified from Long, 1990)

Palaeonisciformes
(Howqualepis, etc.)

CHONDROSTEANS

TELEOSTS

DIPNOI

ACTINISTIA
(coelacanths)

POROLEPIFORMES
(Holoptychius)

RHIZODONTIA
(Barameda, etc.)

OSTEOLEPIFORMES
(Canowindra, Beelarongia, etc.)

TETRAPODA

NEOGENE

PALAEOGENE

CRETACEOUS

JURASSIC

TRIASSIC

PERMIAN

CARBONIFEROUS

DEVONIAN

SILURIAN

ORDOVICIAN

NEOPTERYGII

ACTINOPTERYGII (ray-finned fishes)

OSTEICHTHYES

• Fossil record

CHAPTER 3

ATTACK ON THE LAND

THE EARLY DEVONIAN TO TRIASSIC PERIODS, FROM 410 TO 205 MILLION YEARS AGO

The world's first record of terrestrial vertebrates is in the Early Devonian of Australia — a set of footprints left in the sands of a backwater beach more than 400 million years ago. Such tracks were probably made by labyrinthodont amphibians, a group that dominated Australian land faunas through the late Palaeozoic and early Mesozoic. In contrast, reptiles have left little record, perhaps only an artefact of preservation rather than reality.

The Australian Peninsula of Pangaea may well have been biogeographically distinct from the rest of the world, in part due to its placement at the "end of the Earth". Late Palaeozoic and early Mesozoic times were periods of great climatic change brought about by the shifting position of Pangaea. In the Late Carboniferous and Early Permian, Australia had moved far south of its Devonian equatorial position, and with the Pangaean supercontinent strung from pole to pole, a massive glaciation ensued. Glacial conditions lowered sea levels and pushed cold watermasses towards the tropics, resulting in the spectacular extinctions at the Palaeozoic's end. Terrestrial vertebrates were severely affected, especially those with large body size. Some of the smaller mammal-like reptiles and labyrinthodont amphibians survived to give rise to a rich Triassic fauna. But this fauna, too, met its demise at the end of Triassic times, when yet another series of extinctions all but eliminated the remnants of the Palaeozoic fauna. It was into this ecological void that the dinosaurs and mammals moved and prospered in what remained of the Mesozoic Era.

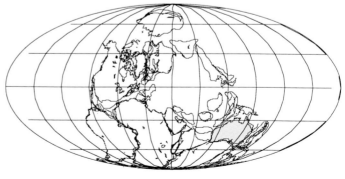

Position of the Earth's continents at 362 million years ago. (Scotese, c. 1990, "terra mobilis" program)

Devonian sandstones in the Grampians of western Victoria. The trackway was found on a paving stone at Glenisla Homestead, and is the world's oldest known record of land-dwelling vertebrates. The trackway is on display in the Dundas Shire Offices in Hamilton, Victoria. x 1.0 (S. Morton, courtesy of the Dundas Shire Council and K. Jenkins)

84

ON BECOMING TERRESTRIAL

Sometime during or before the Early Devonian, vertebrates invaded the land. Undoubtedly, their conquest followed the movement of plants into the terrestrial environment, which for billions of years had lain barren of anything but the tiniest of organic remains. Just what was the spur for such an immigration is a topic of debate. Perhaps vertebrate movement onto land began as an occasional escape mechanism for fishes being pursued by their predatory cousins. Perhaps it was an outright utilization of an area providing a variety of new niches for those capable of dealing with the unique new stresses. Initially, competition in such an environment would have been negligible. Another possibility is that drought may have spurred the development of the tetrapods, and they may have used their ability of navigating the land to move from one waterhole to the next as the water became scarce.

Irregardless of the reason for movement onto land, living there certainly provided a number of challenges. Support was needed in a non-aqueous environment where gravity was not counterbalanced by the buoyancy of water, and land dwellers had to develop effective modes of locomotion in this unforgiving situation. Oxygen had to be gleaned from air rather than water, and water loss had to be controlled lest dehydration cause death. Sensory organs that recorded images and sound had to function in a new medium, and reproduction had to take place in the new environment. Each of these barriers had to be overcome for true terrestriality to be achieved, a process that occurred over the several million years of the middle Palaeozoic and resulted sometime during the Carboniferous in the egg-laying amniotes, vertebrates that were truly capable of surviving away from ponds and streams.

Several changes that plot the development of terrestriality in vertebrate structure can be traced in the fossil record. The osteolepiform crossopterygian fishes, by retaining into adulthood a number of juvenile characteristics, apparently gave rise to the first terrestrial tetrapods ("four-footed" vertebrates). Some workers have suggested that the lungfish, the dipnoans, were the predecessors of tetrapods, but recent work has demonstrated that the crossopterygian ancestry is more likely. Osteolepiforms and tetrapods share a number of specialized characteristics, called synapomorphies, which link them together, more than with any other groups. An example of such features is a single pair of nostrils with both external and internal openings.

Crossopterygian fishes depended on their watery environment for much of their mechanical support. The notochord was still a major structural element in their "skeleton" and it was about all that was needed both for keeping the body from collapsing and for support of the head. *Ichthyostega*, one of the oldest "tetrapod-like" forms known

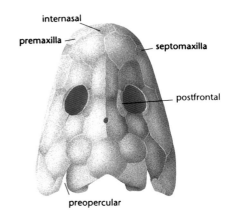

Skull of the Late Devonian primitive amphibian Ichthyostega. (After Carroll, 1988)

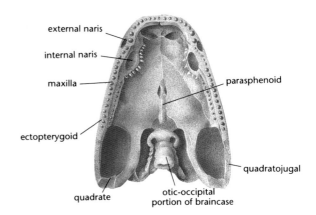

from skeletal remains, is closely related to the osteolepiforms but has some specializations that were acquired by tetrapods. *Ichthyostega* was still dependent on notochordal support but, unlike the osteolepiform fishes, it had developed massive ribs that formed a rather inflexible basket which reinforced the body wall thus protecting the lungs and internal organs as well as supplementing the vertebrae and notochordal column in body support. *Ichthyostega* retained a fin on the elongate tail, which was supported by the neural (dorsal) and haemal (ventral) processes of the vertebrae, suggesting that it used this organ in swimming, so it most likely spent much of its time in the water.

Head support was another necessity for terrestrial living, in order to achieve efficient feeding, locomotion and breathing. Trends in early tetrapods involved much fusion of cranial elements, in both the dermal bones of the skull and the endochondral bones of the braincase and jaws. Mobility of the skull, which had been common in the crossopterygian fishes, was essentially lost in the early tetrapods. In *Ichthyostega*, the skull roof and the cheek regions were firmly attached, as they were in later tetrapods. In early tetrapods bones were lost that had been present in their crossopterygian predecessors, such as the opercular series that had previously covered an area associated with respiration and which had been loosely attached to the shoulder girdle in amphibian precursors. Two remnants of this series were still present in *Ichthyostega*. A further specialization for head support was the development of the atlas-axis complex, comprising the first two vertebrae of the backbone, which attaches directly to the occipital condyles of the skull in a ball and socket arrangement.

Proportions of the skull changed in the fish-to-tetrapod transition. The anterior part of the skull became distinctly elongated and the posterior shortened. What is interesting, however, is that despite these changes related to terrestrial existence there are still remarkable similarities between the earliest tetrapods, such as the amphibian *Ichthyostega*, and members of the fish group that spawned them, the osteolepiform crossopterygians. Many of the bones in the skulls of each of these two groups can be homologized. One remarkable similarity is the labyrinthodont teeth possessed by both. Such

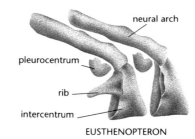

EUSTHENOPTERON

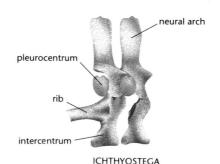

ICHTHYOSTEGA

Comparison of vertebrae of the Late Devonian rhipidistian crossopterygian fish Eusthenopteron and the Late Devonian tetrapod Ichthyostega. (After Carroll, 1988)

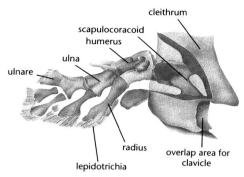

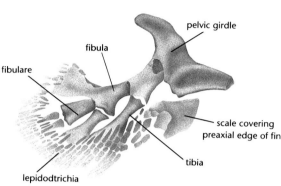

The internal skeleton of the paired fins (pectoral, above, and pelvic, below) of the crossopterygian fish *Eusthenopteron*, perhaps reasonably similar to the ancestor of tetrapods. (After Carroll, 1988)

teeth have a characteristic infolding of the dentine in a labyrinthine array, distinct from the simple circlet of hard tissue as in human teeth and those of the majority of other vertebrates. These labyrinthine teeth were numerous, and many were large and fang-like in early tetrapods just as in the closely related crossopterygian fishes. Both of these groups must have been formidable carnivores.

The ability to move efficiently on a solid terrestrial surface placed many demands on the vertebrate skeleton. The unique arrangement of bones of the paired fins in the crossopterygian group, the rhipidistians, is most similar to the limbs of the tetrapods. But although these bones are comparable, in rhipidistians their function was different, for these animals swam rather than walked. In the transition from fish to amphibian, the shoulder girdle (scapulocoracoids, cleithra, clavicles, interclavicle) became a unit distinct and separate from the skull bones, in contrast to the arrange-ment in fish in which all these bones are interlinked.

In tetrapods a firm connection between the vertebral column and the hind limb developed with a single vertebra, or several vertebrae, fused with the pelvis via a sacral rib. The hind limb in turn articulated with the pelvis. The limb bones and limb joints of the tetrapods were remodelled from those of the rhipidistians with the main change in the more distal parts of the limb where more flexibility developed, allowing the ankle and wrist to twist and hinge the feet. The feet were extended forward and outward, and then musculature pulled the body toward the extended arm or leg. Beyond the wrist and ankle, feet and hands developed which consolidated phalanges into dis-tinctly defined digits. Two patterns arose in the hands (or fore foot) of early amphibians, one having five digits with 2, 3, 4, 5, 3 phalanges and the other having four digits with 2, 3, 3, 3 phalanges. The feet of early amphibians also have one of two arrangements: five digits in both, with either 2, 3, 3, 3, 3 or 2, 3, 4, 5, 4 phalanges. *Ichthyostega* possessed the latter, plus an extra two digits on the hind foot and an extra one on the fore.

Water loss was another stress of the terrestrial environment, both loss through the skin and through the lungs in the breathing process. One of the earliest protective devices of tetrapods was the develop-ment of heavy scales which would have considerably reduced the loss of water.

Yet another difficulty to overcome related to the picking up of information through the sense organs concerning what was happen-ing in the environment. Because there are substantial physical and chemical differences between air and water, sense organs had to be reworked to accommodate the changing conditions. The lateral line system that had served the aqueous fish so well, was no longer useful for helping to sense movement or changing electric fields. The presence of this system in rhipidistians, *Ichthyostega* and most Palaeozoic and some Mesozoic amphibians clearly signals their continuing tie to the water. Increasing terrestriality in Palaeozoic amphibians is signalled by such structures as a nasolacrimal duct, which is associated with producing liquid that bathes the eye and olfactory epithelia. Interestingly, some rhipidistians have this struc-ture, perhaps indicating that they spent some time utilizing these tissues out of water. An eardrum, which may have fitted into a concave space at the back of the skull (the otic notch in many labyrinthodonts), and the associated bone that transmitted sounds to the inner ear, the stapes, developed in early tetrapods and refinement of this system continued in the reptiles, birds and mammals. To begin with, the stapes in early tetrapods may have served as a support for the braincase, but its function in sound transmission was enhanced in more advanced amphibians.

A final problem to be dealt with by "terrestrializing" vertebrates was that of reproduction. The strategy of laying large numbers of small eggs, fertilizing them externally and leaving the larvae with external gills on their own, typical of many fish groups, was not suitable. But for the most part this was a problem that the amphib-ians, the first of the tetrapods, did not solve. It was the reptiles, the first amniotes, which appeared in the Carboniferous that successfully tackled this obstacle.

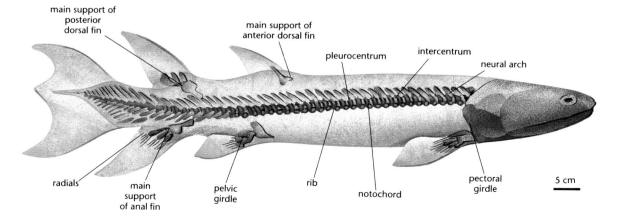

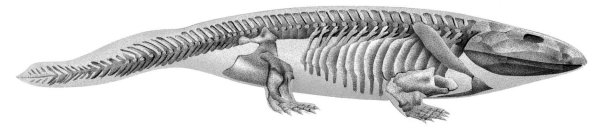

Comparison of the skeletons of the rhipidistian crossopterygian fish *Eusthenopteron* and the primitive amphibian *Ichthyostega*, both from the Late Devonian. (Modified after Carroll, 1988)

5 cm

THE COMING OF THE AMPHIBIANS

Two different kinds of amphibians, the labyrinthodonts and the lepospondyls, developed in the Palaeozoic, and together form a quite distinct evolutionary radiation to that which produced our modern amphibian fauna.

Labyrinthodonts apparently were derived directly from the rhipidistian crossopterygians, most likely with a common ancestor. Some researchers include the ichthyostegids (restricted to the Late Devonian) in the labyrinthodonts as forms most closely related to the rhipidistians, while others place them outside the labyrinthodonts in a separate category, yet involved in the fish-to-amphibian transition. Labyrinthodonts, including the ichthyostegids, ranged from the Devonian to the Early Cretaceous. Labyrinthodonts retained many rhipidistian characteristics, notably labyrinthine infolding of the tooth dentine, large fangs associated with replacement pits on the palate, and vertebrae whose centra are composed of more than one bony ossification. Limbs and the development of the atlas-axis complex of the vertebral column, however, clearly differentiate the labyrinthodonts from the rhipidistians.

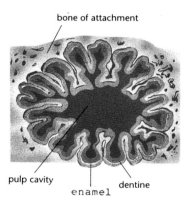

bone of attachment

pulp cavity

dentine

enamel

Cross-section, greatly enlarged, of the tooth of Eusthenopteron, a rhipidistian crossopterygian fish, showing the infolding of the dentine in a very similar fashion to that in labyrinthodont amphibians. This feature, which is the basis of their name, alludes to the labyrinthine complexity of their teeth. (After Carroll, 1988)

Lepospondyls were quite different from the labyrinthodonts. They lacked any labyrinthine complexity of the tooth dentine, and also lacked the fangs and replacement pits for teeth. Generally they had a single ossification for each vertebra, and lacked the otic notch and thus presumably lacked an eardrum. As well, the articulation of the vertebral column with the skull was distinctly different from that in labyrinthodonts. Lepospondyls, too, were generally quite small. So far, no lepospondyls have been found in Australia.

THE LABYRINTHODONTS

The labyrinthine-dentined amphibians are quite a successful group in the late Palaeozoic. They can be split into two major types according primarily to the build of their vertebrae, those structures critical for support in a terrestrial environment: the temnospondyls, ranging from the Devonian to the Cretaceous; and the anthracosaurs, restricted to the Devonian to Permian. (Some scientists classify the earliest temnospondyls as a separate group, the Ichthyostegalia.) Of the two major labyrinthodont types, only the temnospondyls have a record in Australia. Most of this record is restricted to the Early and Middle Triassic, but nevertheless it is the longest and one of the most varied for this group of any place on Earth.

Temnospondyls had a large, crescent-shaped vertebral block (called an intercentrum) associated with a pair of quite tiny blocks (called pleurocentra). They also had a solid skull with the cheek and skull roof interlocked. There are four digits in the hand (with a phalangeal count of 2, 3, 3, 3) and five toes in the foot (with a

85 Reconstruction of a labyrinthodont amphibian that probably impressed the Glenisla trackway in the Early Devonian sandstone of the Grampians.

phalangeal count of 2, 3, 3, 3, 3,). The skull bones had a distinct ornament of grooves and pits separated by sinuous ridges.

Anthracosaurs had vertebrae that were dominated by one bony element, the pleurocentrum. The intercentrum, if present at all, was very much reduced. The skull roof was only loosely attached to the cheek in most anthracosaurs. The ornamentation of the skull bones is characterized by fine radiating grooves, and detailed arrangements of the bones in the skull of this group differ from those in the temnospondyls. The hand had five digits with a phalangeal count of 2, 3, 4, 5, 3, while the foot had one or more extra phalanges in its fifth digit.

THE SILVERBAND TRACKWAY OF THE EARLY DEVONIAN
The oldest trace of vertebrates that walked on Earth comes from Australia, and is impressed on a paving stone slab that once lay in the courtyard of an old Cobb & Co. station. Glenisla Homestead, built for Samuel Carter and his wife in the Western District of Victoria in 1873, was made of a coarse-grained, finely laminated sandstone, quarried at Mt Bepcha in the Grampians of western Victoria. But stone for the pavers in Glenisla's courtyard came from another spot. According to a scrapbook put together by Frederick Carter, son of the property owner, these paving stones had been dug up only about 3 miles away and seem to be from the Silverband Formation in the Grampians Group, most recently dated as Siluro–Devonian age. Across one of these pavers, undulating with ripple marks, is a ladder-like pattern evidently left by a small tetrapod that walked across this water-soaked backwater beach more than 400 million years ago in Early Devonian times. The pattern is quite regular, with 23 positive impressions in all, indicating a stride length of about 193 millimetres. No body or tail trace was left, which probably indicates that the animal making the trail was short-bodied. Extrapolating from the trail, it is probable that the printmaker was about 850 millimetres in length. Unfortunately the details of the Silverband prints were not preserved, most likely because the sediment was so full of water. It is impossible, then, to determine if there were four or five or even six digits on the hands and feet, information which might have made it possible to determine more precisely the relationship of this earliest tetrapod.

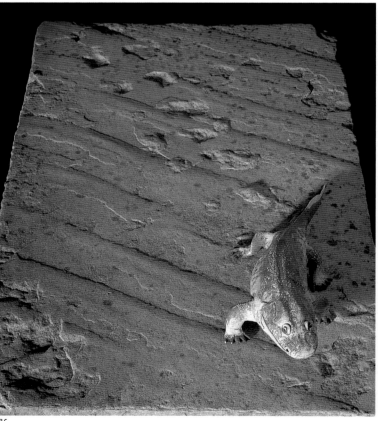

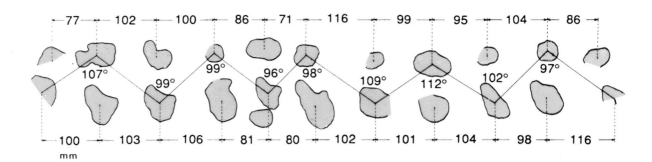

Interpretation and scale of the Glenisla tetrapod tracks. (Warren, Jupp & Bolton, 1986)

In Australia such an early tetrapod stepped ashore onto a land dominated by primitive vascular plants, the lycopods, best exemplified by elements of the *Baragwanathia* flora that appears in the Australian record in the Late Silurian. These early land-plants for the most part were small, but they had many adaptations to survival in a terrestrial environment notably roots, leaves, and a vascular system to transport water and nutrients to all parts of the plant. During the Middle to Late Devonian even more revolutionary changes affected plants, with some developing the ability to produce secondary wood which allowed the attainment of much larger size due to the support offered by such a tissue. Forests emerged as a major land cover in the Middle Devonian—forests of giant lycopods, or clubmosses, and magnificent sphenopsids, the horsetails.

86 One of the oldest land-plants known, Baragwanathia was a large lycopod typical of the Late Silurian and Early Devonian floras of Australia. This specimen, from near Yea in Victoria, represents the sort of vegetation that early tetrapods would have encountered when they invaded terrestrial environments in the middle Palaeozoic. x 0.7 (S. Morton)

87 Baragwanathia from the Late Silurian or Early Devonian of Twenty Mile Quarry on Yarra Track near Yea, Victoria. x 1.2 (F. Coffa, courtesy of the Museum Of Victoria)

86

87

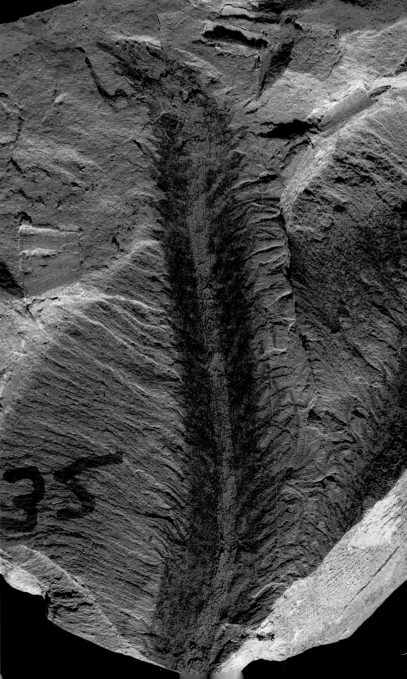

THE LATE DEVONIAN GENOA RIVER PRINTS

Following the footprints in the Grampians the next oldest known tetrapod record is a four-digit print preserved in the Middle to Late Devonian of Brazil. And slightly younger are three sets of tetrapod trackways known from the Late Devonian Genoa River Beds, most likely of Frasnian age, in eastern Victoria, Australia. These tracks are associated with rocks bearing sphenopsid stems and the fern-like foliage of *Archaeopteris howittii*. The trackways vary in preservation, but the best of the three demonstrates that the trackmaker was about 550 millimetres in length. The foot had five digits, none of which had much free length—in other words they were webbed. The hand had at least three digits, maybe more. The foot was at a maximum 3.5 centimetres wide and was slightly larger than the hand. A second of the trackways was accompanied by an undulating tail or body trace. In this trackway, unlike the previous one, the hind feet did not overstep the hands which indicates that the animal was partially using body and tail undulations for locomotion. The pace was about 17.5 centimetres, nearly twice the length of that of the first trackmaker.

The earliest skeletal remains of tetrapods are preserved in sediments of Late Devonian age, the best being *Ichthyostega* from Greenland. *Ichthyostega* was about 1 metre in length, possessed short limbs and had a 3.5-centimetre-wide foot with five short toes. An animal like this could easily have made the Genoa river tracks, but since no more information is available other than the footprints there is no way of knowing for sure just what the creature was like. Perhaps in the future a few bones that can be associated with these traces will be found.

AUSTRALIA'S OLDEST TERRESTRIAL BONE

A road metal quarry into the Late Devonian Cloghnan Shale near Forbes in New South Wales has yielded the oldest bony remains of tetrapods in Australia. This specimen is the fossil of a single jaw thought to belong to an amphibian, *Metaxygnathus denticulus*. It was associated with a rich vertebrate fauna dominated by the placoderms *Bothriolepis*, *Remigolepis* and *Phyllolepis*, and the dipnoan *Soederberghia*. The precise age of the *Metaxygnathus* specimen is somewhat older (Late Frasnian to Middle Famennian) than the Greenland occurrence of *Ichthyostega* (Late Famennian). The Forbes find is the oldest fossil that can be assigned with certainty to this early group.

Rhipidistian crossopterygian fishes have many similarities with early amphibians, especially ichthyostegaliids, but the jaw of *Metaxygnathus* appears to differ from those of its fish relatives in that the dentary bone is more massive and there is a large bony projection on the back of the jaw (the retroarticular process) to which the muscles that opened the jaw were attached. Furthermore, a different set of muscles seem to be involved in opening the jaw (amphibians use the *depressor mandibulae* rather than the hypobranchial musculature), the stresses on the jaw were different than were those in the fishes to which these amphibians were most closely related. This functional difference is reflected in the orientation and shape of the articulation between the lower and upper jaw. The jaw is thickest and heaviest where the *adductor mandibulae* muscles that close the jaw attach (just in front of the jaw articulation) and around the jaw articulation. It is here that the maximum stress occurs. As well, amphibian jaws thin to the front and back because either the muscles that attach in these regions do not deliver great forces (such as the *depressor mandibulae* which attach at the posterior part of the jaw to lower it) or there are no muscle attachments at all.

Metaxygnathus had a variety of conical teeth, with the largest scattered over the middle of the jaw. Some of them were curved towards the inside and were evidently quite effective at grabbing and holding on to small prey.

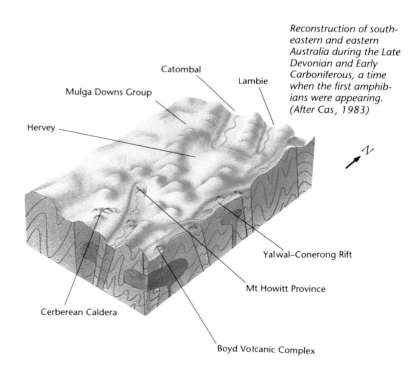

Reconstruction of south-eastern and eastern Australia during the Late Devonian and Early Carboniferous, a time when the first amphibians were appearing. (After Cas, 1983)

Catombal
Lambie
Mulga Downs Group
Hervey
Yalwal–Conerong Rift
Mt Howitt Province
Cerberean Caldera
Boyd Volcanic Complex

88 Trackway of a tetrapod from the Late Devonian of Genoa River in eastern Victoria. It is one of the oldest records of tetrapods on Earth. x 0.25 (S. Morton, courtesy of J. Warren and I. Stewart)

A RARE PERMIAN FIND

Although the earliest known land-dwelling vertebrates occur in Australia, their record between the Devonian and early Mesozoic is paltry. Only one temnospondyl, *Bothriceps australis*, is known. It comes from the Newcastle Coal Measures, from a lake-laid oil shale of Late Permian age) near Airley, New South Wales, and is associated with the palaeoniscoid fishes *Elonichthyes* and *Urosthenes*. *Bothriceps australis* is represented by a small skull that was originally figured by Thomas Henry Huxley in 1859, only the second specimen of a brachyopid amphibian to be described. *Bothriceps* evidently was an inhabitant of, or a visitor to, the lakes and streams of the coal swamps. It had a very flattened head with upwardly directed eyes and was perhaps a small insectivore or small carnivore.

The swamplands around the lake that *Bothriceps* frequented were forested with seed-ferns, not with the giant lycopod (clubmoss) flora of the Late Devonian and Early Carboniferous which had hosted the first Australian amphibians. Seed-ferns were among the first seed-bearing plants, and with their specialised reproductive organs, the equivalent of the vertebrate amniote egg, they were able to prosper in the terrestrial environment and no longer needed free water in some parts of their life cycle. During the Late Carboniferous and some of the Permian these seed-bearing plants dominated the floras. Early in seed-fern history, plant diversity was low in Australia, for as the continent moved southwards during the late Palaeozoic much of it was covered with massive glaciers, especially during Late Carboniferous and Early Permian times. Later in the Permian, by the time *Bothriceps* lived, glaciers had retreated and the seed-ferns, of which the well-known Gondwanan *Glossopteris* was one, began to diversify. It is the remains of these cool temperate plants, after being buried and markedly altered during fossilization, that have provided so much of the energy that fuels Australia's economy—the black coals of New South Wales and Queensland.

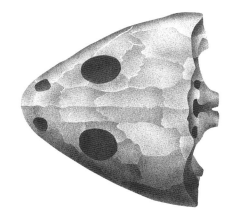

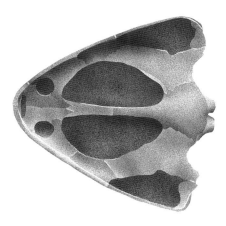

The skull of Bothriceps australis, a brachyopid labyrinthodont of Late Permian age, originally discussed by Huxley in 1859. (Wells & Estes, 1969)

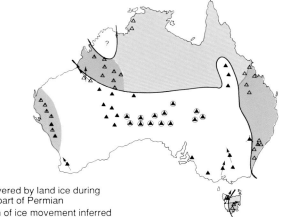

- Area covered by land ice during at least part of Permian
- Direction of ice movement inferred from pavements, exhumed topography, and sedimentary provenance
- ▲ Terrestrial glacial deposits (Sakmarian)
- ▲ Terrestrial glacial deposits (Artinskian - Kazanian)
- ▲? Terrestrial glacial deposits (Permian)
- ▲ Probable terrestrial glacial deposits (Sakmarian)
- △ Marine glacial deposits (Sakmarian)
- △ Marine glacial deposits (Artinskian - Kazanian)

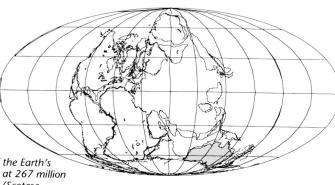

Position of the Earth's continents at 267 million years ago. (Scotese, c. 1990; "terra mobilis" program)

EXTINCTIONS IN THE LATE PERMIAN

In the Late Permian, extinctions of all kinds of life forms were severe and came in several pulses, not a singular event. This scenario was similar to that for extinction patterns of the Late Ordovician and the Late Devonian. Tetrapods of large body size, such as many of the mammal-like reptiles, were dealt the worst punishment, and it was the smaller members of the groups that survived to give rise to Triassic forms. Extinctions were greatest at the lower latitudes, especially affecting organisms of the tropical seas and reef communities. What brought about all this devastation? Although glaciers in the Southern Hemisphere were waning, the poles may still have been cold and there is evidence of a striking regression of seas off the continents in Late Permian times. This fall in sea level certainly would have significantly reduced the ecospace available to nearshore marine organisms and eliminated the shallow seas that had lain on the continents for millions of years. Surely this change was part of the reason for the massive extinctions.

Whatever the detailed explanation, the greatest mass extinction in the history of animal life, with nearly half the known families of the world disappearing forever, occurred at the end of the Permian. On land, 75 per cent of amphibian families and more than 80 per cent of reptile families became extinct.

Remains of fishes preserved both as microfossils and macrofossils are known from the Carboniferous and the Permian of Australia. The Carboniferous record of fish macrofossils is dominated by palaeoniscoids, including specimens from Victoria and Queensland.

Mansfieldiscus sweeti from Victoria is one such example, a fusiform fish about 25–30 centimetres long with the typically long gape and heavy scalation of the palaeoniscoids, distinctly different from modern ray-finned fishes of today. In the same rock unit as *M. sweeti*, the Mansfield Group, a rare crossopterygian *Barameda decipiens* has also been found. This specimen suggests that the group had no internal nostrils, thus was not a choanate. It appears to be closely related to a Middle to Late Devonian form from Antarctica, which is the oldest record for the group anywhere in the world.

Another important specimen of Early Carboniferous age comes from the Utting Calcarenite of Western Australia. It is a single nodule and inside was a partially articulated shark, *Stethacanthus*, showing the association of jaw cartilages, scales on the jaws and teeth. This association of many different elements in a skeleton is critical in allowing interpretation of other isolated bones, teeth and scales.

Occurrences of Permian fish macrofossils are not particularly frequent, some being found in coal deposits such as the Newcastle Coal Measures of New South Wales and others in such units as the Blackwater Shale of Queensland which includes forms like the deep-bodied palaeoniscoid *Ebenaqua*.

Microfossil remains for this time are more abundant and give a better picture of the real diversity of fishes living in aqueous environs on and around the Australian Peninsula of Pangaea. Besides the palaeoniscoids, remains of acanthodians, crossopterygians, lungfish and sharks, including teeth of the freshwater xenacanths, have been recovered from Carboniferous rocks in places such as Mansfield in Victoria and the Narrien Range in Queensland. In New South Wales and Western Australia a variety of shark remains have been found in marine rocks. One of the more interesting assemblages of marine microfossils, however, has been recovered from the Late Carboniferous near Murgon in Queensland. Some of the scales and teeth recovered from this site closely resemble those of modern sharks (neoselachians) and these advanced forms are associated with primitive ray-fins (palaeoniscoids).

89 *Helicoprion davisii*, an elasmobranch from the Permian of the Arthur River, north-east of Carnarvon, Western Australia. (J. Frazier, courtesy of the Western Australian Museum and K. McNamara)

90 *Belichthys magnidorsalis*, a palaeonisciform fish from the Middle Triassic Hawkesbury Sandstone of Beacon Hill Quarry at Brookvale, Sydney, New South Wales. x 2.1 (F. Coffa, courtesy of the Australian Museum, A. Ritchie and R. Jones)

91 *Cleithrolepis* sp., a palaeonisciform fish from the Middle Triassic Hawkesbury Sandstone at Hornsby Heights, Sydney, New South Wales. x 1.0 (F. Coffa, courtesy of the Australian Museum, A. Ritchie and R. Jones)

92 *Ebenaqua ritchiei* from the Late Permian Rangal Coal Measures at Blackwater, Queensland. x 1.2 (F. Coffa, courtesy of the Australian Museum, A. Ritchie and R. Jones)

89

90

THE TRIASSIC HEYDAY OF THE LABYRINTHODONTS

Labyrinthodonts in Australia are best known from Triassic times, and most of the fossil-bearing deposits seem to be of Early Triassic age. There are no endemic families in these Australian Triassic assemblages. As well, no families of labyrinthodonts that have a widespread distribution are lacking in this country. The same situation is not true for reptile faunas of this time in Australia. What is most intriguing, though, is that amphibians dominate the Early Triassic tetrapod faunas in Australia whereas elsewhere in Gondwana reptiles far outdiversify all other land-dwelling vertebrates.

Four areas are well known for Triassic labyrinthodont fossils in Australia: the Arcadia Formation (part Rewan Group) of south-eastern Queensland, the Blina and Kockatea shales of Western Australia, the Knocklofty Formation that crops out near Hobart in Tasmania, and the Sydney Basin which may have material as late as Middle Triassic.

Lake sediments of the Early Triassic Narrabeen Group are thought to be the source of three labyrinthodont specimens collected many years ago. This rock unit also contains a varied fish fauna including sharks, lungfish, palaeoniscoids, and a variety of other actinopterygians. Fossils present in this unit suggest a levee bank scrub dominated by the fern-like pteridosperm *Dicroidium callipteroidium*, cycads and conifers. Towards the upper part of this unit is a forest and heath assemblage. The dominant plant in both the heath and forest fringes was the seed-fern *Dicroidium*.

From a quarry at Brookvale north of Sydney the Middle Triassic Hawkesbury Sandstone, which forms Sydney's cliffs, has yielded a labyrinthodont, the capitosaur *Parotosuchus*, a lungfish and about twenty different genera of actinopterygians, mainly palaeoniscoids and holosteans. Another quarry in this unit, at Somersby, is dominated by the subholostean *Promecosomina*, but also has yielded the eel-like subholostean *Saurichthys*. (The subholosteans are a group somewhat intermediate between the palaeoniscoids and the holostean grade of organization.) Other fishes include a freshwater shark (a xenacanth or pleuracanth) as well as the lungfish *Gosfordia*. The occurrence of the xenacanth is the youngest record of this group anywhere in the world, and so it was a relict form when it lived at Somersby.

The Middle Triassic Wianamatta Group, whose fossils are best known from St Peters Quarry in New South Wales, again includes both fishes and labyrinthodonts. A skeleton of the capitosaur *Paracyclotosaurus* comes from here, as do remains of a freshwater xenacanth shark and a variety of actinopterygians, but holosteans rather than the more primitive ray-fins dominate. This situation is in contrast to the older Narrabeen and Hawkesbury assemblages where the more primitive actinopterygians are most abundant. West of Sydney, a temnospondyl trackway has also been recorded in this same rock unit.

In Queensland's Early Triassic Arcadia Formation, deposited by a braided river complex, 90 per cent of the vertebrate fauna comprises labyrinthodonts. Fishes include lungfish and *Saurichthyes*, and

93 Dorsal view of the skull of Subcyclotosaurus brookvalensis, a labyrinthodont amphibian from the Middle Triassic Hawkesbury Sandstone of Beacon Hill Quarry at Brookvale, Sydney, New South Wales. x 0.9

94 The palaeonisciform fish Cleithrolepis sp. from the Middle Triassic Hawkesbury Sandstone at Somersby, New South Wales. x 0.6

95 Macroaethes cf. brookvalensis, a palaeonisciform fish from the Middle Triassic Hawkesbury Sandstone near Hornsby Heights, Sydney, New South Wales.
(Pictures: F. Coffa, courtesy of the Australian Museum, A. Ritchie and R. Jones)

93

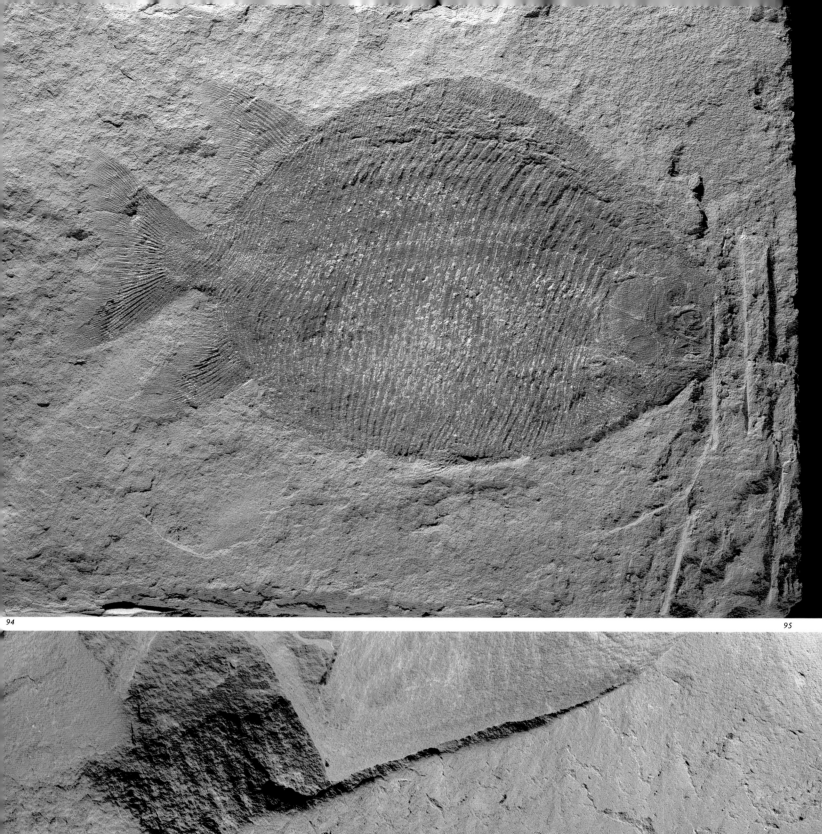

96

99

97

98

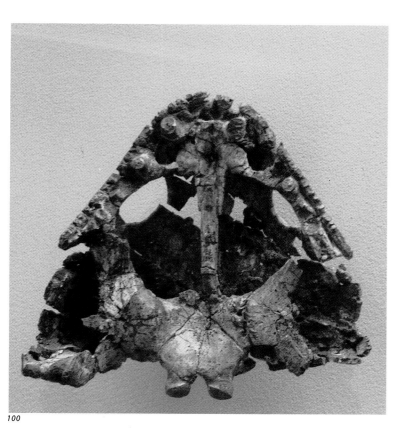

100

The brachyopids and the chigutisaurids are often classified together as the Brachyopoidea, a group which evidently had its centre of diversification in the Australian part of Gondwana. It also had the longest time range for any labyrinthodont group in this country, extending from the Late Permian to perhaps the Early Cretaceous. Amongst the youngest known brachyopoid material is a nearly complete skelekon of *Siderops kehli* from the Late Jurassic Evergreen Formation of Queensland. But the very youngest fossils of this group may be from the Early Cretaceous Strzelecki Group in southern Victoria.

The capitosaurid labyrinthodonts, like the brachyopids, are known from around the world. They, too, had markedly flattened heads. In addition, they had rather small limbs and a reduction in the degree of ossification of the skeleton.

Some of the largest labyrinthodonts were capitosaurs, some with skulls approaching a metre in length. Members of this group fall broadly into three groups, depending on the nature of the otic notch in the back of their skull, which marks the position of the tympanic membrane of the auditory system. The oldest capitosaurs from the Early Triassic had notches that opened broadly to the rear of the

101

reptiles recovered are a rare mammal-like reptile (a dicynodont), small lizard-like forms (*Kadimakara* and *Kudnu*) as well as a procolophonid and a thecodont. Above the Arcadia Formation, the Glenidal Formation, similarly fluvial, includes a long-snouted trematosaurid labyrinthodont.

Tasmania has produced two labyrinthodont-bearing locales, the Knocklofty Formation near Hobart and the less fossiliferous Cluan Formation which crops out further to the north. The Knocklofty is made up of mainly overbank-floodplain deposits that have been cut by higher energy channels containing coarser conglomerates. Together with the labyrinthodonts are preserved both the lungfish *Ceratodus* and ray-finned fishes, plus a proterosuchian reptile, *Tasmaniosaurus*.

Both the Blina Shale and the Kockatea Shale in Western Australia represent estuarine, deltaic or marine deposits. Both formations produce labyrinthodonts, with the former also having lungfish and actinopterygians and the latter having fish and marine invertebrates.

Although Late Triassic coal-bearing sediments are known from Leigh Creek in South Australia, so far only fishes and no amphibians have been recovered from there.

As many as eleven different temnospondyl labyrinthodont groups are known on a world scale but only eight occur in Australia, all of them in the Triassic. The Australian Triassic faunas are the most diversified in the world, so this continent holds a few records: the oldest tetrapods; the most diverse record of labyrinthodonts (when considered at the family level); and the youngest occurrence of labyrinthodont amphibians, which continue into the Early Cretaceous of Australia whereas they became extinct by the Late Jurassic elsewhere in the world, except perhaps in Russia.

The most varied temnospondyl group in Australia is the Brachyopidae, a family made up of five genera unique to Australia. The Australian record is in stark contrast to that of the rest of the world, where brachyopids are generally rare and even absent in South America. They have short, flattened skulls and were one of three labyrinthodont groups to survive beyond the Late Triassic, the other two being capitosaurs and chigutisaurids. In Australia both primitive (*Xenobrachyops*) and advanced forms (*Blinasaurus*) are known, indicating that a considerable amount of evolution occurred on this continent.

Related to the brachyopids are the chigutisaurids which are found only in South America and Australia. One of the better known Australian species is *Keratobrachyops australis* from the Arcadia Formation near Bluff, Queensland.

96 The lungfish *Gosfordia* from the Early Triassic sediments of Gosford, New South Wales. x 0.4 (J. Fields, courtesy of the Australian Museum, A. Ritchie and R. Jones)

97 The palaeonisciform actinopterygian fish *Cleithrolepis sp.* from the Early Triassic freshwater sediments of Hornsby Heights, New South Wales. This fish ranged from 5 to 25 centimetres in total length. (G. Miller, courtesy of the Australian Museum, A. Ritchie and R. Jones)

98 An actinopterygian fish from the Early Triassic freshwater sediments of Hornsby Heights, New South Wales. This fish fauna is dominated by palaeoniscoids, but holosteans, a dipnoan and a cestraciont shark have also been recorded. (G. Miller, courtesy of the Australian Museum, A. Ritchie and R. Jones)

99 *Acrolepis tasmanicus*, a palaeonisciform fish from Triassic rocks of Tinderbox Bay, D'Entrecasteau Channel, southern Tasmania. x 0.6 (S. Morton, courtesy of the Tasmanian Museum and N. Kemp)

100 The labyrinthodont *Xenobrachyops sp.* from the Early Triassic Arcadia Formation, Rewan, Queensland. x 1.0 (F. Coffa, courtesy of the Queensland Museum and R. Molnar)

101 A capitosaur labyrinthodont skull, in top view, from the Triassic Blina Shale, Erskine Range, Western Australia. x 0.6 (A. Warren)

102

skull, and those of the Middle and Late Triassic had closed otic notches. Intermediate forms between these two occur in the Early and Middle Triassic.

It appears that most capitosaurs lived in lakes or rivers. Two genera are known from Australia, *Parotosuchus* and *Paracyclotosaurus*.

Parotosuchus, with an open or semi-open otic notch, is known from a number of sites in Australia and had several species, one of which, *Parotosuchus brookvalensis*, was recovered from a quarry in the Middle Triassic Hawkesbury Sandstone near Brookvale in New South Wales. This amphibian was associated with a rich fish fauna dominated by the palaeoniscoids but also including some holosteans and the lungfish *Ceratodus*.

The only known specimen of *Paracyclotosaurus* is *P. davidi*, a capitosaur with a semi-closed otic notch. It was recovered in a nearly 3-metre-long ironstone nodule in the St Peters Quarry from the Middle Triassic Wianamatta Group in the Sydney Basin. What bone remained was so rotten that it was removed and a cast was made of the impression, thus providing a detailed replica of the amphibian that had left the remains. It was no easy process as this excerpt from D.M.S. Watson's paper (1958) on *Paracyclotosaurus* recounts: "The matrix [surrounding the bone] is an extremely hard and brittle ironstone, quite impossible to work. As the bone was largely rotten, and much of it already lost, Mr. F. O. Barlow of the Museum [British Museum (Natural History)] staff removed all the remains of bone, even from the deepest fissures in the blocks. Many of the blocks are very heavy and irregular in shape, and it was impossible to place them together, run in a flexible casting medium, and so draw out complete casts of individual bones. In practice the cavities in a block and its counter-

part had to be cast separately in glue. From the glue impression a waste mould was prepared from which, in turn, a plaster positive was made. This was then trimmed until it fitted accurately the cast of the same bone similarly prepared from the counterpart. The two parts were then fitted together and cast in a jelly mould which yielded perfect replicas that for all practical purposes are as good as the original bones ... Barlow's work covered many years, and I do not know of any other man who could have done it; it was a technical triumph."

Paracyclotosaurus was a large carnivore for its time, measuring about 225 centimetres in length and probably weighing more than a large human. It would have been a formidable predator prowling the backwaters and rivers of the ancient Sydney environs.

The long-snouted trematosaurids are known from the nearshore marine deposits of the Early Triassic Blina Shale in north-western Western Australia as well as from the fluviatile Rewan Group in south-eastern Queenland. They were rather crocodile-like in their looks and had well developed lateral line systems, suggesting that they spent their time in the water. Trematosaurids are the only amphibian group in the fossil record that is often associated with marine sediments, and perhaps their habits were somewhat like the living Saltwater Crocodile, *Crocodilus porosus*.

The tiny dissorophoids with their rounded heads and enlarged eyes are rare in Australia and South Africa, but relatively common in the Permian and Carboniferous of North America and Europe. One of the Australian forms, *Lapillopsis nana* from the Arcadia Formation of Queensland, is rather similar to the only African dissorophoid, *Micropholis stowi*, known from that continent's *Lystrosaurus* Zone. Both are of Early Triassic age.

Lydekkerinids are, likewise, rare in Australia, with *Chomatobatrachus* from the Early Triassic Knocklofty Sandstone of Tasmania being the only representative described so far. The group is known from only a few other genera in South Africa and Russia, and from one genus in Antarctica.

Yet another group of labyrinthodonts, the rhytidosteoids were somewhat larger, with skulls reaching up to 20–25 centimetres in length. The skulls of this group came in two varieties, rounded and triangular. Digestion in some forms, such as *Acerastea wadeae* from the Rewan Group of Queensland, was aided by gastroliths, stones which helped to grind up the ingested food. An additional point of interest for some rhytiodosteoids found in Australia (such as *Rewana*) was their peculiar vertebral construction: the three singular elements normally making up the labyrinthodont vertebra in these rhytiodosteoids came in pairs — there were two neural arches, two

102 Paracyclotosaurus davidi, a labyrinthodont amphibian from the Late Triassic Wianamatta Series of the St Peters region of Sydney, New South Wales. This specimen, some 225 centimetres long, was preserved as very decayed bone in the sandstone. It was prepared by removing the bone completely, then pouring a moulding material in the holes left behind to produce the skeletal reconstruction. (F. Coffa, courtesy of the Australian Museum, A. Ritchie and R. Jones)

103-104 Dorsal and ventral views of the skull of the labyrinthodont amphibian Blinasaurus townrowi from the Early Triassic Knocklofty Formation at Old Beach, Hobart, Tasmania. x 1.1 (S. Morton, courtesy of the University Of Tasmania and M. Banks)

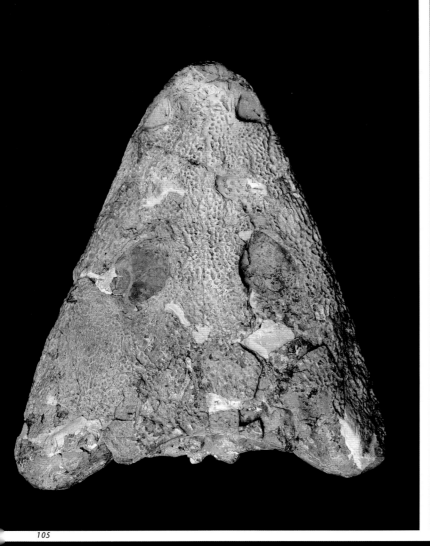

105

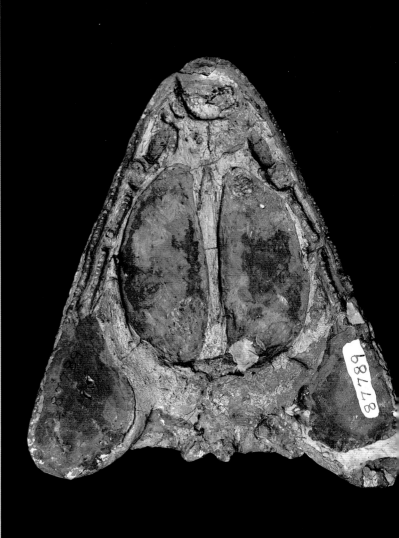

106

107

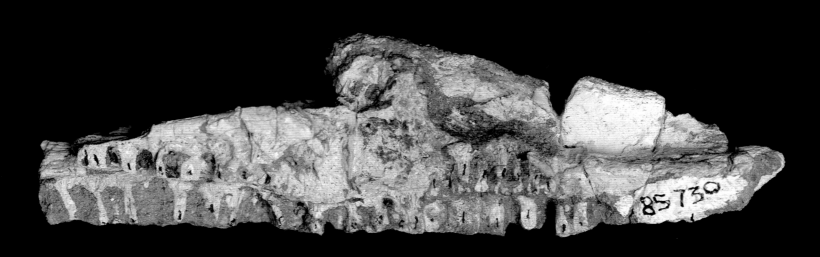

108

109

pleurocentra and two intercentra. In fact, there was much in the way of experimentation occurring in amphibian vertebrae during the late Palaeozoic and early Mesozoic, and many classifications are based on vertebral structure. Such experimentation undoubtedly related to the increasing support provided by the bony column and how much flexibility this support allowed.

Another labyrinthodont group with an Australian record is the plagiosaurs. They were small forms with short, broad, shallow skulls often with a pustular ornamentation and greatly enlarged eye sockets (orbits). Only one plagiosaur has ever been collected in Australia, *Plagiobatrachus australis* from the Early Triassic Arcadia Formation. All other specimens have been recovered from the Late Permian to the Late Triassic of the Northern Hemisphere. Some of these forms actually had external gills, apparently retained into adulthood (a process called paedomorphosis, that is, a retention of juvenile characters in breeding individuals).

LIFESTYLES OF AUSTRALIAN TRIASSIC LABYRINTHODONTS

The kind of life a fossil animal led can often be deduced by the type of skeleton it possessed. Such interpretation has been applied to temnospondyl amphibians based on a number of features of their skeletons, including: overall size and shape of the skull; location of the sockets for the eyes and the size of the eyes relative to the total length of the skull; the degree of development of the lateral line system, which when well developed signals that an animal lived in water; type of teeth; number of vertebrae that lie in front of the pelvic area, and how well ossified they and the rest of the skeleton is; limb proportions and location of limb muscle attachment; and the kinds of sediment that the group is associated with, whether lake, river, floodplain or marginal marine. When all these features are considered, at least five different "ecomorphic" types, that is, body forms that are associated with certain kinds of lifestyle, have been recognized amongst temnospondyls of the world: terrestrial, semi-aquatic freshwater, semi-aquatic euryhaline (with some saltwater tolerance), fully aquatic freshwater, and fully aquatic euryhaline forms.

Of these ecomorphic types, Australia has at least two, perhaps lacking the fully terrestrial, the semi-aquatic salt-tolerant forms (such as the mastodonsaurids), and the fully aquatic freshwater forms.

The capitosaurs, lydekkerinids, brachyopids and chigutisaurids all contain species that apparently were semi-aquatic freshwater ecomorphs. Some of the larger capitosaurids and some brachyopids may have even been obligate aquatics, because of the shortness of the limbs relative to the total body length which would not support activity in a terrestrial environment. The semi-aquatic lifestyle is indicated by the presence of discontinuous lateral line grooves and by the relatively well ossified nature of the skeleton which could efficiently bear the forces generated by an animal actually walking on land. Some lydekkerinids and capitosaurids have dermal armour. All of the families in this ecomorphic type are associated with fluvial, paludal (swamp) and lacustrine deposits. The lydekkerinids were probably small surface-swimming amphibians that ate mainly insects and small fish. Larger forms in this ecomorphic category were probably subsurface feeders on invertebrates, fishes and perhaps smaller tetrapods — feeding by inertia or suction in the aquatic environment. Capitosaurs, for example, may have been a crocodile-analogue.

Fully aquatic forms with broad salt tolerances are known in the Rhytidosteidae (including the Indobrachyopidae), Trematosauridae and Plagiosauridae. Characteristics of this group include continuous, deep grooves for the lateral line system and poor ossification throughout the skeleton, especially in the short limbs. Some have relatively elongate bodies, functionally similar to crocodiles. Skeletons of these labyrinthodonts occur in a range of sediments deposited in rivers, lakes, lagoons and nearshore marine environments. Some of these labyrinthodonts have been found associated with obligate marine species such as ammonites and marine reptiles such as ichthyosaurs. Further evidence for the obligate aquatic nature of the ecomorphic types is the possession by some plagiosaurs of well-ossified branchial skeletons that would have supported their gills for removing oxygen from water, not from air. Feeding styles varied greatly in this group, from the long-snouted

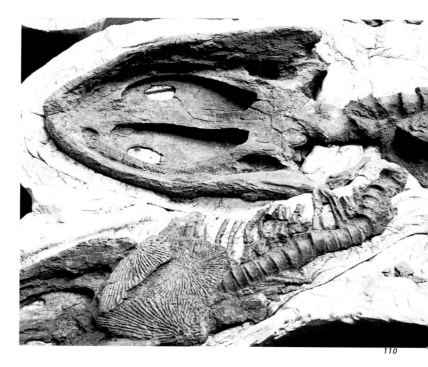

110

trematosaurs that probably had a predilection for fishes to the forms with short and broad snouts that likely preyed on a variety of vertebrates, including the odd tetrapod, and invertebrates.

An invasion of the nearshore marine environment by rhytiodosteids, trematosaurids and plagiosaurids may have occurred in the Early Triassic event, during a marine transgression onto the continental edges which gave rise to a variety of new ecological space. But then the end of the Triassic essentially proved a disaster for these amphibians. In fact, at the end of the Triassic, a major period of extinction affected a broad variety of vertebrates, all but extinguishing labyrinthodonts forever. Only three groups of freshwater ecomorphs survived into the Jurassic, the brachyopids, the capitosaurids and the chigutisaurids. One of these continued into the Early Cretaceous, but its familial identity is uncertain.

105-106 Dorsal and ventral views of the skull of the labyrinthodont amphibian Chomatobatrachus halei from the Early Triassic Knocklofty Formation of Crisp And Gunns Quarry at Mt Knocklofty, West Hobart, Tasmania. x 1.0 (S. Morton, courtesy of the University Of Tasmania and M. Banks)

107 Jaw fragment of the labyrinthodont amphibian Chomatobatrachus halei from the Early Triassic Knocklofty Formation at Midway Point, Hobart, Tasmania. x 2.2 (S. Morton, courtesy of the University Of Tasmania and M. Banks)

108 Jaw fragment of the labyrinthodont amphibian Austrobrachyops from the Early Triassic Lower Fremouw Formation on Coalsack Bluff, Transantarctic Mountains, Antarctica. This specimen was the first Triassic terrestrial vertebrate found in Antarctica. x 1.3 (E.H. Colbert)

109 A humerus (largest bone), ulna and vertebra of the mammal-like reptile Lystrosaurus from the Early Triassic Lower Fremouw Formation of Coalsack Bluff, Transantarctic Mountains, Antarctica. (E.H. Colbert)

110 Two partial skeletons of metoposaur labyrinthodonts from Triassic rocks of Morocco, north Africa. These labyrinthodonts lived at a time when Africa and the Americas were just beginning to split apart. (B. Schaeffer)

AUSTRALIA'S ELUSIVE TRIASSIC REPTILES

Although the first reptiles appear in the Early Carboniferous of Nova Scotia, Canada, their earliest bony remains in Australia are in the Triassic, though footprints may occur in the Permian.

Reptiles represent another step towards true terrestriality over the amphibians, in that they have developed a pattern of reproduction that does not depend on the availability of water at some stage. Reptiles, together with birds and mammals which were to follow, possess an amniote egg that allows the development of the vertebrate embryo inside it, independent of the outside environment. The eggs of many reptiles and all birds are nurtured externally, while those of mammals are generally retained within the body of the mother. These three groups are thought to have come from a common ancestor, and are classed as a distinct group of tetrapods, the amniotes.

Three areas of Australia have produced Triassic reptiles: the Blina Shale in the north-western part of Western Australia; the Knocklofty Formation in Tasmania; and in Queensland, the Arcadia Formation plus two unnamed rock units, one at the tip of Cape York Peninsula.

111 An outcrop of the Triassic Blina Shale in the Erskine Range, Western Australia, which has provided a large collection of reptile and amphibian fossils. (A. Warren)

112 Skeletons of Procolophon sp. from the Early Triassic Lower Fremouw Formation on Kitching Ridge in the Transantarctic Mountains, Antarctica. x 1.4 (E.H. Colbert)

113 Skeleton of the carnivorous archosauriform Chasmatosaurus yuani from the Early Triassic Jiucaiyuan Formation of Kimsar, Xinjiang Province, People's Republic Of China. This form was similar to crocodiles, having a sprawling gait and an elongate skull with an abundance of conical, sharp teeth. The tail was long and flattened, probably an adaptation for swimming. (Courtesy of the Natural History Museum Of Los Angeles County, C. Black and J. Olson; and the Institute For Vertebrate Palaeontology And Palaeoanthropology, Beijing)

114 Side view of the skull and lower jaw of the eosuchian reptile Kadimakara australiensis from the Early Triassic Arcadia Formation of Rewan, Queensland. x 5.8 (B. Cowell, courtesy of the Queensland Museum and R. Molnar)

115 Partial skeleton of the lizard Prolacerta from the Early Triassic Lower Fremouw Formation on Kitching Ridge in the Transantarctic Mountains, Antarctica. x 1.4 (E.H. Colbert)

The Blina Shale, has yielded remains of what might be an ichthyosaur, a marine reptile which was functionally very dolphin-like. Ichthyosaurs have very characteristic vertebrae that look like bi-concave discs and are very similar in shape along the entire length of the column, showing differences only in size. Their limbs were highly modified into flippers, with a great increase in the number of individual bones in each digit. If indeed the Blina Shale bones do belong to an ichthyosaur, it would be one of the oldest occurrences of this group in the world, with other Early Triassic forms known only from Japan, China and Spitzbergen.

The Knocklofty Formation of southern Tasmania has yielded the remains of *Tasmaniosaurus triassicus*, a proterosuchian thecodont. Thecodonts, dominant during the Triassic, were a group of archosaurian reptiles (that is, characterized by having two openings in the posterior part of the skull combined with one just anterior to the orbit, known as the antorbital fenestra). They had a number of skeletal specializations that indicate a more upright posture, evidently associated with more efficient fore-aft motion of the limbs than in more primitive reptilian relatives.

The name thecodont refers to the fact that the teeth of animals in this group are set in sockets, a characteristic which has independently evolved in a few other groups of reptiles. Proterosuchian thecodonts are the most primitive of this group. One of the distinguishing features of the proterosuchians is the downturned front end of the upper jaw, the premaxilla, and in *Tasmaniosaurus* the upper jaw extends well beyond the end of the lower jaw making it quite distinctive.

Tasmaniosaurus was evidently a formidable carnivore, with a long, low skull of nearly 20 centimetres in length. Its jaws were filled with an impressive row of sharp, conical teeth. *Tasmaniosaurus* may have lived much like modern crocodilians, with an aquatic lifestyle and a sprawling stance rather than the more upright stand of the advanced thecodonts. Some proterosuchians, however, may have stood more upright, giving them the ability to breathe at the same time as moving and thus providing an increase in stamina needed for continuous activity. Supporting the interpretation of its rapacious nature, associated with the specimen of *Tasmaniosaurus* were a number of bone splinters and fragments that may represent gut contents. Among them is part of an upper jaw of a small labyrinthodont amphibian, which may have been the last supper of this early Australian reptile.

The Arcadia Formation in south-eastern Queensland has also produced a proterosuchian thecodont, *Kalisuchus rewanensis*, represented mainly by disarticulated fragments of skull containing typical thecodont teeth and ribs with three articulations. Evidently *Kalisuchus* had a broad snout, slender limb bones, long neck and long tail, and like *Tasmaniosaurus* probably resembled a modern crocodile.

Thecodonts are well known and diverse on many other Gondwanan fragments (South Africa, South America, probably India, and Antarctica) and show certain affinities to some groups of dinosaurs to which they might have given rise.

Four additional reptile groups are known from rocks in the Early Triassic Arcadia Formation of south-eastern Queensland: prolacertiforms, procolophonids, lepidosaurs and dicynodonts.

Kadimakara australiensis was a moderate-sized, insectivore, apparently a prolacertiform. Like thecodonts, its recurved (posteriorly curved) teeth were set in sockets. Prolacertiform relationships to other reptiles, whether to the thecodonts or to modern lizards, are not entirely clear, but the group was also quite successful in Permo–Triassic faunas in South Africa and Antarctica as well as in the Northern Hemisphere.

Procolophonid reptiles, also present in the Arcadia fauna, were somewhat lizard-like in appearance and movement. They were rather short and stout, with short tails, a sprawling stance and a skeleton that seems quite primitive within reptiles. They were probably plant eaters, as indicated by their transversely expanded teeth. Their closest relatives may have been the mesosaurs, freshwater reptiles of the Permian that are found only in Brazil and South Africa, and were used as early supporting evidence for continental drift. The procolophonids apparently are an early group of amniotes that developed independently of those types that led to more advanced

112

115

113

114

reptiles, such as turtles and lizards. The Australian procolophonids are quite similar to forms found in South Africa (where the group is reasonably diverse in Permo–Triassic times), South America and Antarctica.

Another of the Arcadia reptiles, *Kudnu mackinlayi*, is a lepidosaur very similar to members of the Paliguanidae from the Late Permian and Early Triassic of South Africa. Paliguanids are the earliest known lizards, a group that specialized in small body size (with a body length of about 150 millimetres or less). They did not have the primitive bony bar along the base of the ventral fenestra of the skull but had developed a joint between the top of the quadrate and the squamosal. With this joint the quadrate was able to move back and forth, a condition called streptostyly, allowing the application of greater forces in the bite.

This jaw specialization was combined with a middle ear structure that was more sensitive to airborne sounds. The functioning of such early lizard ears was in stark contrast to the set up in more primitive reptiles where the massive stapes, the bone connecting the tympanic membrane to the middle ear, had high inertia and would have been more adapted to picking up tissue conducted sounds from the ground. In lizards the stapes is gracile, and the ratio of the tympanic membrane area to the area of the bony footplate connected via the oval window in the skull to the middle ear is 20:1 thus allowing magnification of sounds to the middle ear. Within the tissue of the tympanic membrane is embedded a cartilagenous extracolumella which in turn is attached to the long, thin stapes (also called the columella), with the whole affair acting like a piston to transmit airborne sound vibrations to the middle ear. The middle ear then, magnifies the sound, and thus the ear becomes a much more sensitive instrument for detecting incoming information from the environment than it was in more primitive reptiles. The combination of hearing acuity, more efficient locomotion and a more forceful bite was responsible for the success of the late Palaeozoic–early Mesozoic lizards over the superficially lizard-like groups such as the procolophonids.

Perhaps the most intriguing reptile in the Arcadia fauna is a dicynodont mammal-like reptile, similar to *Kannemeyeria* of Africa. It was first recognized on the basis of a single left quadrate. Dicynodonts of similar morphology are known from several areas of southern and eastern Africa, Argentina, India and China. The dicynodonts were extremely successful on a worldwide scale in

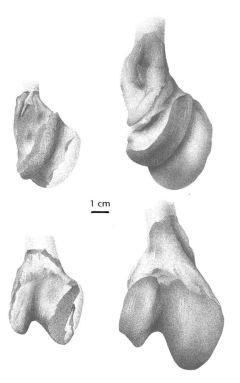

1 cm

The left quadrate of a mammal-like dicynodont reptile from the Early Triassic Arcadia Formation, Rewan, Queensland (two figures on left) compared with an African dicynodont Kannemeyeria (on right). To date, this is about all that is known of mammal-like reptiles in Australia, but future work will likely enrich this record. (From Thulborn, 1983)

Permo–Triassic times, forming up to 80–85 per cent of the total fauna. They tended to be massive reptiles with sprawling stance and short tails. Their heads were large and had a terminal beak, sometimes simple cheek teeth and often tusks made of bone.

Although the dicynodonts themselves are not the group which is most closely related to mammals, this group of mammal-like reptiles often occurs in concert with the cynodonts from which mammals were derived. It is thus quite possible, now that the dicynodonts are recorded on this continent, that their companions are next on the agenda to be found. Only time and further field searching will tell. So far, there is only the barest glimpse of the history of mammal-like reptiles on the Australian continent. Whether their rarity is an artefact of collecting and environmental sampling or a reality of their original abundance is not yet at all clear.

116 *The archosaur Lotosaurus adentus from the Triassic Badong Formation of Sangzhi, Hunan Province, People's Republic Of China. This animal is part of the Pangaean fauna, relatives of which might be expected to be found in Australia when Triassic faunas become better known. The Chinese form had a horny bill and few, if any, teeth. It was about the size of a sheep. (Courtesy of the Natural History Museum of Los Angeles County, C. Black and J. Olson; and the Institute for Paleontology and Paleoanthropology, Beijing)*

A GLOBAL PERSPECTIVE OF TRIASSIC TETRAPODS

Tetrapod faunas are known from a number of different places, especially in the Early Triassic. Furthermore, they were strikingly similar to each other throughout the Gondwanan landmass, and indeed right across Pangaea. There was little to stop tetrapod migration, other than climate. However, while climate had considerable influence in some parts of Pangaea, it seems to have had little effect within the Gondwanan sector.

One example of a form able to traverse long distances at this time is *Lystrosaurus*, a dicynodont mammal-like reptile that historically was used to support the concept of Gondwana. *Lystrosaurus* occurs in such widely separated places as southern Africa, Antarctica, India and China, and has a near relative in Australia, the dicynodont from the Arcadia Formation.

There were some regional differences, such as the reptile-dominated plains fauna, called the Lystrosaurid Empire, that characterized much of Gondwana at the beginning of the Triassic, and to the east the amphibian-dominated Trematosaurid–Rhytidosteid Empire of Australia. One palaeontologist, Anthony Thulborn of the University Of Queensland, has suggested that this difference is due to the peninsular effect, a "rule of thumb" found to be true today and meaning that a fauna is progessively impoverished towards the end of a peninsula. Peninsulas act a bit like islands: the longer the peninsula and therefore the more distant it becomes from other land, the fewer organisms will be able to migrate to it. Perhaps, too, the unusual success of labyrinthodonts in Australia during the Triassic may have been due in part to the evolutionary radiation of forms that were relicts from the Palaeozoic and had essentially become restricted or extinct elsewhere. Protected somewhat by the remoteness of the Australian Peninsula, they prospered in that area. Even this early in time, there was some isolating mechanism at work that set Australia apart from the rest of the world, albeit not as severely as is the case today.

Notable floral and faunal changes occurred globally throughout the Triassic. In the latest Permian and Early Triassic one group of vertebrates that had been successful as medium and large-sized carnivores, the gorgonopsians, not yet known from Australia, was replaced by the thecodonts, which later gave rise to the dinosaurs. At this time, too, there was a gradual change from the *Glossopteris*-dominated plant world to one more dominated by yet another seed-fern, *Dicroidium*. The *Dicroidium* Flora included a variety of other plants such as ferns, ginkgophytes (related to the single living species, the *Ginkgo*), the palm-like cycads, conifers and horsetails as well as other seed-ferns. It was a vegetation that showed arid adaptations at times during the Triassic, for example, thickened cuticle and reduced surface area of the leaves with some even becoming needles or spikes.

The replacement of the gorgonopsians by the thecodonts apparently represents an opportunistic event, for the former were extinct before the latter expanded. Thecodonts similarly replaced cynodont mammal-like reptiles as carnivores in the Late Triassic, even though both had survived side-by-side for millions of years. Rhynchosaurs, diademodontoids and aetosaurs replaced the dicynodonts in the Middle to early Late Triassic. But the prominence of this Rhynchosaur–Diademodontoid Empire was to be short-lived because of the emergence of the dinosaurs in the Late Triassic.

The success of the dinosaurs was quick and striking, and seems to have been associated with equally striking climatic and floral changes. The *Dicroidium* Flora gave way to a conifer-dominated plant assemblage, which was to flourish for much of the history of dinosaurs. Only near the end of the dinosaurs' long and successful reign did the flowering plants begin their spectacular rise to prominence, culminating in their dominance in the Cainozoic.

Dinosaurs became the terrestrial success story of the Jurassic and Cretaceous. Often it has been suggested that they competed outright and won the natural selection battle with other groups of previously successful reptiles. However, rather than outcompeting the losers (that is, the thecodontians, rhynchosaurs and their allies), the dinosaurs probably simply moved into empty niches.

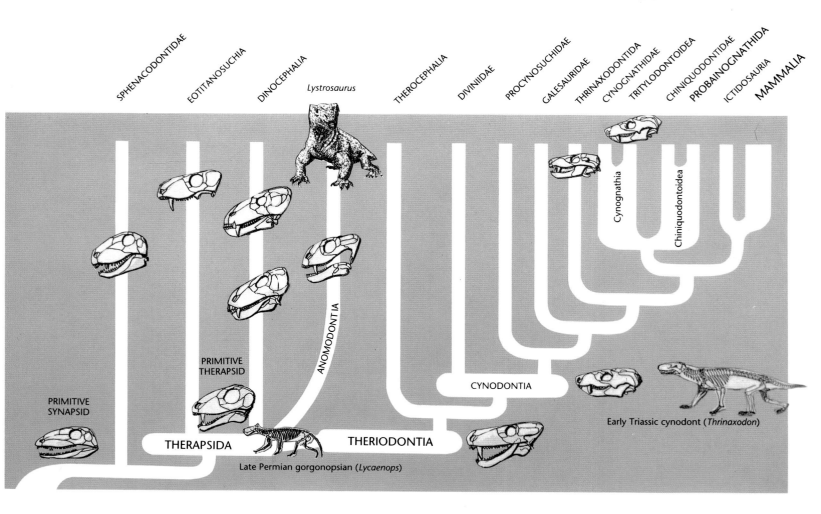

MAJOR EXTINCTIONS AT THE CLOSE OF THE TRIASSIC

Triassic times were warm. Temperatures were higher than those of today. Apparently there were no polar caps, and climates were probably more uniform than they are now. In Australia information on climate during the Triassic is scarce, but the presence of Middle and Late Triassic coals in South Australia (Leigh Creek), Tasmania, the coastal basins of New South Wales and southern Queensland suggests increased precipitation from previous times. In contrast, Early Triassic gypsum deposits in some of these locales indicate more arid conditions. However, Australia lacks the massive Triassic evaporites that signal the intensely arid conditions which characterized many other parts of the world at this time as they were in the grip of a monsoonal regime.

Triassic times were bounded at the beginning and the end by major extinctions affecting the inhabitants of land and sea, swamp and stream. The fauna of the Triassic, thus, is a mixture of sturdy survivors such as the labyrinthodonts and mammal-like reptiles, the ammonites and seed-ferns, and newcomers like the hexacorals and echinoids, the thecodonts, dinosaurs and mammals.

The end of Triassic times brought one of the greatest mass extinctions of all times. About 20 per cent of all marine families died out, including conodonts, conularids and the mollusc-crushing placodonts. The main losers in the terrestrial realm were the thecodonts, the rhynchosaurs and almost all the labyrinthodonts.

There appear to be two pulses to this Late Triassic extinction, one in the last few million years of the Norian Epoch (about 213 million years ago) and another earlier event, maybe two, in the Carnian (about 225 million years ago). They came at a time of floral transition from the seed-fern dominated *Dicroidium* Flora, which had inhabited the lowland habitats of Gondwana for most of the Triassic, to a conifer and other gymnosperm flora. Perhaps this transition reflected a climatic change, on a world scale, to more arid conditions. Or perhaps it was related to some other factor such as temperature or the effects of a meteorite or comet impact on Earth. The cause is yet to be pinpointed, but the effect is reasonably well documented. The Jurassic and younger faunas were vastly different from those which came before: the Eden of the mammal-like reptiles and amphibians was lost, and the Eden of the dinosaurs was just beginning.

117 Triassic terrestrial faunas of Antarctica, like those of many other parts of Gondwana, were dominated by reptiles, mainly by the mammal-like therapsids and the thecodonts. In the foreground a predaceous therapsid, Thrinaxodon, stands atop the anapsid reptile Procolophon that it has just dispatched. The skeleton and skull of Thrinaxodon was very mammal-like and it may have had a covering of hair. The tusked, herbivorous mammal-like reptiles, Lystrosaurus, stand at the water's edge, while in the background the gavial-like thecodont, Chasmatosaurus, bides its time in the water. (P. Trusler, with the permission of Doubleday Inc., New York)

118 Dorsal view of the skull of the labyrinthodont amphibian Blinasaurus townrowi from the Early Triassic Knocklofty Formation, Lime Bay, Tasman Peninsula, Tasmania. x 1.9 (S. Morton, courtesy of the University of Tasmania and M. Banks)

119 Palatal region of the labyrinthodont amphibian Blinasaurus henwoodi from the Early Triassic Blina Shale, Blina, Western Australia. x .85 (J. Frazier, courtesy of the Western Australian Museum and K. McNamara)

120 Ventral view of the skull of the labyrinthodont amphibian Derwentia warreni from the Early Triassic Knocklofty Formation, Old Beach, Hobart, Tasmania. x 1.8 (S. Morton, courtesy of the University of Tasmania and M. Banks)

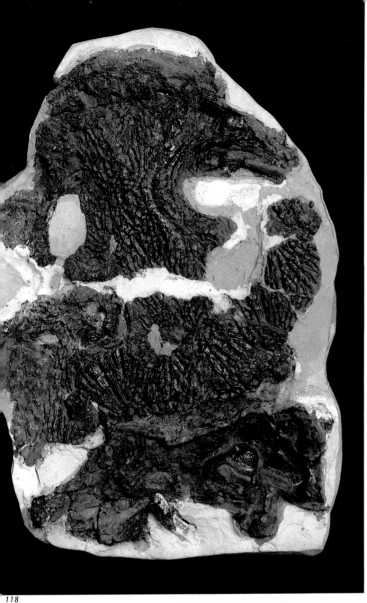

118

119

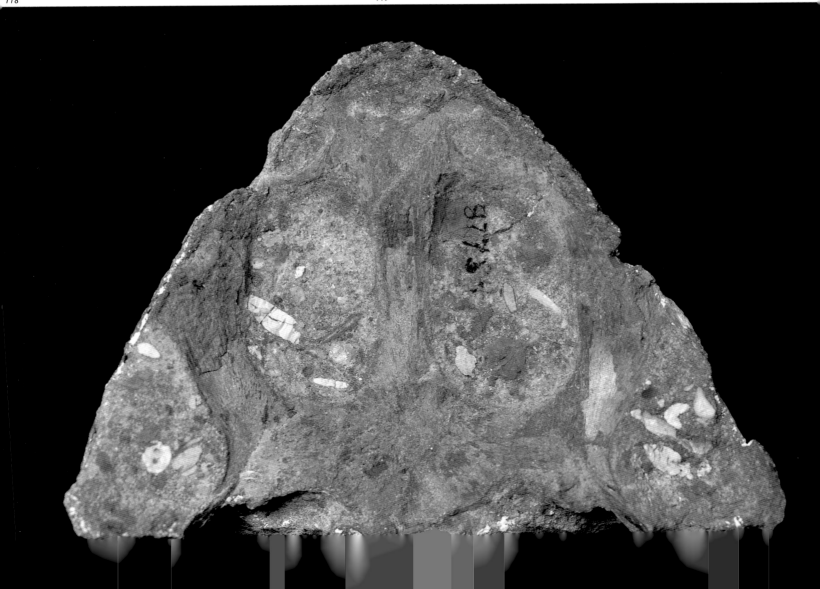

CHAPTER 4

DINOSAURS AND POLAR NIGHTS

THE EARLY JURASSIC TO LATE CRETACEOUS PERIODS,
FROM 205 TO 65 MILLION YEARS AGO

*Dinosaurs were the dominant terrestrial vertebrates from the Late
Triassic to the Late Cretaceous. On no continent is there a continuous
record of this time interval, and, except for Antarctica, Australia has the
most meagre record of all. At the end of the Early Cretaceous, a time
not well represented elsewhere by dinosaurs, Australia exhibits its best
record of them. At that time a number of large dinosaurs are known
from Queensland, while smaller ones of this age are known from
Victoria and the opal fields of South Australia and New South Wales.
Still, it is a paltry record, consisting of less than a dozen partial
skeletons and a few hundred isolated bones. Earlier, in the Late Triassic
and the Jurassic, most of the information on fossil vertebrates consists
of fossil footprints, with a few skeletal remains.*

*Outside of Australia the question of the mechanism for the extinction
of the dinosaurs at the end of the Cretaceous is a dynamic area of
modern research, with much evidence mustered during the 1980s for
and against the suggestion that it may have been owing to the impact of
an asteroid or a comet. Based on the Australian record, there would be
no such debate because the Mesozoic record of vertebrates essentially
ends at the very beginning of the Late Cretaceous. This lack of a record
is not because dinosaurs and other vertebrates became extinct in
Australia at that time but merely because terrestrial sediments for most
of the Late Cretaceous and early Cainozoic of Australia are all but
unknown.*

*Given Australia's poor overall record during the time that dinosaurs
lived and the pronounced bias towards the Early Cretaceous, reconstructions for the Mesozoic are painted with the broadest of brush strokes.
Nevertheless, the impression that emerges is one of a continent on
which fauna strongly reflected two aspects of its geographic setting.
Firstly, as the Mesozoic progressed from the Triassic to the Cretaceous,
Australia was becoming more isolated, initially with the separation of
northern Laurasia from the more southerly Gondwana and subsequently
with the break-up of Gondwana itself. The second aspect mirrored in
the biota is the high latitudinal position of Australia, with the South Pole
either on the continent or close to it.*

121 Aurora would have "lighted" the long winter nights of
Victoria in the Early Cretaceous, when this area lay much
nearer to the South Pole and would have experienced at least
3 months of total darkness annually. (L. Snyder, Geophysical
Institute, University of Alaska)

AUSTRALIA'S OLDEST DINOSAURS

In 1891, Harry Gover Seeley described the first Australian dinosaur, *Agrosaurus macgillivrayi*. No other skeletal remains of dinosaurs as old as it, Late Triassic–Early Jurassic, have been found on the continent. That age estimate was based on the fact that *A. macgillivrayi* is a prosauropod. The species name was in honour of John Macgillivray, a noted collector of natural history specimens who was on board HMS *Fly* when that ship was sent to the Raine Island off Cape York in 1844. By 1890, Macgillivray was long dead, and his notebooks relating to the voyage had been lost. For locality information about this fossil, Seeley had to rely solely on a label that read, "*Fly,* 1844. Jn. Macgillivray, from the N.E. coast of Australia."

Between 1993 and 1995, an attempt was made to relocate the site where the specimen was collected. Just opposite Raine Island on the mainland of Australia are outcrops of the Helby Beds, which are Early Jurassic in age. Though the rocks are the right age and the right type, firsthand examination of them did not uncover any additional fossil bone like that of *A. macgillivrayi*. Nor did examination of the outcrops of the Helby Beds reveal the presence of calcium carbonate, the cement that held together the rock that surrounded the specimen.

It now seems likely that the holotype of *A. macgillivrayi* actually came from Durdham Down, a well-known fossil site near Bristol, England. The trace element chemistry of the holotype is very similar to the bone of a dinosaur from there, *Thecodontosaurus antiquus*, which is probably the correct name for the holotype of *A. macgillivrayi*. In addition, the remains of the teeth of an animal like the living tuatara of New Zealand have been found in rock in which the holotype was enclosed and at Durdham Down as well—and almost nowhere else (Vickers-Rich et al. 1999).

Despite these findings, there is good evidence that dinosaurs did live in Australia during the Late Triassic and Early Jurassic. This evidence takes the form of footprints found in the roofs of coal mines in south-eastern Queensland (Thulborn 1990).

JURASSIC FOOTPRINTS, BUT PRECIOUS FEW SKELETONS

Through the Jurassic, footprints continue to be the principal record of dinosaurs. All are located in eastern Queensland, many of them imprinted in the roofs of mines in the Walloon Coal Measures. One important exception to this generality concerning the Jurassic record is the partial skeleton of *Rhoetosaurus brownei*, a sauropod dinosaur known from a partial hind limb, pelvis and a few vertebrae found on Durham Downs Station near Roma in Queensland. Although far from complete, more is known of this dinosaur than of any of the younger Australian sauropods.

Rhoetosaurus brownei, being Middle Jurassic in age, is one of the oldest sauropods known anywhere in the world. As such, the fact that it is not highly specialized in any way comes as no surprise. Like many dinosaur skeletons the world over, the remains of this individual were discovered alone with no other fossils nearby. Thus, the community to which *Rhoetosaurus* belonged can be only indirectly inferred.

Only a few other Jurassic tetrapods are known from Australia. Like *Rhoetosaurus brownei*, all are from rocks deposited during the early part of the period, about 200 million years ago.

Labyrinthodonts flourished from the Devonian to the end of the Triassic. Until the 1980s it was generally believed that they had become extinct by the opening of the Jurassic. Evidence found more recently, however, suggests that they persisted much longer in western China, Mongolia, south-eastern USSR and eastern Australia. One of the most complete of these labyrinthodonts that "lived beyond their time" (relicts) is *Siderops kehli* from the Early Jurassic Evergreen Formation of south-eastern Queensland. Undoubtedly, it is a brachyopoid, a group well represented in the Early Triassic of Australia. *Siderops* differs from other brachyopoids and other labyrinthodonts, however, in having a smaller body relative to the size of its head and in having an extremely wide skull. This crocodilian-like carnivore may even have been cannibalistic, as a labyrinthodont vertebra of a smaller individual *Siderops* was found in the mouth of the type specimen of *S. kehli*. Also found in its mouth were the remains of a millipede, the only fossil specimens known of this creature in the time span from Carboniferous to Oligocene.

In the same rock unit as *Siderops* other aquatic tetrapods were found, namely the reptilian plesiosaurs, a group better known in the Early Cretaceous in Australia. Plesiosaurs are divided into two groups, the plesiosauroids and the pliosauroids. Plesiosauroids have been characterized as "a snake drawn through the body of a turtle", which does capture the image of them with their small head atop a long neck extending out from their body (without a shell, however!) and four paddle-like limbs. With a slight modification to this image, pliosauroids can also be readily envisioned: simply shorten the neck and enlarge the head.

A freshwater pliosaur is known from the Early Jurassic Razorback Beds of Queensland and is preserved in a most unusual fashion. No bone remains in the hard sandstone where it is preserved. What is "fossilized" is a set of holes left behind when the bone was dissolved, probably by acidic water sometime after the animal was fossilized. When casts of these holes were made, they accurately reflected the morphology of the original skeleton and were as useful for scientific analysis as the bones themselves would have been.

122 Skull fragment of Crocodilus (?Botosaurus) selaslophensis from the Early Cretaceous opal field at Lightning Ridge, New South Wales. x 1.5 (F. Coffa, courtesy of the Australian Museum)

123 Skull and partial skeleton of Siderops kehli from the Early Jurassic of Queensland. When described, this species was the youngest labyrinthodont known. Although younger ones are now known from China, Mongolia, the USSR and Australia, none are as complete as this specimen. x 0.5 (F. Coffa, courtesy of the Museum Of Tropical Queensland)

124 The left hind foot of a sauropod dinosaur, Rhoetosaurus brownei, from Middle Jurassic terrestrial sediments on Durham Downs Station in south-eastern Queensland. This herbivorous quadruped probably reached about 12 metres in total length and utilized its long neck to feed from tall vegetation, much like the giraffe does today. Rhoetosaurus is of special interest as it is one of the oldest sauropods known in the world, but unfortunately only parts of the hind limb, tail, thoracic vertebrae and a fragment of a neck vertebra are presently known. x 0.2 (F. Coffa courtesy of the Museum Of Tropical Queensland)

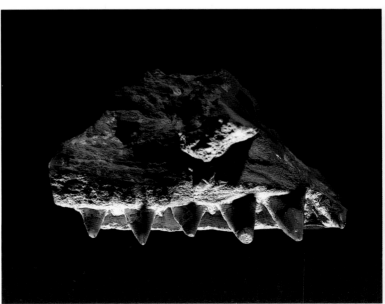

122

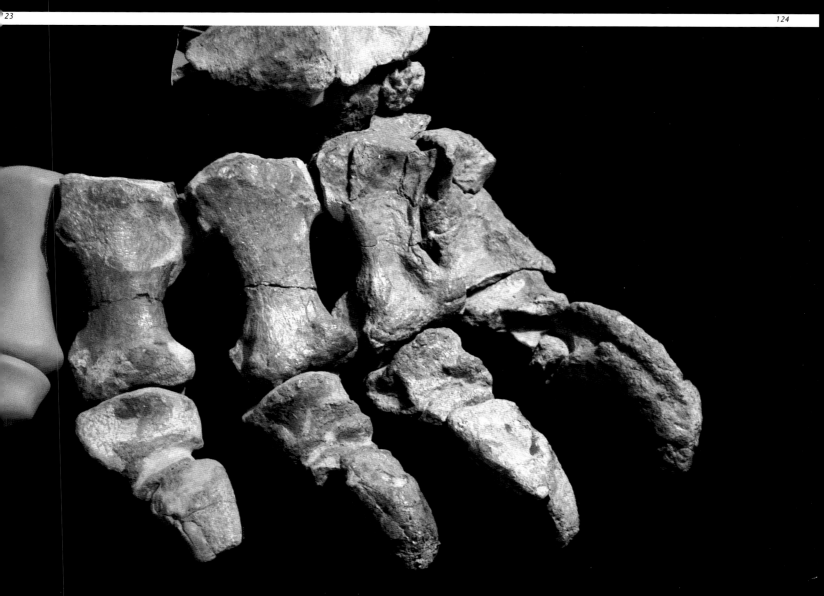

LATE JURASSIC AND EARLY CRETACEOUS VERTEBRATES

THE TALBRAGAR AND KOONWARRA LAKES

During the Jurassic when *Rhoetosaurus* and *Siderops* were alive, eastern Australia was close to or within the Antarctic Circle of the day. Yet there is no indication of any substantial ice build-up. The Kauri Pine and other conifers plus cycads and ferns known from the Talbragar Fish Beds in New South Wales are comparable to those found in modern warm temperate rainforests on the Atherton Tableland of northern Queensland. The freshwater vertebrate fauna from the Talbragar site provides the only glimpse of fishes yet reported from the Jurassic of Australia. All are ray-finned fishes, actinopterygians, and the bulk are teleosts, an advanced group which had appeared in the Triassic and are today the most numerous fishes in the world's oceans and fresh waters. The genus *Leptolepis* is by far the most common of these teleosts in the Talbragar Fish Beds.

The two other, more primitive major divisions of the ray-finned fishes are represented at Talbragar by very few species. One of these species, *Coccolepis australis*, is the only palaeoniscoid, and there were three holosteans, a macrosemionotid (*Uabryichthys*) and two forms in an Australian endemic family Archaeomaenidae (*Archaeomene* and *Madariscus*). In general, the scales of these more primitive ray-fins are thicker and their skeletons are less completely formed of bone than in the case of the teleosts.

Particular layers of the fine-grained, ochre-coloured shale at Talbragar are rich in fossil specimens, and both the good preservation and the concentration of skeletons strongly suggest that catastrophic conditions killed off many fishes at once rather than the fossils representing random deaths in a population over a prolonged period of time.

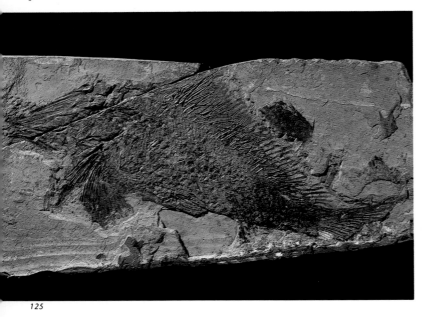

125

Another ancient lake deposit, at Koonwarra in Victoria, has yielded a similar assemblage of plants and animals to that from Talbragar, though it is at least 30 million years younger, having been laid down at the end of the Early Cretaceous. Two of the five genera of fish found at Koonwarra also occur at Talbragar. Both of these genera are widespread during the Jurassic and Cretaceous, *Leptolepis* being known from Spitzbergen and all continents except Antarctica, and *Coccolepis* occuring in Europe and Asia as well as Australia. *Coccolepis* was a late surviving palaeoniscoid. Another late survivor found at Koonwarra was the archaeomaenid holostean, *Wadeichthys*. Lungfish material, although rare, has also been recovered from Koonwarra as well as from isolated locations in the somewhat younger Otway Group rocks to the west, and, where preserved, the dentition consists of tooth plates very like those of modern lungfish, not the isolated denticles that some older members of this group possessed.

But while the flora and the fauna of Koonwarra overlap significantly with those of Talbragar, an even greater similarity exists between the floras of Koonwarra and the Early Cretaceous Rajmahal Series in India. It was the latest time that large scale exchange of terrestrial fauna and flora was possible between Australia and India, as the latter then separated from Antarctica and rapidly moved northwards.

Shortly after the Koonwarra community lived and died, the angiosperms (or flowering plants) began to evolve rapidly and by Cainozoic times came to dominate the terrestrial floras everywhere. Heralding this event is a tiny unnamed flower from Koonwarra, the most ancient yet found anywhere on Earth.

Much effort has been made to understand the mode of accumulation of the fossils at Koonwarra, an endeavour called taphonomy. It is noteworthy that the fossil fishes are of small size, and apparently died owing to some catastrophic event during their first year of life. Amongst freshwater fishes today, it is the young ones which most commonly are found in shallow water, and for this reason it is thought that the Koonwarra site was a shallow part of a lake. The sediments in which the fossils are found occur in alternately light and dark layers. The light ones were evidently laid down during the summer and the dark ones in the winter when the supply of incoming silt and clay was reduced and black organic debris dominated the sediment then being deposited. It is in the darker bands that the fossil fishes are concentrated.

Koonwarra, like the rest of south-eastern Australia at the time, lay within the Antarctic Circle. Although the whole planet was at the warmest it would be during the entire Phanerozoic, it was still quite likely that ice formed in the Koonwarra lake because of Australia's near polar position. A sheet of ice over a shallow arm of the lake each winter could have resulted in a "winter kill", which some palaeontologists have suggested occurred at Koonwarra. The dissolved oxygen in the water would have been exhausted by the creatures respiring beneath the ice, resulting in their deaths if there was a bar or other barrier that prevented interchange with oxygenated waters in the deeper parts of the lake where the adult fishes lived.

Apart from fishes the only vertebrate remains from Koonwarra are a few tiny feathers. They are sufficient only to document the presence of birds in southern Australia towards the end of the Early Cretaceous. It is not possible to say even what order or orders of birds are represented due to the obscured fine detail on these avian remains. The occurrence of these feathers, though, means that the goal of finding more bird fossils is not merely a vain hope but something that might be achieved if enough effort is made. And should a skeleton of a bird ever be found in the fine-grained sediments at Koonwarra, there is an excellent chance that it would be exquisitely preserved, just as are the fish skeletons.

Invertebrates, too, are well represented at the Koonwarra site. Most are insect larvae, but remains of spiders, bryozoans and an ostracod, amongst others, are present. Close living relatives of the two most abundant invertebrates at Koonwarra, siphlonurid mayflies and cantharid beetles, are found today only in the cool montane waters of Australia's south-east.

125 *Uarbryichthys*, a macrosemiiform neopterygian fish from the Jurassic Talbragar Fish Beds near Gulgong, New South Wales. x 0.3 (F. Coffa, courtesy of the Australian Museum)

126 *Aphnelepis australis*, a teleost fish from the Jurassic Talbragar Fish Beds, New South Wales. x 1.3 (F. Coffa, courtesy of the Australian Museum)

127 *Coccolepis australis*, a palaeonisciform fish from the Jurassic Talbragar Fish Beds, New South Wales. x 1.0 (F. Coffa, courtesy of the Australian Museum)

128 *Archaeomene tenuis*, a teleost fish from the Jurassic Talbragar Fish Beds, New South Wales. x 1.6 (F. Coffa, courtesy of the Australian Museum)

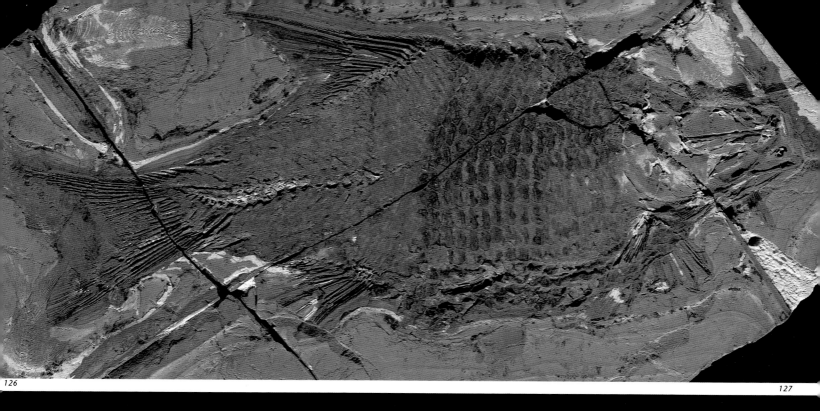

126

127

128

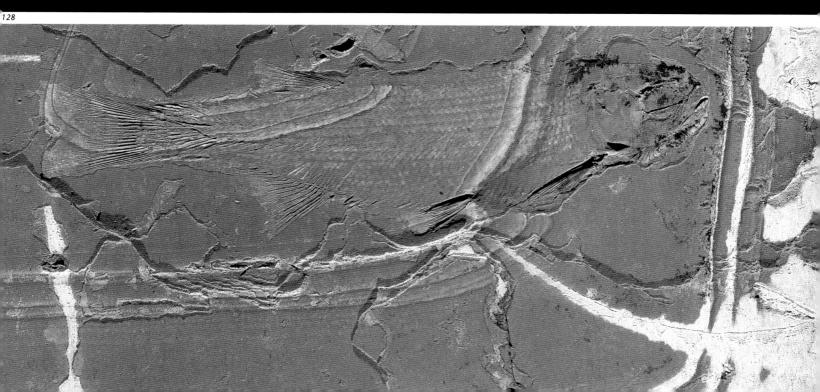

129

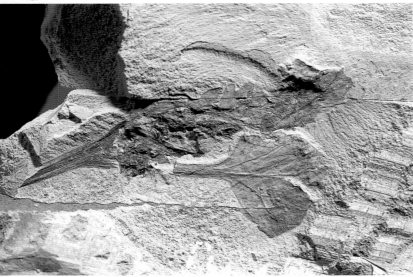

130

131

132

129 *Coccolepis woodwardi,* a palaeonisciform fish
from the Early Cretaceous Koonwarra Fish Beds,
southern Victoria, in the rift valley with Antarctica.
x 1.5 (F. Coffa, courtesy of the Museum Of Victoria)

130 The oldest flower known. An unnamed species
from the Early Cretaceous Koonwarra Fish Beds,
southern Victoria, in the rift valley with Antarctica.
x 3.1 (Courtesy of L. Hickey, Yale University, and the
Museum Of Victoria)

131 A bird feather from the Early Cretaceous
Koonwarra Fish Beds, southern Victoria, deposited in a
lake that accumulated in the rift valley with Antarctica,
x 6.0 (F. Coffa, courtesy of the Museum Of Victoria)

132 *Duncanovelia extensa,* an insect from
Early Cretaceous lacustrine sediments of Koonwarra,
Victoria. The invertebrates from this site have
closest affinities with those inhabiting cool montane
lakes and streams in Tasmania today. x 8.4 (F. Coffa,
courtesy of the Museum Of Victoria)

133 *Wadeichthys oxyops,* a teleost fish from the
Early Cretaceous Koonwarra Fish Beds, southern
Victoria, in the rift valley between Australia and
Antarctica. x 1.8 (Frank Coffa)

134 *Leptolepis koonwarri,* a teleost fish from the
Early Cretaceous Koonwarra Fish Beds in southern
Victoria. x 2.5 (F. Coffa, courtesy of the Museum Of
Victoria)

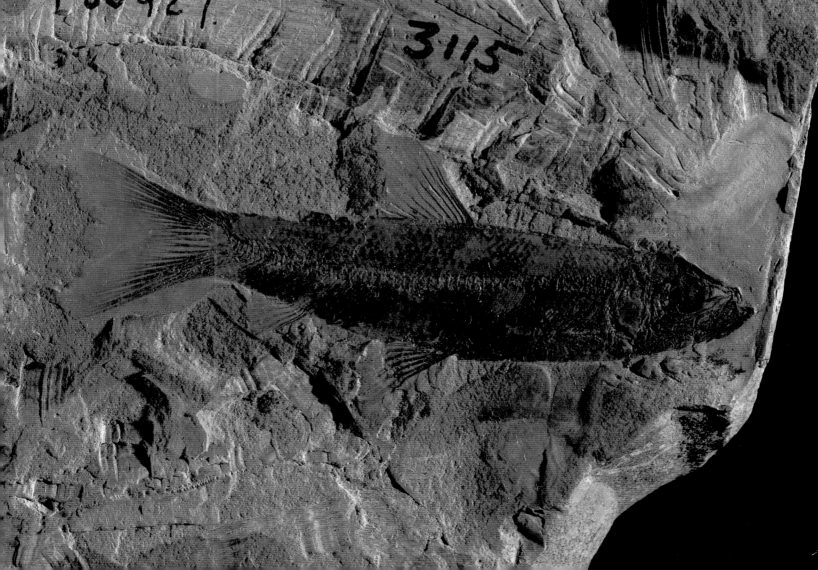

FLOODPLAIN FAUNAS OF THE GREAT SOUTHERN RIFT VALLEY

Lake deposits such as that at Koonwarra are rare in the Early Cretaceous rocks of south-eastern Australia. Much more common are coarser sandstones and siltstones, which represent sediments laid down on the floors of rivers and streams or as overbank deposits resulting when flood waters breached natural levees and spread across the surrounding plains. Although the animals from these channel and flood deposits lived and died close both in time and space to those preserved in the ancient lake at Koonwarra, the fossils from the two environments are quite different. Reptiles, unknown at Koonwarra, are the dominant element in the stream channel and floodplain deposits. On the basis of the animal fossils alone, there would be little basis to surmise that the ages were as close as is in fact thought to be the case. However, several of the same species of spore and pollen fossils were found in facies that represented the two different environments.

Near the former Koonwarra lake there are now heavily overgrown exposures of stream and river channel deposits, however, no fossils have yet been found in them. One must go to the shores of Bass Strait, 25 kilometres away near Inverloch, in order to find bones in that depositional environment. There, extensive, excellent exposures of the channel deposits occur, revealed by the constant pounding of

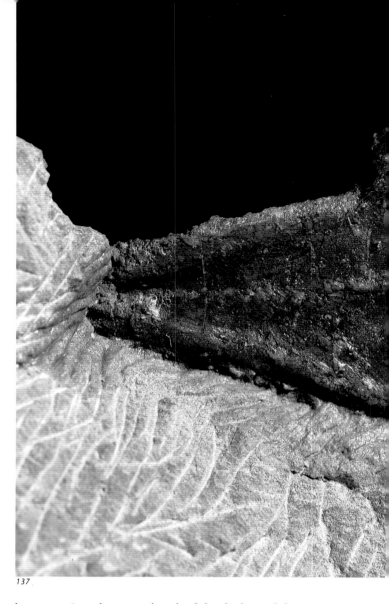

137

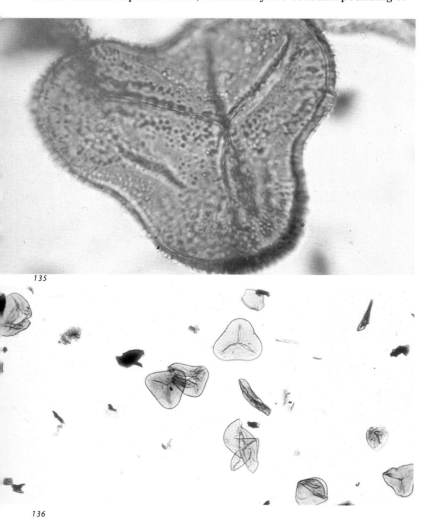

135

136

135 *Pilosisporites notensis, the spore (about 6 microns in diameter) of an extinct fern from the Early Cretaceous floodplain sediments of the Strzelecki Ranges in southern Victoria. Pollen and spores, along with fission track methods, have been important in dating the rocks of this area. (B. Wagstaff)*

136 *A variety of spores from the Early Cretaceous floodplain sediments of the Strzelecki Ranges, Victoria. x 2000 (B. Wagstaff and J. McEwen Mason)*

the sea against the coastal rocks. Inland, channel deposits are not only inaccessible but are often leached by humic acids released in soil formation so that any bones formerly preserved in them have been dissolved. By contrast, on the coast the erosion by the sea outraces such processes, and the bones have a better chance of reaching the surface where the palaeontologist can find them.

The sandstones of these Early Cretaceous channel deposits are so hard that they necessitate rock drills and rock saws to collect the fossils. At one locality, Dinosaur Cove on the flanks of the Otway Ranges, 200 kilometres south-west of Melbourne, fossil stream channels have been found that are rich enough to warrant being quarried. Because the fossil-rich stream channels course into the mountainside at this locality, the excavation of the channel has required tunnelling underground in much the same manner as is employed in small gold mines.

Dominating the fossils found in the stream channel facies at Dinosaur Cove and other locales along the Otway coast, as well as in the Strzelecki Ranges to the east, are hypsilophodontid dinosaurs. These animals would have resembled kangaroos except that they lacked external ears and hair, and the form of the hind feet was different. Like kangaroos, their hind limbs were much larger than their fore limbs, they had a long tail, and they were herbivorous. But of course, their resemblance to kangaroos was only superficial, convergent, for the hypsilophodontids were reptiles not mammals.

Generally, elsewhere in the world, hypsilophodontids were only a minor component of the dinosaur faunas in which they are known, even where collections are thousands of times larger than the Otway and Strzelecki sample. South-eastern Australia in the Early Cretaceous provides an exception to this rule, however. Even though only a few thousand bones are known, from late Early Cretaceous rocks of Victoria, it is clear that at least three and possibly five different species of this family of dinosaurs lived there at this time. Only a few other dinosaur bones, mainly the carnivorous theropods, have been found in these same rock sequences.

AN UNUSUAL ENVIRONMENT FOR DINOSAURS
Although the total quantity of dinosaur material from south-eastern Australia fits comfortably into two museum cabinets, the quality of

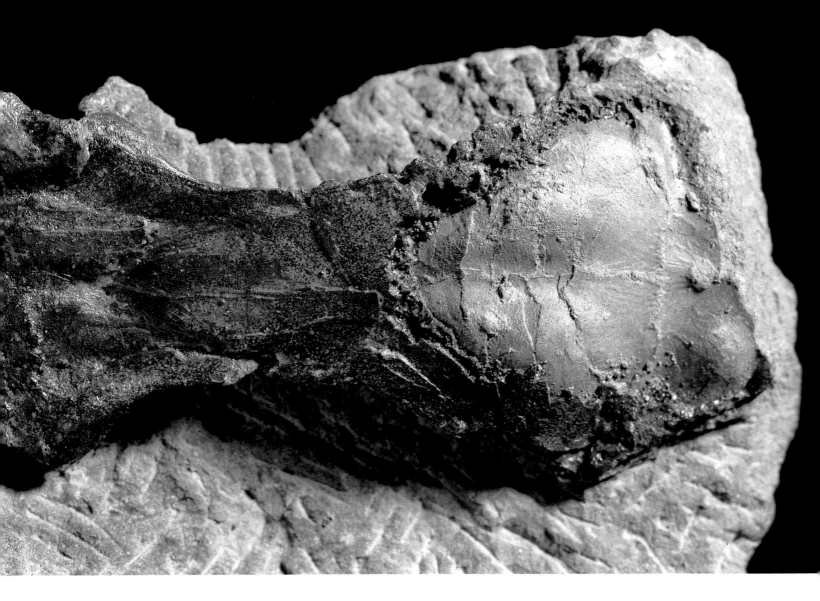

the material is excellent and intriguing aspects of these animals have been brought to light. One of the hypsilophodontids, *Leaellynasaura amicagraphica,* is represented by a skull which preserves a natural cast of the top surface of the brain. Examination of this impression has revealed that the optic lobes were unusually prominent. Undoubtedly, other hypsilophodontids also had optic lobes. However, on the three other fossil specimens in which this area of the brain is known, all non-Australian, the optic lobes were not large enough to have made even a slight impression on the overlying layer of bone. The optic lobes, as their name suggests, are involved with processing nerve impulses from the eye.

What factor or factors might explain the difference in prominence of the optic lobes of *Leaellynasaura amicagraphica* when compared with the other hypsilophodonts from elsewhere in the world? All of the other known hypsilophodonts lived at lower palaeolatitudes, well outside the Arctic and Antarctic circles of the time. In summer, light conditions in the Early Cretaceous of south-eastern Australia would not have been significantly different than at lower latitudes except that day length would have been longer. In the winter, however, the situation would have been much different, with no significant light at any time of day. Because of its enlarged optic lobes, it is tempting to think that the tiny Australian *Leaellynasaura* was capable of remaining active during such dark times. The winter would have been the coldest time of year. Just how cold cannot be measured by the vertebrates, though it has been estimated by studies on the oxygen isotopes preserved in the rocks themselves. The ratio of the isotopes of oxygen (^{18}O to ^{16}O) can be used as a direct measure of ancient temperature, and a study of these isotope ratios preserved in the relatively unaltered carbonate concretions which occur in the dinosaur-bearing rocks of the Otway Ranges reveals some interesting data about environmental conditions 110 million years ago.

Geochemists have estimated the mean annual temperature of the area where dinosaurs were found in south-eastern Australia as somewhere between +5˚C and –6˚C. As the winter would not have been warmer than the mean, for an animal to be active at that time implies functioning at temperatures at or below the freezing point of water. Although some reptiles such as the Tuatara (*Sphenodon*) of New Zealand are capable of being active at air temperatures as low as +6˚C, to have been functional at significantly lower temperatures suggests these animals may have been warm-blooded.

Cold temperatures during the Early Cretaceous in south-eastern Australia may also provide an explanation of another phenomenon. Labyrinthodont amphibians flourished globally from the Devonian to the end of the Triassic. A few specimens are known from the Jurassic of Australia and Asia, but they were not a major element in the terrestrial fauna as they had been in times past. In south-eastern Australia labyrinthodonts (brachyopoids) occur for the last time anywhere in the world. These animals resemble crocodilians in appearance, and their decline at the end of the Triassic corresponds with the arrival of the crocodiles. Crocodilians today are not found in frigid waters, whereas amphibians, such as salamanders, are known to be active in liquid water as cold as –2˚C. In the Late Cretaceous of North America small crocodilians, evidently able to tolerate colder temperatures, ranged further north than larger ones. In the Early Cretaceous of south-eastern Australia, there are no crocodilians from the same localities as labyrinthodonts. What evidence there is for the presence of crocodilians in Australia at all during this time is not conclusive but if they did exist they would have been presumably the most cold-tolerant.

---

137 Top view of the skull of Leaellynasaura amicagraphica *from the Early Cretaceous of southern Victoria, in the rift valley with Antarctica. This small hypsilophodontid dinosaur would have been perhaps 40 centimetres tall. The preserved impression of the top surface of the brain of this specimen shows that the optic lobes were particularly prominent in this individual, much more so than in any other hypsilophodontid known. As this area of the brain interprets the signals received from the eyes, the large size of the lobes suggests high visual acuity. Living as this animal did within the Antarctic Circle of the day, enhanced visual ability would seem to have been advantageous during the winter, a time when the air temperatures were probably below freezing. If* L. amicagraphica *was active at such times, it was probably a warm-blooded animal. x 5.2 (S. Morton, courtesy of the Museum Of Victoria)*

138

It would appear that labyrinthodonts may have persisted in south-eastern Australia after having become extinct elsewhere because they could tolerate colder water than crocodilians. A test of this hypothesis will be to compare the oxygen isotope ratios at the sites where the labyrinthodonts and probable crocodilians occur. To do this, geochemists will measure the oxygen ratio to estimate the mean annual temperatures. If the estimate for the labyrinthodont sites is greater than or equal to that of the possible crocodilian sites, the hypothesis will have been falsified. If, on the other hand, it is less, the hypothesis while not having been proven will at least have been corroborated. Future work may then either support it further or show it to be untenable. The more times such a hypothesis is not falsified by repeated tests, the more confident a scientist becomes of its

correctness, but a hypothesis is never established with absolute certainty no matter how many times tests have been performed. Besides the low temperature, geographic isolation at the end of the Australian Peninsula of Gondwana at this time may also have been a factor in giving some "protection" to labyrinthodonts, allowing them to survive longer here than elsewhere in the world.

The Early Cretaceous stream and river channels of south-eastern Australia come in two sizes, small rivulets only a few metres wide and broad rivers that may be hundreds to thousands of metres across. The rivers flowed much faster than the rivulets, so the sediments deposited in them are much coarser and more massive. Almost no fossil bones have been discovered in these ancient high-energy rivers. Apparently bones that entered these systems were dashed to pieces in the fast-flowing currents, as they collided with other particles in the river's sedimentary load or with other bones and woody remains. It is in the sediments left behind by the rivulets that the vast majority of bones are found. Pools and eddies in this more benign environment provided quiet water where bones came to rest intact.

The majority of vertebrate remains found in these rivulets are those of small animals. This situation is understandable when one considers that only smaller bones could be transported by the slow-moving water courses with widths of only a few metres. Undoubtedly, there were larger animals in the area but for the most part their remains simply could not be moved and concentrated. Evidence of the existence of these larger animals, nevertheless, is seen in the rare occurrences of the smallest of their bones which are identifiable, such as the ankle-bone (astragalus) of a rather large carnivorous dinosaur, an allosaurid.

Allosaurus, a carnivorous theropod dinosaur, is well known in North America and also occurs in Africa. In Australia an astragalus of this animal is known from the Early Cretaceous sediments near Inverloch, Victoria, in rocks slightly older than those from the Otway Ranges. When compared with measurements from 55 individuals from a single site in North America, the Inverloch *Allosaurus* is almost identical in size to the third smallest of that sample. The astragalus is one of the smallest bones in the body of *Allosaurus*, and the smallest one that can be so precisely identified. Thus, a glimpse at one of the larger dinosaurs that was present is afforded by a small bone that may well represent a juvenile individual. Large dinosaurs were certainly in south-eastern Australia during the Early Cretaceous but the factors that determined what skeletal material was ultimately preserved favoured smaller animals, something that must be kept in mind when attempting to reconstruct the real community that lived in this area more than 100 million years ago.

The Victorian record of *Allosaurus* suggests an individual decidedly smaller than most of those in this genus in North America, which reached a height of 5 metres. A few partial limb bones in the southern Victorian Cretaceous sediments reflect the presence of other, smaller carnivorous dinosaurs but, unlike *Allosaurus*, their identification cannot be further refined. Clearly, there were a number of different carnivores that could have preyed upon the herbivorous hypsilophodontids.

No other herbivorous dinosaurs are known in the Early Cretaceous of south-eastern Australia, but again this may be a bias imposed by the small number of fossils so far recovered from the Early Cretaceous rocks of the region.

Stream and lake dwelling plesiosaurs and turtles occur in the Early Cretaceous rivers and rivulets of south-eastern Australia and yet, somewhat surprisingly, both are unknown from Koonwarra. Only a few teeth of plesiosaurs mark their presence in these freshwater deposits, but the turtles, after the hypsilophodontid dinosaurs, are the most common element known.

During the Early Cretaceous when *Allosaurus* and the hypsilophodonts inhabited what is now southern Victoria, a great rift valley was forming along the present margin of south-eastern Australia. It developed as Antarctica began to separate from Australia. Into this valley poured vast quantities of sediment, much of which had a volcanic origin. Just where the volcanoes were located is yet unknown, for no other trace of them exists other than the sediments produced by them. It is from this vast quantity of rocks, which choked the rift valley as it formed, that the Early Cretaceous dinosaurs and other vertebrates from Australia's south-east were recovered.

138 This specimen is the only evidence for the presence of Allosaurus on the Australian continent. In North America this genus is known from literally thousands of bones and a number of skeletons of Late Jurassic age. This one ankle bone came from an individual that lived at least 20 million years later, in the Early Cretaceous rift valley of south-eastern Australia. x 0.6 (F. Coffa, courtesy of the Museum Of Victoria and Monash University)

139 Dinosaur Cove in the Otway Ranges, Victoria, where a diverse Early Cretaceous (Aptian to Early Albian) vertebrate assemblage has been recovered. (F. Coffa)

140 The Otway Group at Dinosaur Cove, Victoria. The cross-bedded, high energy channel sands contain concretions that have been used to study the ratios of oxygen isotopes (16 and 18), which give clues to the past temperatures of the area. (Courtesy of P. Menzel and the National Geographic Society)

141–143 A variety of living plants whose relatives lived in the Early Cretaceous polar floras of southern Victoria.

141 Sphenopsids (horsetails) grow today in the dwarf forest litter near Bettles, Alaska, just south of the Brooks Range and inside the Arctic Circle. Plants related to this group were present in the understorey of the early Cretaceous forest in southern Victoria which then lay inside the Antarctic Circle. (F. Coffa)

142 Ginkgo biloba, the Maiden Hair tree, whose pungent yellow fruits may have been hypsilophodont fodder. In recent times this single species was restricted to a small part of China, however it has now been reintroduced to many parts of the world. (F. Coffa, courtesy of the Royal Melbourne Botanical Gardens)

143 Although cycads are not a dominant member of today's vegetation of Australia, or the world, for that matter, they had a rich and diverse history in the late Palaeozoic and Mesozoic, prior to the expansive radiation of the angiosperms. (S. Morton)

140

41

142

143

144 The Early Cretaceous rift valley of south-eastern Australia is unusual in that it has a variety of hypsilophodontid dinosaurs. Elsewhere, if present at all, the hypsilophodontids tend to be a rare element in the dinosaur assemblages. In south-eastern Australia they dominate the known land vertebrates. Femora are among the sturdiest, and at the same time most characteristic, bones of hypsilophodontids. The evidence for the great diversity of south-eastern hypsilophodontids is to be found in the variety of the femora. Many of these femora types have not been assigned formal scientific names because they are known only in isolation rather than as parts of skeletons. The femora shown here are of: A, as yet unnamed, this 30-centimetre-long femur was from the largest hypsilophodontid known in south-eastern Australia and would have been about the height of a small human adult; B, second unnamed hypsilophodontid femoral type; C, Fulgurotherium australe, also known from Lightning Ridge, New South Wales, in rocks of the same age; D, Leaellynasaura amicagraphica, the smallest of the Australian hypsilophodontids; and E, a third unnamed femoral type probably referrable to Atlascopcosaurus loadsi. x 0.4 (F. Coffa, courtesy of the Museum Of Victoria and Monash University)

145 Jaw of a crocodile-like labyrinthodont amphibian from the Early Cretaceous rift valley of south-eastern Australia. These amphibians may have persisted at higher latitudes, because of their tolerance for colder water, after being displaced elsewhere by crocodilians at the end of the Triassic. x 0.3 (S. Morton, courtesy of Monash University)

146-148 Skull fragments of Atlascopcosaurus loadsi, as seen from the outside (labial), from the Early Cretaceous of southern Victoria, in the rift valley with Antarctica. The single tooth pictured shows how the surfaces wore through the processing of coarse, fibre-rich vegetation. The lateral view shows three unerupted cheek teeth in a juvenile maxilla (upper jaw) fragment. 146, x 2.5; 147, x 2.6; 148, x 9.1 (S. Morton, courtesy of the Museum Of Victoria and Monash University)

149-150 Partial skeleton of a hypsilophodontid dinosaur from the Early Cretaceous rift valley of south-eastern Australia. This individual may belong to Atlascopcosaurus loadsi. Shown are the right hind limb, pelvis and part of the vertebral column; and the lower left hind limb with pathological tibia denoted by a red arrow. Compare this tibia with the normal one indicated by the black arrow on the right hind limb. This individual suffered from chronic osteomyelitis for several years before finally succumbing to the effects of the disease. x 0.6. (S. Morton, courtesy of the Museum of Victoria and Monash University)

OVERLEAF As the flood waters recede, remains transported when a river broke from its main channel lie strewn like jackstraws on this hundred-million-year-old floodplain. The carcass of a chicken-sized hypsilophodontid dinosaur, Leaellynasaura amicagraphica, lies near the cold water's edge where accumulating ice reflects the low-angled polar light of this high latitude area. Other fragments dropped on this sandbar in an ancient rift valley developing between Antarctica and Australia include: a hypsilophodontid dinosaur tooth (B), a lungfish tooth plate (C), a turtle shell fragment (D), and a variety of plant debris — Taeniopteris daintreei (F), Ginkgoites australis (G), Adaniles bifida (H), Sphenopteris warragulensis (I), Phyllopteroides dentata (J), Lycopodium (L) and an araucarian cone bract (M). Pieces of clay — clay galls (K) — ripped up from the floodplain as the waters rushed across it were first rounded somewhat by flood transport and then dropped along with the organic debris. (P. Trusler)

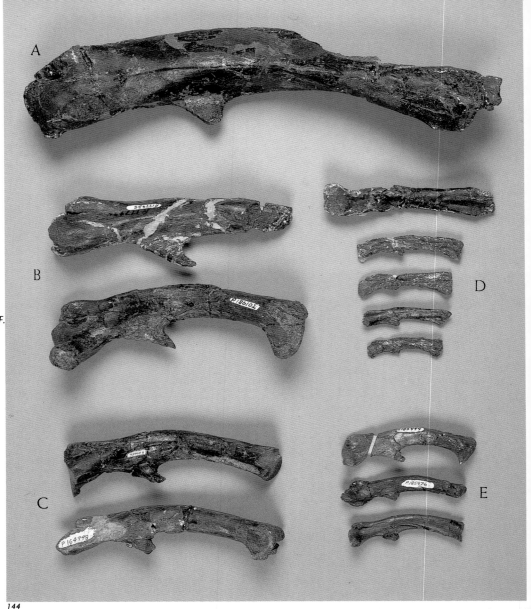

144

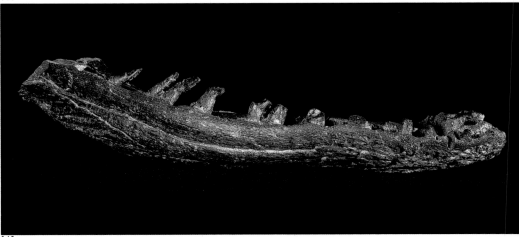

145

149

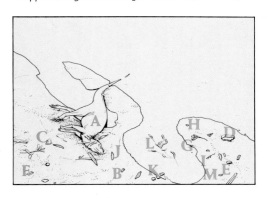

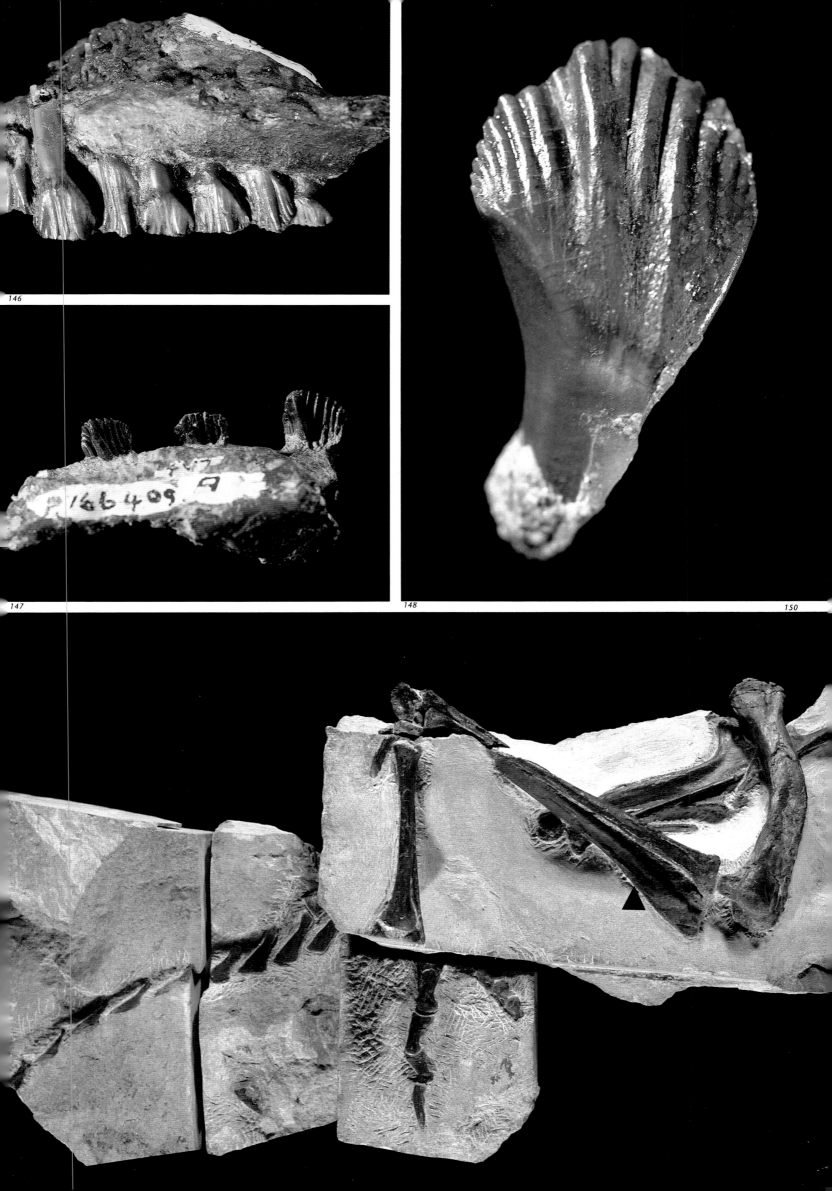

146

147

148

150

Trasler
1990

OPALIZED DENIZENS OF AUSTRALIA'S GREAT INLAND SEA

At the same time as dinosaurs lived along what is now the southern margin of Australia, to the north, in what are today the opal fields of New South Wales and South Australia, there lived lungfish, plesiosaurs, ichthyosaurs, dinosaurs, primitive birds and monotreme mammals, some of an unusual nature when compared to the fauna of the rest of the world. The preservation of many of these vertebrates is unique and fascinating. Originally entombed at the bottom of a shallow inland sea and buried for over 100 million years, such bones were replaced by a natural silica gel which eventually hardened into the fiery and precious stone, opal. This treasure is mined in such places as Andamooka and Coober Pedy in South Australia, and White Cliffs and Lightning Ridge in New South Wales — all of which have produced beautiful fossils.

Of these precious vertebrate fossils, most abundant and most complete are the plesiosaurs. Several partial skeletons of these aquatic reptiles are known, and they demonstrate the variety of these small-headed, long-necked animals. *Cimoliasaurus* from the opal field at White Cliffs is one of the few that has been identified to date. Other, more complete, specimens are known from both there and Andamooka.

Early Cretaceous Australian plesiosaurs are not confined to the opal fields. Numerous specimens are known from marine limestones in Queensland where no gems have ever been found. Not only plesiosaurs but gigantic short-necked, long-headed pliosaurs have also been recovered from these more northerly marine deposits. *Kronosaurus queenslandicus* was one of these, about 14 metres long with a skull about a quarter that length and a neck which was short even for a pliosaur. Its teeth were tall cones, about 40 millimetres in diameter and 250 millimetres tall, well suited for grabbing large fishes as the animal lunged forward with powerful strokes of its flippers. Remains of *Kronosaurus* have been found in a number of different rock formations, and *K. queenslandicus* may be only one of two or three species, though at present the known pliosaur material is not sufficiently studied to allow determination of just how many different kinds are represented.

Remains of ichthyosaurs, dolphin-like reptiles of the Mesozoic, are common in the Early Cretaceous marine deposits of Australia, primarily in Queensland but also in the Northern Territory and South Australia. Conceivably, all the known fossils could belong to a single species, *Platypterygius australis*. The genus *Platypterygius* occurs in many other parts of the world at this time. The structure of the front flippers of the Australian fossils makes it quite clear that *P. australis* is a genuine species, one quite distinct from all others of the genus. Why this should be so is not at first sight obvious, because the inland seas where it lived were directly connected to the world ocean. However, *P. australis* may have been restricted to shallow water in some way, by its diet or some other factor, so that open oceans were a barrier rather than a highway for dispersal.

---

151 Kronosaurus queenslandicus, a pliosaur from the Early Cretaceous inland sea sediments near Hughenden, Queensland. This specimen, in the Museum Of Comparative Zoology at Harvard University, has a total length of about 14 metres. (A.S. Romer)

152 The skull and partial skeleton of an ichthyosaur, Platypterygius australis, from the Early Cretaceous marine Toolebuc Formation on Telemon Station, western Queensland. The skeleton, about 5.6 metres in total length, is that of a young adult. x 0.15 (F. Coffa, courtesy of the Queensland Museum and R. Molnar)

153 The vertical grooves on these conical teeth identify them as those of plesiosaurs. Found in the Early Cretaceous opal fields of New South Wales at Lightning Ridge (longer tooth) and White Cliffs. x 1.5 (F. Coffa, courtesy of the Australian Museum)

154 Vertebrae of Cimoliasaurus leucocephalus, a plesiosaur from the Early Cretaceous opal field at White Cliffs, New South Wales. x 0.9 (F. Coffa, courtesy of the Australian Museum)

155 Teeth from the two genera of lungfish present in Australia during the Mesozoic and Cainozoic, both of which are represented in the Early Cretaceous opal field at Lightning Ridge, New South Wales. The smaller tooth plate is that of Neoceratodus forsteri and the larger, Ceratodus wollastoni. While Neoceratodus is still present in a few rivers near Brisbane, Queensland, Ceratodus became extinct in Australia in the middle Tertiary. x 1.8 (F. Coffa, courtesy of the Australian Museum)

156-157 Opalized marine reptiles from the Cretaceous of Coober Pedy, South Australia (156, a pliosaur) and White Cliffs, New South Wales (157, a plesiosaur). (Courtesy of P. Menzel and the National Geographic Society, Sid Londish and Ken Harris).

152

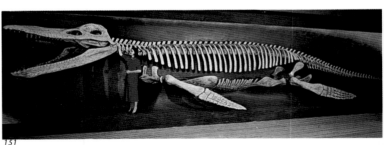

151

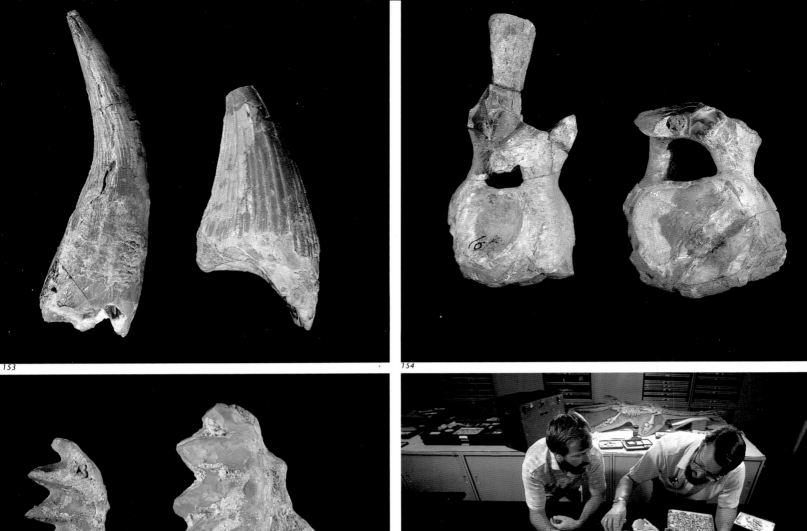

153 154

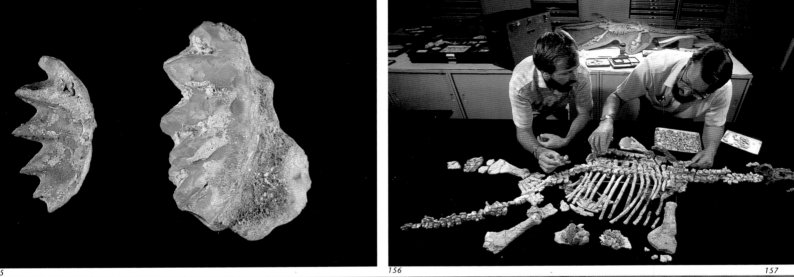

155 156 157

RARE REMAINS OF FLYING VERTEBRATES

Like the bird feathers at Koonwarra, what little pterosaur (or flying reptile) material is known from the Early Cretaceous of south-eastern Australia is so meagre that it shows only that the group was present. Not so with the pterosaur remains from the Toolebuc Limestone exposed along the Hamilton River near Boulia in south-western Queensland. Although no entire skeletons are known from this area, several isolated bones including vertebrae and parts of both limb girdles, which apparently all belong to one species, have been found. The material is characteristic enough to tentatively assign it to *Ornithocheirus*, a genus otherwise known from Late Cretaceous rocks of Europe and Brazil.

Besides being one of the few records of Australian pterosaurs, the Queensland material is important because it is uncrushed. Pterosaur bones, like those of birds, are hollow and the walls are quite thin. Because of this fragility, fossil bones of pterosaurs are often crushed flat, and detail of the shapes of the articular ends of the bones are obscured. The Queensland material, having been found in a limestone that was never deeply buried, is excellently preserved in three-dimensions. Thus, this material can shed light on the general structure of pterosaur skeletons not available elsewhere, despite the fact that the actual number of bones found is quite limited.

Even more rare than the flying reptiles are two specimens of a quite primitive group of birds, the enantiornithines. A single tibiotarsus fragment (a hind limb element) from a bird about the size of a European black-bird or Australian shrike-thrush, named *Nanantius eos*, occurs in the marine Albian Toolebuc Formation on Warra Station near Boulia in western Queensland. One other bone of this group is known, an undescribed tibiotarsus. It is similar to material of this bird subclass known from Mongolia, Mexico and most abundantly from the Cretaceous of Argentina. The Australian occurrence of the enantiornithines is the oldest in the world, for elsewhere it is known only in Late Cretaceous sediments. The enantiornithines were an early group of birds that never acquired many of the specializations shared by modern birds, such as the shapes of the articulating surfaces of the bones of the hind limb and the structure of the pelvis. On the other hand, parts of the skeleton, such as the bones of the wing, are very similar to those of modern birds. By the end of the Mesozoic the enantiornithines were extinct and in their place were birds that gave rise to our modern avifaunas.

158 *The skull of the flying reptile (a pterosaur)*
Cearadactylus atrox from the Early Cretaceous
(Aptian) Santana Formation of Araripe Plateau, Ceará,
Brazil. x 0.8 (G. Borgomanero with the assistance of H.
Alvarenga)

159 *A rare uncrushed scapulocoracoid of a pterosaur*
from the Early Cretaceous Toolebuc Formation of
Queensland. This specimen has been tentatively
assigned to the European genus Ornithocheirus.
x 1.0 (F. Coffa, courtesy of the Queensland Museum)

160 *Partial skeleton of an enantiornithine from the*
Early Cretaceous Rio Colorado Formation of Argentina.
These primitive birds also occur in similar-aged
deposits in Mongolia, Mexico and western Queensland.
x 0.6 (F. Coffa, courtesy of J. Bonaparte and
L. Chiappe)

158

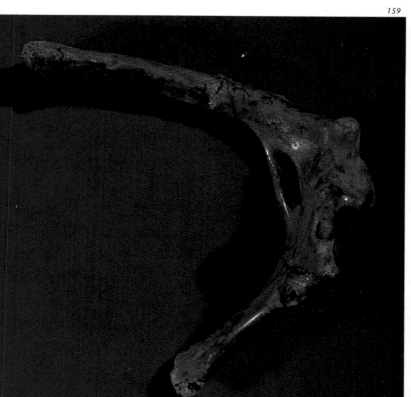

159

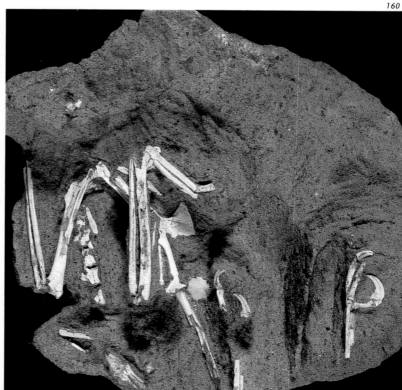

160

MUTTABURRASAURUS AND OTHER NORTHERN AUSTRALIAN DINOSAURS

While plesiosaurs and ichthyosaurs swam in the Early Cretaceous shallow inland seas of Central and northern Australia, and pterosaurs flew over them, dinosaurs occupied mainly terrestrial environs. Yet it is only in marine deposits that Early Cretaceous dinosaurs are found in this part of the continent, because no terrestrial sediments of that age have been preserved. Most of the dinosaurs known from these marine deposits were found as isolated bones, but one complete Australian dinosaur skeleton, that of *Muttaburrasaurus langdoni* (an iguanodontid) has been recovered from near the town of Muttaburra in north central Queensland.

Iguanodontids looked like hypsilophodontids, from which they were descended. Both were mainly bipedal herbivores, with fore limbs smaller than hind limbs and an elongated tail, though a few iguanodontids may have been quadrapeds. The most noticeable difference between the two, however, is that iguanodontids were much larger and, as might therefore be expected, they were also more robust. Iguanodontids flourished in the Early Cretaceous on all the northern continents, but their record is sparse in the Southern Hemisphere.

Like *Iguanodon* itself, *Muttaburrasaurus* probably had a peculiar spike-like thumb, perhaps a defensive weapon, setting it apart from most other dinosaurs. In addition, an inflated area above the nasal region of the skull is unmatched elsewhere. Hadrosaurs, or duck-billed dinosaurs, were a diverse Late Cretaceous group on the northern continents but are also known from South America. Typical of this group were inflated crests on the rear of the skull, but in no hadrosaur did this feature match that of *Muttaburrasaurus*. Many theories have been put forward to explain the function of such structures, one being that they were used to produce sounds, either for mate attraction or dominance displays against other males in the same species or for defence against predators.

The hypsilophodontid *Fulgurotherium australe* was named on the basis of material from the Lightning Ridge opal field in far northern New South Wales. Subsequent to the original discovery, more and better material has been found there and referred to this species. Indistinguishable from it, too, is material from southern Victoria.

Three small to intermediate sized theropods are also known from the marine Early Cretaceous opal field deposits: *Walgettosuchus woodwardi* and *Raptor ornitholestoides* from Lightning Ridge, and *Kakuru kujani* from Andamooka in South Australia. All are represented by single isolated bones, adequate to establish the presence of the different taxa but not sufficient to tell more about them.

Sauropods are known from Early Cretaceous marine deposits of Queensland. Most finds are merely single bones, but one of the more informative records comprises a few associated vertebrae which have been named *Austrosaurus mckillopi*. Sauropods typically have cavities or pleurocoels in the vertebrae, apparently to lighten them, but in *Austrosaurus* these features were rudimentary, perhaps suggesting that this sauropod was rather primitive. Generally, this animal resembles Middle Jurassic forms outside of Australia.

Another group of dinosaurs known from the marine Early Cretaceous of Queensland was the armoured ankylosaurid *Minmi paravertebra*. Less than 4 metres long, it was rather small for an ankylosaur, and it was further unusual in that it was lightly armoured on its back while having reasonably extensive belly armour, a rarity among ankylosaurs. The lack of much back armour might mean that living on an island, as it did, it lacked predators, or that it was agile enough to run away from any predators, rather than rely entirely on body armour for protection.

From the same Queensland marine sediments that produced dinosaur and other reptilian remains, a varied marine fish fauna is known, mainly from the Tambo Series and related late Early Cretaceous rocks. Unlike the freshwater Koonwarra fauna, the marine Tambo fishes are all teleosts, with the exception of one holostean, *Belonostomus*. Most of the teleosts are quite advanced, belonging to the Clupeomorpha, and represent two families, the ichthyodectids (*Cooyoo*, for example) and the pachyrhizodonts (*Pachyrhizodus*), both for the most part large predatory fishes.

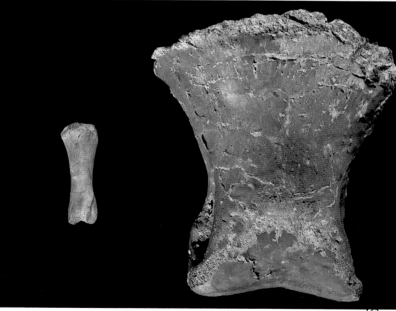

161 Muttaburrasaurus langdoni, an iguanodont dinosaur from the marine Early Cretaceous Mackunda Formation on the Thompson River near Muttaburra, central Queensland. Not many dinosaur skeletons are known from Australia and this one is the most complete of them all. Muttaburrasaurus was about 5 metres tall, and about 10 metres from nose to tail tip. (Courtesy of P. Menzel, the National Geographic Society, and the Queensland Museum)

162 Two dinosaurs' toe bones from the Early Cretaceous opal field at Lightning Ridge, New South Wales. The smaller one may be of a dromaeosaurid, a small carnivorous form. x 0.9 (F. Coffa, courtesy of the Australian Museum)

163 Vertebrae of an as yet unnamed plesiosaur from the Early Cretaceous Toolebuc Formation of Queensland. The Toolebuc Formation is a limestone laid down in a shallow inland sea that covered much of central Queensland during the later part of the Early Cretaceous. Although many fossil vertebrates have been found in this formation, its potential has hardly been scratched as it is quite widespread. x 0.5 (F. Coffa, courtesy of the Queensland Museum)

164 Belly armour impression of an armoured ankylosaur, *Minmi paravertebra*, from Early Cretaceous marine sediments near Minmi in south-eastern Queensland. x 0.3 (F. Coffa, courtesy of the Queensland Museum and R. Molnar)

165-166 A skull (x 0.15) and ribs (x 0.4) with armour, both top views, of an ankylosaurid, or armoured dinosaur, from the Early Cretaceous Toolebuc Formation of Queensland. (F. Coffa, courtesy of the Queensland Museum)

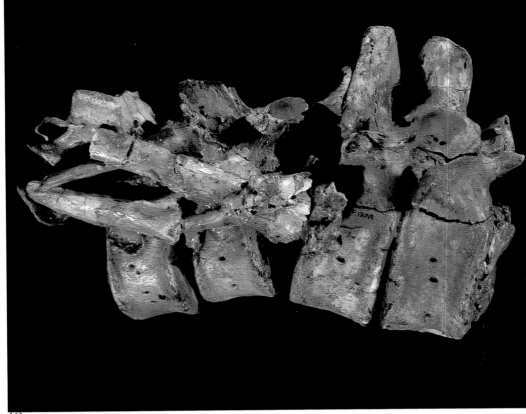
163

164

165

166

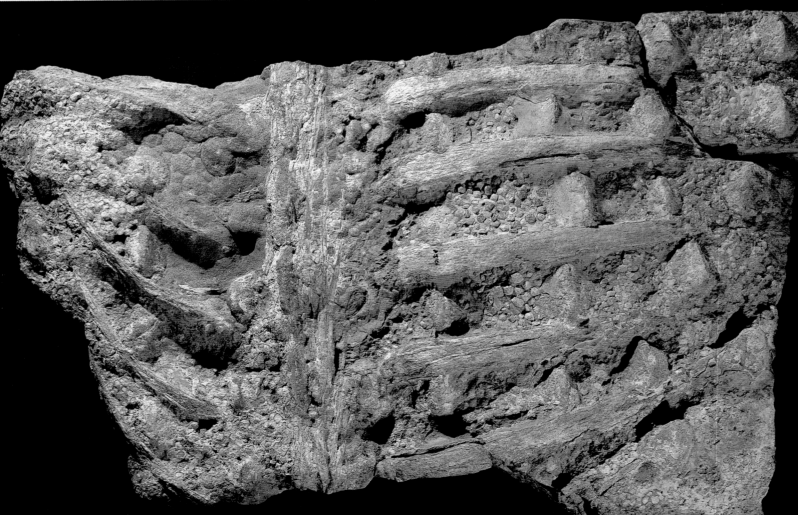

LATE CRETACEOUS VERTEBRATES

THE WINTON FAUNA

The pattern of mountain building, island arc formation and river drainage that had characterized Australia since the Carboniferous came to a close at the end of the Early Cretaceous. Not only did Australia then begin to separate from Antarctica, but the entire continent was uplifted, resulting in the withdrawal of the vast inland seas which until the Late Cretaceous had covered much of the central and northern part of the continent. In addition, rivers over much of the continent began to drain centripetally towards the Ceduna region of the modern Great Australian Bight, a pattern that has been only slightly modified subsequently by the gentle rise along the southern perimeter of the continent. The Lake Eyre Basin remains the final destination for much of the drainage today.

The last image of the Australian terrestrial biota during the Cretaceous comes from the Winton Formation, a widespread rock unit found not only in much of south-western Queensland but in northern New South Wales and north-eastern South Australia as well. The sediments making up this formation were deposited at the beginning of the Late Cretaceous. Unlike the Early Cretaceous units upon which it rests, the Winton sediments were deposited under freshwater, not marine conditions.

A few isolated sauropod bones are known from the Winton Formation, and some have been referred with reservations to *Austrosaurus*. These bones appear to be similar to those of brachiosaurs, a well known group in the Late Jurassic of North America and Africa, which possessed elongated forelimbs.

Another major dinosaur group that may be represented in the Winton Formation is the large ornithopods. Ornithopods include, among other groups, hypsilophodontids, iguanodontids, and hadrosaurs. The reason for the imprecision in the identification of the possible Winton ornithopods is that they are known from only three isolated bones of the ear region. These recently discovered bones are most similar to those of a hadrosaur, *Edmontosaurus*, from Canada. Hadrosaurs, however, are not known to have evolved until later in the Late Cretaceous than the Winton Formation is thought to have been deposited, and up to now they have been unknown in Australia. As there are not many ear regions of most dinosaur groups available for study, it is difficult to evaluate with certainty the significance of the similarity between the Winton bones and those of *Edmontosaurus*. Might not the iguanodontids, particularly *Muttaburrasaurus*, have had a similar ear region? After all, iguanodontids seem to be closely related to the ancestors of hadrosaurs, and the ear region of *Muttaburrasaurus* is unknown.

The three Winton Formation ear regions came from a site near the town of Winton itself, and clearly indicate that at least two skulls were originally preserved. The presence of two specimens at a single site which is littered with bone fragments strongly hints that more might be found. Perhaps an extended excavation would yield a skull that would provide the answer to just what animal these ear regions belong to. However, the exploratory dig which yielded these three ear regions also showed that the fossils which are still buried beneath the surface are quite broken up, having been in the soil zone for sufficiently long that plant roots have caused extensive damage. Because the area where the fossils occur is one of low relief, erosion processes are so slow that before bones become visible the natural destructive agents have had sufficient time to do much damage. This problem is a recurring one in Australia, where much of the land surface we see today is ancient. Indeed, the soil formation processes have had so long to act that frequently fossils which were in rocks are totally destroyed before being exposed.

Iguanodontids and hadrosaurs can be distinguished on the basis of their teeth. Hadrosaurs had a much more specialized set of teeth (with as many as 500 teeth in each jaw) than did iguanodonts. As well, hadrosaurs arose at about the same time that the angiosperms, or flowering plants, came to dominance, and the two events may be causally related. When the iguanodont *Muttaburrasaurus* lived and died towards the end of the Early Cretaceous, angiosperms were extremely rare in Australia. The Winton Formation contains a diversified but as yet undescribed angiosperm flora, and it is

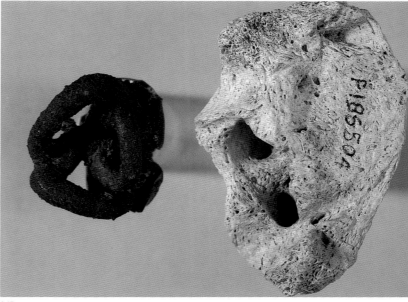

167

tempting to speculate that, just perhaps, these large ornithopod ear regions might belong to animals that, if not hadrosaurs, were evolving in a parallel direction in response to the rise of the flowering plants.

The presence of an even greater variety of Late Cretaceous fossils is mirrored not by bones but by trackways. Thousands of footprints, have been recovered from Lark Quarry, about 100 kilometres south of Winton, Queensland. They provide the only record of large and small theropods plus large and small ornithopods present in the earliest Late Cretaceous Winton Formation.

But much more than simply the identity of fossil types has been gleaned from this site. What is captured in the rocks near Winton is a sequence of events that probably took place in just a few minutes some 100 million years ago. Evidently, a mixed group of about 100 small theropod and ornithopod dinosaurs was drinking from a lake on the edge of a mud flat. A large carnivorous theropod approached them, walking at about 7 kilometres per hour (a speed which can be determined by examining stride length and detail of the individual track impressions). Detecting this potential threat the smaller dinosaurs stampeded, running towards where the large theropod had been only moments before — as evidenced by smaller tracks which can be seen superimposed on the large ones of the big carnivore, going in the opposite direction. The small ornithopods ran at 16 kilometres per hour and the small theropods at 12 kilometres per hour. Whether any of the smaller dinosaurs were seized, or if an attempt to do so was even made, is not known because the preserved area of the trackways do not provide any evidence of such an event.

When deposition of the Winton Formation ceased, the Mesozoic had about 35 million years yet to run. But nowhere on the Australian continent is there a record of terrestrial vertebrates during this interval. Only in Western Australia are there marine rocks that were deposited during that time period. From these rocks have come the only record of mosasaurs, or marine lizards, a group that also flourished only in the Late Cretaceous on other continents. Isolated bones of ichthyosaurs as well as plesiosaurs and a pterosaur are also known from these Western Australian marine deposits, showing that these animals, too, persisted at this time on the Australian continent.

167 The ear region of a large ornithischian dinosaur from the earliest Late Cretaceous near Winton, Queensland. Beside it is a cast of the cavities in this bone for the semi-circular canals. The planes of the three semi-circular canals are mutually perpendicular to one another, a reflection of their function of orienting the individual in three dimensions. x 1.7 (F. Coffa, courtesy of the Museum Of Victoria)

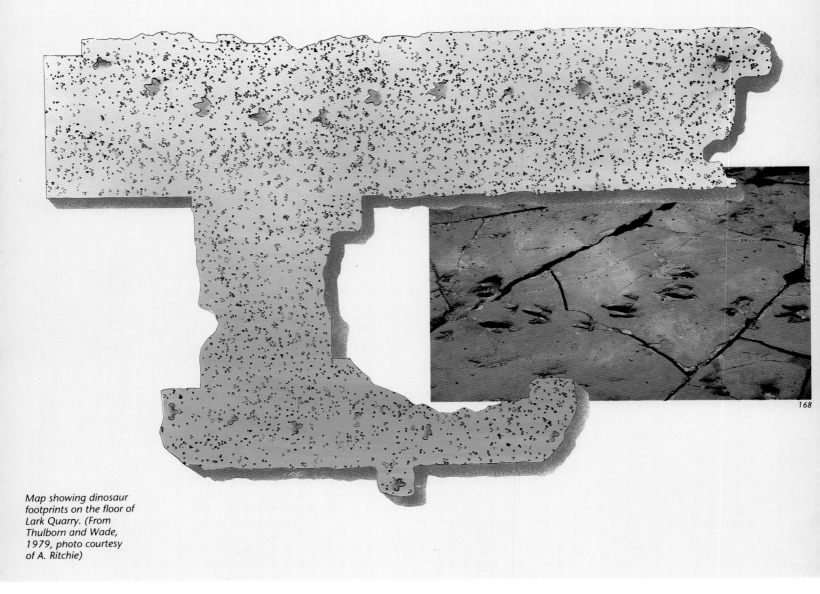

Map showing dinosaur footprints on the floor of Lark Quarry. (From Thulborn and Wade, 1979, photo courtesy of A. Ritchie)

168

168   More than 90 million years have passed since a mixed group of small dinosaurs was suddenly startled into wild flight by the appearance of a lone, large predatory carnosaur. Mute evidence for this brief episode is preserved in footprints at Lark Quarry, south-west of Winton, Queensland. (F. Coffa, courtesy of the Queensland Museum)

169   Tail of a palaeoniscoid fish from the Early Cretaceous Otway Group at the Slippery Rock Locality in Dinosaur Cove, Otway Ranges, Victoria. x 0.9 (F. Coffa, courtesy of the Museum of Victoria and Monash University)

170   Unnamed bony fish with pectoral fins broadly expanded like the living flying fish. From the Early Cretaceous inland sea of Queensland. x 0.9 (F. Coffa, courtesy of the Queensland Museum)

171-172 Skulls of two large, predatory teleost fishes, Cooyoo australis (x0.6) and Pachyrhizodus (x0.4), from the Early Cretaceous inland sea of Queensland. (F. Coffa, courtesy of the Queensland Museum)

173   Side view of a skull of a marine turtle that may belong in the genus Notochelone, from the Early Cretaceous Tooleebuc Formation of Queensland. x 0.5 (F. Coffa, courtesy of the Queensland Museum)

174   Skull of an actinopterygian fish from the Early Cretaceous Otway Group at Dinosaur Cove, Otway Ranges, Victoria. x 1.3 (F. Coffa, courtesy of the Museum of Victoria and Monash University)

169

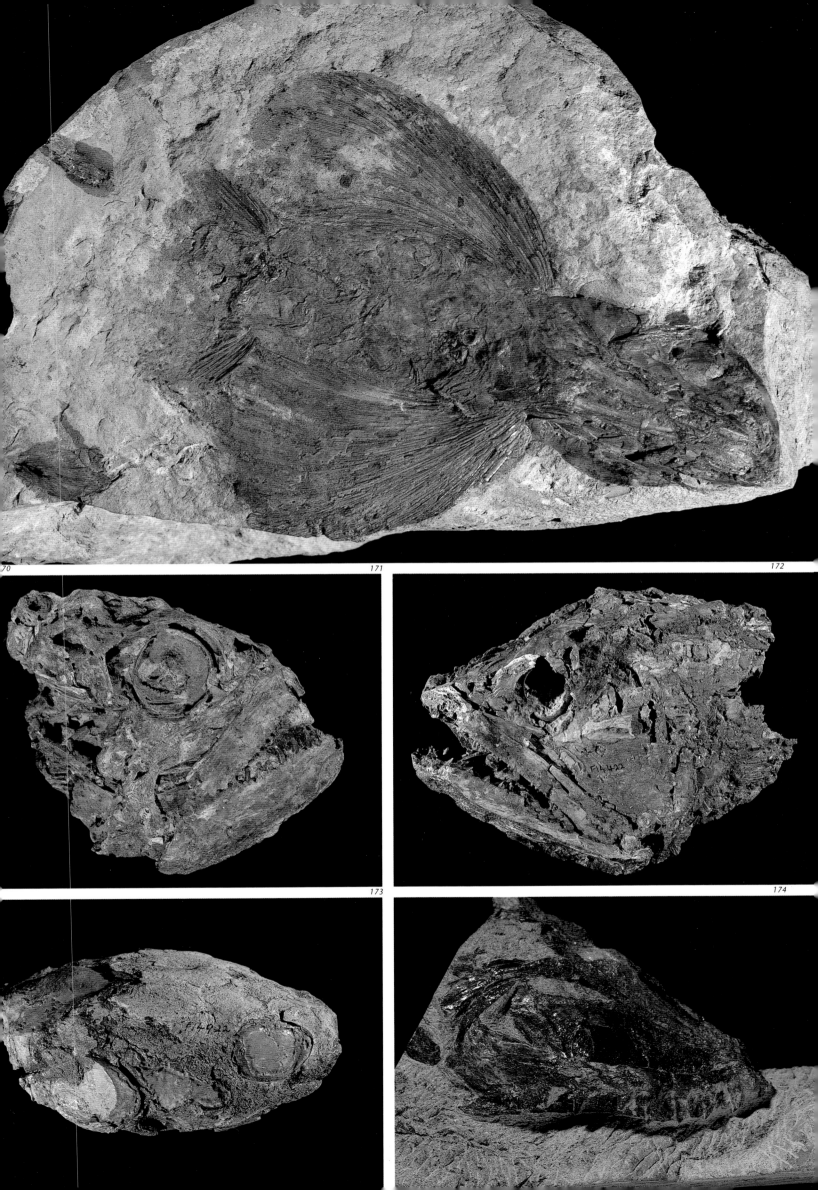

THE DEMISE OF THE DINOSAURS

The absence of fossil vertebrates in terrestrial sediments of Late Cretaceous and early Cainozoic age in Australia means that there is no evidence as to whether the dinosaurs declined gradually or suddenly on this continent, or whether they became extinct at the end of the Cretaceous or perhaps five, ten or even tens of millions of years before or after. The effect on the dinosaur community of the rise of angiosperms in Australia is unknown. Not only hadrosaurs, but also the ceratopsians, or horned dinosaurs, including the well known North American *Triceratops*, arose in the Late Cretaceous and flourished outside of Australia. Both groups may have arisen in response to the appearance of angiosperms. As angiosperms were quite diversified when the "well fossilized" Winton Formation was deposited, a glimpse of the vertebrate response to the beginnings of such a floral radiation in Australia might be gained with further investigation.

So far, our knowledge of Jurassic and Cretaceous terrestrial vertebrates has only just scratched the surface of the Australian continent. Due to the constraints of low topographic relief and deep chemical weathering, the Australian record will never be as good as those of western North America and Mongolia. But, even so, the vertebrates of the late Mesozoic herald both some communication with other Pangaean realms and a distinct isolation, which at the very least produced a series of unique genera (such as *Muttaburrasaurus* and *Leaellynasaura*) that clearly belonged to cosmopolitan families. And then, too, some animals survived in Australia beyond their time elsewhere (such as the *Allosaurus* and labyrinthodonts from the Early Cretaceous of Victoria).

175 *Angiosperm leaves from the Late Cretaceous Winton Formation near Winton, Queensland. x 1.2 (F. Coffa, courtesy of the Museum Of Victoria and Monash University)*

176 *Ginkgoites australis leaves from the Early Cretacous Strzelecki Group near Koonwarra, Victoria. x 1.4 (F. Coffa, courtesy of the Museum Of Victoria and Monash University)*

175

176

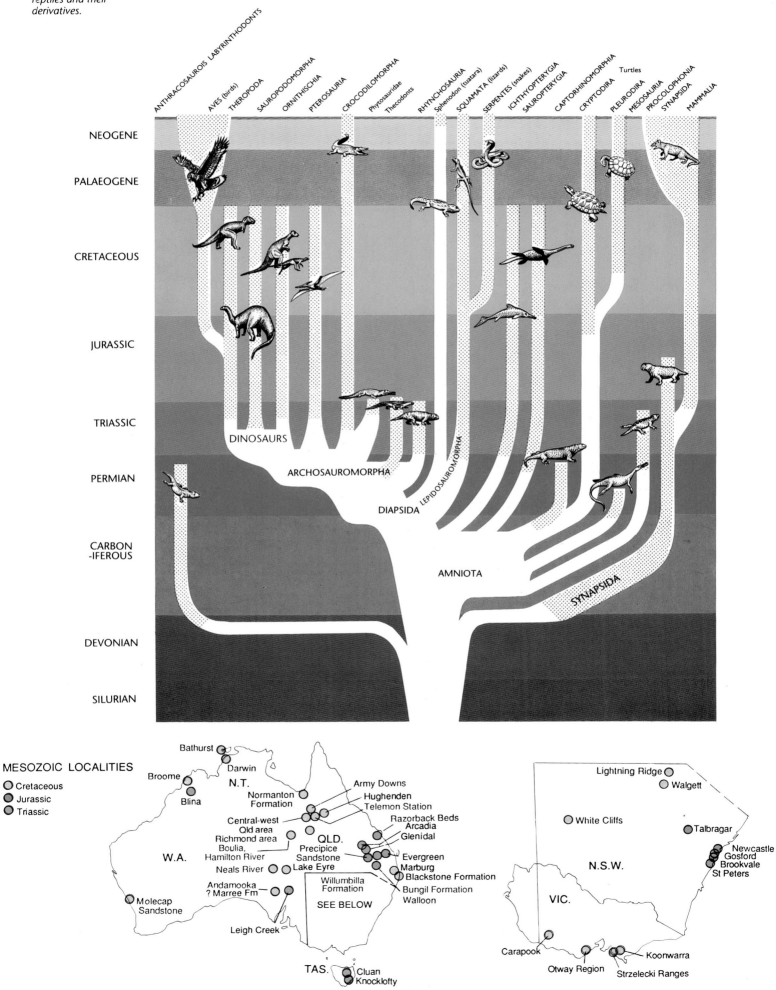

Hypothesis of the inter-relationships of modern and fossil reptiles and their derivatives.

ANTHRACOSAUROIS LABYRINTHODONTS
AVES (birds)
THEROPODA
SAUROPODOMORPHA
ORNITHISCHIA
PTEROSAURIA
CROCODILOMORPHA
Phytosauridae
Thecodonts
RHYNCHOSAURIA
Sphenodon (tuatara)
SQUAMATA (lizards)
SERPENTES (snakes)
ICHTHYOPTERYGIA
SAUROPTERYGIA
CAPTORHINOMORPHA
Turtles
CRYPTODIRA
PLEURODIRA
MESOSAURIA
PROCOLOPHONIA
SYNAPSIDA
MAMMALIA

NEOGENE
PALAEOGENE
CRETACEOUS
JURASSIC
TRIASSIC
PERMIAN
CARBON-IFEROUS
DEVONIAN
SILURIAN

DINOSAURS
ARCHOSAUROMORPHA
LEPIDOSAUROMORPHA
DIAPSIDA
AMNIOTA
SYNAPSIDA

MESOZOIC LOCALITIES

○ Cretaceous
○ Jurassic
○ Triassic

Bathurst
Darwin
Broome
N.T.
Blina
Normanton Formation
Army Downs
Hughenden
Telemon Station
Central-west Qld area
Razorback Beds
Arcadia
Richmond area
Glenidal
Boulia, Hamilton River
QLD.
Precipice Sandstone
Neals River
Lake Eyre
Evergreen
Marburg
Blackstone Formation
W.A.
Andamooka
? Marree Fm
Willumbilla Formation
Bungil Formation
Walloon
Molecap Sandstone
SEE BELOW
Leigh Creek

Lightning Ridge
Walgett
White Cliffs
Talbragar
N.S.W.
Newcastle
Gosford
Brookvale
St Peters
VIC.
Carapook
Otway Region
Koonwarra
Strzelecki Ranges

TAS.
Cluan
Knocklofty

CHAPTER 5

AN ARK TO THE TROPICS

THE TERTIARY PERIOD THROUGH THE PLEISTOCENE EPOCH OF THE QUATERNARY
PERIOD, FROM 65 MILLION TO 10,000 YEARS AGO

*During the Cainozoic Era, the past 65 million years, especially since the
Late Eocene to Early Oligocene, most continents have remained fairly
stable in their latitudinal position. Their animals and plants have been
affected primarily by global climatic trends and the occasional waxing
and waning of continental connections. Australia, however, has broken
this pattern by moving northwards from polar latitudes into the tropics
in less than 50 million years — in near total isolation. Its biota reflects
this move in its own shift from dominantly cool temperate, humidity
loving organisms to those able to cope with aridity and often a great
degree of climatic variability. The* Nothofagus *(southern beech) forests
and dolphins of Central Australia gave way to* Eucalyptus *and kangaroos, most dramatically during the last 5 million years. Australia's biota
also reflects the effects of human intervention and the limitations of the
continent's basic resources, especially its nutrient-starved soils.*

177 *Fossil Bluff near Wynyard, Tasmania. One of
Australia's oldest marsupial fossils,* Wynyardia
bassiana, *was found here, together with marine
invertebrates. The rocks are an Early Miocene marine
unit called the Fossil Bluff Sandstone. (B. Allison,
courtesy of the Queen Victoria Museum And Art
gallery and C. Tassell)*

149

CONTINENTS ON THE MOVE

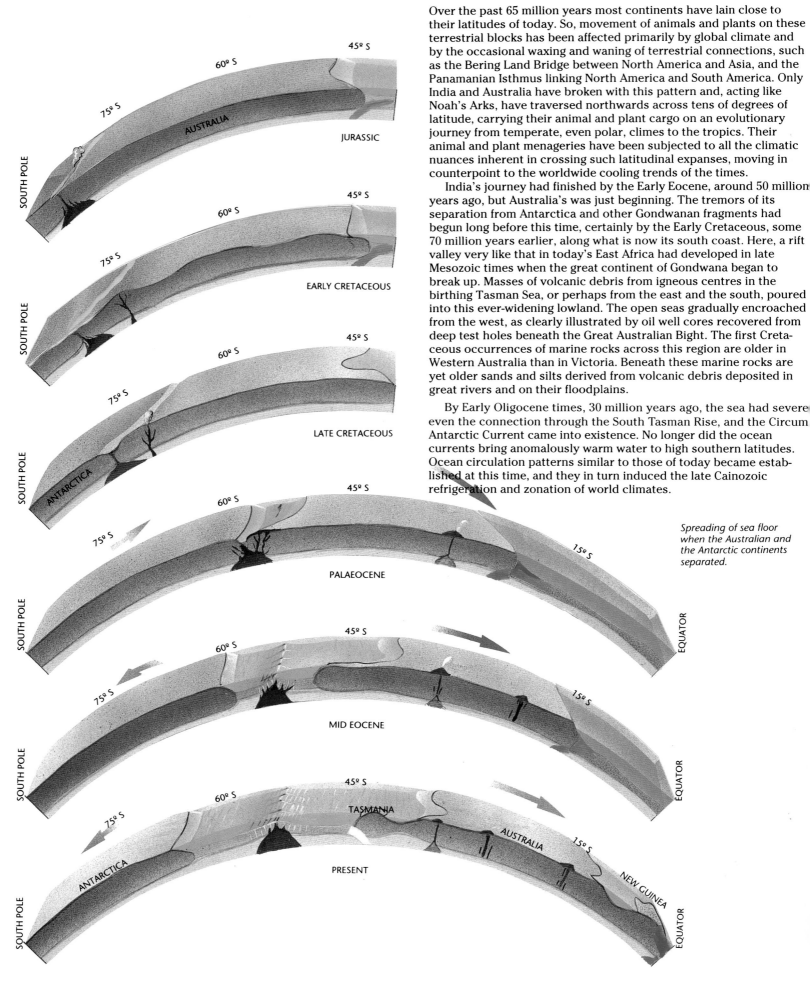

45º S
60º S
75º S
SOUTH POLE
AUSTRALIA
JURASSIC

45º S
60º S
75º S
SOUTH POLE
EARLY CRETACEOUS

45º S
60º S
75º S
SOUTH POLE
ANTARCTICA
LATE CRETACEOUS

45º S
60º S
75º S
SOUTH POLE
15º S
EQUATOR
PALAEOCENE

45º S
60º S
75º S
SOUTH POLE
15º S
EQUATOR
MID EOCENE

45º S
60º S
75º S
SOUTH POLE
TASMANIA
AUSTRALIA
15º S
EQUATOR
ANTARCTICA
NEW GUINEA
PRESENT

Over the past 65 million years most continents have lain close to their latitudes of today. So, movement of animals and plants on these terrestrial blocks has been affected primarily by global climate and by the occasional waxing and waning of terrestrial connections, such as the Bering Land Bridge between North America and Asia, and the Panamanian Isthmus linking North America and South America. Only India and Australia have broken with this pattern and, acting like Noah's Arks, have traversed northwards across tens of degrees of latitude, carrying their animal and plant cargo on an evolutionary journey from temperate, even polar, climes to the tropics. Their animal and plant menageries have been subjected to all the climatic nuances inherent in crossing such latitudinal expanses, moving in counterpoint to the worldwide cooling trends of the times.

India's journey had finished by the Early Eocene, around 50 million years ago, but Australia's was just beginning. The tremors of its separation from Antarctica and other Gondwanan fragments had begun long before this time, certainly by the Early Cretaceous, some 70 million years earlier, along what is now its south coast. Here, a rift valley very like that in today's East Africa had developed in late Mesozoic times when the great continent of Gondwana began to break up. Masses of volcanic debris from igneous centres in the birthing Tasman Sea, or perhaps from the east and the south, poured into this ever-widening lowland. The open seas gradually encroached from the west, as clearly illustrated by oil well cores recovered from deep test holes beneath the Great Australian Bight. The first Cretaceous occurrences of marine rocks across this region are older in Western Australia than in Victoria. Beneath these marine rocks are yet older sands and silts derived from volcanic debris deposited in great rivers and on their floodplains.

By Early Oligocene times, 30 million years ago, the sea had severed even the connection through the South Tasman Rise, and the Circum Antarctic Current came into existence. No longer did the ocean currents bring anomalously warm water to high southern latitudes. Ocean circulation patterns similar to those of today became established at this time, and they in turn induced the late Cainozoic refrigeration and zonation of world climates.

Spreading of sea floor when the Australian and the Antarctic continents separated.

CHANGE IN AUSTRALIA'S CENTRE

Central Australia today does not accurately reflect the environmental conditions typical of this continent over the past 50 or 60 million years. Today, this area is decidedly drier than it has been for most of its past. In fact, we seem to think deserts have always been a major element of the Australian landscape. The fossil record clearly shows, however, that for the most part deserts are relatively young here, perhaps developing only in the last half million years. Certainly Australia has hosted grasslands and grassland savannahs for several million years, but sand ridge deserts, with their erratic rainfall and poor soils, are a comparatively recent phenomenon.

Perhaps the best mirror of past climates is the record left by the pollen, spores, leaves and associated plant parts of fossilized vegetation. Examining fossil floras deposited in the now arid Centre of Australia shows a marked contrast between past and present. By no means is the record perfect. It is, in fact, very sparse, and only a handful of locales provide the data. Sites yielding fossils in this region represent the more moist micro-environments because, of all natural habitats, bogs and swamps preserve pollen and other plant remains best. Therefore, there is a pronounced natural bias in the fossil record against representation of the flora from arid regions. Still, it is true that even the most humid regions in Central Australia today do not host many of the groups of plants that were being fossilized there some 20 or 30 million years ago. Such plants as *Nothofagus* (southern beech) certainly do not grow in Central Australia now but are restricted to humid rainforests.

A complication does arise concerning just how the tolerances of extinct plants can be determined. Some plants preserved in the Central Australian sites have no living counterparts, so are generally useless in interpreting past environments. Other plants are closely related to living forms, perhaps in different species or genera. Specifying their environmental likes and dislikes must be done with caution. Even plants that can be placed in extant genera did not necessarily have the same environmental tolerances as their living progeny. But palaeobotanists look for patterns, and if the majority of the fossil flora points to humid conditions based on the tolerances of their nearest living relatives then hypotheses become more robust. It is only when conflicting tolerances are indicated that major problems arise. The degree of precision of palaeoenvironmental interpretations depends entirely on the composition of the fossil flora being examined and on the types of analyses applied.

Despite such cautions, fossil plants indicate that broad-leaved rainforests were quite widespread in much of what is now the arid Centre of Australia during early Tertiary times (the Palaeocene and Eocene epochs). The dominant gymnosperms, such as *Dacrydium*, are clearly related to temperate rainforest forms today. The pollen of flowering plants that were present likewise reflect high rainfall in Central Australia (such families as the Winteraceae and Gunneraceae).

In the Middle Eocene the Australian rainforest floras underwent a profound change from the typical Mesozoic fern and conifer dominated communities to a Tertiary pattern with an abundance of conifers and *Nothofagus* (southern beech).

Of all the Central Australian Tertiary plant record, that of the Eocene is best. This fact in itself points to higher effective humidity then, which provided widespread conditions suitable for preserving plant material. Eocene floras from the now arid Centre contain abundant *Nothofagus* pollen together with a variety of fossil forms indicating widespread cool temperate rainforests. At Hale River, a locale in the southern part of the Northern Territory, myrtaceous pollen (similar to that of modern *Eucalyptus*) as well as grass pollen first appear. So, even this early in the Cainozoic, there must have been some open areas and drier regions within a generally humid regime vegetated by many familiar forms.

After the Eocene, rainforests persisted for some time in the interior. The Silcrete Flora of the Billa Kalina Basin south-west of Lake Eyre, thought to be of Eocene–Oligocene age, includes impressions of fruits and leaves of Araucariaceae (a relative of the living Norfolk Island Pine) and Podocarpaceae which have strong similarity to forms living in New Guinea today. These taxa are mixed with the more typical Australian forms that survive now in drier areas — *Eucalyptus*,

Melaleuca and *Callistemon*. These specimens most likely include the oldest record of true *Eucalyptus*.

During the Oligocene only indirect evidence suggests slightly cooler and drier conditions, but by Miocene times there is a clear indication of aridity in Central Australia. The Etadunna Formation, a series of often greenish clays, limy silts and white sands exposed in northern South Australia, contain floras dominated by the humidity-loving *Nothofagus* in its lower part. Present, too, in the lower part of the Etadunna is a sufficient proportion of grass pollen to suggest that grasslands were present and not far away, perhaps on the interfluves of the river channels where the rocks of this formation were laid down.

As one collects and examines clays in the upper part of the Etadunna it is clear that there are no *Nothofagus* pollen grains. Instead, the Black Oak, *Casuarina*, and *Acacia*, a Tertiary arrival from South-East Asia, together with grass pollen dominate. The more moisture-loving *Dacrydium* is there. Thus, conditions still were wetter than at present. But, quite clearly, Central Australia had begun to dry during the time that the Etadunna rocks were being deposited. Not surprisingly, it was at the same time that the Antarctic ice sheet had expanded considerably. Temperatures decreased, as did precipitation, and the global air circulation, no longer sluggish, became quite intense.

From the Middle Miocene to the present, aridity intensified, but this change has not occurred without perturbations. Oscillations back to more humid conditions occurred, for instance, in the Early Pliocene. Even so, the takeover by grasses, chenopods and the composites, with their colourful splash across the countryside, continued afterwards.

Long cores of sediments taken in the last decade from Lake George in New South Wales, not far from Canberra, have provided perhaps the best mirror of events over the past 2 million years. Fossil plant sites of this age, Quaternary and very latest Tertiary, are rare in the arid zone. But this rarity is not at all surprising. Although the Quaternary Period is not all that long ago and usually boasts a good fossil record, conditions conducive to preserving plant material were rapidly deteriorating from the moister Tertiary. Whereas lakes and swamps are ideal preservers, sand dunes and dry uplands are not. So, in this case, the very conditions that were being investigated are those hampering the investigation!

178 Middle Tertiary outcrops at Lake Palankarinna, east of Lake Eyre, South Australia. From these rocks came the first extensive collections of Australian Tertiary land mammals. The vertical thickness of the Tertiary rocks here is as great as that of any Tertiary land mammal-bearing sequence in Australia. This entire middle Tertiary sequence could span a time interval as brief as 1 million years. (N. Pledge)

178

THE AGE OF MAMMALS IN GONDWANA

The global climate changed dramatically during the Cainozoic, a fact reflected by fossil floras and reinforced by fossil vertebrate faunas.

A suite of Cretaceous and possibly earliest Tertiary faunas in South America taken together suggest an intriguing history for mammals on the entire Gondwanan landmass. In the undoubted Cretaceous assemblages from the La Amagra (Early Cretaceous) and the Los Alamitos (Late Cretaceous) formations of Patagonia, several archaic groups of mammals are present: pantotheres, symmetrodonts, triconodonts, paratheres and possibly multituberculates. (The placental paratheres, or edentates, include modern forms such as the armadillo and tree sloth.) These assemblages have been dubbed the *Gondwanatherium* Fauna for one of its most prominent and unusual members. Prominent by their absence

in the *Gondwanatherium* Fauna are the marsupials and advanced placental mammals. Both groups are widespread and abundant during the Late Cretaceous in the Northern Hemisphere.

Two somewhat younger mammalian assemblages are known, the Tiupampa Local Fauna from Bolivia and the Laguna Umayo Local Fauna from Peru. Their exact age is in dispute, some workers regarding them as latest Cretaceous and others as Early Palaeocene. However, all are agreed that they postdate the Los Alamitos by about 5 to 10 million years. In these younger faunas, the only mammals present are marsupials and placentals. All of the groups represented in the La Amagra and Los Alamitos formations are absent. Presumably, except for the paratheres which are present in somewhat younger faunas and are still alive today, these La Amagra and Los Alamitos groups had become extinct in South America, for they are absent in all younger fossiliferous rocks there.

This dramatic turnover of the South American mammalian fauna near the Cretaceous–Palaeocene boundary suggests that, towards the end of the Jurassic, interchange between the Northern Hemisphere and the Southern Hemisphere became restricted. Groups such as the symmetrodonts and triconodonts that had evolved on much of Pangaea prior to that time were common to both, but those that appeared later, such as the paratheres, would be confined to one hemisphere or the other during the Cretaceous. Then, at about the time of the Cretaceous–Tertiary boundary, placentals and marsupials reached the Southern Hemisphere, having orginated in the Northern Hemisphere during the Cretaceous.

It was the comparison of the abundant South American record at such a critical juncture in time with that of the Northern Hemisphere which fostered this picture of mammalian evolution on the Gondwanan continents during the Cretaceous. A single tooth in the Cretaceous of Malawi is concordant with this picture of Africa and similarly India. In Australia, a single, exquisitely opalized jaw from the Early Cretaceous opal fields at Lightning Ridge, New South Wales, shows that monotremes, too, were part of this late Mesozoic Gondwanan suite. An isolated monotreme tooth has recently been discovered in Early Palaeocene deposits of Patagonia.

Conventionally, paratheres are classified as placentals, those mammals in which the young are born in an advanced state of development. For some time it has been thought that the paratheres are so different from other placentals that they should be split off as a separate group. Although paratheres share the advanced developmental condition of their offspring at birth, they differ from other placentals in that they maintain a lower average body temperature, and their temperature fluctuates through a greater range. So, the geographic distribution of paratheres in the Cretaceous supports the view that the paratheres evolved in the Southern Hemisphere at that time, having split off by the end of the Jurassic from the stock that would subsequently give rise to marsupials and other, unquestioned placentals in the Northern Hemisphere.

179 The skeleton of the parathere Ernanodon antellios from the Late Palaeocene of China. When the interconnection between North America and South America was established near the Cretaceous–Tertiary boundary, there was a brief two-way interchange of mammals, the ancestors of this species having reached Asia presumably through North America. Oddly enough, there is more evidence of mammals with a South American origin in the Palaeocene record of Asia than in that of North America. It was about the size of a sheep. (Courtesy of the Natural History Museum of Los Angeles County and C. Black)

180 Skull and jaws of Vincelestes neuquenianus, a pantothere from the Early Cretaceous La Amagra Formation in Patagonia, Argentina. This animal is a typical mammalian element of the older Gondwanatherium Cretaceous fauna of Argentina and probably most, if not all, of the Gondwanan continents of the time. Like all the other mammals of this assemblage, Vincelestes neuquenianus was neither a marsupial nor a placental mammal but a member of a group which had separated from their ancestors by the end of the Jurassic. x 1.5 (F. Coffa, courtesy of the Museo Argentino De Ciencias Naturales "Bernardino Rivadavia" and J. Bonaparte)

181-182 Skull and partial skeletons of male and female Pucadelphys andinus didelphid marsupials found in the latest Cretaceous or earliest Palaeocene Tiupampa Local Fauna from Bolivia. These two individuals were apparently buried in their dens, in the resting position. 181, x 0.85; 182, x 1.25 (L. Marshall, courtesy of R.S. Soruco)

183 Jaw of Steropodon galmani, an Early Cretaceous platypus with well developed teeth, unlike those of the living platypus. Found in the opal fields at Lightning Ridge, New South Wales. Top, in relfected light; bottom, in transmitted light. x 3.8 (J. Fields, courtesy of the Australian Museum and A. Ritchie)

179

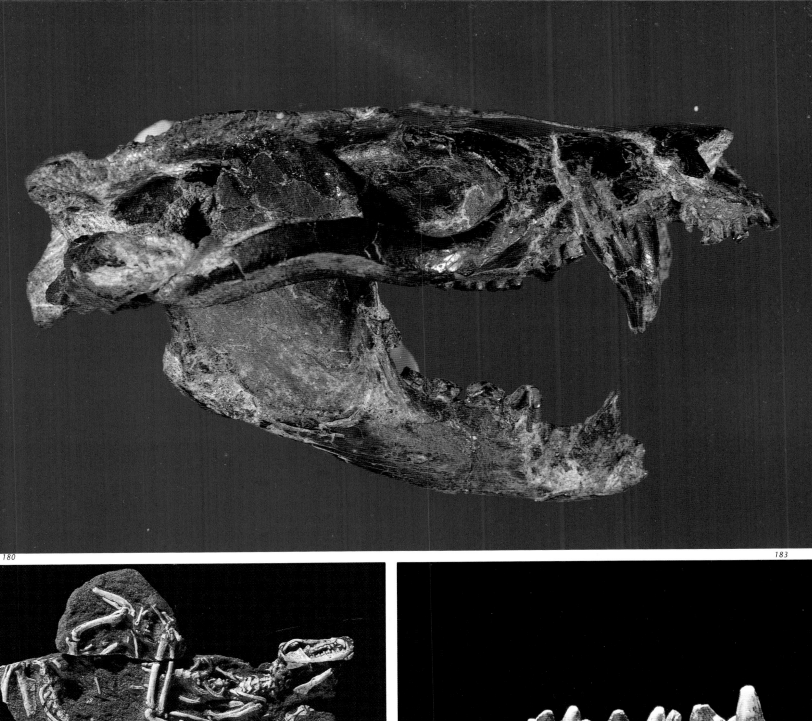

180

181

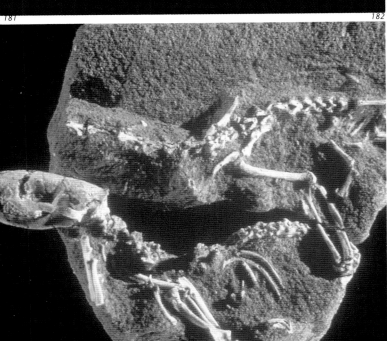

182

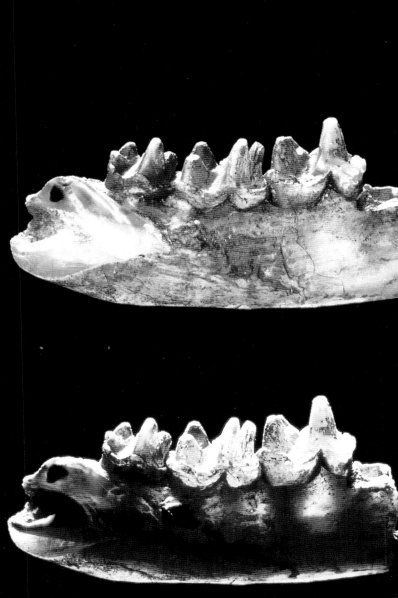

183

MARSUPIALS OF THE AMERICAS AND AUSTRALIA

On the basis of the shape of bones in the ankle, marsupials can be divided into two major groups, the ameridelphians and the australadelphians, the former centred in the Americas and the latter in Australia. Indeed, their distinctiveness is remarkably clear: every Australian marsupial is an australadelphian and all but one small group of the American marsupials is an ameridelphian. The intriguing exception is the living *Dromiciops australis* and its fossil relatives in the Microbiotheridae.

D. australis is one of the few marsupials that inhabits the temperate rainforests of southern Chile, an environment not unlike that which existed in much of Australia and Antarctica at the beginning of the Cainozoic. Few modern marsupials live in such habitats, which suggests that although it may well have been physically possible to cross from South America through Antarctica to Australia through such forests at the opening of the Cainozoic, few marsupials would have done so because the environment of the route was not generally suitable.

This picture suggests that Australian marsupials migrated from South America. But what is it about the evidence that militates against the alternative — that migration occurred in the reverse direction? Could not marsupials have originated in Australia during the Cretaceous and spread out from there? After all, modern marsupials are far more diverse in Australia than elsewhere. And the fossil record is poor, being represented in the Australian Cretaceous by merely a single jaw and a cranial fragment of a primitive platypus, *Steropodon galmani*. In the early Tertiary, too, Australian mammals are poorly known, the available record consisting of a handful of marsupial teeth. Is this idea of a South American origin simply the product of an all too human failing of reading the fossil record literally, not taking into account its biases, and assuming that because marsupials first appear in the fossil record of the Northern Hemisphere then that is where they originated? The short answer to that question is, "No".

Given the late Mesozoic connection across Antarctica between Australia and South America, and the relative isolation of Australia from North America at that time, it is difficult to imagine how marsupials would have evolved in Australia and reached North America first, only subsequently arriving in South America. Only if future discoveries show that the Late Cretaceous fauna from the Los Alamitos Formation is atypical of South America at the time, or if Cretaceous marsupials are eventually found in Australia, will this alternative be a viable hypothesis.

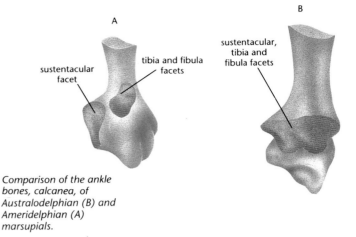

Comparison of the ankle bones, calcanea, of Australodelphian (B) and Ameridelphian (A) marsupials.

Another piece of evidence supporting the idea that the marsupials migrated from South America towards Australia rather than the reverse is the structure of the ankle. The australadelphian condition is structurally more advanced than the ameridelphian one. It is easier to imagine that the advanced condition arose in South America among the ancestors of *Dromiciops*, and it was members of that group which subsequently reached Australia. The alternative hypothesis would be that australadelphians reached South America from Australia and from them evolved the more primitive ameridelphian condition, the ameridelphians then radiating to form the majority of the American marsupials.

184 *The small rainforest-dwelling Dromiciops is the only South American marsupial that has a tarsal structure nearly identical with that of all Australian marsupials. All other South American marsupials, except Dromiciops, have a homogeneous ankle structure. Many other features of Dromiciops, including the lack of pairing of its sperm just as in all Australian marsupials, suggest that one group of New World marsupials could have given rise to the entire Australian metatherere fauna. x 1.3 (A. K. Lee)*

184

THE EARLIEST AUSTRALIAN MAMMALIAN FAUNAS, *AT TINGAMARRA AND GEILSTON BAY*

The first glimpse of the marsupials in Australia is to be found in south-eastern Queensland. In sediments at least 54 million years old (Early Eocene), preserved until modern times by an armour of overlying basalt, is an accumulation of mainly crocodilians and turtles. But amongst these vertebrate remains are occasional fragments of birds and tantalizingly rare dental remains of Australia's oldest marsupials, the Tingamarra Fauna. These few teeth are quite unlike those of younger Australian marsupials, and represent our first look at a group slightly removed from its South American relatives, although changed enough from them to obscure precise ancestry.

A hiatus of nearly 30 million years separates the oldest Australian marsupials of the Tingamarra Fauna from the next oldest assemblage, another handful of isolated teeth and a jaw fragment found at Geilston Bay near Hobart, Tasmania. This assemblage contains marsupials quite clearly allied to recognized Australian groups typical of the latter half of the Cainozoic, a phalangerid, a burramyid, a diprotodontoid and a possible dasyurid.

First discovered in the 1860s when the site was an active quarry, the specimens were sent to Sir Richard Owen, who dismissed them as a few insignificant Pleistocene scraps. For a century they languished in a drawer in the British Museum (Natural History). Only then did Jack Mahoney of the University Of Sydney happen to open that drawer and recognize the primitive nature of the specimens. By the time Richard Tedford of the American Museum Of Natural History in New York was able to revisit the Geilston Bay site in 1973 and confirm the date, it was no longer a quarry but had been filled in years before to become a playing field for a school. Had the true age been recognized in Owen's day, undoubtedly more fossils could have been collected there.

The Geilston Bay site is overlain by a basalt which has been dated at a minimum of 22 million years old, thus occurring near the Miocene–Oligocene boundary. It is unfortunate that the collection from the site is so small, because if it were larger, more meaningful comparisons could be made between its marsupials and those from a number of sites in central and northern Australia which are as yet imprecisely estimated to be of Late Oligocene to Middle Miocene age, and probably best referred to as middle Tertiary.

---

185 Top view of the carapace of Emydura macquarii from the probably Oligocene sediments of the foreshore near Taroona High School, Taroona, Tasmania. x 0.8 (S. Morton, courtesy of the University Of Tasmania and M. Banks)

186 Unidentified reptilian bone from a well core taken in Oligocene sediments at the 256-foot (75-metre) level at the Narrows, Queensland. x about 1.3 (F. Coffa, courtesy of the Queensland Museum and R. Molnar)

185

186

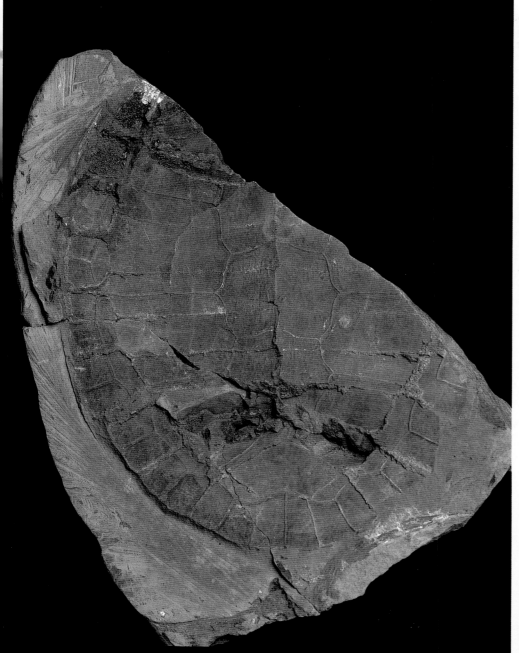

AN EXPLOSION OF VARIETY IN THE MIDDLE TERTIARY

It is at sites approximately contemporaneous with Geilston Bay that the first broad look at the terrestrial mammalian fauna of Australia is obtained. These sites are of two kinds, based on their location and the kinds of rocks that enclose the fossils: the playa lakes and dry stream beds of northern South Australia, and the limestone plateaus of north-western Queensland and the northern part of the Northern Territory.

LAKES AND STREAMS OF CENTRAL AUSTRALIA:

THE LAKE EYRE AND TARKAROOLOO BASINS

To the east of Lake Eyre and Lake Frome in South Australia are a number of fossil vertebrate sites on the margins of small playas or dry lake beds. One advantage of some of these sites, especially those at Lake Palankarinna, is that a number of fossiliferous horizons are actually stacked one on top of the other, or, in geological terms, they exhibit superpositional relationships to one another. So, the sequence of evolutionary events can be reconstructed relative to each other. This reconstruction method is not possible for many fossil vertebrate locales in Australia because they lie isolated from other sites.

The Tertiary-aged fossil-bearing rocks of the Lake Eyre and Lake Frome regions are typically white sands and pale green to black clays that were deposited in the channels of small streams and on the floors of lakes when the area was quite different from the desert conditions of today. Flanking these watercourses when they were full were lush rainforest communities, while between them were developed woodlands and even possibly grasslands. The lakes, where salinity increased through the Cainozoic, were suitable habitats for the flamingoes and the flamingo-like palaelodids. Dwelling there, too, were a variety of mammals, many of which would have looked much like their modern descendants except for their generally smaller size. There were others, however, that are rather unexpected. In the Lake Frome area, evidently due to a direct opening to the Southern Ocean in the south, freshwater dolphins made their way into the streams and lakes, certainly a reflection of the humid conditions of the time. Here, too, as well as in the more northern parts of South Australia, fully-toothed platypuses, Obdurodon, inhabited the permanent waters, while in the forests fringing these watercourses an arboreal assemblage including both familiar and exotic forms provides testimony of the differences between this time and today.

Diprotodontoids were for the most part the dominant mammalian herbivores. They were sheep-sized marsupials, rather than bullock-sized creatures like their Quaternary descendants. The diprotodontoids were part of an extensive radiation of wombat-related marsupials (Vombatiformes) that appear to have reached their peak in these middle Cainozoic terrestrial faunas, the first diverse assemblages of mammals on the Australian continent. Koalas and wombats, the only vombatiforms alive today, are represented in these assemblages by a few teeth. Other vombatiforms, the wynyardiids and the ilariids are known only in these associations. Ilariids were the largest marsupials of their day. At present little more is known about them than their molar teeth, which are quite distinct from those of all other mammals in that the crowns possess a unique, complicated pattern.

As the name suggests, the first wynyardiid specimen was discovered near Wynyard, Tasmania, in the latter half of the nineteenth century. It was found on a beach in a loose block of rock that had broken free from the nearby sea cliffs. In those same cliffs are millions of shells of marine invertebrates, which make possible the dating of the wynyardiid specimen as Early Miocene. At the time it was collected it was one of the very few Tertiary terrestrial mammalian specimens known from Australia and, therefore, received a great deal of scientific attention over the years. The analyses of the specimen led to varied interpretations, but generally Wynyardia was regarded as close in form to the ancestors of the larger diprotodontian marsupials. The diprotodontians form the bulk of the Australian herbivorous marsupials and are characterized by having a single pair of prominent lower incisor teeth. They include modern animals as diverse as wombats and koalas (the vombatiforms), kangaroos, and the many varieties of possums (phalangeriforms).

Most bizarre of the diprotodontians were the carnivorous thylacoleonids. Generally, amongst mammals, the evolutionary trend is from carnivore to omnivore to herbivore. However, thylacoleonids, the marsupial lions, have defied this "rule". Quite clearly, they are diprotodontians and were regarded as phalangeriforms until investigations of the base of the skull, the basicranium, suggested that they were more closely allied to the vombatiforms. In any case, their ancestors lost the canine tooth and developed a gap in the tooth row between the incisors in front and the last premolars behind, useful to herbivores in manipulating their food. As they shifted into the carnivorous lifestyle, thylacoleonids enlarged their most anterior incisors, which were used in piercing and grasping prey just as the canine tooth does in most carnivorous mammals.

The earliest known marsupial lions from the middle Tertiary deposits of South Australia, are the possum-sized Priscaleo and the Dingo-sized Wakaleo. Both had the typical highly specialized pattern to their teeth which characterizes this group, implying that this evolutionary pathway had been followed for some time before it first appears in the fossil record near the Oligo–Miocene boundary. The thylacoleonids, therefore, were probably a separate, distinct group of diprotodontians from much earlier in the Tertiary.

Early browsing kangaroos are represented by fossils found in middle Cainozoic sediments east of lakes Frome and Eyre. Grazers, which were to become the dominant Pliocene and Quaternary kangaroos, had yet to appear. The potoroos in these fossil faunas, although not as outwardly different from living species as were the kangaroos, were clearly more primitive.

The contemporaneous possums of middle Cainozoic interior Australia include a mixture of extant families such as pseudocheirids (ringtailed possums) and petaurids (gliders plus both Leadbeater's and striped possums) as well as some which make their only appearance at this time, the phalangerid-like Miralinidae and the petaurid-like Pilkipildridae. All are known solely on the basis of their dentitions, many only by isolated teeth. On this meagre evidence alone, there is no indication that any major structural changes have occurred amongst the survivors since this time, unlike the case with the kangaroos. They seem to have been primarily arboreal animals then, just as their descendants are today.

Notable by their late appearance were the phalangerids, which today include the highly successful brushtail possum, or Trichosurus. Phalangerids are an exception to the generality that possums were all present and much as we see them today when this diverse land mammal group first flooded the Australian fossil record in the middle Cainozoic.

187 Prearticular and anterior medial calvarial bone of the living lungfish, Neoceratodus forsteri, from Enoggera Reservoir in south-eastern Queensland. x 1.2 (F. Coffa, courtesy of the Queensland Museum and A. Kemp)

188 A prearticular and anterior medial calvarial bone of a middle Tertiary lungfish, Neoceratodus sp. from Ericmas Quarry at Lake Pinpa, Tarkarooloo Sub-basin, South Australia. x 0.9 (F. Coffa, courtesy of the Queensland Museum and A. Kemp)

189-190 Palatal view of platypus skulls. Obdurodon, on the top, is from the Middle Tertiary limestones on Riversleigh Station in north-western Queensland. Both anterior teeth and sockets for the rear teeth are clearly visible. Ornithorhynchus, on the bottom, is the modern platypus. x 1.6 (F. Coffa, courtesy of the University Of New South Wales and M. Archer)

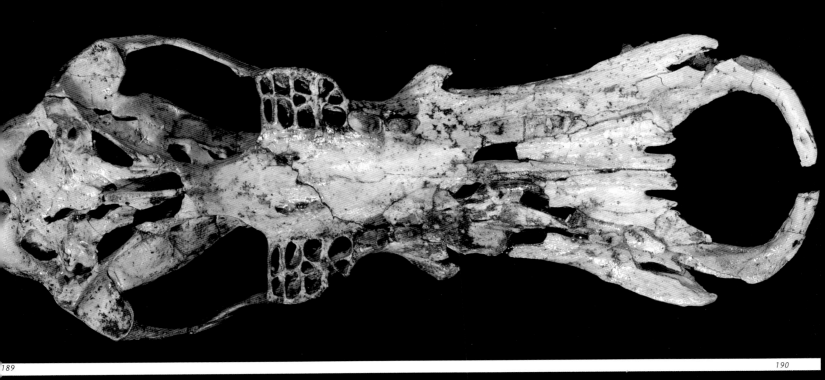

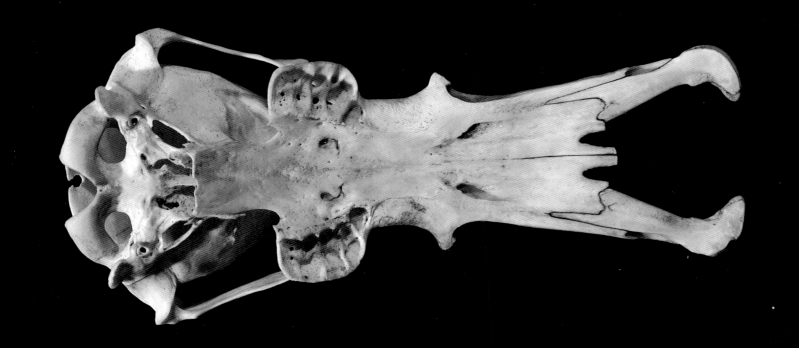

191 The holotype of *Wynyardia bassiana* from Early Miocene marine deposits exposed on the coast near Wynyard, Tasmania. For the better part of the twentieth century, this was the most complete specimen of a Tertiary terrestrial mammal known from Australia. x 1.3 (S. Morton, courtesy of the Tasmanian Museum and N. Kemp)

192-193 *Ngapakaldia tedfordi,* a diprotodontoid specimen from middle Tertiary rocks of South Australia. This animal is not only common from South Australia but is found at Riversleigh in north-western Queensland as well. On the basis of the structure of the base of its skull, *N. tedfordi* appears to have been related to the peculiar, sloth-like diprotodontoid *Palorchestes* of the Late Miocene to Late Pleistocene. x 0.5 (S. Morton, courtesy of the South Australian Museum and N. Pledge)

194-195 The lower jaw of *Ilaria lawsoni,* a member of an enigmatic group of large marsupials from the middle Tertiary known only from a few dentitions. This specimen was found at Lake Palankarinna, east of Lake Eyre, in the Etadunna Formation. x 0.7 (S. Morton, courtesy of the South Australian Museum and N. Pledge)

196-197 Fragment of lower jaw of *Purtia,* a middle Tertiary potoroid, or rat-kangaroo, from the Etadunna Formation at Lake Palankarinna, South Australia. x about 0.5 (S. Morton, courtesy of the South Australian Museum and N Pledge) About 0.5x.

198 The skull and lower jaw of the holotype of *Muramura williamsi,* another possible wynyardiid from the middle Tertiary Etadunna Formation at Lake Palankarinna, east of Lake Eyre, South Australia. x 0.6 (S. Morton, courtesy of the South Australian Museum and N. Pledge)

199 This fragmentary skull from middle Tertiary deposits east of Lake Frome, South Australia, was the first specimen of *Namilamadeta snideri* to be discovered. Because its teeth appeared intermediate between a number of diprotodontians, the name *Wombaroo* was originally proposed for this animal to emphasize its apparent structural position mid-way between two groups. Fortunately or unfortunately, reviewers objected and the name *Namilamadeta* was substituted. While a single species of this genus is known from South Australia, several others have been subsequently recognized in the middle Tertiary limestones at Riversleigh in north-western Queensland. *Namilamadeta* may be a wynyardiid, but its precise relationship to *Wynyardia* is not firmly established. x 1.3 (S. Morton, courtesy of the South Australian Museum and N. Pledge)

200 The lower jaw of *Raemeotherium yatkolai,* a rare diprotodontid from middle Tertiary deposits east of Lake Frome, South Australia. The form of the known molars suggests this animal may be the earliest branch of the diprotodontids, which eventually gave rise to larger forms such as the Late Miocene *Kolopsis* and the Late Miocene to Late Pleistocene *Zygomaturus.* However, this relationship is not certain yet because the animal is so poorly known. Besides the specimen pictured here, only a few isolated teeth of *R. yatkolai* are known. x 0.8 (S. Morton, courtesy of the South Australian Museum and N. Pledge)

201 *Wakaleo oldfieldi,* a Dingo-sized marsupial lion found in middle Tertiary deposits in South Australia. x 1.7 (S. Morton, courtesy of the South Australian Museum and N. Pledge)

191

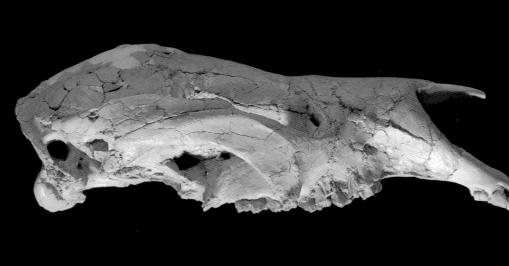

192

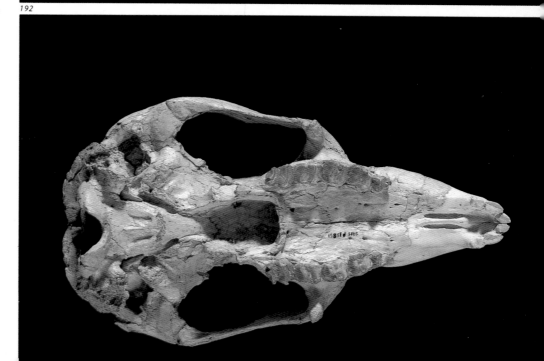

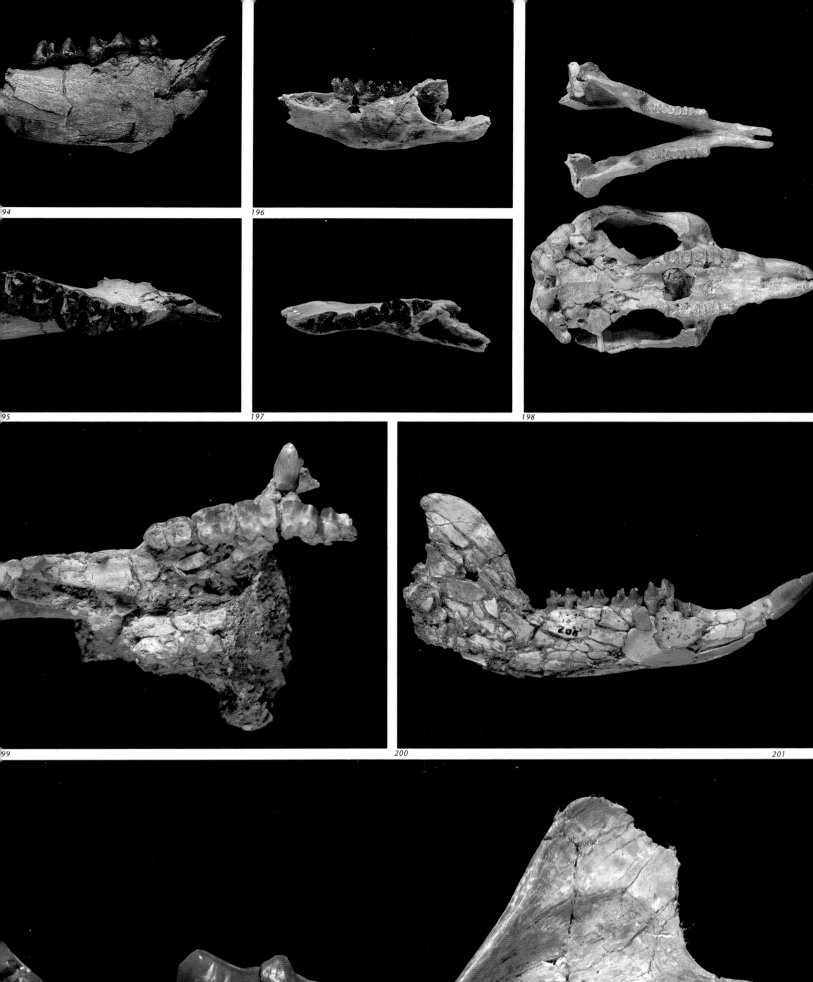

194

196

95

197

198

99

200

201

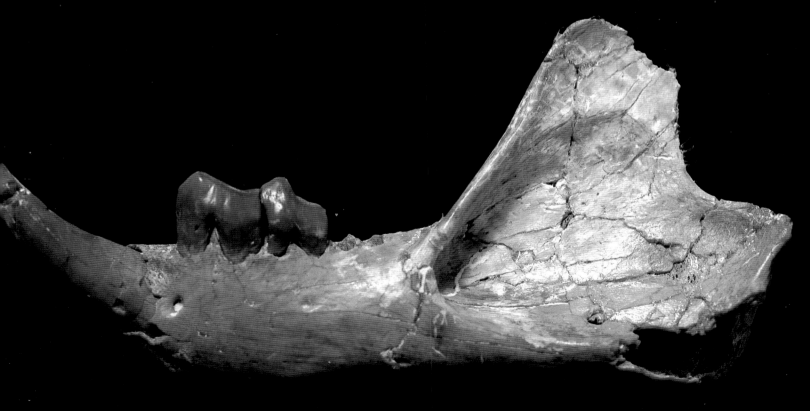

Dentally, at least, an extremely peculiar group of possums were the ektopodontids. When first discovered in 1962 their teeth were considered to be so odd that they were tentatively identified as multituberculates, a group of mammals widespread in the Mesozoic and early Cainozoic of the Northern Hemisphere. But it was soon recognized that their teeth were oriented in a manner altogether unlike those of multituberculates. By 1967 when the first published scientific description of these animals appeared they were, again tentatively, referred to the monotremes. With the discovery of a more primitive member of this family in 1972, finally their phalangerid affinities were recognized. The teeth consist of two or three rows of cusps possibly adapted to seed-eating. Their eyes were directed forward, giving them overlapping fields of view and probably the ability to see in three dimensions. Such an adaptation would favour a life in the trees. The group was never common in the fossil record, but persisted until the Early Pleistocene. Just what these animals ate is open to speculation, and ideas have ranged from seeds and nuts to insects. Their decline and extinction may have been brought about as a consequence of the entry of rodents into Australia during the Pliocene or may be due to the contraction of rainforests during the Pleistocene (T.F. Flannery, pers. comm.).

Although there are differences in the composition of these South Australian Tertiary faunas from one site to the next, it is difficult to interpret what they mean. Do these differences represent the evolutionary changes to be expected over a significant length of geological time, perhaps 5 or 10 million years, or are they simply due to dissimilarities in the habitats sampled? The inability to choose between these alternatives is a seemingly intractable problem for the inland Australian Tertiary mammalian sites as they are extremely difficult to date. Unlike some sites on the periphery of the continent, there are no volcanic rocks which permit radiometric dating. Nor are there marine deposits containing shells and microfossils interbedded with the rocks that yield the terrestrial vertebrates. If there were, such marine fossils could be correlated with similar marine organisms elsewhere in the world, and a relative age determination could be made in that manner.

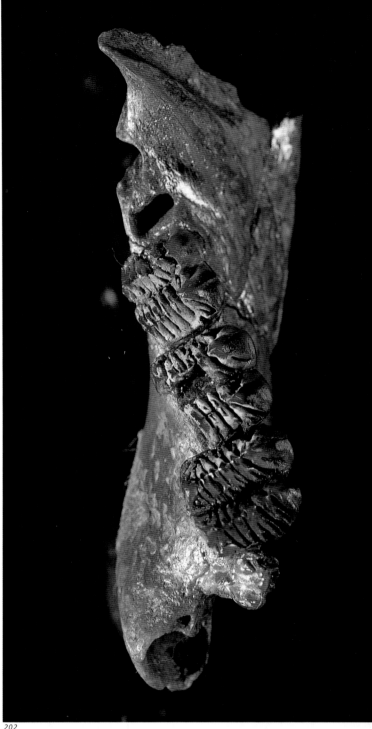

202

202 *Ektopodon stirtoni, a middle Tertiary phalangeroid related to the common Brush-tailed Possum. Equipped with uniquely structured molars in which there were a multitude of cusps evidently adapted for seed-eating, it was found in the Etadunna Formation at Lake Palankarinna, east of Lake Eyre, South Australia. x 3.5 (S. Morton, courtesy of the South Australian Museum and N. Pledge)*

203 *Quipollornis koniberi, a primitive owlet-nightjar (Family Aegothelidae) from Miocene diatomite deposits near Bugaldi, Warrumbungle Mountains, New South Wales. x 0.6 (F. Coffa, courtesy of the Australian Museum, A. Ritchie and R. Jones)*

204–209 *A record of the oldest and most primitive known macropodoids, a collective name for kangaroos and rat-kangaroos, is found in a small assemblage of isolated teeth in the middle Tertiary Namba Formation at Lake Tarkarooloo, east of Lake Frome, South Australia. Shown are: an upper premolar of Nambaroo tarrinyeri, x 12.3, an upper molar of Nambaroo novus, x 10.8, an upper molar of Palaeopotorous priscus, x 12.4, and a partial upper premolar, x 9.7, an upper molar, x 8.5, and a lower molar, x 13.2, of Gumardee. Nambaroo is the oldest and one of the smallest known macropods, or kangaroos, and Palaeopotorous is the most primitive known rat-kangaroo. (F. Coffa, courtesy of the Museum Of Victoria)*

210 *This avian foot from the Redbank Plains on the outskirts of Brisbane, Queensland, is probably of Eocene age. It is one of the very few bits of known evidence about bird evolution in Australia during the early Cainozoic. x 1.2 (F. Coffa, courtesy of the Queensland Museum and R. Molnar)*

211 *The partial hind leg of a very primitive, middle Tertiary emu, Emuarius gidju, from the Wipijiri Formation of Lake Ngapakaldi, east of Lake Eyre, South Australia. x 1.3 (F. Coffa, courtesy of the South Australian Museum and N. Pledge)*

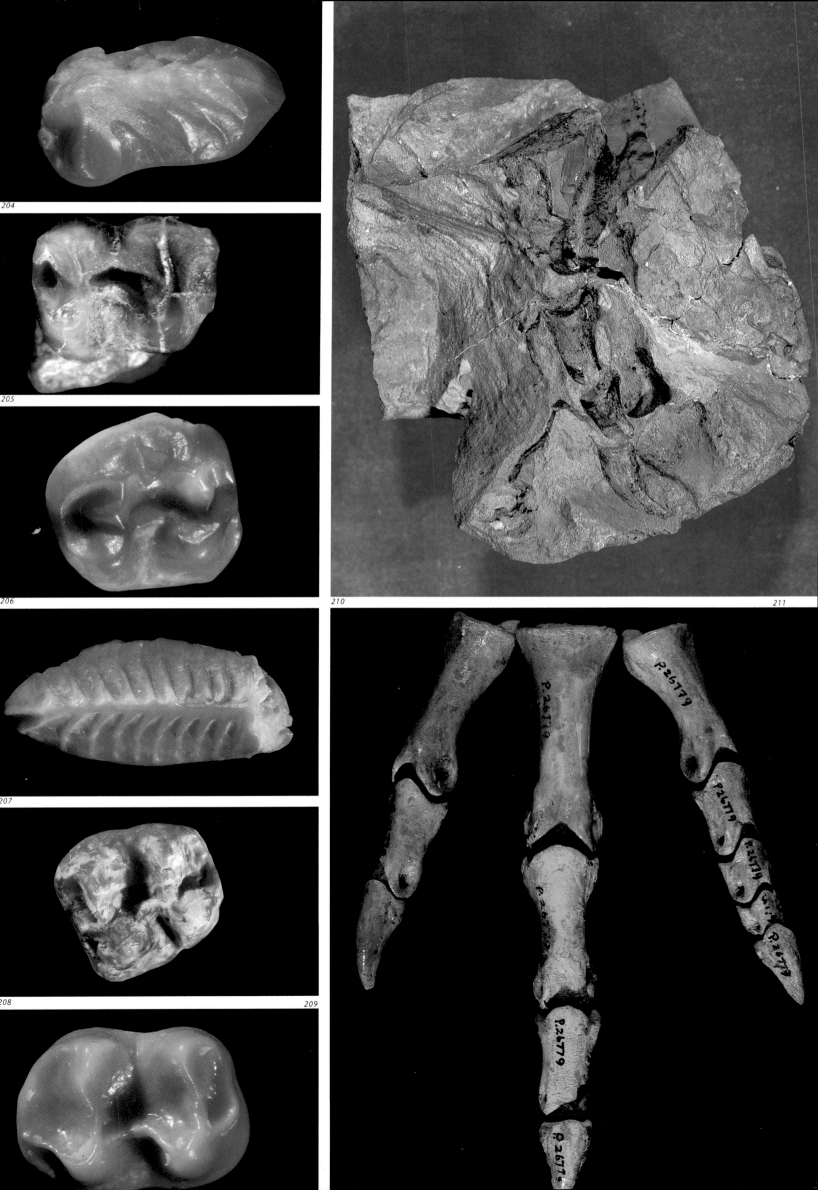

204

205

206

207

208   209

210   211

THE NORTHERN LIMESTONE COUNTRY:

RIVERSLEIGH AND BULLOCK CREEK

The second major suite of middle Tertiary Australian terrestrial vertebrate sites is found principally on Riversleigh Station in north-western Queensland and at Bullock Creek on Camfield Station in the north-western Northern Territory. Extensive deposits of freshwater limestones occur at these locales and are probably contemporaneous with the fossil-bearing white sands and green claystones of South Australia. Despite being formed at more northerly, tropical latitudes and in a different environment from that of the more southerly sites, significant similarities in the mammalian faunas and, to a more limited extent, the avifaunas of the two regions have been observed.

Freshwater limestones of the middle Cainozoic are widely developed in the Northern Territory as well as in westernmost Queensland and the eastern part of Western Australia. Typically, they are incised into Cambrian limestones and appear to represent deposits of a moist climatic phase when the older marine limestones were dissolved and subsequently redeposited by the lime-charged streams and lakes that flooded and cut into the countryside.

Of the two principal freshwater limestone fossil-bearing areas, by far the greatest amount of collecting effort has gone into Riversleigh. As in South Australia, different sites appear to represent different times within an interval nebulously characterized as middle Tertiary. Again, the poorly dated nature of individual localities makes it difficult to assess just how many years passed between the deposition of any two of them. Amongst the sites at Riversleigh, those that are regarded as the older rather than the younger ones appear more

likely to be of similar age to the South Australian middle Tertiary sites. The various sites at Bullock Creek have not been so well studied as yet, but all appear to be of much the same age: indeed, the known outcrops of the Tertiary fossil-bearing limestone in this area seem to occur along a solitary, somewhat sinuous east-west band, suggesting that all sites may represent deposition in a single river channel active during one brief moment of geological time.

In the younger assemblages at Riversleigh but still within the middle Tertiary interval, as well as at Bullock Creek, there is a suite of mammals that were more advanced than those that occur in the middle Tertiary South Australian faunas. This northern limestone country boasts larger diprotodontoids and a greater variety of kangaroos and potoroos than are found in the South Australian assemblages. Amongst the kangaroos, larger ones had evolved, and grazing forms made their first appearance. Here, too, are unique and peculiar mammals whose relationships to other marsupials are not clearly evident. Some of them are strange enough that they may not be marsupials at all.

One such unusual animal is *Yalkaparidon coheni*. First known informally as "Thingodon", it was originally thought to be a placental mammal. Then, more material was found, including a lower jaw with the typically inflected angle (a flange on the bottom and rear part) of marsupials. *Yalkaparidon* has molars rather like those of the marsupial mole incisors, which are diprotodontian-like, and a skull base resembling that of the bandicoots. This mammal does not fit comfortably into any one of the major marsupial groups, and for this reason it has been placed in an order of its own. Perhaps more complete material will one day allow this strange and intriguing animal's relationships to be resolved.

A single tooth tentatively assigned to the rhinolophids or "horse-shoe bats" is all that is known about the Chiroptera in the middle Tertiary South Australian deposits. In stark contrast, the Riversleigh sites have yielded tens of thousands of fossil bat specimens. About 25 species have been recognized so far, all clearly members of modern Australian families and most belonging to living genera. Far more than just isolated teeth are known. In some cases, it has been possible to reconstruct nearly the entire skeleton for particular species.

212 Riversleigh, north-western Queensland, is today an area of contrasts: the dry plains are dotted with fossiliferous limestone hills, but these arid lands are incised by rivers, such as the Gregory, along whose banks grow a lush tropical vegetation. Riversleigh has produced a rich array of fossils entombed in the ancient rivers, lakes and caves that characterized this area for at least the last 30 million years or so. (P. Vickers-Rich)

212

The Riversleigh bats are most similar to Early Miocene species from Europe. Unfortunately, this similarity is not one of species that are shared in common, but rather a structural similarity between different species of one genus, *Brachipposideros*. Because of the constraints imposed by the requirements for flight, the earliest known bats, which are recorded in Eocene sediments of North America, are not markedly different from modern ones. As a result, their subsequent rate of evolution has been amongst the slowest for any mammalian group. Thus, only the most subtle differences are found between bats of different geological ages, a fact that limits their usefulness in dating rocks. Far more research has been carried out on the European than on the Asian fossil record of bats. Therefore, it is likely that the apparent greater affinity of the Riversleigh bats to European rather than Asian forms may be more a reflection of the amount of study than of real similarity. Only further discovery and study, especially of the Asian faunas, will resolve this issue.

The Riversleigh sequence is characterized by a wide variety of mammalian species, and reflects greater biotic diversity than is found in the region today. Comparison of the diversity patterns in the Riversleigh fossil mammalian faunas with those in modern environments suggests that during the middle Tertiary the Riversleigh area, with its abundance of arboreal forms and the high total number of species, probably hosted a tropical rainforest rather than a woodland community that is typical of the area today. The occurrence of fossils such as those related to the marsupial moles, or notoryctids, at Riversleigh suggests that some of the modern desert-adapted marsupials of Australia arose from species which inhabited strikingly different environments as recently as the middle Cainozoic.

The limestones at Riversleigh and Bullock Creek preserve individual bones and skulls in exquisite detail. This situation is in marked contrast to the sites in South Australia, where contemporaneous fossil vertebrates are frequently crushed. On the other side of the coin, however, articulated skeletons are found in those green South Australian clays but are so far unknown from the limestone tombs of northern Australia. The two types of preservation reinforce one another in providing a picture of what Australian middle Tertiary land vertebrates were like.

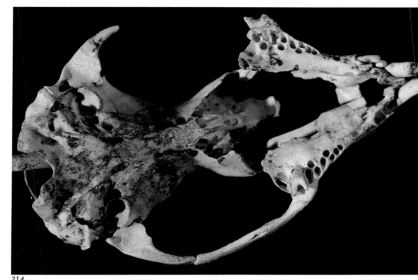

214

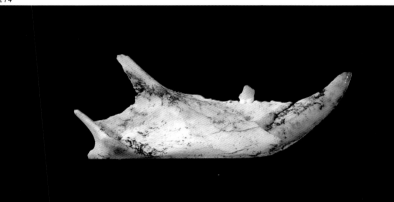

215

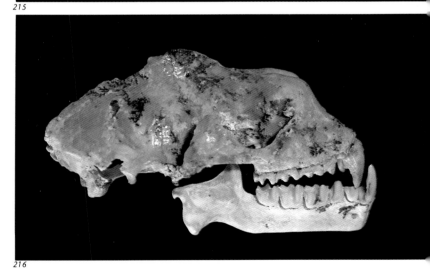

216

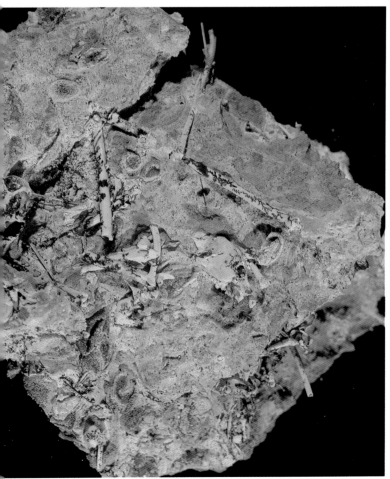

213

213 A block of limestone containing a mass of middle Tertiary bat bones, stacked like jackstraws, collected at Riversleigh, north-western Queensland. x 1.4 (F. Coffa, courtesy of the University Of New South Wales and M. Archer)

214–215 *Yalkaparidon coheni,* a mammal from the middle Tertiary limestones at Riversleigh, north-western Queensland, with a mosaic of characters suggesting relationships with a diversity of marsupials including bandicoots, marsupial moles, and diprotodontians. x 2.1 (F. Coffa, courtesy of the University Of New South Wales and M. Archer)

216 The skull of a hipposiderid bat extracted from the middle Tertiary limestone at Riversleigh, north-western Queensland. x 3.4 (F. Coffa, courtesy of the University Of New South Wales and M. Archer)

217 Palatal view of the skull of *Neohelos tirariensis* from the Camfield Beds at Bullock Creek, Northern Territory. A few teeth of this species occur in the youngest part of the South Australian middle Tertiary Wipijiri Formation. However, this diprotodontid is best known from a large sample of skulls and jaws found at Bullock Creek. It is intermediate in size between the earliest well known diprotodontoids, such as *Ngapakaldia tedfordi* from the middle Tertiary of South Australia, and the late Miocene ones from Alcoota in the Northern Territory. x 0.5 (S. Morton, courtesy of the Museum Of Victoria)

218 Mandible of a marsupial lion, *Wakaleo vanderleueri*, from the middle Tertiary Camfield Beds at Bullock Creek, Northern Territory. This genus of primitive thylacoleonids also occurs in the middle Tertiary of South Australia. x 1.0 (F. Coffa, courtesy of the Spencer And Gillan Museum and P. Murray)

219 *Propalorchestes novaculacephalus*, a rare diprotodontid from the Camfield Beds at Bullock Creek, Northern Territory. Note the crocodile tooth marks on this specimen. x 1.0 (F. Coffa, courtesy of the Spencer And Gillan Museum and P. Murray)

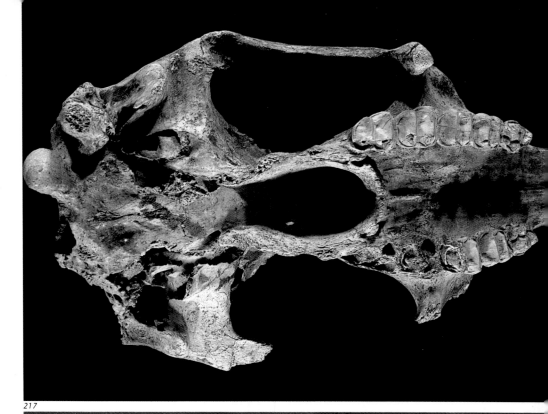

217

218

219

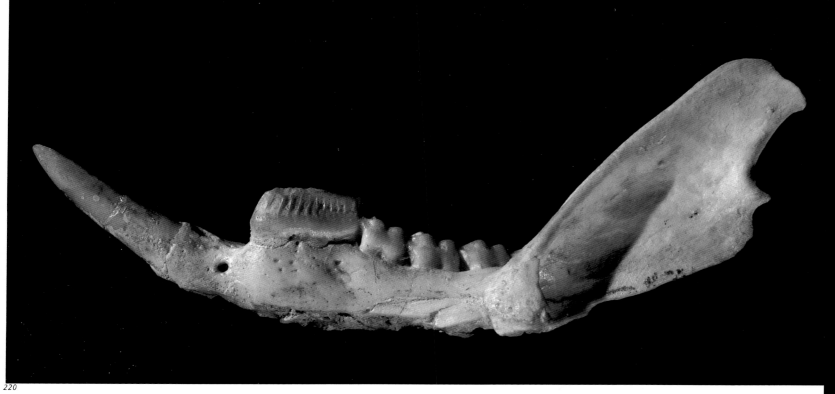

220

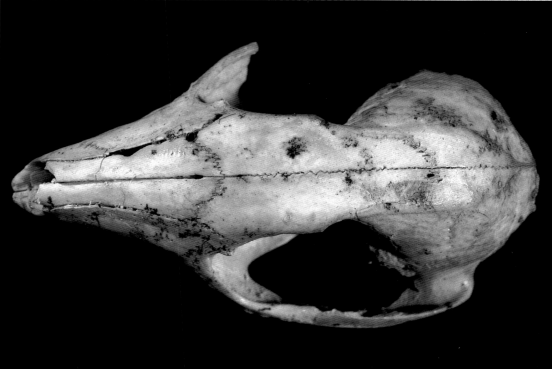

220–222 Although the species is extinct, the genus of this rat-kangaroo, *Bettongia moysei*, is extant. This specimen was collected from the middle Tertiary limestone at Riversleigh, north-western Queensland, in deposits thought to be younger than the middle Tertiary mammal-bearing deposits in South Australia. Jaw x 3.9; skull x 1.8 (F. Coffa, courtesy of the University Of New South Wales and M. Archer)

221

222

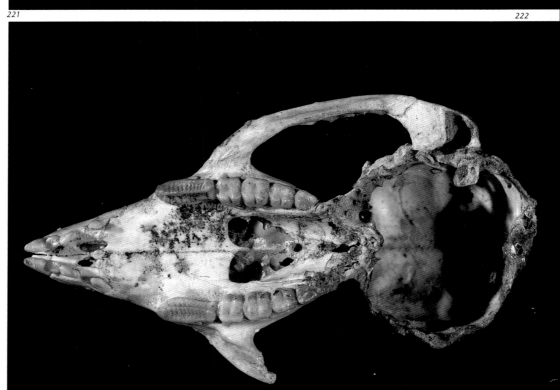

MIDDLE CAINOZOIC VERTEBRATES ASSOCIATED WITH THE MAMMALS

Other vertebrates besides mammals have left their remains in many middle Cainozoic sites.

Virtually all of the non-passeriform bird families that have an Australian fossil record are represented in the South Australian middle Tertiary deposits. With two prominent exceptions, all these families are alive today. One of the two extinct avian families is the large, ground-dwelling Dromornithidae (the Mihirungs) which radiated broadly in Australia during the middle Tertiary only to dwindle to a single species by the Pliocene. The second now extinct avian family consists of the flamingo-like palaelodids, which, like the dromornithids, persisted into the Pleistocene. Palaelodids are also known in the Neogene of Europe and North America but are now completely extinct. Flamingoes, although living today, do not now occur in Australia. In the middle Tertiary avifaunas of Central Australia they were diverse, and remained so well into the Pleistocene. In addition to true flamingoes, another prominent and varied bird group in Australia at this time was the ducks, and to a lesser extent the charadriiforms, or wading birds. At this time, too, are recorded the earliest emus, which possessed many features not only of living emus but also of cassowaries. These primitive "emus" of the late Palaeogene and early Neogene lacked the excessive elongation of the hind limb that is characteristic of the highly cursorial, or swiftly running, living Emu, *Dromaius novaehollandiae*.

The picture painted by the avifauna is one of permanent lakes that provided not only a food resource but a vast nesting area for such colonial species as the flamingoes and probably the palaelodids. The record is skewed towards waterbirds, such as the thicknees (Burhinidae), waders, ducks and cormorants, but the rarer upland representatives, such as the dromornithids and the emus, give a hint that forested environments abounded near the lakes and streams, and somewhere not far away must have been some more open country, particularly in later Neogene times. In the later Miocene both emus and dromornithids evolved some cursorial forms with elongated hind limbs, which were speed enhancers and energy savers, and thus reflect that vertebrates were moving into more open environments.

In central and northern Australia a diverse actinopterygian fish fauna (dominated by siluriform and perciform teleosts) along with a variety of lungfish and crocodiles reinforce the picture of a much wetter and more predictable environment in the middle Cainozoic than exists in those regions today. Up to nine different species of lungfish may have been present in middle Tertiary times. Aquatic vertebrates filled both herbivore and carnivore roles, and included the broad-toothed ceratodonts and the neoceratodonts. Ceratodonts survived into the Pleistocene in Australia, having lived long beyond their time elsewhere in the world. The neoceratodonts radiated broadly in the Miocene, but the group was dramatically restricted during the Pliocene, leaving only the living *Neoceratodus forsteri*

confined to the Burnett, Mary and Brisbane rivers in south-eastern Queensland today. Despite their decrease in diversity at this time, the group remained more varied in Australia for far longer than anywhere else in the world. Only two other genera of living lungfish remain of this once varied group, both occurring on Gondwanan fragments to which they were probably restricted with the break-up of Gondwana and Pangaea in the late Mesozoic. *Lepidosiren paradoxa* still survives in South America and several species of *Protopterus* persist in Africa.

Tree frogs (from the family Pelodryadidae, and including *Litoria*) and leptodactylid frogs (*Crinia, Kyarranus, Limnodynastes* amongst others) are known from South Australian and central and northern Australian middle Tertiary sites. Amongst the oldest known Australian frogs is *Australobatrachus ilius* which comes from the middle Tertiary central Australian sites at Lake Palankarinna as well as from those in the Lake Frome area of South Australia. Its relationships to other frogs are not clear, but it is well represented in these oldest diverse Cainozoic vertebrate assemblages.

The reptile fauna of the middle Tertiary locales is varied, and particularly diverse from the northern limestone sites. Crocodiles, turtles, lizards and snakes are present. Crocodiles include not only familiar forms but also ziphodonts, which were probably terrestrial crocodilians having deep, laterally compressed snouts and steak-knife serrated teeth, quite resembling those of carnivorous dinosaurs. The ziphodonts survived in Australia for millions of years beyond their time elsewhere, becoming extinct in Miocene times in other parts of the world but persisting into the Pleistocene of Australia in the form of *Quinkana fortirostrum*.

Turtles are present in both the central and northern Australian sites and include the chelids and the horned turtles, or meiolanids. They seem prosaic to us, while dinosaurs do not. Were turtles totally extinct, they would undoubtedly fascinate people more than they now do. But a turtle with large, recurved horns on its skull would make even the most blasé people sit up and take notice. *Meiolania* was just such an animal. It roamed Australia from at least the middle Tertiary to the end of the Pleistocene. In Queensland, during the Pleistocene, there were probably two different species of these giant horned turtles.

The record of *Meiolania* on mainland Australia is confined to the eastern half of the continent and consists of horn cores plus skull and skeletal fragments. It is to Lord Howe Island that one must turn for complete skeletons of these animals. The fossils there occur in soft sandstone formed of bits of shell laid down between 100,000 and 120,000 years ago. The Lord Howe Island meiolanids were up to 2 metres long. An indication of the much larger dimensions of the mainland *Meiolania* is the fact that some horn cores from Queensland are twice the size of those found on Lord Howe Island.

Of the lizards, the agamids, the skinks and the varanids are known from the middle Tertiary northern South Australian sites and the geckoes from the northern limestone sites of Riversleigh and Bullock Creek. Snakes, too, are quite varied, especially in the northern limestone sites, and include the pythonids, the madtsoiids with their South American affinities, the typhlopids and the fanged elapids. Boids are also known from the northern South Australian sites. The early to middle Cainozoic reptile assemblages consist of forms with a mixture of origins. Some, like the ziphodont crocodilians, meiolanid turtles and madtsoiid snakes hark back to a Gondwanan or perhaps Pangaean origin, later becoming isolated on the continental break-up fragments. Other forms, such as the varanid lizards, the geckoes and the elapid vipers most surely were amongst the earlier colonizers from the north as Australia began to dock with southern Asia.

223 Lord Howe Island. The most extensive collections of the horned tortoise Meiolania platyceps from anywhere in the Australian region have been found here. (P. Vickers-Rich)

224 Skeleton of the horned tortoise Meiolania platyceps from the Late Pleistocene of Lord Howe Island. x 0.3 (J. Fields, courtesy of the Australian Museum, A. Ritchie and R, Jones)

225 A clutch of eggs of the horned tortoise Meiolania platyceps from the Late Pleistocene of Lord Howe Island. x 0.5 (F. Coffa, courtesy of the Australian Museum and A. Ritchie)

226 The skull of a horned tortoise Meiolania platyceps from Lord Howe Island, partially prepared at the Australian Museum. x 0.4 (G. Miller, courtesy of the Australian Museum, A. Ritchie and R. Jones)

223

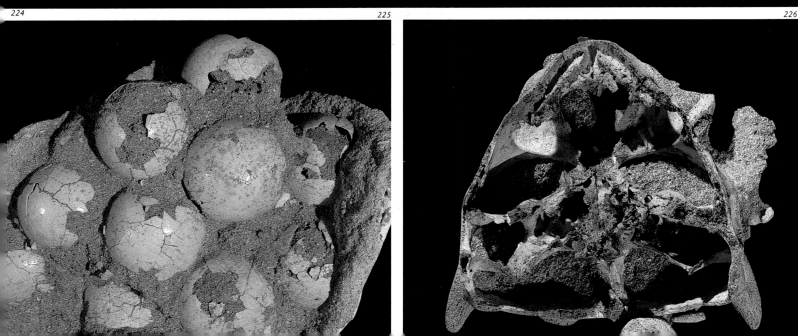

227

228

229

230

ALCOOTA, A FAUNAL SWITCH-POINT IN THE LATE MIOCENE

Only one inland assemblage, that of Alcoota, provides a glimpse of Australian terrestrial vertebrate life in the Late Miocene. It lies less than 200 kilometres north-east of Alice Springs, about mid-way geographically between the middle Tertiary South Australian sites and those of Bullock Creek and Riversleigh on the northern limestone plateau. The fossil fauna of Alcoota indicates that the rainforest communities, all but ubiquitous during the Cainozoic until this time, were being replaced by more arid-adapted woodland and savannah communities. Grazing diprotodontians are an important component of the mammalian fauna for the first time, indicating that more open habitats were now existent, but the variety of crocodilians present shows that there was still an abundance of permanent water in the area. Unfortunately, no plant macrofossils, spores or pollen have yet been recovered from this site.

The diprotodontoids and macropodids both show a profound size increase over their middle Tertiary predecessors. A number of new genera put in their earliest appearances in the geological record and continue on for some time, for example, the diprotodontoids *Palorchestes* and *Kolopsis*. Other forms are known only from this one locality: the diprotodontoids *Alkwertatherium*, *Plaisiodon* and *Pyramios*, and the kangaroos *Dorcopsoides* and *Hadronomas*. The trend in both of these marsupial groups towards larger size in the late Cainozoic may have been the result of selection of animals that could survive on plants with ever-decreasing nutritive value. Larger animals have an advantage over smaller ones because their metabolic rates operate more slowly, and thus less nutritive value per unit mass of fodder processed is required. Perhaps, owing to the competitive advantages that similar-sized kangaroos had over diprotodontoids, at any one site in the late Cainozoic, the diprotodontoids are consistently larger than the kangaroos. The kangaroos were more efficient utilizers of the environment. Possibly they were faster breeders and/or were more efficient in moving about in search of ever-scarcer fodder. Eventually they displaced diprotodontoids of comparable size.

While the diprotodontoids and kangaroos reached their maximum size in the Late Pleistocene, the flightless dromornithid birds were at their largest early in their history, in the Miocene. Their first appearance in the geologic record is a few tracks in Late Oligocene sediments mined for tin at Pioneer, Tasmania, and fossil bones at Bullock Creek and Riversleigh. These large herbivores reached their acme of size and diversity in the Late Miocene of Alcoota. The largest of these birds was *Dromornis stirtoni*, which outstripped gargantuan Malagassy elephant birds in size and was thus the largest bird ever known. In each of two later epochs the dromornithids were reduced to but two single, smaller species, *Dromornis australis* in the Pliocene and *Genyornis newtoni* in the Pleistocene.

Alcoota marks a time of fundamental change in the terrestrial mammalian fauna of Australia. Only one genus, *Wakaleo*, the marsupial lion, is known from older deposits. Four other genera are found at Alcoota and at younger sites: *Thylacinus*, the Tasmanian wolf; the diprotodontoids *Kolopsis* and *Palorchestes*; and the ringtailed possum *Pseudochirops*. The remaining mammalian fauna is unique to the site or known only at a few other localities of equivalent age. Thus, Alcoota stands apart, with a few representatives of the groups that were to dominate in the future and only one holdover from the past.

---

227 Replicas of the skull of three meiolaniid turtles from the Late Cretaceous of Patagonia (Niolamia, left), the Pleistocene of Lord Howe Island (Meiolania platyceps) and the Pleistocene of the Darling Downs, Queensland (Meiolania oweni). (E. Gaffney)

228 The horned tortoise from the Tertiary Camfield Beds at Bullock Creek, Northern Territory. Although Meiolania specimens from this site are among the oldest remains of Meiolania from Australia, the distribution of meiolanids suggests that they were on this continent long before this time. x about 0.4 (F. Coffa, courtesy of the Spencer and Gillan Museum and P. Murray)

229 Underside of the skeleton of the horned tortoise Meiolania platyceps from the Late Pleistocene of Lord Howe Island. x 0.1 (J. Fields, courtesy of the Australian Museum, A. Ritchie and R. Jones)

230 A sparid perciform teleost, Chrysophrys sp., from the Early Miocene Mannum Formation near Overland Corner, South Australia. x 0.4 (S. Morton, courtesy of the South Australian Museum and N. Pledge)

231 Fossil deposits at Alcoota Station in the Northern Territory occur in the grassy flats in front of the low red hill (P. Vickers-Rich)

231

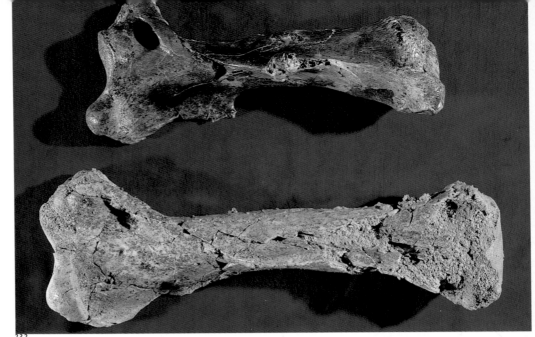

232 Humeri of diprotodontoids *Neohelos* (small specimen) and *Propalorchestes*. *Neohelos* is from the Late Oligocene to Middle Miocene Camfield Beds of Bullock Creek in the Northern Territory, while *Propalorchestes* is from the Late Miocene Waite Formation of Alcoota Homestead, also in the Northern Territory. x 0.4 (F. Coffa, courtesy of the Spencer And Gillan Museum and P. Murray)

233 Skull of *Palorchestes painei*. This species is the earliest known member of this diprotodontoid genus which persisted until almost the end of the Late Pleistocene. It has been reconstructed as a marsupial ground sloth, utilizing enlarged claws as a feeding aid. The specimen was found in the Late Miocene Waite Formation at Alcoota, Northern Territory. x 0.4 (F. Coffa, courtesy of the Spencer And Gillan Museum and P. Murray)

234 Skull of *Alkwertatherium* from the Late Miocene Waite Formation at Alcoota, Northern Territory. This diprotodontoid combines features of *Plaisiodon* and *Pyramios* in a perplexing manner that makes its position within the family uncertain. x 0.3 (F. Coffa, courtesy of the Spencer And Gillan Museum and P. Murray)

235-237 Specimens of *Kolopsis torus*. The diprotodontid genus *Kolopsis* provides the principal basis for assigning the fossil deposits in the Waite Formation at Alcoota, Northern Territory, to the Late Miocene. The same genus occurs at Beaumaris, Victoria, in the marine Black Rock Member of the Sandringham Sand. 235, partially prepared lower jaw, x 0.4; 236, top view of skull, x 0.7; 237, side view of lower jaw. (F. Coffa, courtesy of the Spencer And Gillan Museum and P. Murray)

232

233

234
235

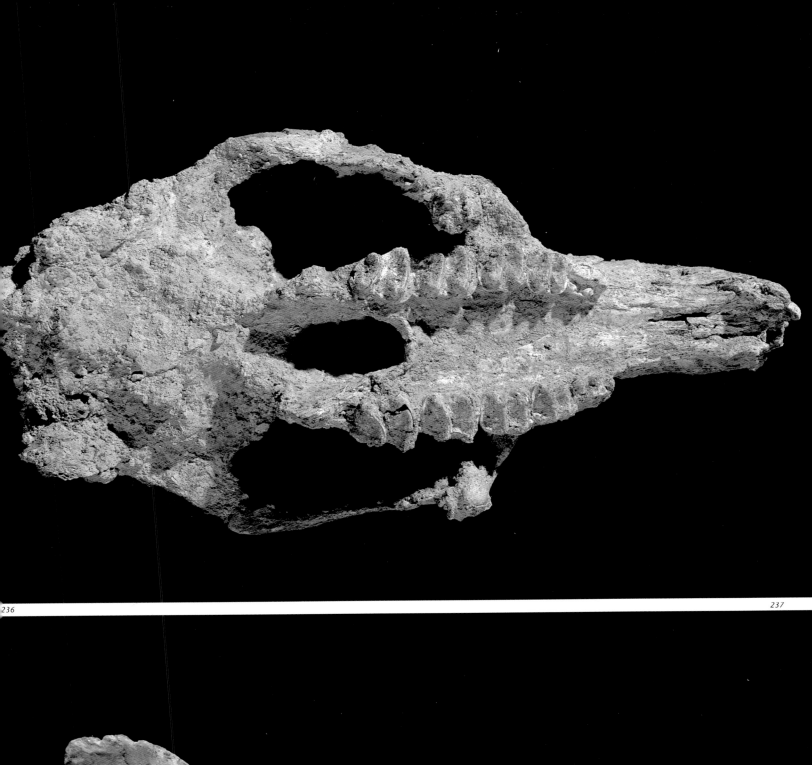

238 Mandible of *Pyramios alcootense*, the largest diprotodontid from the Late Miocene Waite Formation at Alcoota, Northern Territory. x 0.4 (F. Coffa, courtesy of the Spencer And Gillan Museum and P. Murray)

239 Palatal view of a kangaroo, *Hadronomas puckridgi*, which is of uncertain position within the family of Macropodidae. Although common in the Late Miocene Waite Formation at Alcoota, Northern Territory, it is known nowhere else. x 0.6 (F. Coffa, courtesy of the Spencer And Gillan Museum and P. Murray)

240 Mandible of the kangaroo *Dorcopsoides fossilis* from the Late Miocene Waite Formation at Alcoota, Northern Territory. The generic name is in reference to the similarity to the living *Dorcopsis*, the New Guinea forest wallaby. x 1.8 (F. Coffa, courtesy of the Spencer And Gillan Museum and P. Murray)

241 Vertebrae, in dorsal or top view, of the largest ground bird that ever lived, the dromornithid *Dromornis stirtoni*, from the Late Miocene Waite Formation at Alcoota, Northern Territory. x 0.5 (F. Coffa, courtesy of the Spencer And Gillan Museum and P. Murray)

242 Comparison of the tibiotarsi of two of the large flightless ground birds found in the Late Miocene Waite Formation at Alcoota, Northern Territory. From bottom to top: *Emuarius gidju* and *Ilbandornis* x 0.5 (F. Coffa, courtesy of the Spencer And Gillan Museum and P. Murray)

243 Tarsometatarsi (distal hind limb bones) and scapulocoracoids (shoulder girdles) of *Dromornis stirtoni* (large specimens) and *Ilbandornis* from the Late Miocene Waite Formation of Alcoota Homestead in the Northern Territory. These large herbivorous birds formed much of the plant-eating biomass of Central Australia for most of the middle to late Cainozoic. x 0.25 (F. Coffa, courtesy of the Spencer And Gillan Museum and P. Murray)

244 Hind limb of *Dromornis stirtoni* from the Late Miocene Waite Formation at Alcoota, Northern Territory. x 0.18 (F. Coffa, courtesy of the Spencer And Gillan Museum and P. Murray)

245 Lower hind limb of *Ilbandornis woodburnei* from the Late Miocene Waite Formation at Alcoota, Northern Territory. x 0.3 (F. Coffa, courtesy of the Spencer and Gillan Museum and P. Murray)

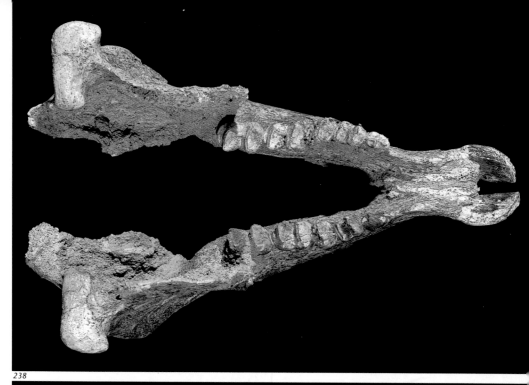

238

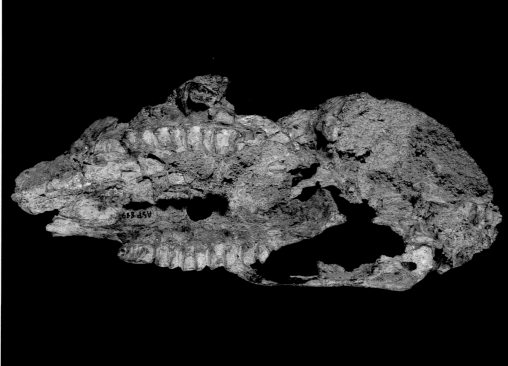

239

240

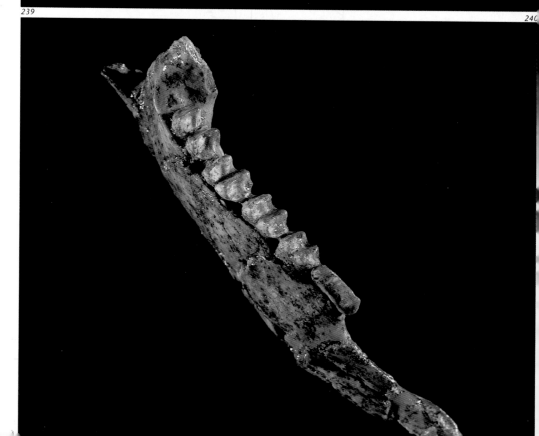

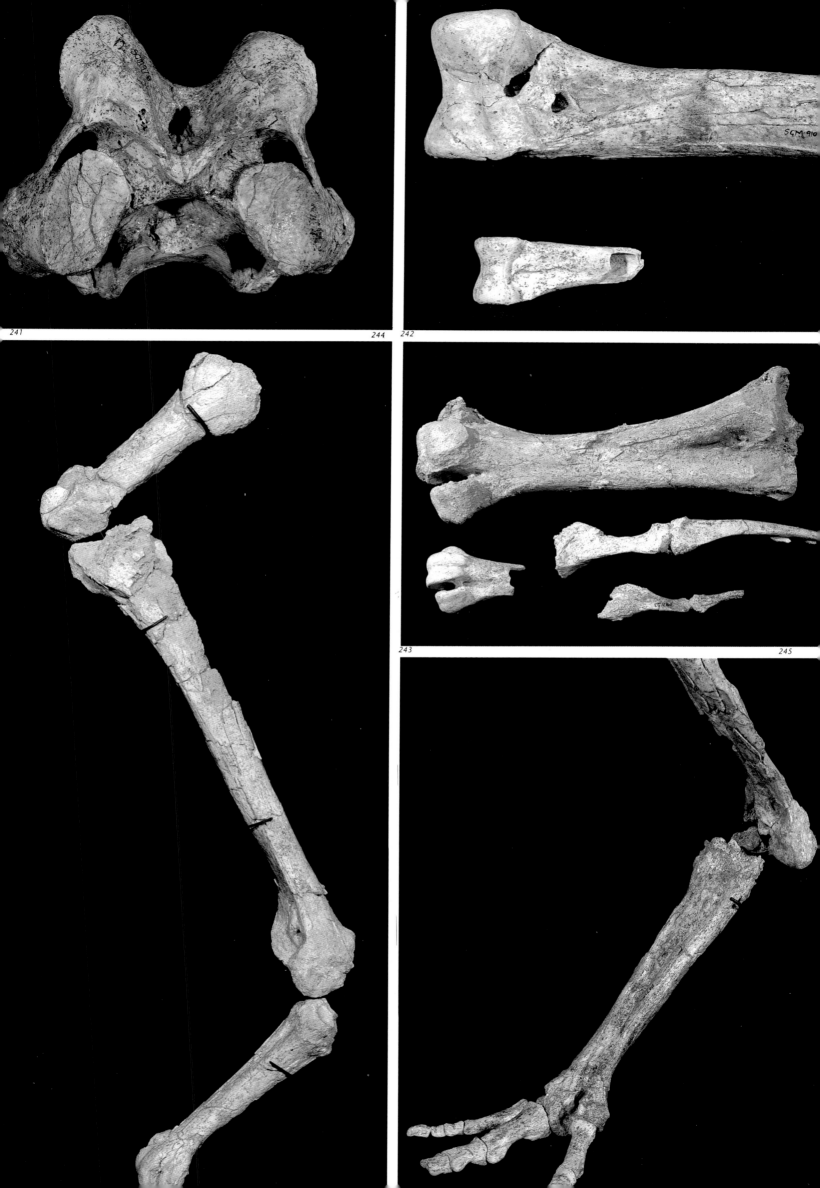

241

244

242

243

245

INCREASING ARIDITY IN POST-MIOCENE AUSTRALIA, 5.2 *MILLION YEARS TO THE PRESENT*

The later part of the Cainozoic reflects the effects of increasing aridity. In the Pleistocene the climate takes on a cyclic nature, with the advance and retreat of the distant continental ice sheets in the Northern Hemisphere manipulating increases and decreases in humidity on the Australian continent.

Early Pliocene Australian terrestrial mammalian sites can be divided into those representing relatively moist environments on the eastern and southern margins of the continent and those restricted to the drier Centre.

The principal sites along the continental margin are Bluff Downs near Charters Towers in Queensland, Bow in New South Wales, and Hamilton in Victoria. Faunas from all of these locales reflect well-watered environments with luxuriant vegetation. Except for Hamilton, the fossil communities recovered from these sites are mixtures of animals related to those from modern woodlands, savannahs and rainforests with crocodilians and large predaceous varanid lizards much in evidence. An unusual crocodilian, *Pallimnarchus*, had a broad, flattened snout (even broader than the living Saltwater Crocodile, *Crocodylus porosus*) and upwardly directed eyes. It inhabited the inland waterways of eastern Queensland during this period and into the Pleistocene, most likely not frequently overlapping geographically with *Crocodylus*. Grazing kangaroos dominated the mammalian faunas at all of these localities, with many living genera making their earliest appearances; for example, *Macropus*, which includes the Grey Kangaroo together with the Euro, some wallabies, wallaroos and the Red Kangaroo; *Petrogale*, the rock wallabies; *Dendrolagus*, the tree kangaroos; *Dorcopsis*, the New Guinea Forest Wallaby; and *Thylogale*, the pademelons. Specimens from Bluff Downs of the diprotodontid *Zygomaturus*, which first appeared in the Late Miocene at Beaumaris near Melbourne, show the general late Cainozoic pattern of size increase. Kangaroos also mirror this trend throughout the Pliocene and into the Pleistocene. *Euryzygoma*, a diprotodontid with flaring cheekbones in the males, first appears at Bluff Downs, as well.

By the Early Pliocene, diprotodontian marsupials (including possums and wombats) are quite "modern", most belonging to living genera. There are a few tantalizing exceptions, however. One of these is *Koobor*, which was thought to be a koala when first discovered at Bluff Downs and at the Late Pliocene or possibly Pleistocene Chinchilla locality of south-eastern Queensland. With the recognition that *Ilaria* in the South Australian middle Tertiary assemblages represented a totally new vombatiform family, the Ilariidae, the relationships of *Koobor* were suddenly not so clear. It resembles ilarids, and if it is closely related to them then the late Tertiary decline of the vombatiforms was less abrupt than previously thought. Most vombatiform family groups may well have persisted until the end of the Pliocene, although the total number of genera declined markedly during that epoch.

Hamilton in western Victoria is another Early Pliocene locality. In contrast to the northern marginal sites of similar age, it has produced a fauna typical of temperate rainforests. Oddly enough, only mammalian fossils are preserved at this site, a reflection of bias in the fossil record. Not a scrap of the lizards, turtles, frogs, fish or even birds that must have been living with them has been found. Easier to understand is the rarity of mammals whose adult body weight would have exceeded 5 kilograms. Although *Palorchestes*, a few of the kangaroos such as *Macropus* and *Protemnodon*, and an indeterminate diprotodontid would have exceeded this mass, the majority of the 26 species present at Hamilton were small-to-intermediate-sized forms. This ratio of larger to smaller forms is probably close to what the living community structure would have been in such a temperate rainforest environment.

The smaller forms in the Hamilton fossil assemblage include the Brushtailed Possum, *Trichosurus*, and the Mountain Pygmy Possum, *Burramys*. The tiny *Burramys* stands out because of its present-day habitat: it is now restricted to a small area of the Australian Alps, far from any modern rainforest, temperate or tropical. Other living mammalian genera represented in the Hamilton fossil assemblage

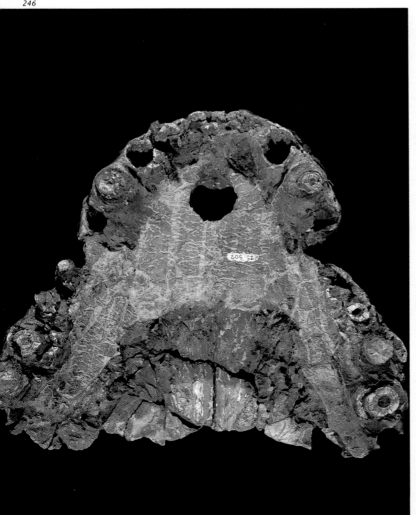

today occupy either a broad variety of habitats (such as *Trichosurus* which is now almost ubiquitous in Australia) or are most at home in tropical rainforests (for example, *Strigocuscus*).

Hamilton is not the only place where a Pliocene or Pleistocene fossil assemblage contains terrestrial mammalian genera which do not today, nor at any time in the Holocene (or last 10,000 years), occur together but rather are found in markedly different habitats. A persuasive explanation for this phenomenon, termed "disharmonious associations", points to a more predictable climate during the Pliocene compared to that of modern times — a climate which lacked swings through the year at a given place. Such faunal associations are now impossible because variability has increased significantly during the Holocene. Imagine two species, one a warmth-loving form and another cold-loving that overlap in some small part of their geographic ranges. If the mean annual temperature were to remain the same but the winters became colder and the summers warmer in the area of overlap, it might well be that neither species could survive there. For the warmth-loving form, the winters might be too cold and for the cold-loving form, the summers too hot — although on average the temperature may remain the same as before. Because the entire Earth is now experiencing greater annual variations in environmental conditions than was the case in either the Pliocene or most of the Pleistocene, there are most likely few habitats today that exactly match the ones of earlier times. Such disharmony may also be in part due to human-induced change (T.F. Flannery, pers. comm.).

Poorly-dated and with a small number of known taxa, the Palankarinna assemblage from northern South Australia gives the best glimpse available of what the Central Australian Early Pliocene mammalian fauna looked like. Genera such as the bandicoot *Ischnodon* and the diprotodontid *Meniscolophus* are unique to this assemblage. Others such as the diprotodontid *Zygomaturus* and the kangaroo *Prionotemnus* are widespread, both in space and time. That the Centre had not dried out to anything remotely like its present condition is attested to by the presence of bony fishes, lungfish and a variety of crocodiles. On the other hand, *Ischnodon* is regarded as a rabbit-eared bandicoot, or thylacomyid, and the two modern species of thylacomyids are restricted to arid or semi-arid regions. Like the living members of the family, *Ischnodon* had high crowned molars, seemingly adapted for a diet of grass seeds, a rich food source in an arid environment. Also in this assemblage is *Dromaius ocypus*, an emu with a clearly elongated hind limb — intermediate in proportions between the more cassowary-like Miocene relative and the living, highly cursorial Emu, *Dromaius novaehollandiae*. Certainly, by Pliocene times, there had been some selection for open country living. Aquatic elements still persisted in this now arid area, with *Neoceratodus*, a close relative of the present-day lungfish, as well as a variety of teleost fish and crocodiles abundant in the fluviatile and lacustrine sediments. Besides the modern *Crocodylus*, a more unusual crocodilian perhaps related to the South American sebecosuchian crocodiles or the eosuchian pristichampsines is also represented by isolated teeth in this assemblage. Further study will be needed to resolve the relationships of these reptiles, but no doubt they were top carnivores of their times with their rapacious laterally-flattened and serrated battery of teeth.

Stratigraphically higher and more securely dated than the Palankarinna assemblage is the Late Pliocene Kanunka assemblage, located east of Lake Eyre. With the aid of palaeomagnetic dating, these fossils are known to have been deposited just after the reverse Matuyama Chron at about 2.48 million years ago. Except for a dasyurid (native cat), all other marsupials in this assemblage are large diprotodontians. They include the earliest *Diprotodon* as well as *Zygomaturus*, a plethora of kangaroos dominated by grazing forms, and wombats, both the modern *Vombatus* and the extinct giant *Phascolonus*, which persisted until the end of the Pleistocene. The construction of the teeth of the kangaroos suggests that open grasslands must have been widespread in the Kanunka area. Present too, was permanent water, flanked by woodlands, for lungfish and crocodiles were elements of this Central Australian fauna. Flamingoes, too, were still very much in evidence, and in fact quite diverse. As many as three different forms, *Phoenicopterus*, *Xenorhynchopsis* and *Ocyplanus*, may have been present, rivalling the diversity of flamingoes anywhere on Earth today, and demanding the presence of permanent water to ensure their ability to feed and reproduce.

248

In the Late Pliocene of Kanunka and other localities, rodents appear in Australia for the first time. These rodents are the "Old Endemics", murids particularly well-suited for the drier habitats of the Centre. They include many forms found nowhere else. By the Late Pleistocene, rainforest-adapted rodents had developed, particularly members of the highly successful genus *Rattus*. Despite this further immigration of rodents, all belong in a single subfamily, the Murinae, which appears to have emigrated, perhaps several times, from South-East Asia. Evidently, as the earliest rodents in Australia are arid-adapted forms, their route to this continent may have been through the drier regions of Indonesia, such as parts of Timor, rather than through the rainforest-dominated areas of New Guinea.

In South America, the arrival of rodents triggered an evolutionary explosion. Their earliest record on that continent is in the Early Oligocene, when seven families unique to South America were present. All are placed within a single suborder, the Caviomorpha, a group essentially restricted to South America. The difference in the fortunes of the rodents on the two continents has suggested that compared with the situation in the Early Oligocene of South America, the evolutionary opportunities in Australia in the Late Pliocene and Pleistocene were fewer. The physical environment may have played a role in this respect, but so too may have competition. Niches that rodents might have filled in Australia may have been already occupied by parrots and possums.

246-247 Pallimnarchus sp., a common crocodilian of the Australian late Cainozoic. Both the snout and scutes, or dermal armour, shown here were found in Pleistocene deposits on Cooper Creek, South Australia, indicating that the area did not dry up to its present state until much of that epoch had passed. The teeth are from Late Pliocene or Early Pleistocene deposits from Chinchilla in south-eastern Queensland. Snout (x 0.5), armour (x 1.0) (S. Morton, courtesy of the South Australian Museum and N. Pledge; F. Coffa, courtesy of the Spencer and Gillan Museum and P. Murray)

248 Basalt from above the fossiliferous layer at Hamilton, Victoria. The Pliocene mammals were found in the green sandy clays seen immediately beneath the basalt. It was possible to obtain radiometric dates on the basalt and invertebrate fossil "dates" on the marine sediments immediately beneath the fossil-bearing layer. Hamilton is one of the most securely dated Tertiary terrestrial mammal sites in Australia. (P. Vickers-Rich)

Another factor may be important here, too. Rodents have not been in Australia for longer than about the last 4 million years, whereas they are first known in South America some 30 million years ago. The Tertiary mammalian record in South America, just as in Australia, is not continuous, and before the Early Oligocene when the rodents first appeared there, a gap of at least 8 million years occurs in the South American record. Although seven families of indigenous rodents are recognized in South America, they were not all that different from one another when the order first appeared there. It is primarily because scientists have traced back the more recent and highly diverse living South American rodents to their roots in these seven groups that their status as separate families in the Early Oligocene has been acknowledged.

As rodents formed about a quarter of the total Australian mammalian fauna at the time of European contact, even though confined to a single subfamily, it is interesting to speculate what the South American rodent fauna looked like 4 million years after having reached that continent. Was it the assemblage seen in their earliest known record in the Early Oligocene or was it a much less diverse suite of rodents, unknown to us because no mammalian fossils have been found during the Middle to Late Eocene gap in the South American mammalian record? If the latter is the case, perhaps Australian rodents are diversifying at about the same rate as they did in South America during their early history there, and the apparent differences are related not to differing evolutionary rates but to our own perspective.

A further sequence of sites that straddle the Pliocene–Pleistocene boundary occurs along the Murray River. Vertebrate assemblages from these locales make up the Fishermans Cliff and Bone Gulch

249

faunas. They are assemblages typical of the Australian inland, with an abundance of kangaroos and the constant presence of the aquatic elements — lungfish, both ceratodonts and neoceratodonts, and bony fishes. Reptiles are represented mainly by chelid turtles. These assemblages do not differ markedly from that at Lake Kanunka in Central Australia, except for the lack of crocodilians. They are composed of animals that were living on the edge of Lake Bungunnia, formed in the Late Pliocene when uplift on the lower Murray River created a natural dam that held until Middle Pleistocene times.

The pattern of inland faunas set at the beginning of the Pliocene thus persisted until the Middle Pleistocene, with grazing kangaroos the dominant herbivores but with the larger herbivores *Diprotodon* and *Zygomaturus* still a part of these assemblages. The smaller diprotodontids such as *Meniscolophus* had disappeared by the end of this phase. The rodents, dasyurids and possums for the most part are extant genera. Many other forms, such as koalas, are today extinct in the desert areas where these inland fossil faunas occur. Such fossil occurrences, then, are compelling evidence that significant environmental changes have occurred in the Centre during the past 1 million years.

Likewise, on the periphery of the continent, the Late Pliocene and Early Pleistocene fossil assemblages are a continuation of those known since the beginning of the Pliocene. New species arose, but the community structure of assemblages, such as the Late Pliocene to Early Pleistocene Chinchilla fauna in south-eastern Queensland, is remarkably similar to that at Early Pliocene Bluff Downs or Bow.

A site in south-eastern Australia is an example of yet another late Cainozoic environment. At some time in the Pliocene, a cave system developed near the modern town of Geelong, Victoria. Over time, the cave gradually filled up with sediment, which was washed into it, until there was no opening at all. Rarely in these sediments were buried isolated teeth and, even less often, jaw fragments of a variety of mammals. Subsequently, a flow of basalt sealed off this former cave. Such a cave deposit in which there is no air space remaining is called a fissure-fill. As is the case with this site near Geelong, such fissure-fillings are generally found only due to quarrying operations. Otherwise, because the sediments filling the ancient caves are typically softer than the surrounding rock, generally limestone, they are removed first when exposed by natural erosion.

The small collection of mammalian fossils from this site, which constitutes the Dog Rocks Local Fauna, includes rodents that did not arrive in Australia prior to the Pliocene. So, the age of this particular locality lies somewhere in that epoch. Study of the sediments in the fissure show that when they accumulated, the Earth's magnetic field was the reverse of what it is at present. Therefore, within the Pliocene, the age is further constrained either between 2.03 and 2.48 million years ago or 3.40 and 3.48 million years ago, times when, had a modern compass existed, its present magnetic North would have pointed South. Although the collection from this site is meagre, because it can be dated within the Pliocene it is important in establishing that many modern Australian mammal species such as *Macropus giganteus*, the Grey Kangaroo, extended back to that epoch.

249 Koalas have an extremely meagre Tertiary record, as might be predicted by their arboreal habits. This *Perikoala palankarinnica* mandible is from the Early Pliocene deposits at Lake Palankarinna, South Australia. x 2.2 (S. Morton, courtesy of the South Australian Museum and N. Pledge)

250-251 Mandible of *Phoberomys burmeisteri*, a Late Miocene or later descendant of the caviomorph rodents which began to radiate by the Early Oligocene in South America. From the Acre region of Brazil. x 0.6 (F. Coffa, courtesy of the Museo De La Plata and R. Pascual)

253 Partial mandible of *Ischnodon australis*, the only fossil bilby or rabbit-eared bandicoot of Tertiary age. From Pliocene deposits at Lake Palankarinna, South Australia. x 3.6 (S. Morton, courtesy of the South Australian Museum and N. Pledge)

252 Mandible of *Troposodon kenti*, just one of numerous kangaroo species found in the Kanunka assemblage east of Lake Eyre, South Australia. x 0.7 (S. Morton, courtesy of the South Australian Museum and N. Pledge)

254 This jaw of *Kolopsis* from the Black Rock Member of the Sandringham Sand at Beaumaris, Victoria, has an unusual history. The rear half was found in the beach shingle at Beaumaris in 1967. An alert museum technician remembered another specimen from Beaumaris that had been collected more than half a century earlier. When he put them together, they fitted perfectly. Shells in the Black Rock Formation enable that unit to be correlated with similar shells at overseas localities, which in turn can be correlated with the type Miocene localities in Europe. x 1.0 (F. Coffa, courtesy of the Museum Of Victoria)

255 *Zygomaturus gilli* from the Late Miocene marine Black Rock Member of the Sandringham Sands Formation at Beaumaris, Victoria. It is the oldest species of this genus of diprotodontid which persisted until the Late Pleistocene. During its history, *Zygomaturus* quadrupled in mass, a typical pattern found in both diprotodontoids and kangaroos during the late Cainozoic. *Zygomaturus* appears to have favoured moister habitats than did *Diprotodon*. Although the two overlap in their geographic ranges, *Zygomaturus* alone occurs on Tasmania and in south-western Australia while it does not occur in the more arid Centre where *Diprotodon* is well known. The maxilla shown here was found by a diver who was searching for old bottles, not old bones. After the specimen was measured, photographed, and a mould and cast made of it, it was returned to the discoverer and owner, Dwayne Gates. x 1.7 (F. Coffa, courtesy of D. Gates)

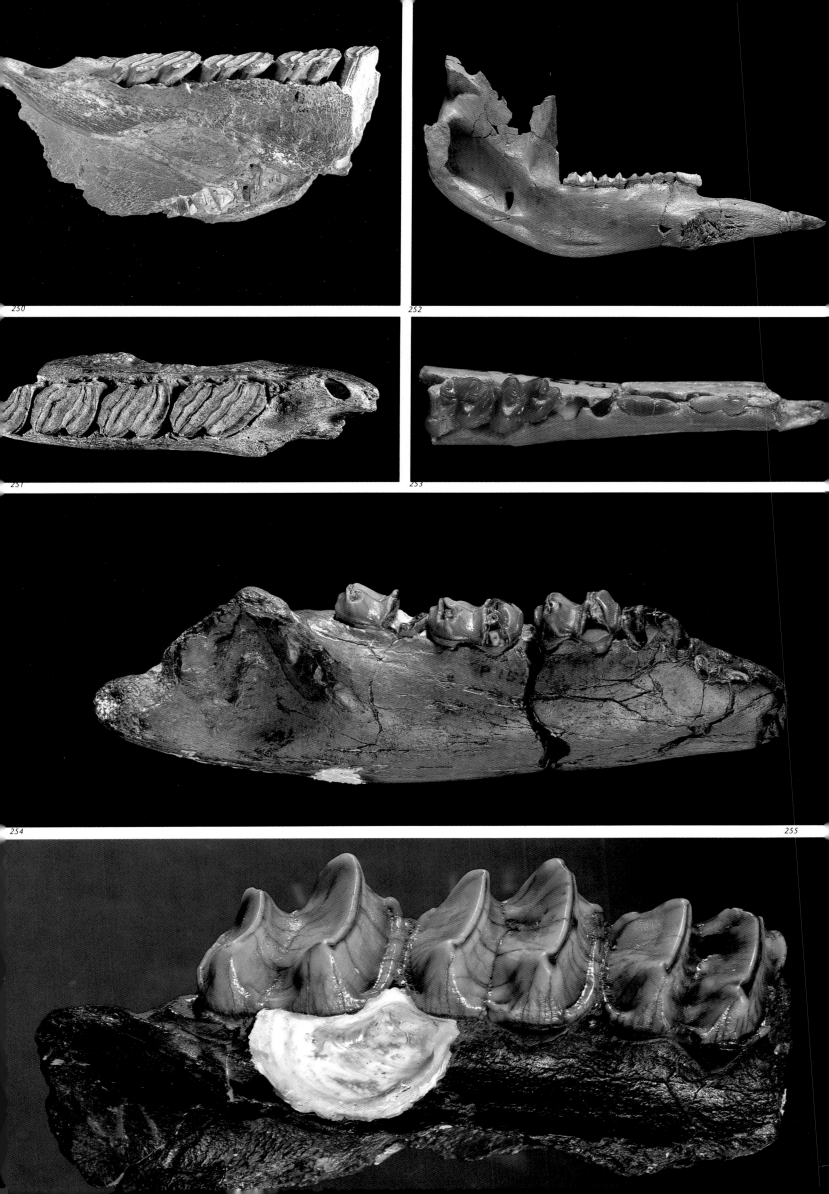

250

252

251

253

254

255

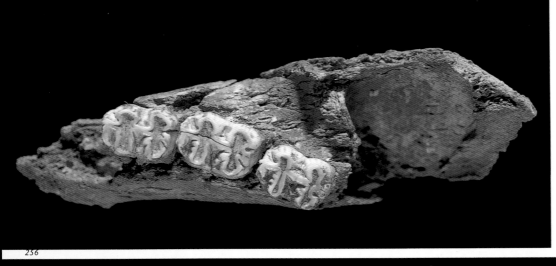

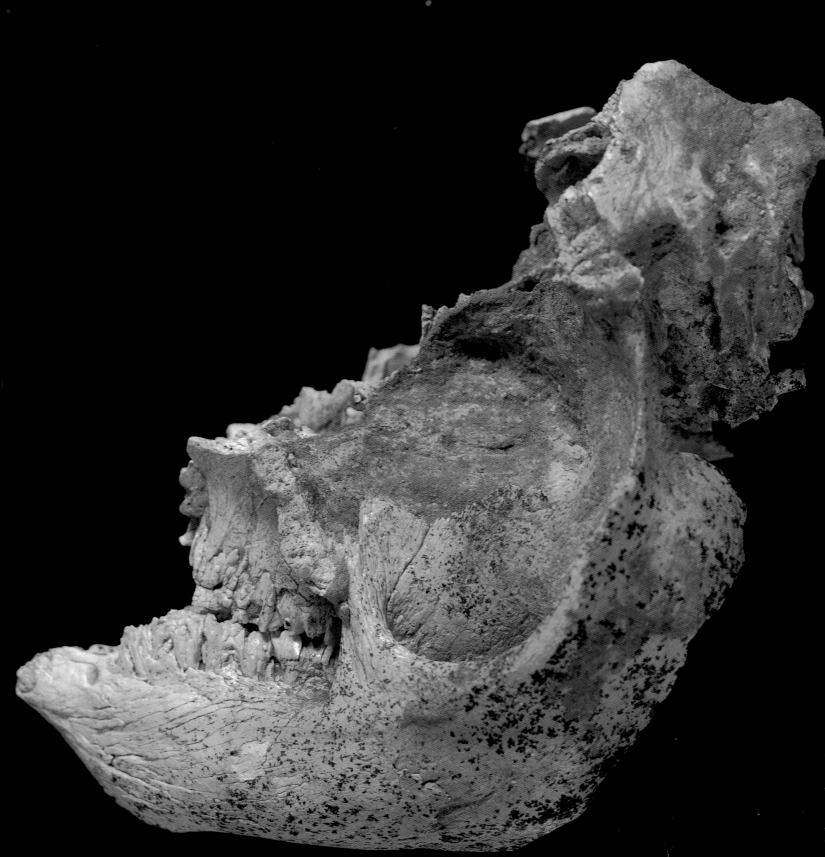

256-257 Jaw and skull fragments of *Procoptodon goliah* which stood 2.6 metres tall and was the largest kangaroo that ever lived. An animal of such stature must have approached the size limit set by the strengths of biological materials for a hopping mode of locomotion, if it indeed moved in that manner. Lower jaw, 256, from Pleistocene deposits at Henschke's Fossil Cave, south-eastern South Australia, x 1.0; skull and lower jaw, 257, collected in Pleistocene sediments at Lake Victoria in south-western New South Wales. X 0.8 (S. Morton, courtesy of the South Australian Museum and N. Pledge)

256

257

UNPREDICTABILITY IN
THE LAST 500,000 YEARS

The replacement of those terrestrial vertebrate communities typical of the Early Pliocene to Middle Pleistocene interval by those characteristic of the Late Pleistocene occurred through the last 500,000 years, and most dramatically in the last 25,000 years. It was during the last half million years that the present sand ridge deserts of the Centre became established. While dune development had begun earlier, it is not clear just how widespread such dunes were nor how severe the environment was when they first became active, sometime in the Pleistocene. Although the waxing and waning of the desert during the past half million years does appear to be finely correlated with the advance and retreat of the continental glaciers in the Northern Hemisphere, again, with the exception of the last glacial cycle, the evidence is scanty indeed. The best that can be done now is to look at what happened during the last glacial cycle and attempt to determine whether one can extrapolate backward in time to the Middle Pleistocene when the cyclic activity began in the Northern Hemisphere.

One of the best studied areas offering insights into this question lies in western New South Wales, upstream of the confluence of the Murray and Darling rivers. Here, 100,000 years ago, conditions were much as they are today. It was an area of dry lakes surrounded by saltbush and other species of the modern mallee flora. As today, this was a time 10,000 to 20,000 years after a glacial maximum in the Northern Hemisphere.

These conditions of 100,000 years ago persisted for at least another 50,000 years. Then a wet phase began, which increased the run-off into the region. Lakes filled, and an abundant community of vertebrates inhabited both the lakes and their surrounds. Large marsupials such as *Diprotodon* and the kangaroo *Protemnodon* mostly became extinct sometime during this phase, which lasted for 30,000–35,000 years. Their fossils are known only in those deposits in the Murray–Darling region that are not associated with human artefacts or remains. Virtually all the mammals known in such human association from this time interval belong to extant species.

Beginning about 25,000 years ago, the lakes which had been such bountiful resources to the animal communities began to dry out. Just 7000 years later, they were mere barren, dusty claypans, far harsher environments even than they are today. The diverse mammalian community of grazing kangaroos was reduced to only a few species, like the Red Kangaroo. This species could thrive in the desert conditions; many others could not. Such evolution towards more arid conditions was not confined to the Murray–Darling region. Desert dunes were also building on the north-east tip of Tasmania. When maximum aridity was reached 18,000 years ago, the continental glaciers in the Northern Hemisphere were extended their furthest south. Consequently, because of the enormous amounts of water locked into the glaciers, the sea level dropped more than 80 metres below its present level. Bass Strait was dry and the Tasmanian dune field may well have been linked with those in western Victoria.

Although it had taken about 100,000 years from the previous glacial maximum for the glaciers in the Northern Hemisphere to build up to their great size of 18,000 years ago, their retreat was startlingly rapid. By 10,000 years ago, the Earth was decidedly warmer than at present and glaciers of continental proportions were gone. The pattern of Pleistocene glacial activity thus seems characteristically to be a slow build-up towards a maximum ice advance (80,000–100,000 years) followed by a quick melting off. As the melting occurred, conditions ameliorated in the Murray–Darling convergence area and over much of Australia. The longitudinal dune fields that had been active during the harshest phases of aridity became stabilized once again, frequently covered with composite wildflowers, grasses and other still arid-adapted vegetation.

Presumably, this cycle of events was repeated at least five and perhaps up to seven times during the latter half of the Pleistocene. In Central Australia, the conditions in areas such as the Simpson Desert and the Great Sandy Desert probably followed a similar path. These presumptions are, of course, based on extrapolation of data from a restricted time interval and area, but unfortunately the direct data bearing on such generalities are as yet much too scanty to assert without equivocation.

258

259

260

258 Marcus Locality at the north end of Kalamurina Waterhole on the Warburton River, South Australia. Sediments, mainly of Quaternary age, form the cliffs here and are often richly fossiliferous. J.W. Gregory investigated sediments like these in the Lake Eyre Basin early in the twentieth century, and they were later more thoroughly studied by R.A. Stirton, R.H. Tedford and R. Wells, their work beginning in the 1950s. (R.H. Tedford)

259 Carapace fragments of the turtle Chelodina insculpta from the Quaternary Katipiri Sands along the Warburton River, South Australia. x 0.9 (S. Morton, courtesy South Australian Museum and N. Pledge)

260 The lower jaw, inside view, of a Zygomaturus trilobus from probably Pleistocene sediments in the Murchison River bed, Billabong Station Crossing, Western Australia. x 0.3 (J. Frazier, courtesy of the Western Australian Museum and K. McNamara)

261 A petrel, *Pterodroma solandri*, still encased in Quaternary sediments on Old Settlement Beach, Lord Howe Island. x 1.5 (F. Coffa, courtesy of the Australian Museum, A. Ritchie and R. Jones)

262 The distal ends of several tarsometatarsi ("drumsticks") of flamingoes known from middle to late Cainozoic lacustrine and fluviatile sediments of Central Australia. Australia probably had the most diverse flamingo fauna of any place on the globe. The oscillatory and unpredictable climate since then, which brought about the demise of the permanent interior lakes, was the death knoll to these birds on this continent, even though they survived elsewhere in the world. x 0.9 (P. Vickers-Rich)

263 Pelvis of a fossil emu, *Dromaius sp.*, from Quaternary paludal sediments of Mowbray Swamp near Smithton, Tasmania. x 0.5 (S. Morton, courtesy of the Queen Victoria Museum and C. Tassell)

264 Hind limb bones of a fossil emu, *Dromaius sp.*, from Quaternary paludal sediments of Mowbray Swamp near Smithton, Tasmania. x 0.5 (S. Morton, courtesy of the Queen Victoria Museum and C. Tassell)

265 Mummified partial limb of a fossil emu, *Dromaius sp.*, from Quaternary deposits in Tasmania. x 0.6 (S. Morton, courtesy of the Queen Victoria Museum and C. Tassell)

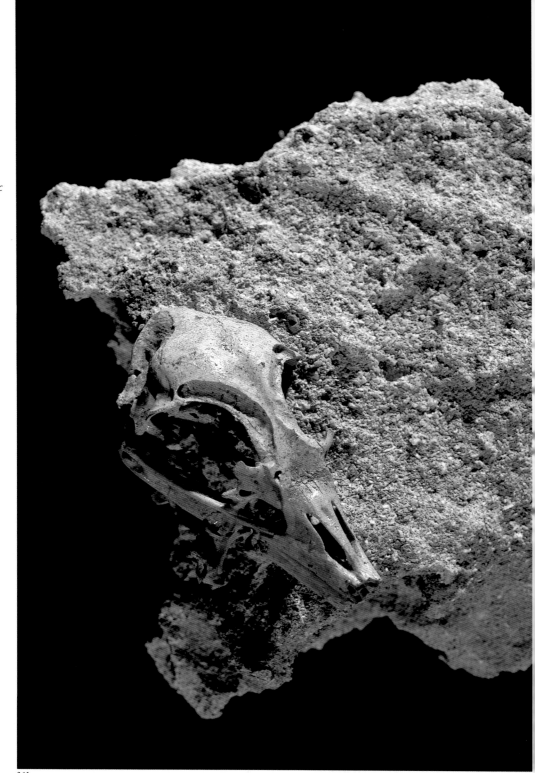

261

262

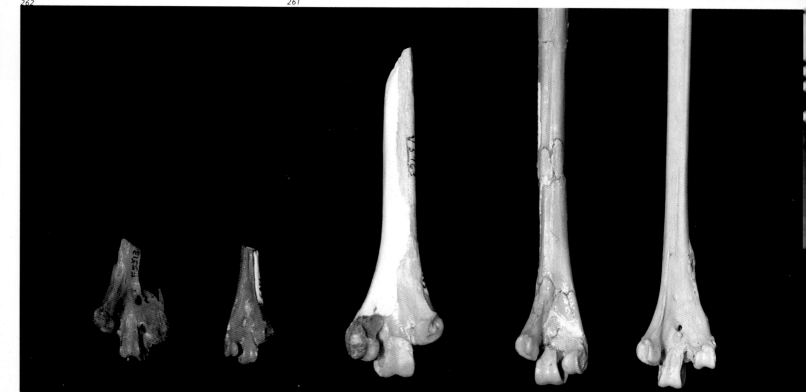

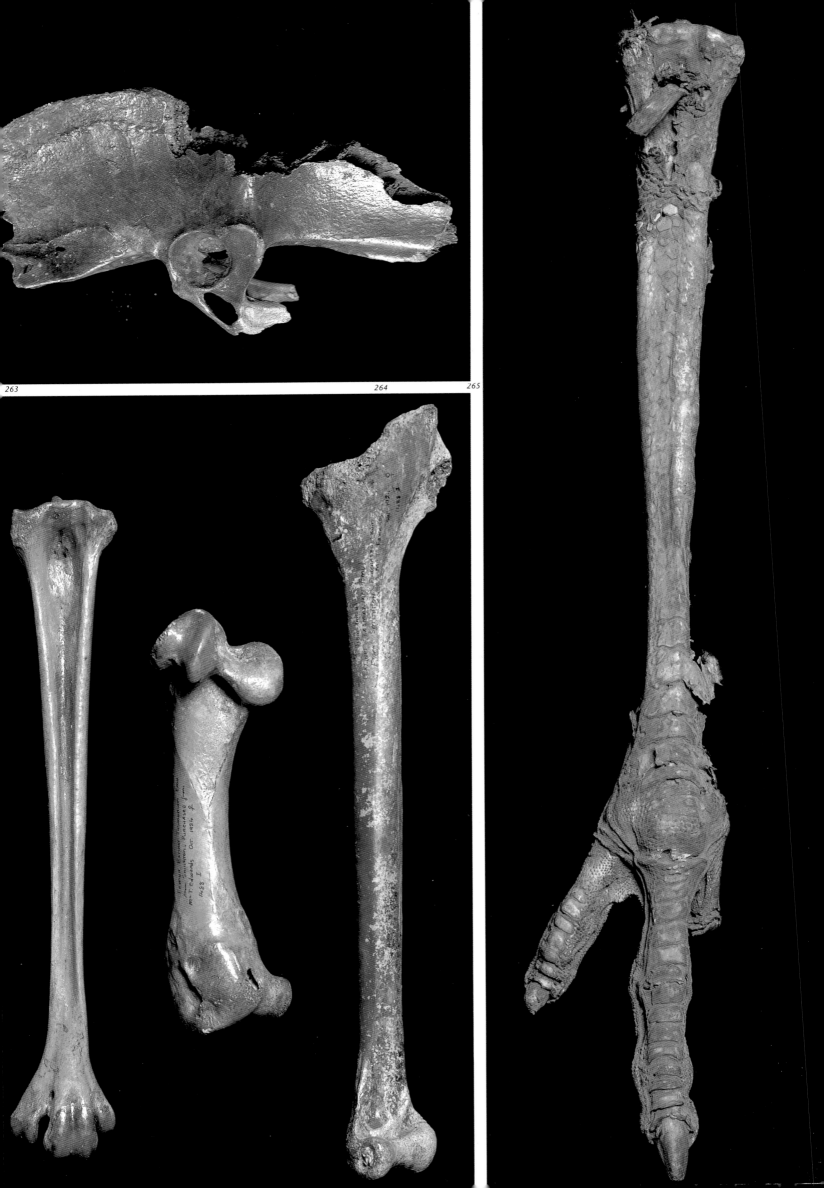

EXTINCTION OF THE QUATERNARY MEGAFAUNA

Uncertainty about events and their timing in the Pleistocene (the last 2 million years) has a direct bearing on hypotheses about the cause of the extinction of the larger land vertebrates in Australia during the past 35,000 years, and it is possible that such extinctions occurred as recently as within the last 5000–10,000 years. One fossil site, Lancefield, offers significant clues for unravelling this interpretive dilemma. Located just 50 kilometres north of Melbourne, Lancefield is a small country town with a large recreational reserve. On one edge of that reserve is a swamp, only a few hectares in area. Buried beneath it are the remains of at least 20,000 individuals of the Grey Kangaroo, *Macropus giganteus*. These Grey Kangaroo fossils were decidedly larger than their modern descendants, perhaps twice as massive. Many of the mammals buried along with them belong to genera that later became extinct rather than merely declining in size. Together with the large and now extinct ground-dwelling Mihirung, *Genyornis*, these genera include the large diprotodontids *Diprotodon* and *Zygomaturus*, the kangaroos *Protemnodon*, *Propleopus* and *Sthenurus*, the Marsupial Lion, *Thylacoleo*, and the Giant Wombat, *Phascolonus*. Such forms are referred to collectively as the "megafauna". In contrast, the few smaller fossil mammals known from Lancefield, such as the Tasmanian Devil, *Sarcophilus*, the wombats *Vombatus*, and the Broad-toothed Rat, *Mastacomys*, are all still extant.

When these animals lived and died at Lancefield 26,000 years ago (a date based on Carbon 14 analysis), the conditions were much more arid than they are at present, just as they are known to have been at that time in the Murray–Darling convergence. The age structure of the fossil kangaroo population at Lancefield reveals that most were young adults, individuals with the highest reproductive potential. In modern times, when kangaroos have been surveyed during periods of prolonged drought, the older individuals die off first and reproduction declines or ceases altogether in the population. If these conditions persist for a few years, the last individuals to die are those just entering breeding age. They usually die in the vicinity of the rare water sources. What kills the last survivors is not lack of water but lack of fodder. Under these conditions, the animals are "tethered" to their water supply. They have eaten all the fodder as far away from the waterhole as they dare to go. In these circumstances, they return hungry to their water supply, take a drink, and collapse as if poisoned. The corpses thus accumulate both in and around the waterhole. Disease in such a highly stressed population and water poisoning also play roles in accelerating mass death.

It seems likely that the Lancefield accumulation represents the predictable result of such a prolonged drought. What it also demonstrates is that those large mammals and birds which have subsequently become extinct survived until that time of high environmental stress. If the extrapolation of the Late Pleistocene environmental cycles for Australia is valid, particularly suggestions that the magnitude of the last episode of aridity was similar to that of earlier cycles, the survival of these large animals until at least Lancefield time tells us something about the cause of their demise.

For more than a century it has been known that there were large land mammals and birds in Australia during the recent past, which are now extinct. Except for Antarctica, the same can be said for all the other continents as well as large islands such as Madagascar and New Zealand. Explaining what has brought about this extinction of large forms has led to the spilling of much ink. In various guises, two general theories have been offered in explanation. These are: firstly, that these animals succumbed because of the increasingly harsh environmental conditions brought about by the climatic cooling during the Pleistocene ice ages; or secondly, that humans directly (hunting) or indirectly (by altering the environment) brought about their demise.

Extinctions worldwide occurred during the Pleistocene, but each continental mass has its own history (for example, North America exhibited mass extinctions 11,000 years ago, as did New Zealand 1000 years ago).

In Australia, the likelihood that climatic deterioration alone was responsible for the disappearance of the megafauna hinges mainly on the validity of extrapolating the cycles of wet and dry phases back into the Middle Pleistocene. If they do continue back that far and their magnitude was similar to the present cycle, it is difficult to see why the latest harsh conditions alone would have brought about this extinction. There is even tantalizing evidence that some of the megafaunal elements such as *Diprotodon* survived until as recently as 6000 years ago, well past the harshest climatic conditions of 18,000 years ago.

If human activity alone was responsible for this extinction event, then what evidence is there? One of the eminently curious aspects of the human fossil record in Australia is that there is little evidence suggesting direct interaction between Aboriginal populations and the megafauna. There are as yet no known *Diprotodon* butchering sites analogous to the mammoth or bison kill sites on other continents. This may be owing simply to the fact that such sites are yet to be found or it could be due to the human groups not being large enough to warrant a task of the magnitude of hunting such large animals when smaller, more easily dealt with prey was about. But there is little evidence allowing decisions on this issue to be made as yet.

However, despite the lack of evidence for direct killing, it does not mean that humans had no effect on the megafauna. Radiometric dates on fossil sites suggest that humans and megafauna overlapped in Australia for a minimum of 15,000 years, perhaps as much as 100,000 years. Even a minimal overlap would have been enough time for the subtlest environmental changes to have had their effect. One of the important ways that humans could have altered the environment was with fire. Annual burning of large areas to improve the conditions for hunting was a common Aboriginal practice in historical times. Sustained over centuries, this burning would have fundamentally altered the vegetation and consequently the entire community structure of animals based upon it.

Perhaps it was the combination of both climatic change and human intervention that led to loss or change of the megafauna — or maybe one factor was omnipotent. Still, we are faced with an unanswered dilemma.

266-268 Mandibles (x 1.0) and adult humeri (x 0.3) of Macropus giganteus, the Grey Kangaroo. The black specimens are from Lancefield Swamp, Victoria, and are of Late Pleistocene age. The white ones are from individuals that lived within the last century. Most of the larger Late Pleistocene mammals of Australia that survive to the present day underwent a marked diminution in size during the past 26,000 years. Dwarfing among the mammalian and avian survivors of the Late Pleistocene was a common phenomenon worldwide. (F. Coffa, courtesy of the Museum Of Victoria)

269 A skull of Diprotodon sp., from Bacchus Marsh, Victoria. In the late Cainozoic, there is a trend towards ever increasing size in many groups of diprotodontoids and macropods (kangaroos). On the basis of this specimen's smaller size, it is probably older than the typical Late Pleistocene species, Diprotodon optatum, such as is found at Tambar Springs in New South Wales and Lake Callabonna in South Australia. x 0.2 (F. Coffa, courtesy of the Museum Of Victoria)

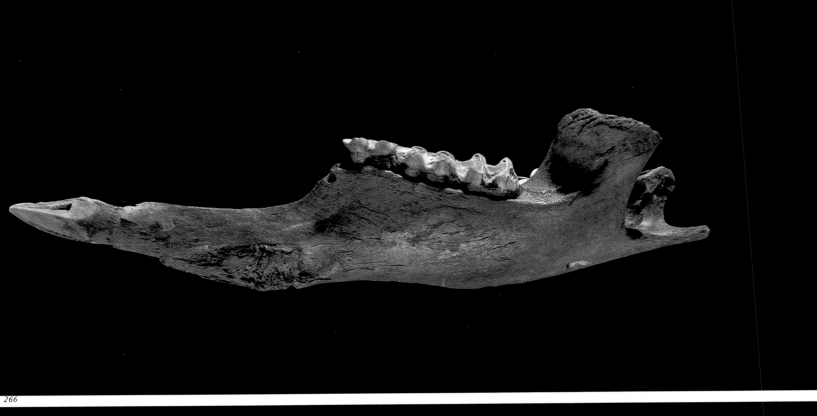

266

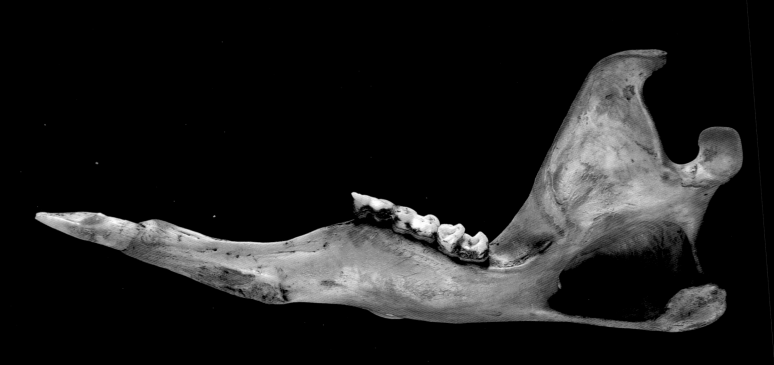

267

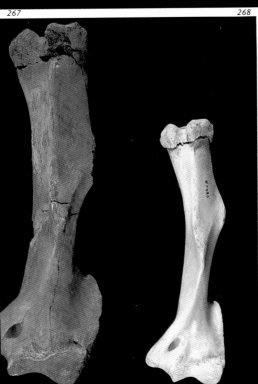

268

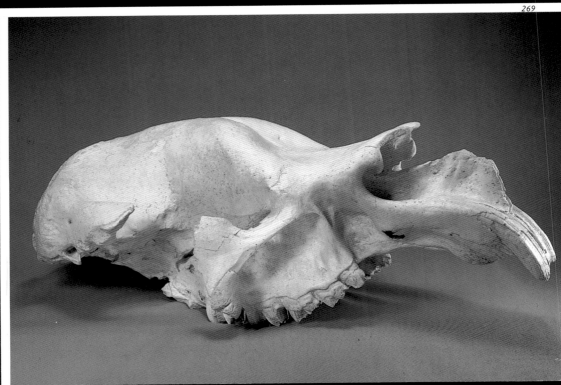

269

270 Mandible fragment of *Zygomaturus trilobus* from the late Pleistocene at Lancefield Swamp, Victoria. *Zygomaturus* was a diprotodontid of the wetter regions of Australia, while *Diprotodon* favoured the drier areas. The areas of overlap are only a small part of the total range of the two genera. x 1.0 (F. Coffa, courtesy of the Museum Of Victoria)

271 Not only did the larger mammals and birds produce "giant" forms towards the end of the Cainozoic, which then disappeared before modern times, but smaller ones did as well. The lower specimen is the jaw of a large koala from Cora Lynn Cave, South Australia, and it is compared with a fossil jaw of the still living Koala, *Phascolarctos cinereus* from Henschke's Fossil Cave. South Australia. Both specimens are Quaternary in age. x 1.2 (S. Morton. courtesy of the South Australian Museum and N. Pledge)

272 Juvenile mandible of *Diprotodon optatum* from the Late Pleistocene at Lancefield Swamp, Victoria. About half the individuals of *D. optatum* found at Lancefield are juveniles, in marked contrast to *Macropus giganteus*, the most abundant species in this fauna, in which juveniles are all but unknown. Whether it was drought or some other factor that caused the lack of juvenile kangaroos, if it affected *Diprotodon* at all, it did so in a quite different manner. x 0.5 (F. Coffa, courtesy of the Museum Of Victoria)

273 A hind limb element (the tarsometatarsus) of the New Zealand Moa, *Eurapteryx gravis*, from Quaternary sediments. These birds were a diverse group in New Zealand, with only a Late Pliocene and mainly Pleistocene record. They were probably the remnants of a Gondwanan dispersal into the area, and their earlier record is unknown because of the great rarity of fossiliferous terrestrial sediments older than the Quaternary. Moas were totally extinguished by the arrival of humans in New Zealand, a scenario very familiar on most Pacific islands. x 0.8 (F. Coffa, courtesy of the Queensland Museum and R. Molnar)

274 Jaw of *Corracheirus curramulkensis*, a "giant" ring-tailed possum about twice the size of the living *Pseudocheirus*. This specimen was found in Mio-Pliocene cave deposits at Curramulka, South Australia. x 1.8 (S. Morton, courtesy of the South Australian Museum and N. Pledge)

275 Occlusal (dorsal) view of fig. 274. x 1.8 (S. Morton, courtesy of the South Australian Museum and N. Pledge)

276 Distal end of the tarsometatarsus of a dromornithid, probably *Genyornis*, from the Late Pleistocene at Lancefield Swamp, Victoria. x 1.0 (F. Coffa, courtesy of the Museum Of Victoria)

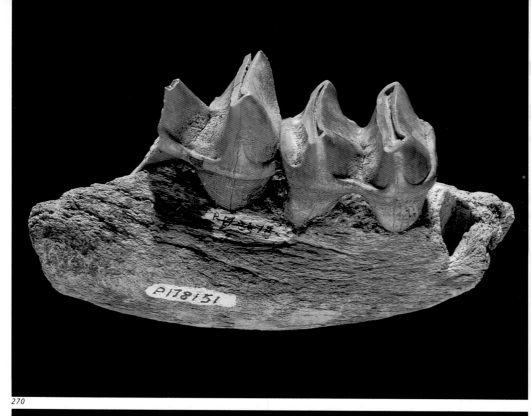

270

271

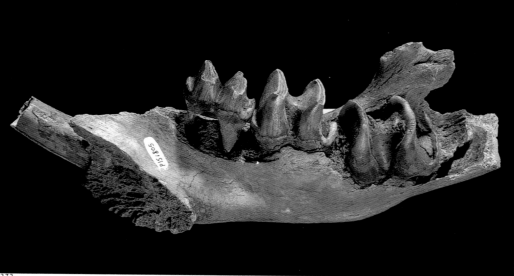

272

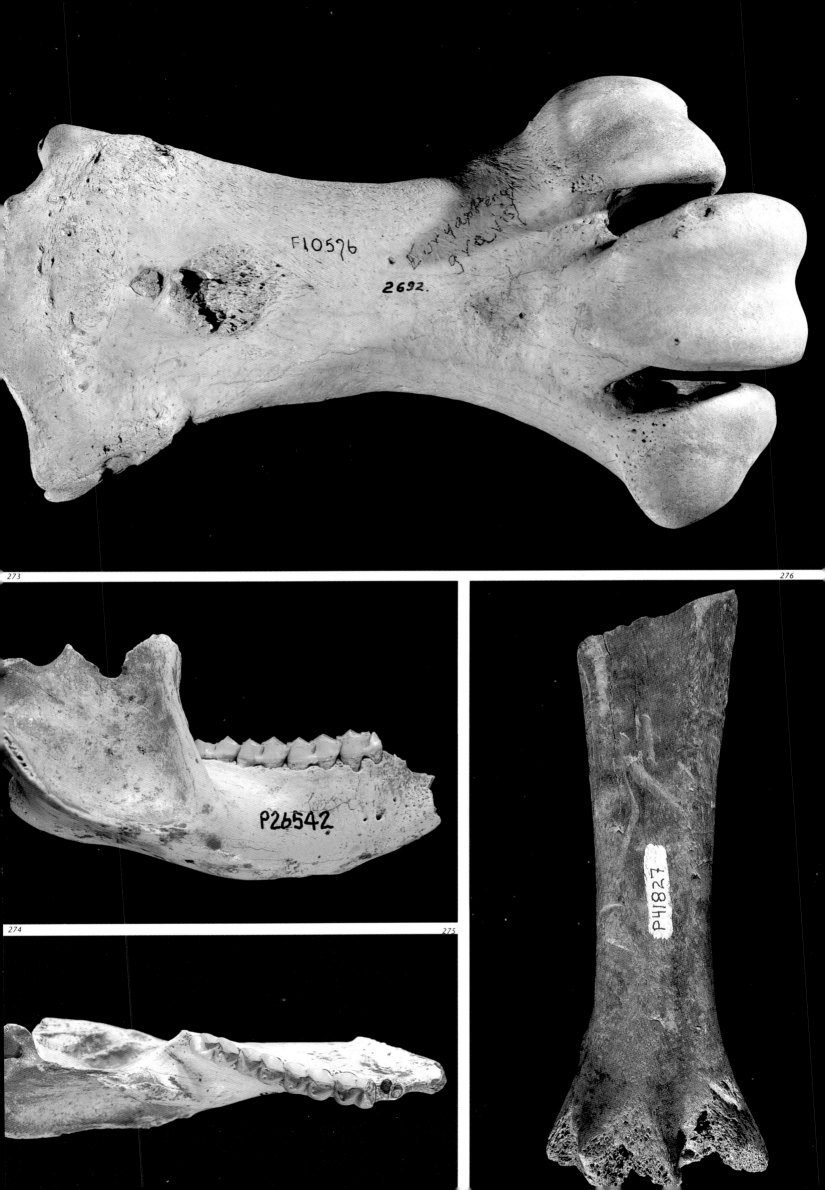

273

276

P26542

274

275

AUSTRALIAN QUATERNARY CAVES

Although Pliocene cave deposits are known, it is in the Quaternary Epoch that this mode of fossil accumulation assumes major importance in telling the story of the Australian terrestrial vertebrate fauna. Accumulations of fossil bones in caves are one of the richest sources of specimens, particularly of smaller vertebrates, and such deposits occur in Tasmania and on all margins of the mainland, except the extreme north-west. None are known in the Centre. Caves seem to have only a limited life span when they are open and accumulating fossil bones. There are no early Tertiary or older fossil deposits laid down in caves that are still open today anywhere in the world. Where fossils are found in pre-late Tertiary cave deposits, the caves have filled up completely (and are called fissure-fills).

The first Australian fossil vertebrate localities to attract European attention were caves in the Wellington district, north-west of Sydney. Beginning in the late 1820s, work on these cave faunas has continued intermittently. The fauna recovered is diverse and includes both large and small forms. From these caves came the first remains of many now well-known Australian fossil mammals. Among them are *Diprotodon* and the kangaroo *Sthenurus*.

Typical of many caves, those at Wellington were formed in limestone by the gradual dissolution action of mildly acidic groundwaters flowing through natural cracks and joints.

Both Pliocene and Pleistocene fossils have been found in the Wellington caves system. Unfortunately, although the deposits are rich, within the Pleistocene deposits it has not been possible to work out a sequence. This same difficulty besets many other Australian caves such as Victoria Cave near Narracoorte, South Australia, which produces abundant, beautifully preserved material. In both these instances, it appears that the Pleistocene faunas are probably from a time late in that epoch.

Fossil remains from Victoria Cave at Narracoorte, South Australia, provide classic examples of the three different types of fossil accumulation that can occur in caves. The first type occurred because Victoria Cave was a natural pitfall trap. As the cave chamber ex-

277

panded, a vertical shaft developed in the limestone, which breached the surface. Animals were prone to fall into this opening and become permanently trapped in the cave, unable to escape. The second type of accumulation occurred because the cave provided a roost for predatory birds such as owls, which regurgitated the bones of small animals after having separated the surrounding flesh. And after the bones got into the cave, many were redeposited by the movement of water within the cave system, this being the third method of fossil concentration in subterranean deposits. The Victoria Cave fossils are some of the most exquisitely preserved of any cave in Australia, or for that matter in the world, and due to their abundance and the quality of preparation after collection, they serve as one of the very best collections of Pleistocene vertebrates for the Australian continent. Study of these fossils has allowed detailed anatomical reconstructions of whole skeletons of many of Australia's now extinct forms, known elsewhere only by unassociated fragmental remains.

In Western Australia, there are cave deposits where sequences within the Late Pleistocene and Holocene have been established. The best understood are Devils Lair near Perth and Madura Cave on the Nullarbor Plain. A favourable geological setting plus meticulous collecting practices has made it possible to document a sequence of events in the Late Pleistocene and Holocene at both of these sites.

At Madura Cave, the bulk of the fossil mammals are small forms, with the largest being the Grey Kangaroo, *Macropus giganteus*. Dated between 16,000 and 38,000 years old the assemblages from this site include several small dasyurids, plus a koala and a potoroo, which do not occur in the area today but are found in south-western and eastern Australia. These specimens are associated with other species that are still occupying the area — a disharmonious association as seen in the older Pliocene locality at Hamilton. Prominent by its absence from the fossil record, but present in the region today, is *Macropus rufus*, the Red Kangaroo, which is at home in the most arid modern habitats. Based on the composition of this fauna a picture emerges of a moister environment on the Nullarbor Plain prior to 16,000 years ago than is characteristic of the region today.

At Devils Lair, the fossil mammalian succession suggests a different pattern of climatic change than that mirrored by Madura Cave. In the Late Pleistocene, conditions at Devils Lair were drier than at present in the oldest records, in this case at about 35,000 years. The driest conditions occurred at the very end of the epoch, some 10,000 years ago, which is about 6000–8000 years later than is the case in southern Australia. During the Holocene, open vegetation gave way to more heavily forested conditions, suggesting a wetter environment than previously. Today, Devils Lair is about 5 kilometres from the sea. At the time of the maximum drop in sea level, about 18,000 years ago, the seashore would have been approximately 40 kilometres away from Devils Lair. This local factor may, thus, have played a role in making the late Quaternary climatic history at Devils Lair distinctly different from that of other Australian sites.

277 Caves, such as this, in the Buchan area of eastern Victoria have yielded large collections of Pleistocene vertebrates which have allowed detailed study of the climatic changes that have occurred in this part of Australia over the last few tens of thousands of years. The climate has dried and moistened, warmed and cooled in concert with the growth and decline of the massive continental glaciers in the Northern Hemisphere. (F. Coffa)

278 The skull of Sthenurus atlas, in bottom view, from the Quaternary Henschke's Fossil Cave at Naracoorte, South Australia. x 0.4 (S. Morton, courtesy of the South Australian Museum and N. Pledge)

279 Tooth of the primitive six-gilled shark Hexanchus agassizi from the marine Oligocene Ettrick Formation of Wellington, South Australia. x 5.0 (S. Morton, courtesy of the South Australian Museum and N. Pledge)

280 Humerus of the giant coucal, Centropus collosus, from Quaternary sediments in Green Waterhole Cave at Tantanoola, South Australia. x 2.3 (S. Morton, courtesy of the South Australian Museum and N. Pledge)

281 Skull and jaws of the Tasmanian Tiger, Thylacinus cynocephalus, from Quaternary cave sediments near Loongana, Tasmania. x 0.4 (S. Morton, courtesy of the Queen Victoria Museum and C. Tassell)

282-283 Tooth plate of Diodon formosus, a mollusc-crushing teleost fish from marine Late Miocene rocks of Victoria x 2.0 (S. Morton, courtesy of D. Jeffries)

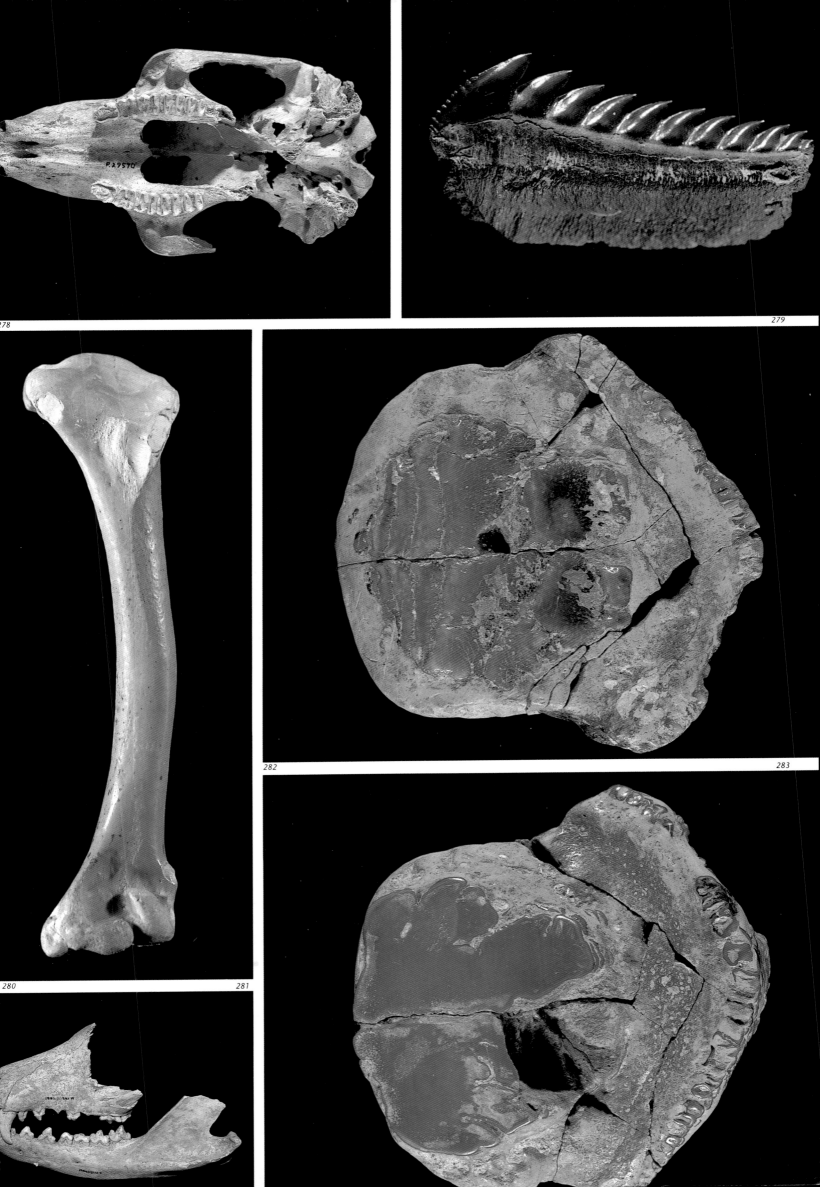

278

279

280

281

282

283

284-285 Skull of a sthenurine kangaroo, *Simosthenurus sp.*, from the Pleistocene of Green Waterhole Cave, South Australia. x 0.6 (S. Morton, courtesy of the South Australian Museum and N. Pledge)

286 Lower jaw of a pouch young of the giant one-toed kangaroo, *Procoptodon goliah* in top (occlusal) view, showing a new molar erupting and minimal wear on the other teeth. From Quaternary deposits of Bairstow's Sand Pit, Port Pirie, South Australia. x 0.7 (S. Morton, courtesy of the South Australian Museum and N. Pledge)

287 *Thylacoleo carnifex*, a marsupial lion, with a wombat femur in its mouth, from the Late Pleistocene Victoria Cave, Naracoorte, South Australia. x 1.3 (F. Coffa, courtesy of Flinders University and R. Wells)

288 A femur of *Vombatus sp.*, showing tooth marks of *Thylacoleo carnifex*, from the Late Pleistocene Victoria Cave at Naracoorte, South Australia. x 1.0 (F. Coffa, courtesy of Flinders University and R. Wells)

289 Articulated manus, or hand, of *Diprotodon optatum* at Lake Callabonna. This sight eluded Sir Richard Owen, who gave *Diprotodon* its name, throughout his life. x 0.4 (R. Jones, courtesy of the Australian Museum)

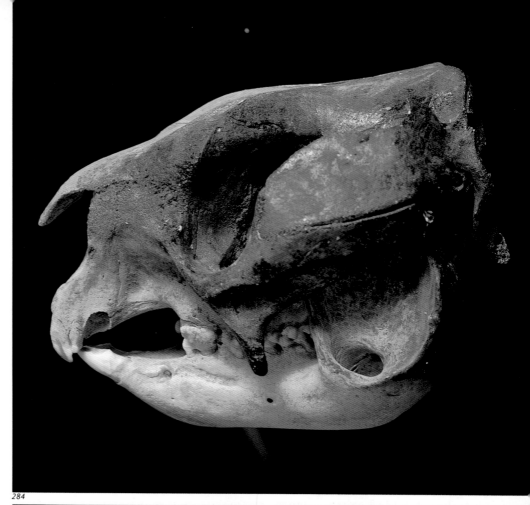

284

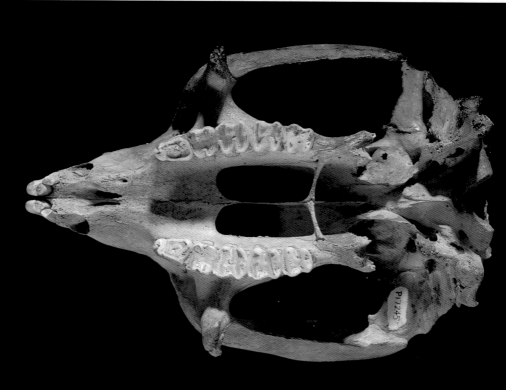

285

286

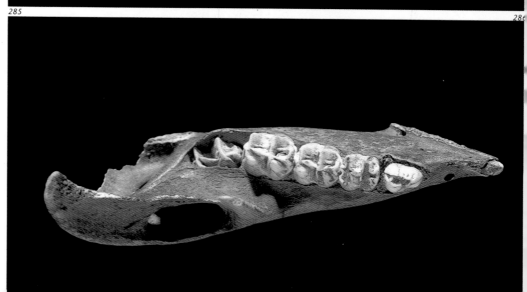

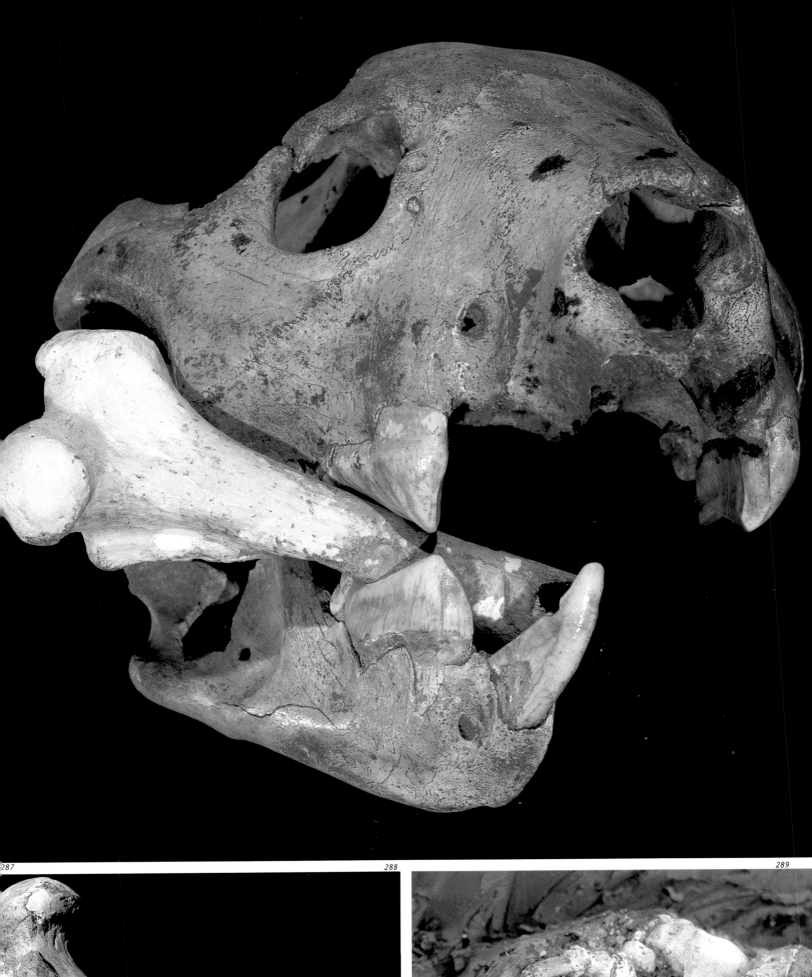

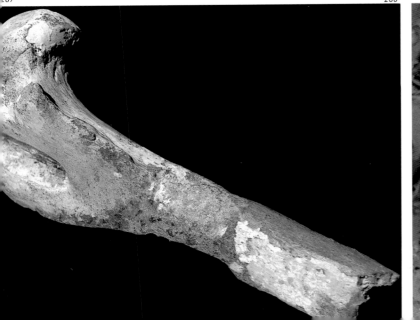

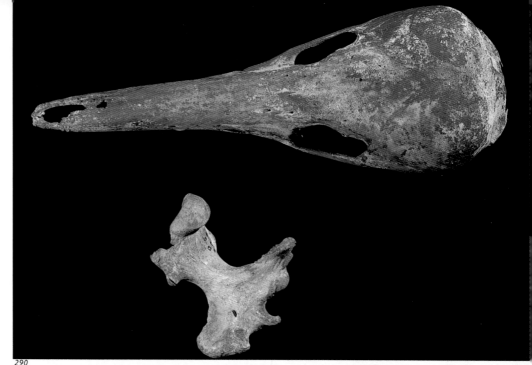

290 The skull and humerus of a giant echidna, *Zaglossus robusta* from the Late Pleistocene Victoria Cave at Naracoorte, South Australia. x 0.8 (F. Coffa, courtesy of Flinders University and R. Wells)

291 Two vertebrae of the snake *Wonambi naracoortensis* from Pleistocene cave sediments at Naracoorte, South Australia. x 3.1 (F. Coffa, courtesy of the Queensland Museum and R. Molnar)

292 A skeleton of *Wonambi naracoortensis* from Quaternary sediments of Henschke's Fossil Cave at Naracoorte, South Australia. x 0.3 (S. Morton, courtesy of the South Australian Museum and N. Pledge)

293-294 Skull and lower jaws of *Sthenurus gilli* from Quaternary sediments of the Ossuary of Victoria Cave at Naracoorte, South Australia. x 0.5 (F. Coffa, courtesy of Flinders University and R. Wells)

295 The skeleton of a sthenurine kangaroo, *Sthenurus sp.*, reconstructed by R. Wells from Pleistocene sediments of Victoria Cave at Naracoorte, South Australia. x 0.2 (F. Coffa, courtesy of Flinders University and R. Wells)

290

291

292

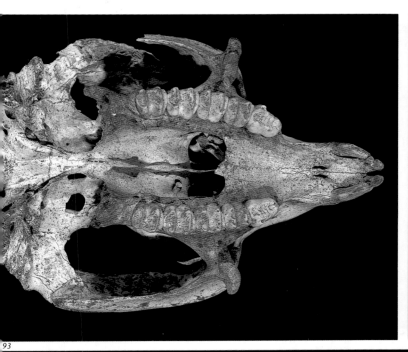

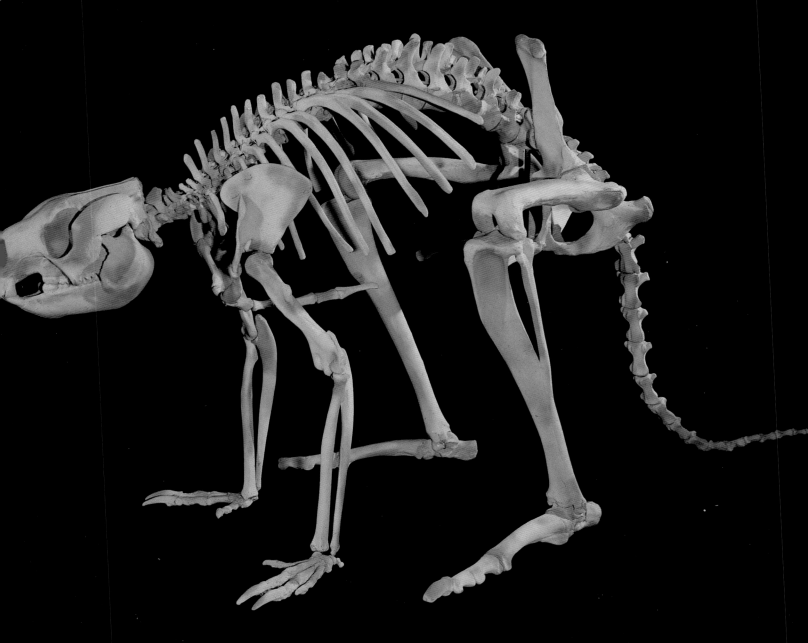

296 The skull and mandibles of a short-nosed bandicoot, *Isoodon*, from the Late Pleistocene Victoria Cave at Naracoorte, South Australia. x 2.0 (F. Coffa, courtesy of Flinders University and R. Wells)

297 Skulls of *Thylacinus cynocephalus* and a pseudomyine rodent encrusted in cave calcite from Quaternary deposits in Mair's Cave, South Australia. x 0.7 (S. Morton, courtesy of the South Australian Museum and N. Pledge)

298 Lower jaw of the "Tasmanian" Tiger, *Thylacinus cynocephalus*, from Quaternary sediments of Mairs Cave in the Flinders Ranges of South Australia. x 0.7 (S. Morton, courtesy of the South Australian Museum and N. Pledge)

299 Skull of the giant kangaroo, *Procoptodon*, from the Late Pleistocene Victoria Cave at Naracoorte, South Australia. (F. Coffa, courtesy of Flinders University and R. Wells)

300 Lower jaws of a juvenile (top) and an adult marsupial lion, *Thylacoleo carnifex*, from Quaternary sediments of Henschke's Fossil Cave at Naracoorte, South Australia. x 0.8 (S. Morton, courtesy of the South Australian Museum and N. Pledge)

301 Lower jaws of Quaternary "Tasmanian" Devils, *Sarcophilus* sp., from Henschke's Fossil Cave (bottom) and Big Cave at Naracoorte, South Australia. The large specimen has been affected by a disease known as "lumpy jaw". x 1.0 (S. Morton, courtesy of the South Australian Museum and N. Pledge)

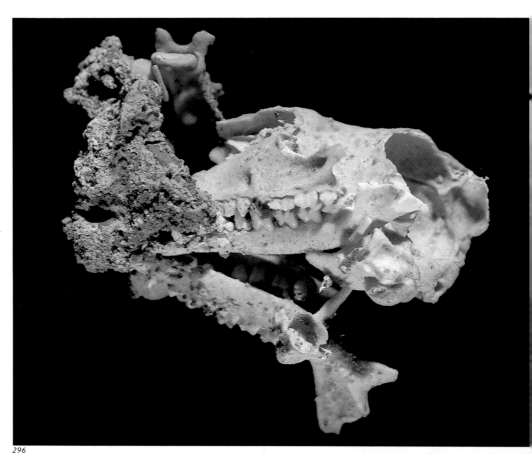

296

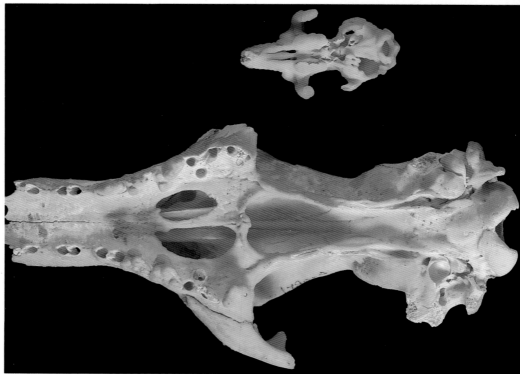

297

298

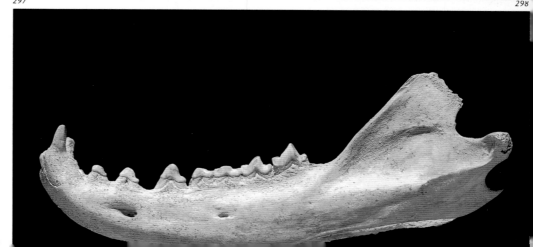

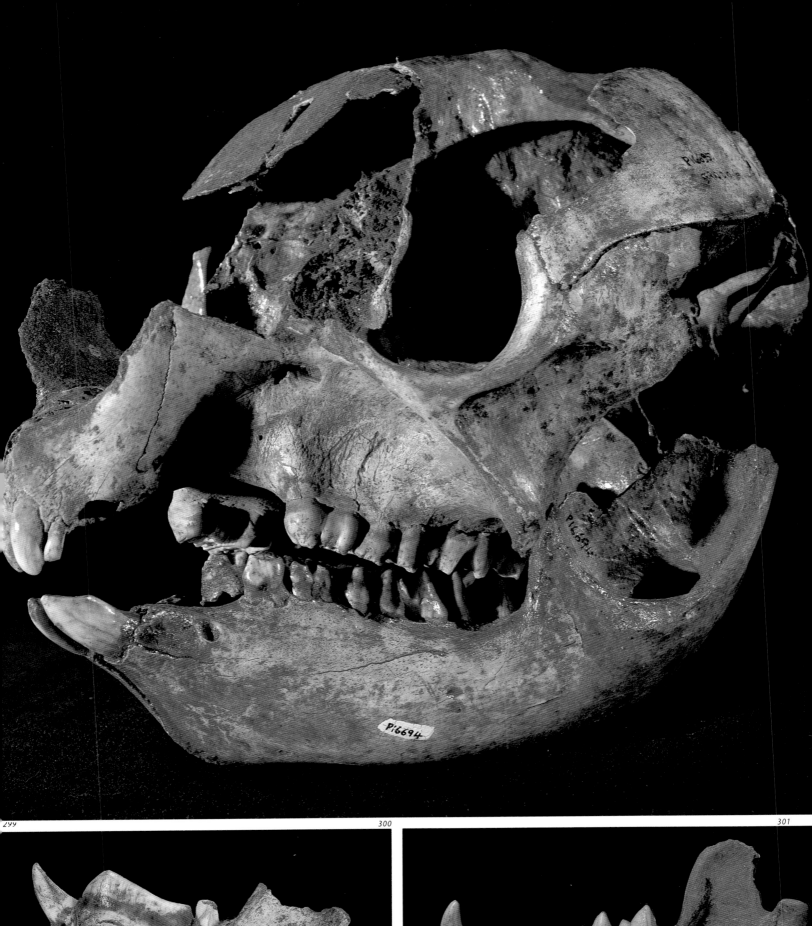

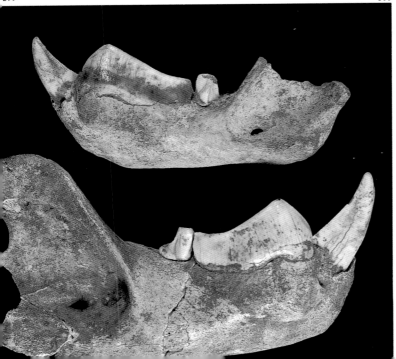

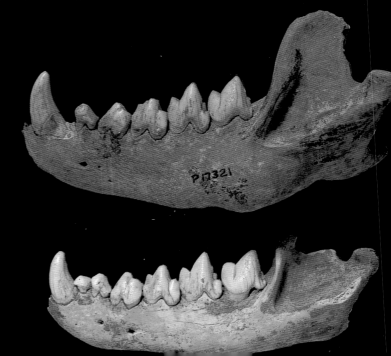

LAKE CALLABONNA

Although Quaternary terrestrial vertebrate fossils are far from a rarity in Australia, their preservation as articulated skeletons rather than as jumbled clusters of bones is not at all common. A prominent exception to this generality was discovered by an Aboriginal stockman in 1893 at Lake Callabonna in South Australia. There, on the floor of this giant playa, literally hundreds of skeletons of the Late Pleistocene megafauna have been found: *Genyornis, Diprotodon, Phascolonus* and *Sthenurus*, to list but a few. These animals evidently became bogged in the unctuous clays surrounding waterholes and mound springs in the region, the only sources of water in that severe landscape. Skeletons are typically found with their feet vertical and the body lying on its side, above. Because the feet are the deepest part of the skeleton in the clay, they are generally the best preserved of all the bones. The skull and upper part of the body were buried in the clay closest to the surface and thus are often weathered and fragmentary. This poor preservation of skulls and uppermost skeletal parts has resulted from the constant wetting and drying of the Callabonna near-surface sediments over thousands of years since the animals' deaths. When the lake dried, salt crystallized within the fossilizing bones, causing them to break apart. During wet phases, the salt would dissolve only to crystallize during the next dry episode.

For many years the skeletal construction of *Diprotodon* was a mystery. Sir Richard Owen, who named and described much of the animal, restored it in 1877 with its feet hidden behind tussocks of grass because he did not know what they looked like. It is ironic that it was the year in which he died that the feet were finally identified. It turned out that foot bones were actually known to Sir Richard, however they were so small and peculiar that no one was able to make their association with the rest of the animal until articulated skeletons were found with the feet in place at Lake Callabonna.

303

302

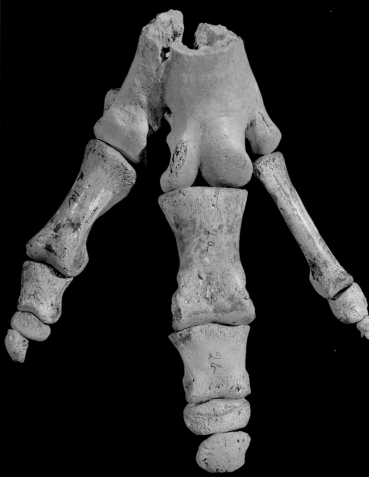

304

302 *Excavation at Lake Callabonna. The fossils occur on the floor of this vast dried up lake, far from the shore. Once fossil remains are discovered, a trench is dug around them. The hessian, or burlap strips, soaked in plaster-of-Paris is utilized to encase each specimen. Once the plaster-of-Paris is hardened, the pedestal is undercut and then the moment of truth arrives: the specimen is rolled over. If it does not collapse at that* moment, the points on what was the underside, where the pedestal attached, are covered in hessian and plaster-of-Paris, and the specimen appears, in essence, like a large egg. It can then be readily packed for transport to a museum where it can be carefully prepared for exhibition and scientific study. (Courtesy of the Australian Army)

303 *Fossil eggshell fragments of Genyornis and another bird (top left), possibly the duck Tadorna tadornoides, from Pleistocene sands of Dempsey's Lake, South Australia. x 0.7 (S. Morton, courtesy of the South Australian Museum and N. Pledge)*

304 *A right foot of the giant flightless bird Genyornis newtoni, found upright in the clay where the animal became trapped on Lake Callabonna, South Australia, in the Late Pleistocene. x 0.5 (F. Coffa, courtesy of the South Australian Museum and N. Pledge)*

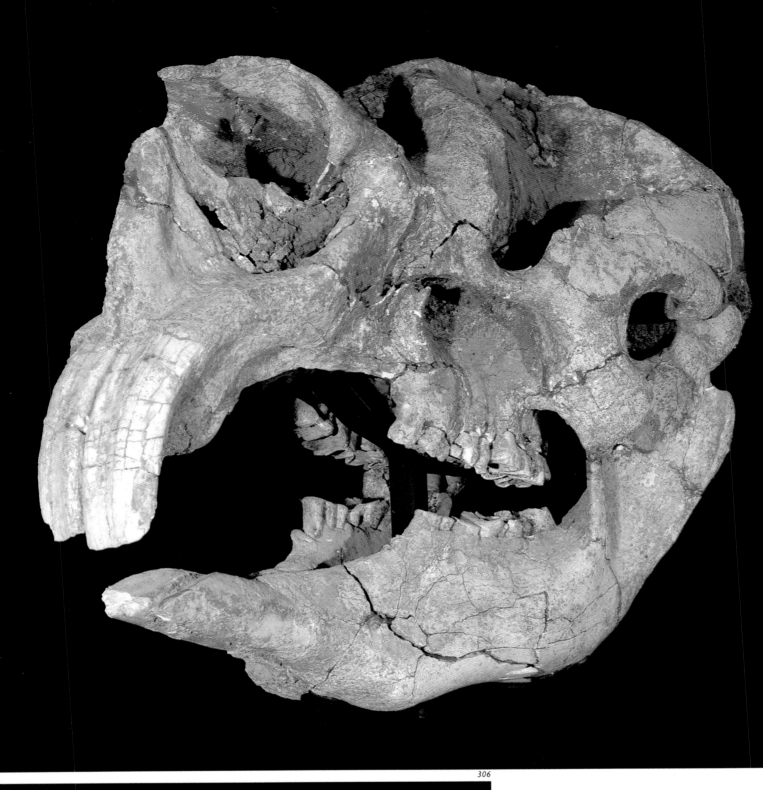

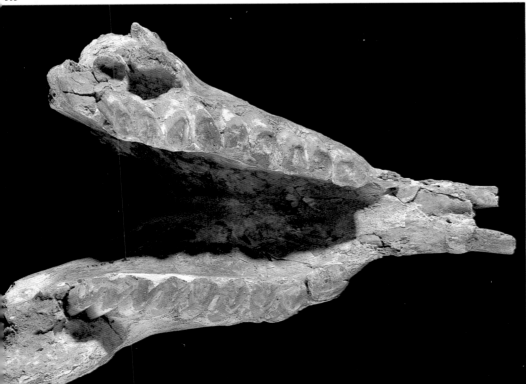

305 A skull and lower jaw of Diprotodon optatum from Late Pleistocene deposits at Tambar Springs, New South Wales. Although only this single specimen of this species is known from Tambar Springs, the skull is much better preserved than any of the numerous ones from Lake Callabonna, all of which have been damaged by salt. x 0.3 (F. Coffa, courtesy of the Australian Museum, A. Ritchie and R. Jones)

306 Lower jaw of the Giant Wombat, Phascolonus gigas, from the Late Pleistocene deposits on the floor of Lake Callabonna, South Australia. Although Diprotodon is commonly referred to as a "giant wombat", it was not, but Phascolonus was. The proportions of Phascolonus's forelimbs suggest that, like the living wombat, it was a good digger. If so, it was probably the largest burrowing animal to have ever lived. x 0.6 (S. Morton, courtesy of the South Australian Museum and N. Pledge)

DWARFING AND LOW DIVERSITY ON ISLANDS

One of the fascinating vertebrates collected by Nicolas Baudin on his scientific travels through Bass Strait was a Dwarf Emu, *Dromaius ater*, which he obtained on King Island. Dwarfing is not an uncommon happening on islands, or, indeed, amongst mainland populations where resources become restricted either by climate or some other factor. On islands plant resources for herbivores can be in short supply and can also show much less diversity than they do on larger landmasses like continents. The environmental quality can thus be lower.

Many theories have been proposed attempting to explain the details of dwarfing. One theory relates dwarfing to a decrease in plant diversity and a subsequent deterioration in environmental quality. This theory suggests that the main growth of animals, somatic growth (that is, body frame growth when protein is being added in the form of connective tissue, muscles, nerves, bones, skin), generally occurs during a limited period each year: it is not a continuous process, and there are times of somatic growth dormancy. The "growth period" corresponds to the time each year when the volume and nutrient level of forage exceeds the capacity of the herbivores to assimilate it. If this period is short, or in some years essentially non-existent, dwarfing can occur. The period of optimum protein availability can also be shortened by increasing levels of plant defences against herbivory, such as the production of fibre, resins, tannins and the like. Young shoots on plants generally contain the lowest amounts of these defences, while at the same time protein components, nitrogen, phosphorous, easily digestible energy resources, easily fermentable fibres and soluble carbohydrates are at their peak. If plant diversity is greater, then different plants may show growth peaks at different times, but if the vegetation is depauperate or homogeneous, the optimum period for high value forage may be shortened. Such a theory can explain the dwarfing observed on continental islands, such as King Island and Kangaroo Island, and even on Tasmania, and likewise, it can explain the dwarfing of modern species relative to their Pleistocene counterparts. The environmental stresses imposed by fluctuating climates during the last 25,000 years or more not only brought about outright extinction, but favoured dwarfing as well.

Islands impose another stress that favours dwarfing, and that is the necessity to maintain genetic variability within a population of animals living in a restricted area. Smaller body size allows the maintenance of larger populations and thus the necessary genetic variability which favours survival.

Other theories regarding dwarfing suggest that hunting pressures, either by humans or some other predator, can favour dwarfing. This is a direct result of smaller forms being taken less often due to such factors as their being able to escape notice more frequently, or, in the case of the human hunter, a preference for a larger prey.

On the Bass Strait islands which have been studied, changes in species composition through the Pleistocene have been recorded, and they seem to reflect both the swings in climate through this time and the arrival of both Aboriginal and European humans. On Kangaroo Island, for example, a sequence of stratified sediments at Seton Rock Shelter shows the faunal composition between 10,000 and 16,000 years ago. Human occupation is recorded in the two upper units of this sequence. Twenty-eight species of mammal and nearly 40 species of bird are known from this site, though today only seven of the mammals and few of the birds still inhabit the island. Most of the forms that became extinct were grassland, open country species. Such a natural habitat has almost completely gone on Kangaroo Island today, a result of changing climate, a rising sea level and decreased fire frequency in the most recent times due to the decline of the local Aboriginal population. Kangaroo Island, like many of the Bass Strait islands, has not always been a small landmass. Many times during the Pleistocene, when sea levels were markedly lowered due

to maximum glacial build-up, these islands were part of a broad landbridge that linked mainland Australia with Tasmania. This landbridge was severed most recently about 9000 years ago when rising ocean waters finally flooded right across the area, establishing Bass Strait and leaving only the highest peaks as islands.

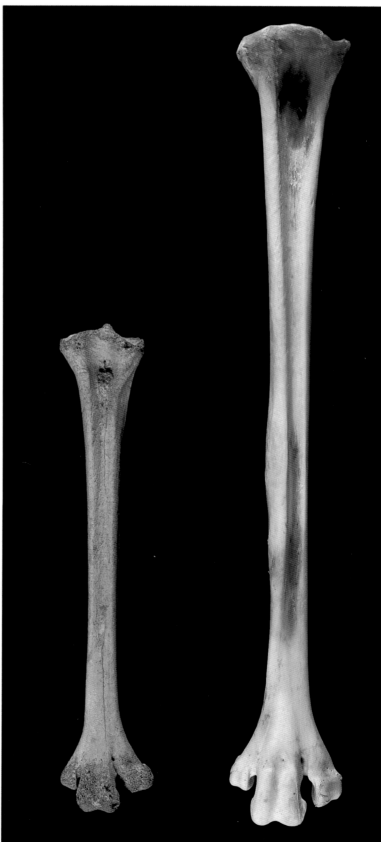

307 Tibiotarsi of the living Emu, Dromaius novaehollandiae, alongside the dwarfed King Island Emu, Dromaius ater, which survived until the beginning of the nineteenth century. x 0.5 (F. Coffa, courtesy of the Museum Of Victoria)

307

THE MISSING PREDATORS

Large mammalian herbivores in the Cainozoic had large mammalian predators making meals of them everywhere except Australia and perhaps Antarctica. In Australia there were such large herbivores as *Diprotodon* and *Zygomaturus*, both the size of a hippopotamus or a bullock, but predators such as *Thylacoleo*, the Marsupial Lion, or *Thylacinus*, the Tasmanian Tiger, would not have been able to attack even half-grown individuals, much less adults.

However, this does not mean that the big herbivores were immune from predation, only that the predators were not mammalian; instead they were probably reptiles. During the Pliocene and Pleistocene, *Megalania prisca*, the Giant Goanna, grew to lengths of at least 5 metres and perhaps 7. At about 600 kilograms, it was ten times as heavy as the largest living goanna, the Komodo Dragon of Indonesia, so attacks on *Diprotodon* and other herbivores would not have been deterred by size.

A somewhat smaller Pleistocene predator, but formidable nonetheless, was *Quinkana fortirostrum*, a 3-metre-long terrestrial crocodilian known only from eastern Queensland. Its teeth were laterally flattened and sabre-shaped, reminiscent of carnivorous dinosaurs such as *Tyrannosaurus*. Such teeth are called ziphodont and crocodilians with them became extinct elsewhere in the world during the Miocene. Only teeth and two skulls are known of *Q. fortirostrum*; however, related forms in Europe had hooves instead of claws on their feet and rounded rather than flattened tails, indicating that they were more adapted for life on land than are any living crocodilians.

Large-sized predators are not the only meat-eaters missing, or apparently missing, from the Australian late Cainozoic mammalian faunas. Fleet-footed carnivores are also absent in the groups where they might be expected. The limb proportions of both *Thylacoleo* and *Thylacinus* indicate that these animals could not run fast. No four-footed marsupial as large as a thylacine appears to have been capable of extremely rapid, prolonged running like a wolf. This may be related to the inability to carry young in a pouch while moving in such a manner. It seems that for a large marsupial to be capable of swift movement over the ground, the kangaroo's solution may be the only one available.

It is in this group of animals that a marsupial cheetah counterpart may be found. *Propleopus*, from the Pliocene and Pleistocene, and the middle Tertiary *Ekaltadeta* are kangaroos known only from skull, teeth and jaws. Their premolars are robust blades suitable for grasping and cutting very tough sinew and muscle. The stout molars behind have been utilized for crushing parts of the skeleton, as they have flanges on their margin that could have prevented bone splinters from entering the gums — just as are present in modern bone-crushing carnivores. Lower incisors, suitably modified by evolution for stabbing, complete the dental adaptations of these "killer kangaroos". If and when post-cranial skeletal material of these kangaroos is discovered, it is to be expected that carnivorous adaptations will be found there as well, particularly in the forelimbs. That these animals were, in fact, carnivores is supported by the way their teeth were actually utilized. Seen under a microscope, the teeth have coarse scratches on their enamel wear surfaces strikingly similar to those found on the teeth of carnivores such as dogs and thylacines but not present on those of herbivores.

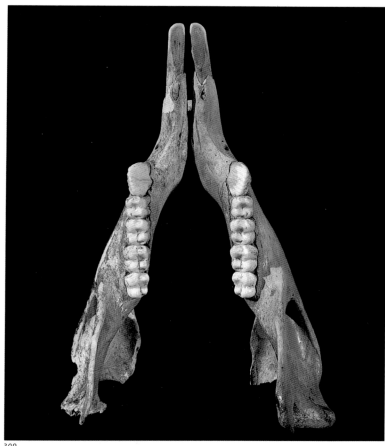
309

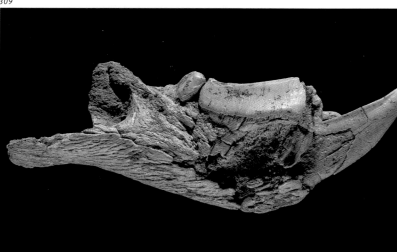
310

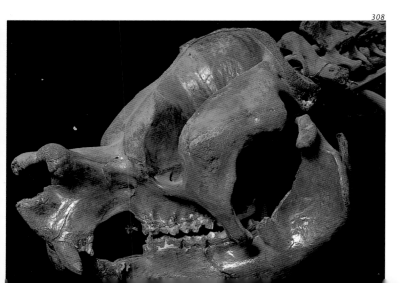
308

308 A Quaternary skull of *Zygomaturus tasmanicus* from Mowbray Swamp at Smithton, Tasmania. The structure of the lower incisor of this species is quite distinct from the mainland Quaternary form, *Zygomaturus trilobus*, suggesting a prolonged period of isolation x 0.2 (S. Morton, courtesy of the Tasmanian Museum and N. Kemp)

309 Lower jaw of *Propleopus*, the "killer kangaroo", from the underwater Green Waterhole site at Tantanoola, South Australia. x 0.6 (F. Coffa, courtesy of Flinders University and R. Wells)

310 Lower jaw fragment of a marsupial lion, *Thylacoleo carnifex*, from Late Pleistocene sediments at Lake Victoria, New South Wales. x 0.8 (F. Coffa, courtesy of the Museum Of Victoria)

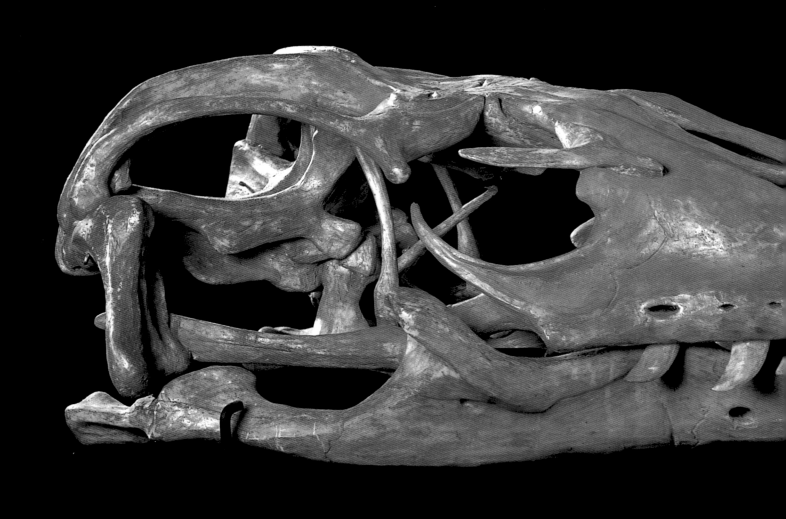

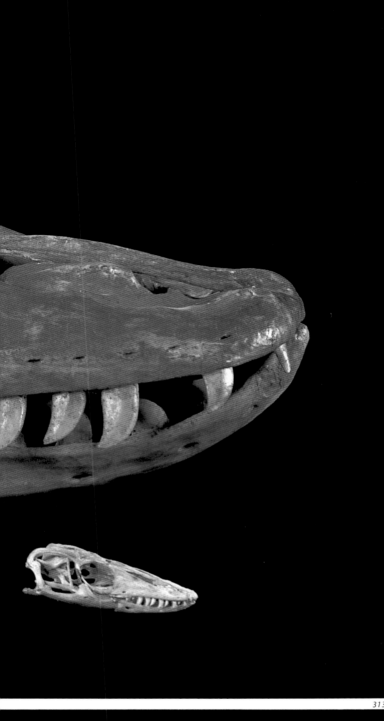

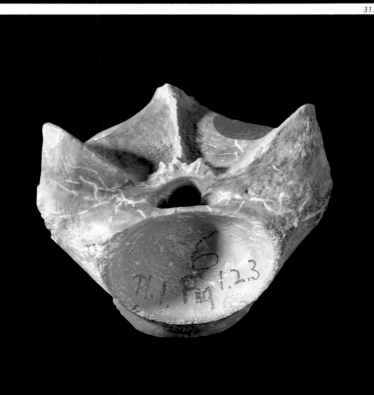

311 Reconstruction of the 80-centimetre-long skull of *Megalania prisca* compared with the 5-centimetre-long skull of a living goanna. (F. Coffa, courtesy of the Museum Of Victoria)

312 An upper jaw fragment of the giant varanid lizard *Megalania prisca* from Quaternary sediments near Waterford on the Darling Downs of south-eastern Queensland. x 1.0 (F. Coffa, courtesy of the Queensland Museum and R. Molnar)

313 Dorsal vertebra of a juvenile *Megalania prisca* lizard from the Quaternary Katipiri Sands along the Warburton River, South Australia. This specimen was collected by the early South Australian geologist H.Y.L. Brown. x 1.1 (S. Morton, courtesy of the South Australian Museum and N. Pledge)

314-316 Sketches for a restoration incorporate the ideas of artist and scientist. Then details must be worked out to produce an accurate restoration. Shown are an early skeletal drawing of *Megalania prisca* and *Genyornis newtoni*, and a later sketch overlaying the skeleton with muscles. (P. Trusler)

OVERLEAF   The giant goanna, *Megalania prisca*, approaches a nest of the dromornithid bird, *Genyornis newtoni*, in Pleistocene Central Australia. Both species are now extinct. *Megalania*, known only from fragmentary material, has been reconstructed using its nearest living relative, the Komodo Dragon Lizard, *Varanus komodensis*, from Indonesia, as a model. *Megalania* may have reached up to 7 metres in total length, and, like the Komodo Dragon Lizard, it was probably an ambush killer and scavenger. (P. Trusler)

THE DEVELOPING PSYCHROSPHERE AND VERTEBRATES AT SEA

Whales arose from archaic terrestrial carnivores called mesonychids. Extremely primitive whales with many of the features from their terrestrial past are known from the Early and Middle Eocene of Pakistan. By the Late Eocene, a variety of primitive archaeocete whales had developed in the Northern Hemisphere, but none have yet been found in Australia despite the occurrence of several marine rock units of this age.

Worldwide, the Early Oligocene record of whales is almost non-existent, and Australia is no exception. Evidently, it was during this time, so poorly documented with fossils, that the major living groups of whales evolved — the odontocetes, or toothed whales, and the mysticetes, or baleen whales. The evolution of these two groups out of the less specialized archaeocetes may have been triggered by two events that occurred during the Oligocene. At the beginning of the Oligocene, the psychrosphere became established as Antarctica became colder and more isolated from the rest of the world. (This cold, nutrient-rich bottom current today originates along the edges of frigid Antarctica to flow north into all major oceans of the world.) Then, in the Late Oligocene, with the final separation between Australia and Antarctica, the Circum-Antarctic Current became established.

Both the mysticetes and the odontocetes evolved feeding adaptations that took advantage of the new food sources made available as a consequence of psychrosphere development — the upwelling of nutrient-rich waters. The development of baleen in mysticetes enabled them to become filter-feeders, and may have been a response to the abundant plankton that thrived in these cold southern ocean waters because of the psychrosphere. The development of echo-location in the odontocetes enabled them to find and capture larger prey in a manner unavailable to their archaeocete ancestors.

By the Late Oligocene, when the cetacean record markedly improves, the odontocetes and mysticetes were fully established and quite distinct from their archaeocete forebears. It was a time of experimentation and differentiation for the whales, with the appearance of many short-lived families that would disappear by the end of the Oligocene or sometime during the Miocene.

Mammalodon colliveri from Torquay, Victoria, is one of the most intriguing of these Late Oligocene whales. It is probably the most primitive mysticete known, retaining well-developed teeth and lacking the vascular foramina on the palate which generally marks the point of attachment for the baleen. However, because other undoubted mysticetes also lack such foramina, this absence does not militate against their belonging amongst the baleen whales. It is possible that baleen was present in the spaces between the teeth without leaving any trace. In rocks of the same age as those containing *M. colliveri* much more advanced mysticetes occur; thus, when *Mammalodon* was alive, it was a living fossil.

The somewhat younger *Prosqualodon davidis* was found at Table Cape in Tasmania, the same site at which *Wynyardia bassiana* was discovered. Unfortunately, the one known skull of *Prosqualodon* was lost in about 1961, but photographs and casts provide much information about this toothed whale. Unlike Northern Hemisphere members of the family Squalodontidae, this animal had a short, broad rostrum rather than a long, narrow one. The bones in the face were pushed backwards and scooped up in the fashion similar to those of modern echo-locating odontocetes, so this fossil form was probably quite efficient at finding its prey using its own sonar. Very similar animals, perhaps even closely related species, occur in rocks of comparable age from Patagonia and New Zealand, an instance of a circum-Antarctic distribution of a cetacean species, in this case at about the time of the Oligocene–Miocene boundary and probably signalling the establishment of the Circum–Antarctic Current.

Squalodontids became extinct worldwide by the end of the Miocene. Other odontocetes replaced them and still survive today. Most of the fossil specimens are partial skulls or even less complete remains.

At the end of the Miocene occurs a rock unit only a few tens of centimetres thick that can be recognized at a limited number of widely scattered areas across the entire length of Victoria. Called the Nodule Bed, it is a lag deposit in which sands and silts were winnowed away and larger rocks and fossil bones were left behind. When this winnowing process was operating, the rocks and bones were frequently rolled, and consequently it is only the most durable parts of the skeleton such as the ear bones and the vertebrae that are frequently preserved in this unit.

The Nodule Bed is evidence of a widespread erosional phase in the marine environment, the cause of which is uncertain. At the time it formed, there was a pronounced lowering of sea level owing to the filling of the Mediterranean Basin. A few million years earlier, the Mediterranean Sea had dried up due to the blocking of the Straits Of Gibraltar. As a result, it had become a vast desert. When the natural dam at the Straits Of Gibraltar broke, world sea levels dropped precipitously. Perhaps that event, half a world away, found expression in Australia in the development of the Nodule Bed.

Among the most common vertebrate fossils preserved in the Nodule Bed are the remains of both mysticete and odontocete whales. Although the majority of these fossils are the durable vertebrae and ear bones, a rostrum of *Mesoplodon*, a beaked whale which still lives in Australian waters, managed to survive. Another marine mammal group which has its oldest Australian record in the Nodule Bed is the seals.

Seals, fur-seals and sea lions are all members of the placental order Carnivora.

Phocidae, or earless seals, appear to have originated in the Atlantic Basin at about the Middle Miocene and entered the Pacific at the end of that epoch. It is their remains which occur in the Nodule Bed. Because these specimens comprise only vertebrae and ear regions, little more can be said about them than that they are phocids.

Fur-seals and sea lions, the Otariidae, originated in the temperate waters of the North Pacific at the end of the Middle Miocene and reached the Peruvian coast by the end of that epoch. They have an even shorter Australian record than the earless seals, not appearing until the beginning of the Pleistocene. None of the Australian fossil specimens are clearly from extinct species. For example *Neophoca cinerea*, the living Australian Sea Lion, is known from the Late Pleistocene at Queenscliff, Victoria.

317 Tooth of an odontocete (toothed) whale from Quaternary gravels on Flinders Island, Tasmania. x 0.4 (S. Morton, courtesy of the Queen Victoria Museum and C. Tassell)

318 Tooth of the cetacean Scaldicetus macgeei from the Late Miocene Black Rock Sandstone of Beaumaris, Victoria. x 0.8 (F. Coffa, courtesy of the Museum Of Victoria)

319 Tooth of the cetacean Scaldicetus lodgei from the Pliocene sediments of Clifton Bank on Muddy Creek near Hamilton in western Victoria. x 1.0 (F. Coffa, courtesy of the Museum Of Victoria)

320 A skull and lower jaw of Mammalodon colliveri from the Late Oligocene at Torquay, Victoria. With a full dentition, this species is the most primitive baleen whale known. x 0.25 (E. Fordyce)

321 Cetacean (whale) vertebrae in matrix from the Early Miocene marine Fossil Bluff Sandstone at Fossil Bluff, Wynyard, Tasmania. x 0.4 (S. Morton, courtesy of the Tasmanian Museum And Art Gallery and N. Kemp)

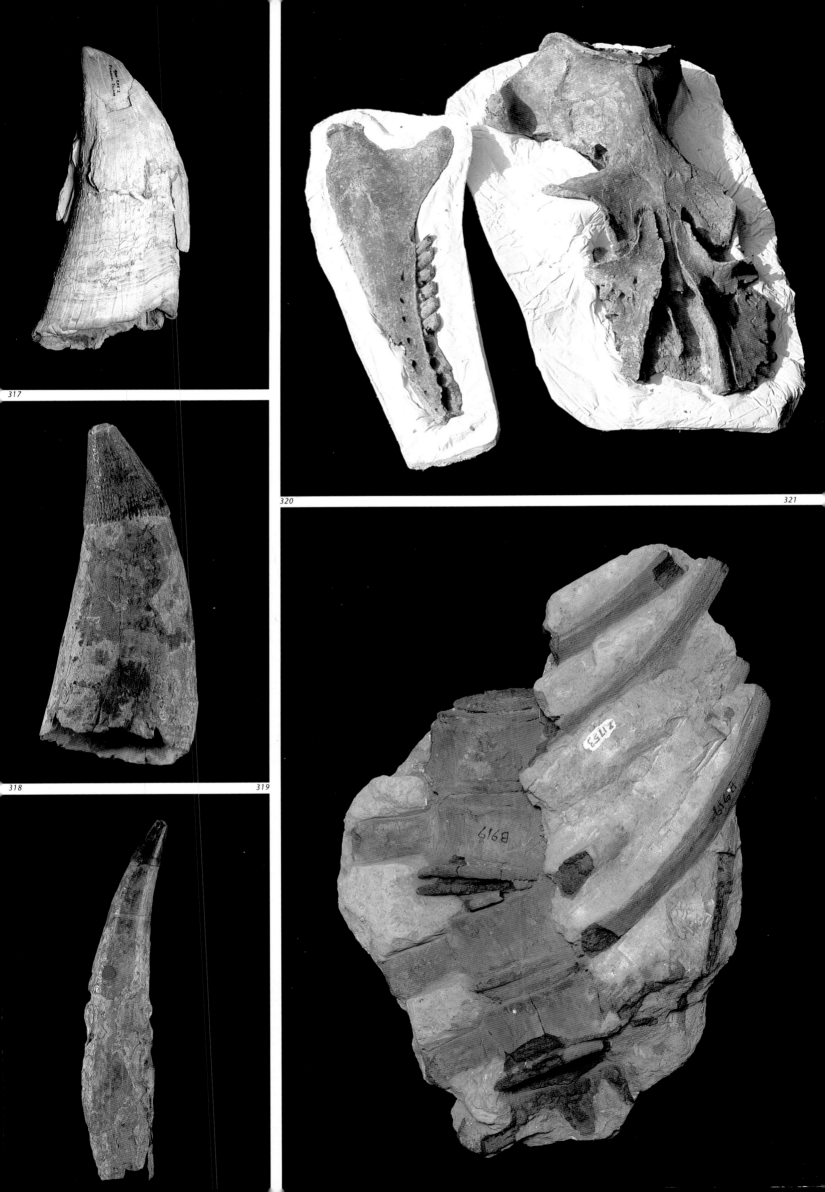

317

318

319

320

321

322 Tooth of a primitive whale, Squalodon gambierense, and an ear region (otic bulla) of an unidentified whale. The tooth comes from the Oligocene marine Gambier Limestone at Mt Gambier, South Australia, and the ear region from the Pliocene Loxton Sands of Sunlands Pumping Station west of Waikerie, South Australia. x 1.0 (S. Morton, courtesy of the·South Australian Museum and N. Pledge)

323 Cetacean petrosals (ear bones) from Tertiary sediments exposed in the seabed off Flinders Island, Tasmania. x 0.5 (S. Morton, courtesy of the Queen Victoria Museum and C. Tassell)

324 A skull and lower jaw of Prosqualodon davidis, a toothed whale, from the Early Miocene Fossil Bluff Sandstone at Fossil Bluff, Wynyard, Tasmania. x 0.3 (S. Morton, courtesy of the Tasmanian Museum And Art Gallery and N. Kemp)

325 White Pointers were common in the Late Miocene coastal waters of Victoria. Teeth of Isurus hastalis from the Black Rock Member of the Sandringham Sands at Beaumaris, Victoria, are shown in the background while an even larger and now extinct shark, Carcharodon megalodon, is represented by the single enormous tooth in the foreground. x 0.6 (F. Coffa, courtesy of the Museum Of Victoria)

326 A tooth of the large shark Carcharodon angustidens from Miocene marine sediments of the Cape Range, Western Australia. x 1.2 (J. Frazier, courtesy of the Western Australian Museum and K. McNamara)

327 Fossils from the marine Pliocene Loxton Sands at Sunlands Pumping Station, west of Waikerie in South Australia. Included are: a variety of sharks, rays, Diodon (a teleost) and invertebrates (bivalves, echinoids, bryozoans and crustaceans). This site is especially important because of the occasional occurrence of terrestrial vertebrates, for which the marine assemblage provides a globally related dating. x 1.2 (S. Morton, courtesy of the South Australian Museum and N. Pledge)

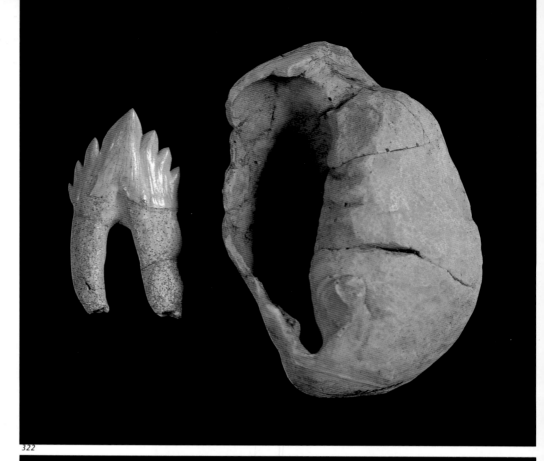

322

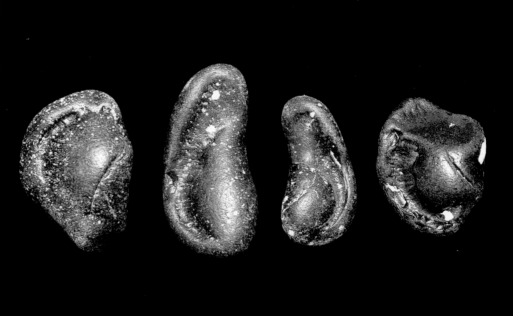

323

324

MARINE BIRDS OF THE CAINOZOIC

Besides mammals, birds are also known in several Cainozoic marine deposits. They include both penguins and albatrosses from nearshore marine deposits along southern Australia, and a variety of mostly extant forms in assemblages from Quaternary deposits on continental islands and oceanic islands such as Norfolk Island and Lord Howe Island off the east coast of Australia.

Diomedea thyridata from the Late Miocene of Beaumaris, near Melbourne, is the oldest albatross from Australia. The older *Manu antiquus* from the Early Oligocene of New Zealand, a species defined on a furcula (wishbone) fragment, has been called an albatross, but is too incomplete to assign to any family. The Australian record of albatrosses is considerably younger than the oldest record for the family, from the Late Oligocene of North America. Other Southern Hemisphere records include one species from the Late Miocene of Argentina, and another from the Early Pliocene of southern Africa. Clearly, the record is too sparse to allow reconstruction of the southern ocean history of the group other than to simply say it existed from Miocene times onwards.

Penguins have a longer record, having frequented the Australian coast from Palaeogene times. They, like the whales, seem to have appeared and expanded as the psychrosphere developed. At least three different penguins have been recovered from the Late Eocene Blanche Point Formation in south-eastern Australia. One form, *Anthropornis nordenskjoeldi*, was truly gigantic, and like most other penguins of this age is closely related to contemporaneous forms in New Zealand and West Antarctica. *Anthropornis* may have reached 135 centimetres in height and 90 kilograms in weight, significantly exceeding the size of the largest living penguin, the Emperor Penguin, *Apetnodytes forsteri*, which inhabits Antarctic waters. A bill, which was preserved in one specimen of a giant penguin from contemporaneous Antarctica, was long and dagger-like, very unlike

any living penguins, and much more loon-like. This and other characters of these earliest penguins seem to suggest a relationship with the loons. Olson (1986) is one proponent that some of the more ancient penguins, possibly including *Anthropornis*, fed more by spearing their prey, whereas modern penguins do not. Most living penguins, in fact, are specialized for feeding partly or entirely on plankton.

The penguin record continues through the Miocene to the present, with Australian fossils restricted to the south-east of the continent including a Quaternary record in Tasmania. Quaternary forms appear to be still extant species, but only the Little or Fairy Penguin, *Eudyptula minor*, breeds on Australian shores today, the remainder of living forms breeding essentially on the Antarctic continent, sub-Antarctic islands, coastal islands of South America (including the Galapagos) and Africa, or New Zealand. The larger species of penguins seem to disappear in the Early Miocene, which coincides with the expansion of the whales and pennipeds — and perhaps these two events are related.

To date, what knowledge there is about fossil Australian marine vertebrates has been acquired by accidents of discovery. No systematic attempt has yet been made to locate and subsequently excavate the fossils needed to document this evolutionary story. Fortunately, the major barrier that retards progress in unravelling the history of Australian land mammals, namely the lack of sufficient rock outcrops of the proper age and type, is not a problem in this instance, at least at the present time. Particularly in southern Australia, there are widespread outcrops of marine rocks of all ages in the Cainozoic exposed on the present coastline that could be prospected. Just as is the case for the dinosaur sites on the Victorian coast, this potential source of information about Australia's past will be available only for as long as there is not a significant rise in sea level.

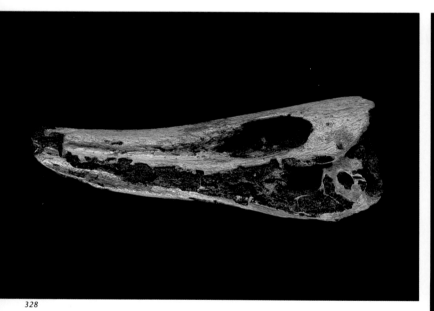

328

328 Bill fragment of Diomedea thyridata, the oldest known Australian albatross, from the Late Miocene Black Rock Sandstone of Beaumaris, Victoria. x 1.2 (F. Coffa, courtesy of the Museum of Victoria)

329 Humeri of Palaeeudyptes antarcticus (left specimen) and that of an unidentified penguin with tooth marks possibly from a shark. The Palaeeudyptes specimen was recovered from the Oligocene Gambier Limestone of Mt Gambier, South Australia, and the penguin specimen from the Late Eocene Blanche Point Formation at Christies Beach, South Australia. The giant penguin Anthropornis, whose range in the Eocene extended to Seymour Island in Western Antarctica, is also known from the Blanche Point Formation. x 0.7 (S. Morton, courtesy of the South Australian Museum and N. Pledge)

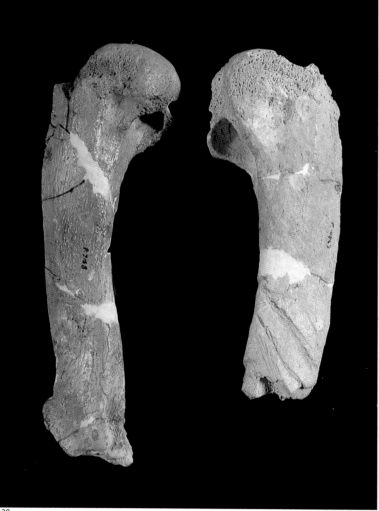

329

TOWARDS OUR MODERN WORLD

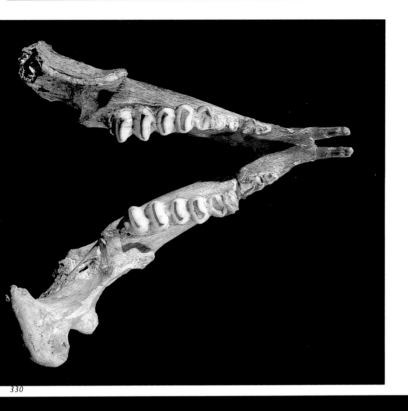

The journey of Australia during the Cainozoic has been a truly epic one — and it is still in progress. It has brought Australia from a far southerly position dominated by a cool temperate climate to the tropics with concomitant development of hot, steamy rainforests in the north-east through to stark deserts in the continental interior. All of this change can be clearly reconstructed from the fossil fauna and flora that lie locked in the rocks and sediments left behind by ancient streams and swamps. The future for Australia appears to be just as changeable. With the appearance of humans of various cultures and commitments the biota of this antipodean land has changed dramatically over the past 40,000 years, most rapidly in the last 200. And the continent continues to slowly and persistently move northwards, pushing up mountains in its bow wave. In 50 million years, whether humans are still living or extinct, it will join with Asia and cease to exist as a separate biological entity. Its plants and animals will be an amalgam with the great northern landmasses of Europe and Asia, and success will depend on *survival of the fittest* on the grandest of scales.

330 The lower jaws (in occlusal view) of a zygomaturine diprotodontoid from Pleistocene sediments of Dani Valley, Irian Jaya. x 0.4 (F. Coffa)

331 The skull and lower jaws of a zygomaturine diprotodontoid from Pleistocene sediments of the Dani Valley, Irian Jaya. Vertebrate fossils are still quite rare from the island of New Guinea, but the area must have a rich history, at least since the Miocene. x 0.7 (F. Coffa)

330

331

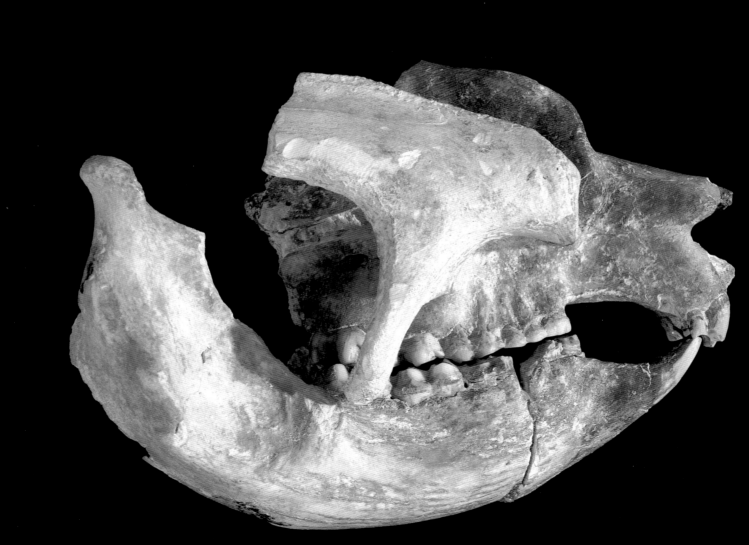

332

Trusler

332 Australia in the middle to late Cainozoic may have had the richest flamingo fauna known anywhere, along with a variety of aquatic vertebrates. The flamingoes inhabited the myriad of lakes that were then dotted across Central Australia. Two species of flamingoes are known to have lived in the shallow waters of northern and eastern South Australia during the Late Oligocene to the Middle Miocene—a more robust form, Phoeniconotius (dark legs), and the more gracile Phoenicopterus novaehollandiae (middle foreground), which was about the size of the largest living flamingo, the Greater Flamingo. The flamingo-like palaelodids, depicted lower right, show some affinities with middle Tertiary species in Europe. Also resident in these lakes were a variety of cormorants, pelicans, rails and ducks as well as freshwater dolphins (rhabdosteids) in the more southerly lakes near today's Lake Frome. (P. Trusler, with the permission of Doubleday Inc., New York)

333 While flamingoes crowded onto the inland lakes, the rainforest vegetation surrounding these waterways hosted a rich fauna of browsing and arboreal silvaphiles. A tree-dwelling marsupial squirrel, Ektopodon, feeds undisturbed as its joey emerges from the pouch, while three diprotodont marsupial herbivores, Ngapakaldia (about the size of a sheep), cautiously approach the water's edge. In the ferns behind them a primitive rat kangaroo, a browser, pauses momentarily, deciding whether to bound away. (P. Trusler, with the permission of Doubleday Inc., New York)

334 The early Cainozoic vertebrate fauna of Asia was for the most part quite distinctive from that of Gondwana. Typical of the time was the carnivorous Cimolesta, shown in the foreground. Common in this fauna, too, were the pika-like anagalids shown grazing on the right. Approaching the water in the background, however, are two sheep-sized edentates, Ernanodon, which probably arose from Gondwanan stock that migrated northwards from South America sometime in the early Cainozoic and moved into Eurasia. All of these vertebrates were placentals. (P. Trusler, with the permission of Doubleday Inc., New York)

TERTIARY LOCALITIES
- Pliocene
- Miocene
- Oligocene
- Eocene
- Palaeocene

Mogorafugwa

Awe

Riversleigh, Nooraleeba, Rackham's Roost

Bullock Creek
Quanbun
Floraville
Tara Creek
Bluff Downs

Alcoota
Kangaroo Well
Cape Range
Lake Ngapakaldi,
Kanunka
Tingamarra
Chinchilla
Redbank Plains

Lake Palankarinna,

Ian's Prospect
Pinpa, Yanda,
Tarkarooloo,
Ericmas, Wadikali
Currambulka, Town Well
Sunlands

Christies Beach
Wynyard
Geilston Bay

Warrumbungles
Krui River, Bow

Talyawalka
Bone Gulch,
Fisherman's Cliff

Canadian Lead
Big Sink

Great Buninyong
Estate Mine, Smeaton
Coimadai

Mt Gambier
Orbost
Jemmy Pt
Lake Tyers
Bunga Creek
Morwell

Forsyth's Bank,
Hamilton,
Grange Burn
Balcombe Bay
Beaumaris
Batesford Quarry,
Dog Rocks,
Waurn Ponds
Janjuc, Torquay

333

Trusler

334

Trusler

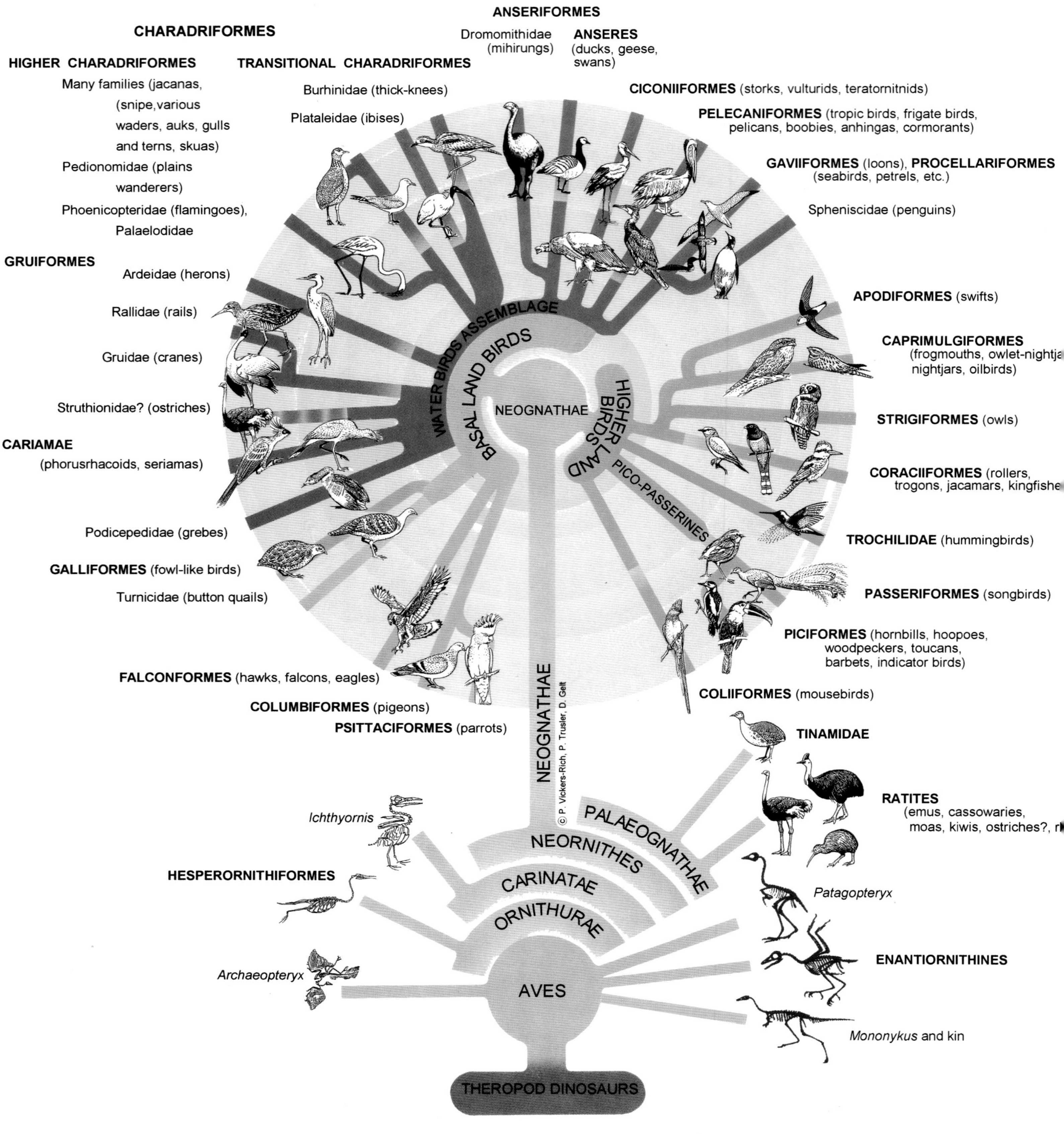

ANSERIFORMES

Dromomithidae (mihirungs)

ANSERES (ducks, geese, swans)

CHARADRIFORMES

HIGHER CHARADRIFORMES

Many families (jacanas, (snipe, various waders, auks, gulls and terns, skuas)

Pedionomidae (plains wanderers)

Phoenicopteridae (flamingoes), Palaelodidae

TRANSITIONAL CHARADRIFORMES

Burhinidae (thick-knees)

Plataleidae (ibises)

CICONIIFORMES (storks, vulturids, teratornitnids)

PELECANIFORMES (tropic birds, frigate birds, pelicans, boobies, anhingas, cormorants)

GAVIIFORMES (loons), **PROCELLARIFORMES** (seabirds, petrels, etc.)

Spheniscidae (penguins)

GRUIFORMES

Ardeidae (herons)

Rallidae (rails)

Gruidae (cranes)

Struthionidae? (ostriches)

CARIAMAE

(phorusrhacoids, seriamas)

Podicepedidae (grebes)

GALLIFORMES (fowl-like birds)

Turnicidae (button quails)

FALCONFORMES (hawks, falcons, eagles)

COLUMBIFORMES (pigeons)

PSITTACIFORMES (parrots)

APODIFORMES (swifts)

CAPRIMULGIFORMES (frogmouths, owlet-nightja nightjars, oilbirds)

STRIGIFORMES (owls)

CORACIIFORMES (rollers, trogons, jacamars, kingfishe

TROCHILIDAE (hummingbirds)

PASSERIFORMES (songbirds)

PICIFORMES (hornbills, hoopoes, woodpeckers, toucans, barbets, indicator birds)

COLIIFORMES (mousebirds)

TINAMIDAE

RATITES (emus, cassowaries, moas, kiwis, ostriches?, r

WATER BIRDS ASSEMBLAGE

BASAL LAND BIRDS

NEOGNATHAE

HIGHER LAND BIRDS

PICO-PASSERINES

NEOGNATHAE

PALAEOGNATHAE

Ichthyornis

NEORNITHES

CARINATAE

ORNITHURAE

Patagopteryx

HESPERORNITHIFORMES

AVES

ENANTIORNITHINES

Archaeopteryx

Mononykus and kin

© P. Vickers-Rich, P. Trusler, D. Gelt

THEROPOD DINOSAURS

Hypothesis of the relationships of modern and fossil birds (based in large part on Olson, 1986 and Chiappe, in press).

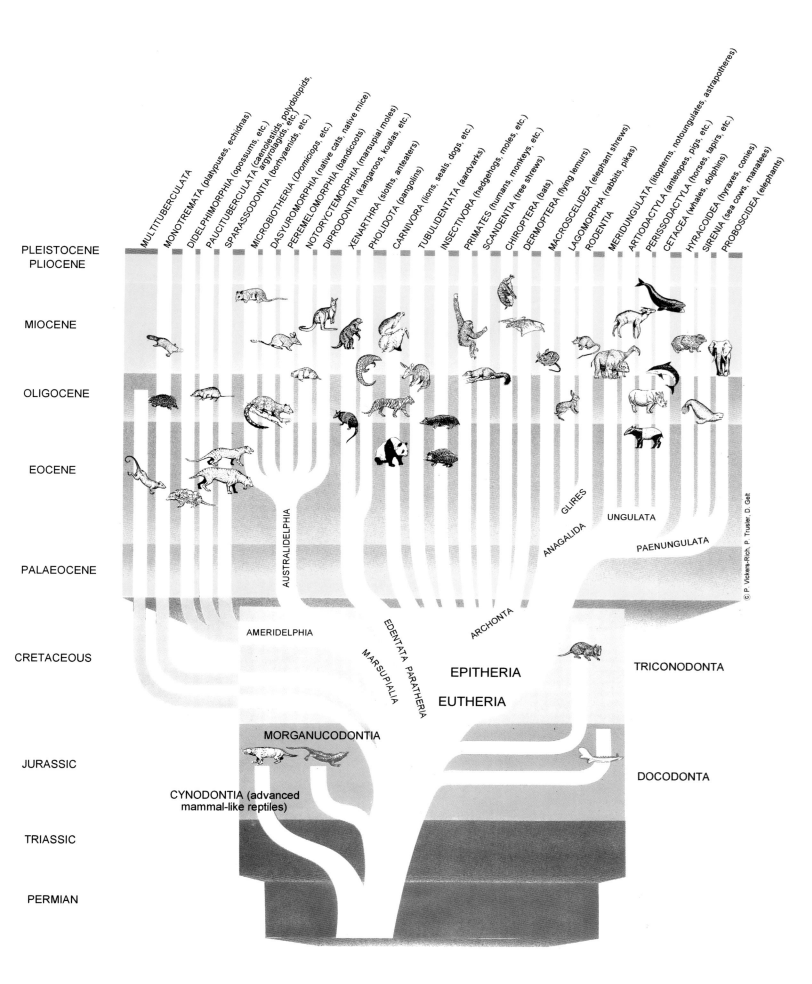

MULTITUBERCULATA
MONOTREMATA (platypuses, echidnas)
DIDELPHIMORPHIA (opossums, etc.)
PAUCITUBERCULATA (caenolestids, polydolopids,
SPARASSODONTIA (borhyaenids, etc.)
MICROBIOTHERIA (*Dromiciops*, etc.)
DASYUROMORPHIA (native cats, native mice)
PEREMELOMORPHIA (bandicoots)
NOTORYCTEMORPHIA (marsupial moles)
DIPRODONTIA (kangaroos, koalas, etc.)
XENARTHRA (sloths, anteaters)
PHOLIDOTA (pangolins)
CARNIVORA (lions, seals, dogs, etc.)
TUBULIDENTATA (aardvarks)
INSECTIVORA (hedgehogs, moles, etc.)
PRIMATES (humans, monkeys, etc.)
SCANDENTIA (tree shrews)
CHIROPTERA (bats)
DERMOPTERA (flying lemurs)
MACROSCELIDEA (elephant shrews)
LAGOMORPHA (rabbits, pikas)
RODENTIA
MERIDUNGULATA (litopterns, notoungulates, astrapotheres)
ARTIODACTYLA (antelopes, pigs, etc.)
PERISSODACTYLA (horses, tapirs, etc.)
CETACEA (whales, dolphins)
HYRACOIDEA (hyraxes, conies)
SIRENIA (sea cows, manatees)
PROBOSCIDEA (elephants)

PLEISTOCENE
PLIOCENE

MIOCENE

OLIGOCENE

EOCENE

PALAEOCENE

CRETACEOUS

JURASSIC

TRIASSIC

PERMIAN

AUSTRALIDELPHIA

GLIRES
ANAGALIDA
UNGULATA
PAENUNGULATA

AMERIDELPHIA

EDENTATA
PARATHERIA
MARSUPIALIA

ARCHONTA

EPITHERIA
EUTHERIA

TRICONODONTA

MORGANUCODONTIA

DOCODONTA

CYNODONTIA (advanced
mammal-like reptiles)

© P. Vickers-Rich, P. Trusler, D. Gelt

Hypothesis of the relationships of
living and fossil mammals.

CHAPTER 6

THE LIVING VERTEBRATES OF AUSTRALIA

THE QUARTNERARY PERIOD, FROM 1.78 MILLION YEARS AGO TO THE PRESENT

The living biota of Australia is the result of a pot-pourri of past events, and reflects the dynamism of the Australian landmass itself. Today's fauna inhabit a variety of modern environments, ranging from the vast areas in the grip of intense aridity in the continental Centre, to the humid tropical lands of the north and north-east, and the temperate country in the south. Overprinted on such environmental controls are the effects, very often horrific, of humans over at least the last 40,000 years, especially within the last 200.

In some ways, the modern fauna is not at all typical of most of Australia's past, for during the vast majority of vertebrate history on this continent it has been a wetter place, a more benign environment. There have been exceptional times, however, such as those of severe glaciation at the end of the Palaeozoic, when vast ice sheets covered up to half of what is today Australia. As well, Australia's present isolation from the rest of the world, and so the moulding of today's fauna, is an anomaly with the situation in the past. More often, this continent has been part of a much larger landmass, for the most part Pangaea, and for a brief time in the late Mesozoic and a longer period in the early to middle Palaeozoic — important when considering vertebrates — a southern continent, Gondwana, had some reality.

To a large extent, the future of the Australian biota, and for that matter the biota of the Earth, is very intertwined with human actions, especially related to burgeoning population size. Unless that problem is addressed, and brought under control, all other issues shrink to insignificance. It seems impossible to imagine how forests can be preserved, species diversity maintained, quality of life upgraded and pollution controlled if humans continue to breed and increase their population size at the astounding rate that presently characterizes our species. Much of the Australian vertebrate biota, and, indeed, that of the world, especially our own species, has little hope for a lengthy future unless humans face this very real problem — now.

335 The Koala, Phascolarctos cinereus, occurs over the eastern one-third of Australia, being restricted to those areas with the acceptable food trees, predominantly the River Red Gum and Forest Red Gum in the north and the Grey, Manna, Swamp and Blue gums in the south-east. Animals in the southern part of the range are larger than those from the north, with total lengths varying from 69 to 82 centimetres and weights ranging from 4 to 13.5 kilograms. (F. Coffa, courtesy 335 of Healesville Sanctuary)

THE UNIQUENESS OF THE AUSTRALIAN BIOTA

The biota of Australia, its plants and animals, is quite different from those elsewhere in the world. Kangaroos and koalas, lyrebirds and emus, *Eucalyptus* and waratahs — all are distinctly Australian. Other parts of the biota are shared with other continents, for example, *Acacia* and rabbits, but they are only rather recent arrivals, rabbits within the time of European settlement and *Acacia* sometime in the last few tens of millions of years, on their own, from the Old World.

The uniqueness of the Australian vertebrates, as well as other parts of the biota, can be viewed in many ways. One of the most striking aspects of this fauna is that *many forms are totally unique to Australia*, in other words, they are *endemic* to this continent, not only at the species level but at higher taxonomic levels as well, such as at family and ordinal levels.

336

336 *The Eastern Grey Kangaroo, Macropus giganteus, is a marginal occupant of interior Australia, entering the arid regions in good seasons but absent in times of drought. (F. Coffa)*

337 *Emus, Dromaius novaehollandiae, are obligate land dwellers endemic to Australia. Closely related to Cassowaries, their relatives outside Australia and New Guinea are not generally agreed upon. The fossil record of Emus extends into the Oligo–Miocene. (F. Coffa)*

338 *Mallee Fowl, Leipoa ocellata, on its nesting mound. The megapodes as a group may have had a Gondwanan origin but some biogeographers still favour a northern derivation. (F. Coffa)*

339 *The Tawny Frogmouth, Podargus strigoides, is in the family Podargidae and occurs Australia-wide. It ranges in total length from 33 to 47 centimetres. (F. Coffa, courtesy of Healesville Sanctuary)*

340 *Cape Barren Geese, Cereopsis novaehollandiae. As well as occurring on Cape Barren Island in Bass Strait, this species is known all along the southern coast of Australia. Its total length ranges from 75 to 90 centimetres. (F. Coffa, courtesy of Healesville Sanctuary)*

Marsupials are notable, but by no means the only, examples of these endemic animals, with major marsupial groups (orders) such as the native cats and mice and their relatives (Dasyuromorphia), the bandicoots (Peramelemorphia), the marsupial moles (Notoryctemorphia) and the vast array of kangaroos, koalas, possums and all their relations (Diprotodontia) being entirely restricted to the Australian region. Marsupials are still present in South America in some variety, and have had a rich past there, but the groups on this other Gondwanan fragment are different from those in Australia, with only one group showing close affinity to antipodean forms. The living Chilean *Dromiciops australis*, a microbiothere marsupial, of all its South American compatriots shares a number of specialized (synapomorphic) characters with Australian marsupials. *Dromiciops* has an ankle, a tarsus, which is constructed in the same fashion as that of Australian marsupials and not like that of other South American marsupials. The articular faces of the astragalus and calcaneum, both bones of the ankle, form a unique pattern (the continuous lower ankle joint pattern, or CLAJP) which defines the *Australidelphia* — all Australian marsupials plus the New World *Dromiciops* and fossil relatives — as distinct from the *Ameridelphia* which comprises most all of the New World marsupials.

Emus and cassowaries are yet another of the Australian (including New Guinean) groups that are endemic at a higher taxonomic level, as are an array of other avian forms including several of the subfamilies of parrots (Loriinae, Cacatuinae) and many families of songbirds, for example, the lyrebirds (Menuridae) and mudlarks (Grallinidae).

Another aspect of the Australian vertebrate fauna is that *many groups which have high diversity elsewhere in the world have low diversity within Australia*, or may be entirely absent. Placental mammals, those that develop a placenta between mother and offspring so that the embryo can be retained within the mother's body for a relatively long period of gestation, are the dominant mammalian group on a world scale. Not so in Australia. Marsupials, and even parrots, fill many of the niches in Australia which elsewhere in the world are occupied by placental mammals.

The only placentals that have occupied Australia without the assistance of humans are bats and rodents. Bats are represented in Australia's modern fauna by six families, the earliest of which had arrived by at least the Early Eocene. The rodents are not nearly so diverse, being represented by only one group, the family Muridae. Their oldest record at the antipodes is Early Pliocene, and they seem most closely related to forms in south-eastern Asia. In the cases of both the bats and the rodents, there seem to have been several invasions of the Australasian area, not simply a single event, and some rodents, such as the Plague Rat (*Rattus rattus*), have even been introduced, albeit inadvertently, by humans, perhaps initially from Dutch shipwrecks.

With humans came an entourage of placental companions, including the Dingo (*Canis familiaris dingoensis*), whose oldest record is about 3500 years before the present and was brought in by Aboriginal people, as well as domestic dogs, cats, house mice and rats, sheep, horses, cows, deer and the ubiquitous rabbits, to mention but a few, all European innovations of the past two centuries.

Yet a third aspect of the Australian biota is the *survival of a number of groups on this continent long after their extinction elsewhere*. One of the oldest records of this phenomenon was Nicolas Baudin's recovery of the living *Trigonia*, a small clam that had been extinct elsewhere in the world for millions of years. Other examples of this relictual austral distribution of forms, long gone elsewhere, are such primitive mammals as the platypus (*Ornithorhynchus*) and the echidnas (*Tachyglossus* and *Zaglossus*). Moreover, this aspect of the modern biota is reflected time and time again in the fossil record: the survival of labyrinthodont amphibians and the carnivorous dinosaur *Allosaurus* into the Early Cretaceous of Australia and the survival of ziphodont crocodiles late into the Cainozoic of this continent are but three examples.

The living terrestrial vertebrate fauna is also somewhat unique in that it *lacks certain functional elements which are usually present on other continents*. Missing, for example, are open country carnivores that would be functional analogues of the African Cheetah and Lion.

337

338

339

340

341 The Short-beaked Echidna, *Tachyglossus aculeatus*, occurs Australia-wide. It is a toothless anteater which uses its forepaws to dig into termite nests and then extends its long snout and tongue into the exposed galleries. The total length of this species ranges from 30 to 45 centimetres, and its weight varies from 2 to 7 kilograms. (F. Coffa, courtesy of Healesville Sanctuary)

342 The Platypus, *Ornithorhynchus anatinus*, is found along the well-watered east coast of Australia, and as far west as eastern South Australia. It eats a wide variety of adult and larval invertebrates, and occasionally small vertebrates. The Platypus ranges in size from 40 to about 54 centimetres in total length, and from 700 to 2200 grams in weight. The largest forms occur in New South Wales, west of the Great Divide. (F. Coffa, courtesy of Healesville Sanctuary)

343 The Dingo, *Canis familiaris*, south of Darwin, Northern Territory. The "native dog" was introduced to Australia by Aboriginal people only a few thousand years ago. (F. Coffa)

344 The Spinifex Hopping Mouse, *Notomys alexis*, is known throughout much of arid Central and Western Australia, where it prefers sandy soils. It ranges from about 225 to 270 millimetres in total length and from about 27 to 45 grams in weight. (F. Coffa, courtesy of Healesville Sanctuary)

345 The Eastern Barred Bandicoot, *Perameles gunnii*, is known only from south-western Tasmania and a few relict colonies around Hamilton, Victoria. It prefers grasslands and feeds opportunistically on a wide range of foods including earthworms, insects and berries. This animal ranges in total length from 34 to 46 centimetres and weighs from 950 to 1800 grams. (F. Coffa, courtesy of Healesville Sanctuary)

346 The Long-footed Potoroo, *Potorus longipes*, known from only a small part of eastern Gippsland, Victoria. It inhabits open forest of mixed eucalypts associated with a little understorey of shrubs, dense wiregrass, ferns and sedges. This species ranges from about 70 to 75 centimetres in total length and from 1.6 to 2.2 kilograms in weight. (F. Coffa, courtesy of Healesville Sanctuary)

347 The Brown Antechinus, *Antechinus stuartii*, is widespread in forested areas of south-eastern Australia and in several pockets of wet dense forest of north Queensland. It is primarily an arthropod eater. This animal has a short life span of about 1 year. It ranges from 14 to 25 centimetres in total length, with a weight range of 17 to 71 grams. (F. Coffa, courtesy of Healesville Sanctuary)

348 The Feather-tailed Glider, *Acrobates pygmaeus*, is found in well-watered eucalypt forests of the eastern coast of Australia and extends into the drier regions of more stunted sclerophyllous forest and woodlands. It feeds on nectar and sugary sap and on small insects found on flowers and foliage. This animal's total length is 13 to 16 centimetres and it weighs between 10 and 14 grams. (F. Coffa, courtesy of Healesville Sanctuary)

349 The Mountain Pygmy Possum, *Burramys parvus*, was first known to Europeans as a fossil, from skull and jaw fragments found in the Wombeyan Caves of New South Wales. In 1966 a living specimen was collected in a ski hut on Mount Hotham, Victoria, and this species is now known from two small areas, one near Mount Hotham and a second in the Kosciusko National Park of New South Wales. It ranges in size from 25 to 27 centimetres, and weighs from 40 to about 44 grams. (F. Coffa, courtesy of Healesville Sanctuary)

350 The Squirrel Glider, *Petarus norfolcensis*, is known from eastern Australia, inland from the coast in dry sclerophyll forest and woodland. Its total length ranges from 40 to 53 centimetres and it weighs from 200 to 260 grams, (F. Coffa, courtesy of Healesville Sanctuary)

Perhaps the living fauna of Australia, like that of North America, is somewhat biased because of the impoverishment of the Late Pleistocene extinctions which cut down, for example, the North American "big cats" (*Felis atrox* and *Smilodon*). Perhaps the Pleistocene loss of the Australian *Propleopus*, almost certainly a carnivorous kangaroo, and *Megalania*, the big carnivorous varanid lizard, is a parallel event. *Megalania*, however, was at best only an ambush killer, and neither the Marsupial Lion, *Thylacoleo*, nor the Tasmanian Tiger, *Thylacinus*, the latter of which once had a wide distribution on the Australian mainland, were truly cursorial animals. Only *Propleopus* could have occupied the open country cursorial (in this case "saltatorial") carnivore niche. If that were the case, the saltatorial carnivore aspect is an example of a rather unusual method of occupying such a niche. It may have been the only possible way to hold such a feeding style, however, because of genetic limitations imposed by the marsupial founder population on the Australian continent. Whatever the explanation, this lack of big carnivores amongst the Australian vertebrate fauna sets it apart from faunas on other continents of the world today and in the past.

341

342

343

344

345

346

347

348

349

350

THE EFFECTS OF DRIFTING CONTINENTS AND CHANGING CLIMATES

Two interrelated factors have markedly affected the Australian vertebrate fauna of today and, for the most part, offer explanations for its unique nature when compared to the faunas of other continents. One factor is that the arid climate of today is a relatively recent mask donned by the Australian continent, and is not typical of much of Australia's past. The second factor is that Australia has been isolated on an unprecedented scale from the rest of the world for more than 50 million years, which has led to the evolution of its biota quite independently of the rest of the world throughout that time.

These two factors are wedded by the behaviour of the Earth's crust: its division into rigid plates that grow and are consumed, driven by the plastic behaviour of the deeper, molten parts of the Earth. Parts of these crustal plates are continents, where most vertebrates play out their lives, leaving their offspring and eventually their fossil bones. Their ability to move over the Earth's surface is controlled by the positions of such continental masses, and ultimately the restless interior of our planet. Thus, vertebrate biogeography and the current explanation of the Earth's crustal behaviour, the *Theory Of Plate Tectonics*, are undeniably intertwined.

Australia, of course, has not always occupied its present geographic position. In fact, it has been a long distance voyager for hundreds of millions of years. At the beginning of vertebrate history it lay near the palaeoequator. With the advent of fishes and even the first tetrapods, the continent straddled or moved only slightly south of the Equator. But with the expansion of labyrinthodonts and the rise of reptiles, Australia made an incredible and rapid excursion southwards, and by the end of the Palaeozoic it lay very close to the South Pole. There it remained for much of the Mesozoic, during the dominance of reptiles, through the rise of birds and mammals, and during the time of origin of modern amphibian groups such as frogs and salamanders. Only near the close of this era did its movement northwards begin, this time a slower, more gentle waltz towards the Equator, which only now does it approach in the north.

As a result of this chequered history, the modern fauna of the Australian continent is polytypic: it still holds many old endemic elements, reminders of a more southerly, wetter, more predictable past when Australia as a continental entity did not exist. Such organisms as the southern beeches (*Nothofagus*) and tree-ferns, lungfish, leptodactylid and microhylid frogs, the extinct meiolaniid and chelid turtles, perhaps some boid snakes (related to the extinct *Wonambi*), the extinct Mihirungs (Dromornithidae), emus and cassowaries (Casuariidae), many old endemic passeriform bird groups such as the scrub-birds (Atrichornithidae), lyrebirds (Menuridae), mudlarks (Grallinidae) and woodswallows (Artamidae), platypuses and echidnas (Monotremata) and marsupials are but a few of these Mesozoic and early Tertiary survivors. They were derived both as direct leftovers from groups with a much wider dispersion on the larger landmasses of Gondwana or Pangaea, or as their progeny, which developed a unique identity distinct from ancestral stock as the isolated Australian ark set sail, alone, for Asia. *Eucalyptus* was one such *derived* Australian endemic as were a variety of the families of birds, such as the parrots and painted quail (Pedionomidae), and endemic marsupial orders (marsupicarnivores, bandicoots and the varied diprotodontians), all so characteristic of this continent.

During Australia's northward trek, as it moved into the drying Horse Latitudes, water came in shorter supply. In the southern parts of the continent rains were confined to the winter, and the water supply, even during those cooler times, was neither predictable nor dependable. More and more the biota was stressed to cope with such harsh conditions. Some forms were able to retreat to such humid enclaves as the south-west of Western Australia, the Eastern Highlands and Tasmania. Others were able to partially or fully adapt to the new opportunities: a significant segment of the Australian biota exhibits arid adaptations, from the burrowing marsupial moles (notoryctids) and nocturnal bilbies (thylacomyids) to the cursorial emus (dromaiines) and the needle-leafed *Casuarina*, most derived from old endemic stock. More recent arrivals have also adapted to the rapidly drying continent, such as the acacias, the small-leaved and salt-tolerant saltbush and bluebush (the chenopods) which cling close to the ground of the dry interdune valleys in Australia's Centre, some varanid and agamid lizards, skinks, some colubrid, elapid and boid snakes, the Stubble Quail (*Coturnix pectoralis*), and many murid rodents.

Still another part of this final overprinting are a number of forms that adapted to life in the blossoming humid tropics which graced the more northerly parts of Australia. Some were immigrants from the north, like the hylid frogs, carettochelyid and trionychid turtles, many elapid, colubrid, typhlopid and boid snakes, various varanid, agamid, scincid and gekkonid lizards (and their Australian endemic progeny, the legless fossorial pygopodids), and such forms as the Dollar Bird (*Eurystomus*) and fruit bats. Some derivitives of the old Gondwanan–Australian endemics moved opportunistically into this new environment, including the cassowaries and many parrots and pigeons, as well as a variety of marsupial groups.

The reactions of old natives and several waves of new immigrants from the Old World, over 50 million years or more, to the increasing aridification of a once humid continent and the encroachment of warm tropical conditions in the north has produced the Australian biota of today. The composition and distribution of the elements of this biota reflect those changing environmental conditions and the dynamic tectonic features which shaped them.

351 The Estuarine Crocodile, Crocodylus porosus, one of the two species of crocodile found in Australia, both restricted to the coastal and near-coastal part of the northern tropics. This species averages under 5 metres in total length, but rare individuals have been recorded that are in excess of 6 metres. (M. Deller)

352 The Lace Monitor, Varanus varius, a common tree-climbing goanna of eastern Australia and the Outback. It reaches up to 2 metres in length. (F. Coffa, courtesy of Healesville Sanctuary)

353 One of the brown snake group of elapids derived from an old world stock sometime during the Cainozoic. This group occurs across most of Australia, successfully occupying the small vertebrate predator niche. They are highly venomous and aggressive. (F. Coffa)

354 The Common Scaly-foot Lizard, Pygopus lepidopodus, is a member of the family Pygopodidae, a group restricted to Australasia. These lizards are limbless, and often mistaken for snakes, but retain remnants of the hind limbs. (F. Coffa)

355 The Jew Lizard, Pogona vitticeps, one of the agamid lizards which are diverse in the Australasian region. (F. Coffa)

Rainfall >3500 mm
Rainfall <1000 mm
Uniform rainfall (>250 mm/yr)
Winter rainfall (>250 mm/yr)
Summer rainfall (>380 mm/yr)
Arid zone
Alpine grassland and heath
Freshwater swamp and swamp forest
Dunefields and sandplains
Playa or salt lake
Coastal clay plains

Mean discharge to sea >150 m³/sec
Coral reefs
Cold ocean current
Warm ocean current

Morphoclimatic map of Australia and New Guinea today (modified after Veevers, 1986).

351

354

352

353

355

AUSTRALIA'S MODERN BIOGEOGRAPHIC PROVINCES

Australia's most modern face, defined in part by the kinds of vertebrates that live on this continent and their geographic distribution, is
the result of the continent's movement and the changes in global
climate during the last 400,000 years. Australia has been moving
slowly northwards during much of the Cainozoic, and simultaneously
global climate has become more and more narrowly zoned.
Geomorphologist and palaeoclimatologist Jim Bowler (1982) has
explained the resultant present-day climatic profile for the Australian
continent as the end result of a northwards race between the Australian continent and the STHP (subtropical high pressure) belt, which
comprises areas of global scale dominated by dry, descending air
masses. Places on Earth located beneath the STHP are deserts. The
aridity produced in such areas shows up in the absence of surface
water, and thus in the disappearance of lakes, the presence of saline
features in soils and temporary watercourses (such as those active
only during times of flash-flood), widespread development of alkaline
geochemical environments in the soils, and development of aeolian
sedimentary regimes and landforms such as sand dunes and other
wind-deposited features and sediments. Of course, flora and fauna
reflect such aridity by their development of adaptations to cope with
water loss (for example, thick cuticle and small leaves in plants, and
nocturnal activity in vertebrates), water scarcity (the ability to gain
moisture from food rather than standing water, or by utilizing dew),
heat (nocturnal activity or use of shade, lowered activity during
times of maximum temperature, posture) and increased salinity in
the soils.

356 Rainforest such as this, which is typical of south-
eastern Australia today, was much more widely spread
across the continent during the Cainozoic. (S. Morton)

The result of the more rapid northwards movement of the STHP
when compared to that of the Australian continent was that the STHP
won the race about 400,000 years ago. The first noticeable effect was
about 6 million years ago when the STHP overtook the southern part
of the Australian continent in the Nullarbor region. There is then a
widespread "synchronization in the disappearance of lakes and
contraction of mesothermal floral elements, followed by a type of
geochemical weathering that produced widespread secondary
opaline silica in sediments from Lake Eyre to Kerang" (Bowler, 1982:
42). By 1 million years ago the STHP had taken up its present geographic position relative to the Australian continent, and conditions
would have been much as they are today. However, during the last
million years events occurred that made times even more stressful
for the Australian biota, and these hard times were what most likely
led to widespread extinctions of some of the larger vertebrates, the
megafauna. It was into this harsher environment, too, that humans
first moved into the antipodes.

During the past 500,000 years cyclicity of climate seems to have
been the norm, with at least four or five repeats of the pattern observed during the last 100,000 years. This cycle involved a period
when lakes were much expanded over their current area, closely
approaching their Miocene extents, but temperature regimes were
quite different. "This [last] episode [lakes expanded] in south-eastern
Australia dated from about 50,000 to 30,000 years ago, known as the
Mungo lacustral phase . . . was followed by a period of major water
deficit extending from at least 25,000 to 14,000 years ago with the
apparent maximum aeolian activity located near 18,000 to 16,000
years ago . . . which, in global terms, is taken to coincide with the
period of maximum extent of Northern Hemisphere ice sheets. During
this period, characterized by accelerated strength of global circulation, the Australian desert expanded on its poleward margin with the
construction of longitudinal dunes as far south as Kangaroo Island . . .
and even on to the north-eastern tip of Tasmania, then joined to

mainland Australia during maximum sea-level lowering." (Bowler, 1982:43.)

During the last 500,000 years of this cycle of unpredictability, many vertebrates succumbed. In times of stress often there is initial selection for larger size in herbivores that allows the processing of food, which, in this case, as aridity developed was of poorer quality. The increase in size carried a risk, however, for it also demanded a certain level of free-standing water. Thus, as the water became scarce, big herbivores may have become tethered to waterholes — and if the waterholes eventually disappeared, so, too, did the vertebrates. Such could be the scenario explaining the demise of much of the megafauna. For some species, such as the Grey Kangaroo, *Macropus giganteus* (which evidently underwent as much as a 30 per cent size reduction in hard times), dwarfing and thus lowered water requirements during times of stress seem to have guaranteed survival. But for other groups, such as *Diprotodon* and *Genyornis*, failure to scale down led to their extinction.

Perhaps it was not only climate that had its effects on Pleistocene faunas. Aboriginal peoples could also have played some role by firing the savannahs and thus altering the environment, perhaps even by direct killing of megafaunal elements. Thus far, however, there is not enough evidence to sensibly evaluate this issue. Maybe, in time, there will be, but to pass judgement just now would not be warranted.

HEAT AND WATER IN AUSTRALIA TODAY

"The concentric zonation of rainfall in 'dry' continental Australia and the contrast with 'wet' New Guinea focuses attention on the water regime as a major determinant of biogeographic pattern and evolutionary process. But the thermal regime is no less significant." (Nix, 1982: 49–50.) In addition, the amount of light available plays a role. As Nix so aptly points out, both temperature and the availability of water are important in determining the distribution of animals and plants today on the Australian continent, and to consider one without the

other is naive. Nix has proposed several distinct bioclimatic regimes that are controlled by temperature, moisture and light: megatherm (with an upper limit of 46°C, a lower limit of 10°C and an optimum range of 26–33°C), mesotherm (upper limit 33°C, lower limit 5°C, optimum range 10–22°C), microtherm (upper limit 25°C, lower limit 0°C, optimum range 10–14°C) and hekistotherm (upper limit 20°C, lower limit –10°C, optimum range 6–8°C). Australia has been subdivided into five biogeographic areas based on biotic content, and they can be better understood in terms of the temperature-moisture-light factors:

Irian (*Megatherm, Non-seasonal*) with temperature optima of 26–33°C, and a range of extremes of 10°C to 46°C.

Torresian (*Megatherm, Seasonal*) with the same characteristics as the Irian, except seasonal.

Tumbunan (*Mesotherm/Microtherm, Non-seasonal*) with temperature optima of 10–22°C, and a range of extremes of 0°C to 33°C.

Bassian (*Mesotherm/Microtherm, Seasonal*) with the same characteristics as the Tumbunan, except seasonal.

Eyrean (Eremaean) (*Megatherm/Mesotherm, Arid*) with temperature optima of 19–33°C, and a range of extremes of 5°C to 46°C.

Such physical controls on the distribution and survival of different groups of vertebrates have combined to shape today's Australian biota and will, in their dynamic way, strongly influence the future. As well, the changing nature of factors that are clearly tied to the tectonic forces which have driven Australia both south and north during the 450-million-year history of vertebrates on this continent will continue to exert their influence on the future of this mobile landmass.

357

357 Small coral island in the Capricorn Group, Great Barrier Reef, Queensland. Much of this area was subaerial in parts of the Cainozoic, but is now shallow marine shelf on which the great reef has developed during the last few million years of the Cainozoic as Australia drifted into the tropical latitudes. (P. Vickers-Rich)

358 Screw Palm, Darwin, Northern Territory. (F. Coffa)

359 The Sturt Pea, Clianthus formosus, a legume, is a prostrate annual or biennial species highly adapted to living in the arid Centre. It can be self-fertile in times of stress. (P. Vickers-Rich)

360 Rainforest at Berry Springs in the Territory Wildlife Park south of Darwin, Northern Territory. Pockets of rainforest like this exist because of water availability from springs. Such "moist" areas are surrounded by drier grassland savannah country. (F. Coffa)

361 Lake Mungo, western New South Wales, where some of the oldest remains of humans have been found in Australia, dated at 38,000–40,000 years before the present. (F. Coffa)

362 The Macdonnell Ranges, near Alice Springs, Northern Territory, in the heart of Australia's arid Centre. (F. Coffa)

363 Termite nest, south of Darwin, Northern Territory. The grasslands of northern Australia abound with these nests. The Termites inside are grazers of the northern plains and once may have "competed" with Diprotodon though now they are dominant herbivores. (F. Coffa)

364 Mallee dunes, Sunset Country, north-western Victoria. Depending on the availability of water such dunes may be active, as in times of drought, or stable when a moister regime prevails and they are vegetated. (F. Coffa)

365 Gibber plain in Central Australia, on the Birdsville Track north of Marree, South Australia. Gibber is a residual accumulation of silica-rich rocks that remain after almost everything else has been weathered, washed or blown away. (P. Vickers-Rich)

358

359

360

361

362

364   363

365

AND WHAT OF MAN'S DALLIANCE?

Environment, tectonics and biotic interactions have made their impression on Australia's vertebrates over hundreds of millions of years. Perhaps no one factor, however, has so rapidly modified the vertebrate make-up of this part of the globe as has human intervention. It is difficult to assess the speed of impact of those climatic changes at the end of the Mesozoic, perhaps set off by extraterrestrial events, on the Australian biota. But what is quite clear, is that European human presence on this antipodean continent has made a dramatic impression in a geologic instant in time — just 200 years. It is more difficult, presently impossible, to accurately estimate pre-European human effects, although certainly there have been some.

Europeans have affected Australian vertebrates through their alteration of the environment, their direct killing of certain species for entertainment, profit and in order to diminish direct competition, and their introduction of exotic animals and plants. Clearing and burning have either reduced or completely obliterated habitats of a myriad of Australian and New Zealand vertebrates. Although the figures are not clearcut for Australia, since 1840 something like 66 per cent of the original forest cover of New Zealand has been removed or in some way altered. And the forest removal, whilst now becoming a contentious issue in need of review by all parties concerned, is still occurring significantly in both Australia and New Zealand. Open grasslands and savannah areas have also been greatly altered due to the introduction of European grazing animals and grain crops. In New South Wales, for example, since settlement, upwards of 86 per cent of terrestrial grassland mammal species have either become extinct or are exceedingly rare, caused most likely by overstocking with domestic vertebrates, invasion by the human-associated House Mouse, *Mus musculus*, which displaced endemic granivores, and replacement of the native grasses by exotic grasses to which the domestic stock was accustomed and the European population used to culturing. On a continent-wide scale since the coming of humans to Australia, at least 70 of about 260 species of land mammal have been driven to extinction, that is, about 25 per cent of the total terrestrial mammal fauna — and 23 of these species have disappeared in just the last 200 years. On some of the islands near Australia the destruction has been even more horrific. For example, on Timor the entire native land mammal fauna has been destroyed, save one species of shrew. Only a few thousand years ago this island had a diverse fauna of rodents, pygmy elephants and, amongst the reptiles, a giant carnivorous lizard, a relative of the extant Komodo Dragon of the Flores. But when man and his domestic fauna arrived, all that changed, and fast. The story is repeated again and again in the Pacific Basin, even in the Hawaiian islands which have been ironically viewed as primaeval paradises but are in fact "among the most devastated ecosystems on Earth!" (Flannery, 1989: 20.)

Bounties on the Tasmanian Tiger, *Thylacinus cynocephalus*, open seasons on fur-bearing marsupials (such as the Koala), sporting chases with dogs for the fleet-footed Toolache Wallaby, *Prionotemnus greyi*, eating of the King Island Emu, *Dromaius ater*, removal of "pest" species (such as the Kangaroo Island Emu, *Dromaius baudinianus*), and introduction of both sporting species (like rabbits and hares) and so-called beneficial species (such as the Cane Toad, *Bufo marinus*, to control sugar-cane pests) — all have taken their appalling toll of native fauna and genetic variation on this continent.

But, in some ways, all of these pale to insignificance when one final human action is considered, that of phenomenal population growth.

As Paul and Anne Ehrlich (1990) point out, since 1968 the human population of the world has increased from 3.5 billion to 5.3 billion — nearly 2 billion more people on the face of the Earth in 22 years! Every hour of the day 11,000 more people enter the world, making 95 million in a year. This population explosion is happening in a world where most people are quite "unaware of the role that overpopulation plays in many of the problems oppressing them" (Ehrlich & Ehrlich 1968), such as outright famine or higher food prices or the death and human misery caused by flooding of areas where people shouldn't be living in the first place (but are because there are no other places to live). Australia is spared many of these problems, in part because of its physical isolation and, perhaps, in part because of the limitations of some natural resources such as water and its poor quality soils. But we, who demand so much more of world resources than do those peoples from the many desperately poor nations of the world, have no call to be smug. We, who demand so much of this world, cannot allow our own desires for more sons and daughters than just enough to replace ourselves numerically to override our responsibility to maintain a decent world for all, not just us, our local community and our country. The whole world is our community. We, as citizens of that community, must change our behaviour if we want to make decisions about our children's and grandchildren's future, and give them the chance to make those decisions too. Otherwise, Nature will be the decider, and the consequences will no doubt be unimaginably horrendous, not just in Rio De Janeiro or Bangladesh, but in Sydney and Melbourne as well.

We can only hope that humans will come to cherish the variety of life that has developed in the world, such as the unique biota presented in this book — the vertebrates that have evolved on the Australian continent over the past 450 million years. We can only hope that in understanding the global forces that have shaped the land, the animals and the plants, in understanding the finality of extinction that has repeatedly affected a great variety of vertebrate life on this continent, we can learn something about ourselves, and the fragility of our own existence. Unlike other vertebrates which have come and gone through geologic time, we humans may be able to do something to prevent our own extinction, that is, if we are wise enough to make the hard decisions. And if not, then . . .

366

366 Male pouch young of the Tasmanian Tiger, *Thylacinus cynocephalus*, in the spirit collection of the Tasmanian Museum. It was taken before 1910. x 0.5 (S. Morton, courtesy of the Tasmanian Museum And Art Gallery and N. Kemp)

367 A bounty skin of the Tasmanian Tiger, *Thylacinus cynocephalus*, taken sometime before 1933 and now in the collections of the Tasmanian Museum. (S. Morton, courtesy of the Tasmanian Museum And Art Gallery and N. Kemp)

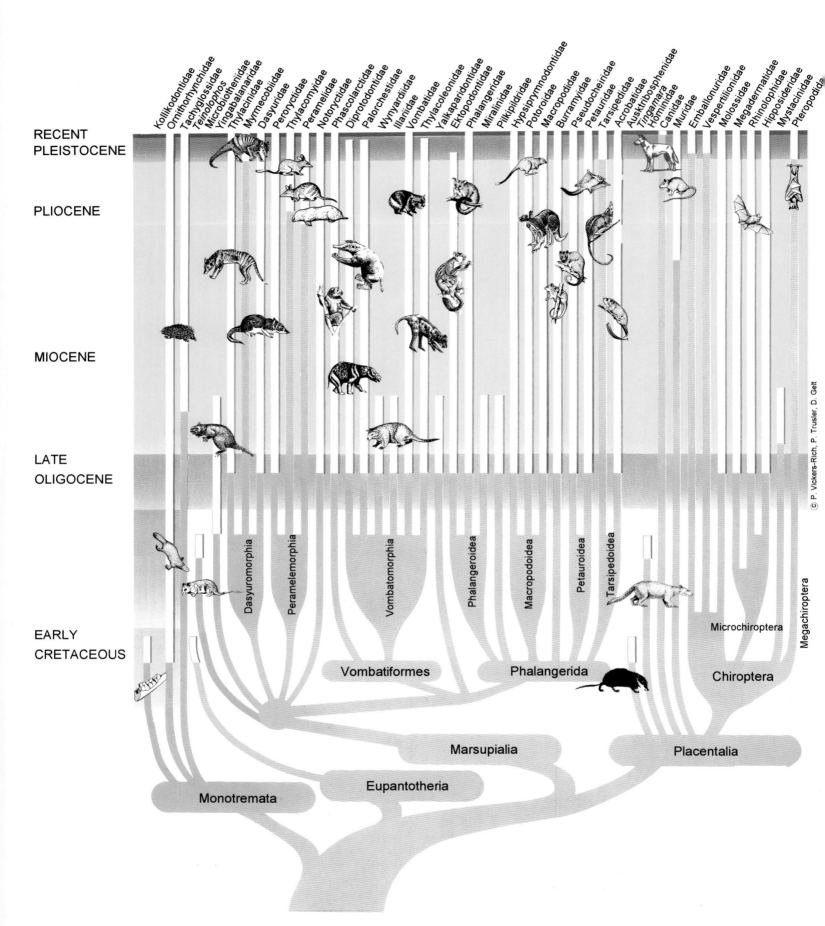

RECENT
PLEISTOCENE

PLIOCENE

MIOCENE

LATE
OLIGOCENE

EARLY
CRETACEOUS

Kollikodontidae
Ornithorhynchidae
Tachyglossidae
Teinolophos
Microbiotheriidae
Yingabalanaridae
Thylacinidae
Myrmecobiidae
Dasyuridae
Peroryctidae
Thylacomyidae
Peramelidae
Notoryctidae
Phascolarctidae
Diprotodontidae
Palorchestidae
Wynyardiidae
Ilariidae
Vombatidae
Thylacoleonidae
Yalkaparidontidae
Ektopodontidae
Phalangeridae
Miralinidae
Pilkipildridae
Hypsiprymnodontidae
Potoroidae
Macropodidae
Burramyidae
Pseudocheiridae
Petauridae
Tarsipedidae
Acrobatidae
Ausktribosphenidae
Tingamara
Hominidae
Canidae
Muridae
Emballonuridae
Vespertilionidae
Molossidae
Megadermatidae
Rhinolophidae
Hipposideridae
Mystacinidae
Pteropodidae

Dasyuromorphia
Peramelemorphia
Vombatomorphia
Phalangeroidea
Macropodoidea
Petauroidea
Tarsipedoidea
Microchiroptera
Megachiroptera

Vombatiformes
Phalangerida
Chiroptera

Marsupialia
Placentalia

Eupantotheria

Monotremata

© P. Vickers-Rich, P. Trusler, D. Gelt

Hypothesis on the interrelationships of Australian mammals, primarily marsupials.

PART FOUR

GONDWANAN
FAUNAS
IN
GLOBAL CONTEXT

368

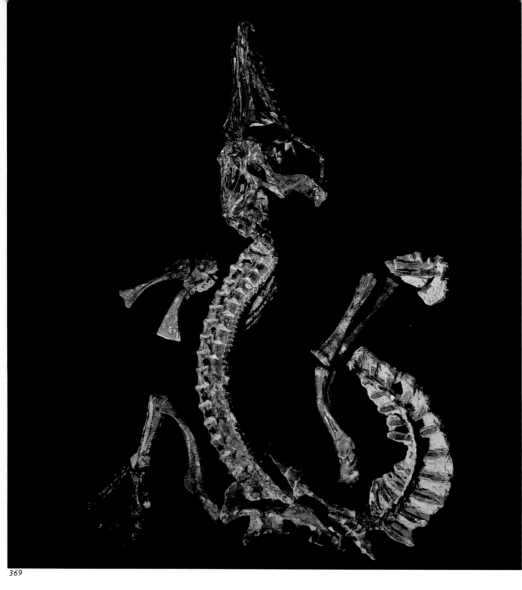

368 *A skeleton of the aetosaur thecodont* Neoaetosauroides engaens *from the Late Triassic Los Colorado Formation of La Rioja Province, Argentina. Aetosaurs are exceptional thecodonts in that for the most part they were apparently carnivorous, yet they lacked teeth in the front of the snout and on the lower jaw. This, together with the shape of the teeth in certain species, suggests that some of them may have been herbivorous. Their entire bodies were heavily armoured. x 0.2 (F. Coffa, courtesy of the Instituto Miguel Lillo and H. Powell)*

369 *A skeleton of the primitive proterochampsid thecodont* Chanaresuchus sp., *from the Middle Triassic of Los Chañares, Nodulo No.1, La Rioja Province, Argentina. This primitive thecodont may be related to phytosaurs or to crocodiles, both formidable predators, but further work is needed to sort out close ties to any group. Thecodonts are a prominent element of Triassic faunas on a world scale. x 0.3 (F. Coffa, courtesy of the Instituto Miguel Lillo and H. Powell)*

369

The fossil remains of plants and animals have been important clues in reconstructing past geographies, as their similarities and dissimilarities through time reflect the proximity of continents and oceans to each other. Gondwana, or Gondwanaland as it was formerly known, was originally recognized from such information. And, besides the fossil relics, the rocks of the continents that once comprised this great southern landmass reflect a similar past cohesiveness of its components: Africa, Madagascar, India, South America, Antarctica, New Zealand and Australia, and even parts of eastern Asia.

During the early Palaeozoic, from the Cambrian until the Devonian, Gondwana existed as a physical reality, separate from other continental fragments. North America, Europe and most of Asia, excepting India and parts of China, were separated from the megacontinent of Gondwana by ocean expanses. The exact arrangement of all of these continental fragments is not absolutely certain. During early Palaeozoic times terrestrial biotas could have spread across Gondwana, being hindered only by the presence of mountain ranges and perhaps climate.

From Cambrian Times, beginning about 545 million years ago, until the Middle to Late Devonian, about 354 million years ago, Gondwanan freshwater, terrestrial and nearshore marine faunas, where known, show some similarity, and often a corresponding distinctness from faunas on non-Gondwanan landmasses. The placoderms from Australia and south China, for example, in the earlier parts of the Devonian show marked similarity, and some groups appear to originate in this area, only later becoming cosmopolitan when Gondwana fused with the northern continents of North America, Europe and parts of Asia towards the end of the Devonian, thus forming the largest continuous landmass on the Earth, Pangaea.

From the Late Devonian until sometime in the Cretaceous Period, that is, from about 354 until around 130 million years ago, Gondwana and the now northern continents remained a single great megacontinent. Terrestrial organisms seemingly should have been able to move around on this gargantuan terrestrial mass, as there were essentially no seaways barring their movement. But it is during this very time, especially from the Carboniferous Period, which began about 362 million years ago, until the Triassic Period, which ended 208 million years ago, that the plants and rock sequences were most

similar across Gondwana — and the plant assemblages, characterized by such forms as the Permian seed-fern *Glossopteris*, were extremely distinct from assemblages elsewhere in the world. Without doubt, the Gondwanan continents of Africa, India, South America, Antarctica, New Zealand and Australia were closely juxtaposed; that, alone, explains the similarity of their rocks and biotas. But close, too, were the continents of North America, Europe and parts of Asia. The distinctiveness of the Gondwanan floras at this time, from the Carboniferous to the Triassic, seems very dependent on climate, not upon physical isolation. The Gondwanan continents of this time were for the most part in a far southerly position and the cool temperate, even glacial climate that prevailed dictated the distinctiveness of the plants. The tropics of the time acted as an effective barrier for migration either north or south of the latitudinally controlled floral assemblages.

Pangaea began to break apart in the Cretaceous Period, some 130 million years ago. North America, Europe and Asia were all separated from Gondwana by seaways, some quite narrow. These seaways were between North America and Europe on the one hand and South America and Africa on the other. For a few million years Gondwana was again an isolated physical entity, and a distinctive terrestrial vertebrate fauna developed. Then Gondwana itself began to break apart, first India from Africa, then Africa from South America, and New Zealand from Antarctica and Australia in the later parts of the Cretaceous, and then Australia from Antarctica in the latest part of the Cretaceous and the early Tertiary. Each of these once cohesive fragments subsequently began to develop unique faunas. Some continents experienced repeated northerly contacts — South America with North America, Africa with Europe — but others, once sundered, experienced profound isolation — New Zealand, New Caledonia and Australia. And one fragment, Antarctica, not only experienced profound isolation, but an ever deteriorating climate which eventually severely limited its biotic diversity.

The following is a brief review of the major terrestrial vertebrate faunas known from the Gondwanan landmasses exclusive of Australia, and is intended to point out both what is known and, perhaps even more importantly, what is yet to be explored in order to gain an understanding of the evolutionary patterns that have characterized the past 500 million years.

THE PALAEOZOIC RECORD

Pre-Devonian vertebrates on Gondwana are not abundant and have been mentioned in earlier sections, especially the Ordovician fauna of Bolivia. In Early and Middle Devonian times, the freshwater fish faunas together with the invertebrates and plants support division of the world into northern and southern biogeographic realms, but by the end of the period the biota becomes cosmopolitan. This homogeneity at the end of the Devonian suggests that while Gondwana and other continental fragments had previously been separate entities, by this time Pangaea was established, and terrestrial forms were able to move about, almost unhindered, on this larger landmass.

It is the flora of the Carboniferous and the Permian periods which provides the compelling biotic support that Gondwana existed, *not* the terrestrial vertebrates. Characterized by the presence of the seed-fern *Glossopteris*, this flora is associated with glacial deposits as well as with vast coal fields on the southern continents. What may have been critical in the geographic restriction of this flora was its high latitude position rather than water barriers separating it from the contemporary *Pecopteris*, *Angaridium* and *Gigantopteris* floras of the northern continents.

The record of terrestrial vertebrates is so scanty towards the end of the Palaeozoic, that differences seen between hemispheres are as likely to be owing to the chances of preservation as to real physical barriers between faunas living in different areas at the time. This situation is well established with the synapsids (mammal-like reptiles). In the Late Carboniferous and Early Permian periods, the synapsids were restricted to North America and Europe. They overlap in time with early Late Permian forms known from Russia. The Russian sequence of synapsids continues upward and in turn overlaps in time with the oldest of the South African Late Permian forms. There, in southern Africa, synapsids survived until the end of the Triassic. Where such time overlaps occur, as might be expected, there seem to be similar or closely related species or genera in both regions. Their absence in other places may be simply due to the lack of rock records of the right age in those places.

---

370 A skeleton of the mesosaur Stereosternum tumidum from the Permian Irati Formation of Sao Paulo, Brazil. This small freshwater reptile is very similar to a contemporary mesosaur in Africa and has been used as evidence suggesting South America and Africa were once much closer together. x 0.8 (G. Borgamanero with the assistance of H. Alvarenga)

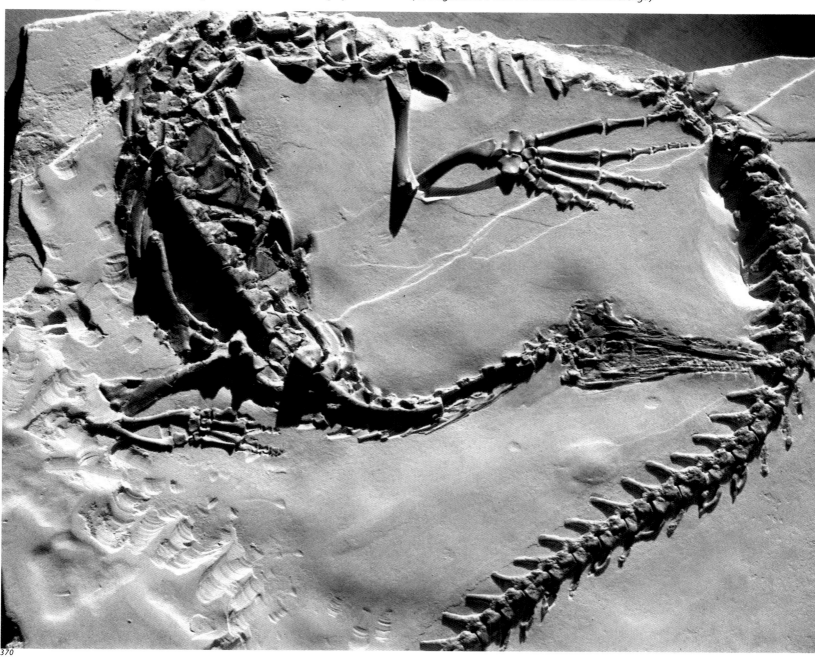

370

THE MESOZOIC RECORD

In the Early Triassic, resemblances are seen between the synapsid faunas of South Africa and those of India, Antarctica, the USSR and China as well as with the meagre representation from Australia. Although these resemblances have been often cited as evidence that tetrapod distribution supports the existence of Gondwana at that time, an examination of the Early Triassic faunas from the undoubted Laurasian continents, which lack synapsids, suggests that facies may have been more important in controlling faunal composition (that is, controlling what was preserved as fossils) than were geographic barriers. The Laurasian fossil assemblages are generally of a more aquatic nature and can be characterized as fish and labyrinthodont dominated, just as are those of Australia. The mechanisms that account for the accumulation of certain types of fossils are not distributed evenly through the Geological Time Column, hence the inherent biases of the record must be taken into account when drawing any biogeographic conclusions based on it.

A diversity of synapsids is known from the Late Triassic of North America. Either they were able to reach that continent from the Gondwanan continents at about that time, although continental separation is thought to have been increasing then, or their ancestors were present in North America during the Late Permian and Early Triassic in North America but simply not recorded in the fossil record. As Late Permian and Early Triassic tetrapods of any kind are all but unknown in North America, the absence of evidence of synapsids is not compelling for inferring the real absence of these animals in North America during that time period.

The idea that there are significant differences between the typical Gondwanan tetrapods and those of Laurasia has been challenged in recent decades by the discoveries of rare remains in one of these areas which were previously thought typical of and exclusive to the other. For example, one specimen of the thecodont *Ticinosuchus* was found in the Middle Triassic of Switzerland in a marine rock unit, not the sort of environment where such animals are to be expected. But, that one fossil is sufficient to demonstrate a close relationship of the Swiss fauna with the much richer and better known assemblages of the same age from Brazil and South Africa where *Prestosuchus* and thecodonts similar to *Mandasuchus* occur in quite different settings considered to have been deposited in lakes on broad floodplains.

Although a significant Early Jurassic terrestrial vertebrate fauna is known from India and an extensive Middle Jurassic one from China, there are no others of those ages of comparable magnitude on any other continent. Thus, useful comparisons of faunas of these ages are simply not possible. In the Late Jurassic, to the contrary, the exten-

sive assemblages from the widespread Morrison Formation of western North America and that from Tendaguru in Tanzania, East Africa, overlap in time extensively and provide strong evidence that interchange of dinosaurs and other vertebrates was quite possible between these two regions. *Brachiosaurus*, the largest sauropod dinosaur known from a complete skeleton, was common to both areas, as was the smaller sauropod *Barosaurus*, the dryosaurid *Dryosaurus*, and possibly the theropods *Allosaurus* and *Ceratosaurus*. Other dinosaurs, although assigned to different genera, are quite similar to each other, such as the North American *Stegosaurus* and the African *Kentrosaurus*. A detailed comparison of the Late Jurassic dinosaur fauna from China with that from North America or Africa has not yet been made, but examination of published faunal lists suggests that such an investigation might reveal a striking degree of similarity there as well.

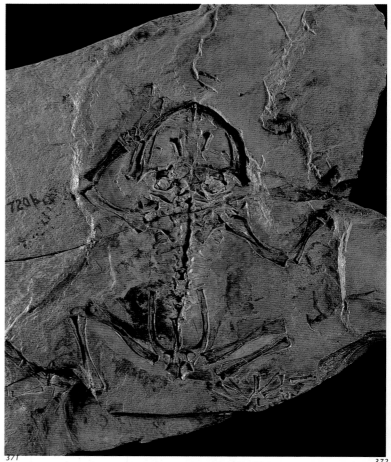

371 A skeleton of the frog *Notobatrachus degiustai* from Middle Jurassic rocks near Santa Cruz, Patagonia, Argentina. x 0.6 (F. Coffa, courtesy of the Museo Argentino De Ciencias Naturales "Bernardino Rivadavia" and J. Bonaparte)

372 The skull of a tiny prosauropod dinosaur, *Mussaurus*, from the Late Triassic El Tranquilo Formation at Laguna La Colorada, Estancia Canadon Largo, Santa Cruz Province, Argentina. The skull length is 31.4 millimetres. (F. Coffa, courtesy of the Instituto Miguel Lillo and H. Powell)

373 An actinopterygian fish, "*Tharrias*" ferugloi, from the Middle Jurassic Conodon Astolto Formation at Cerro Condor, Chubut Province, Argentina. It is closely related to *Leptolepis*. x 2.0 (F. Coffa, courtesy of the Museo De La Plata and R. Pascual)

374 The aspidorhynchiform neopterygian *Vinctifer comptoni* from the Early Cretaceous Santana Formation of Chapada Do Araripe, Brazil. x 1.0 (F. Coffa, courtesy of the Museo De La Plata and R. Pascual)

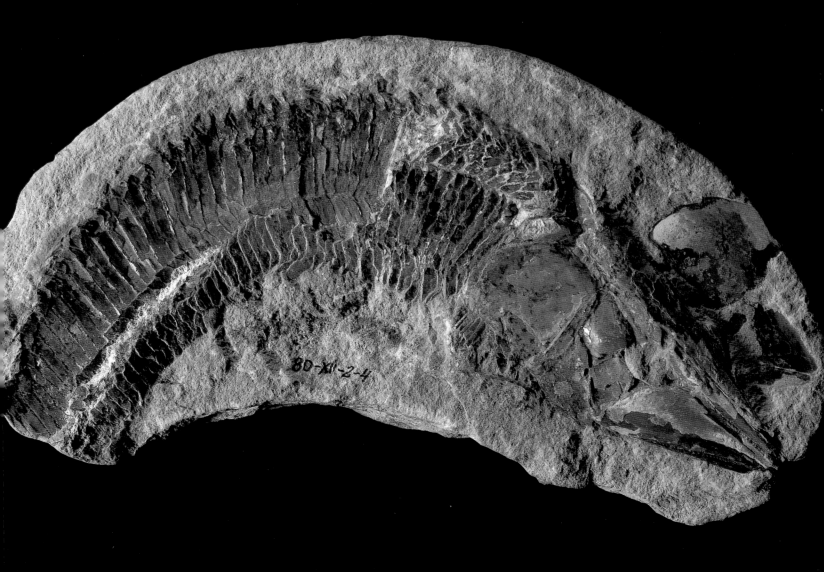

375 A skeleton of the tiny prosauropod dinosaur *Mussaurus* ("mouse reptile") from the latest Late Triassic El Tranquilo Formation at Laguna La Colorada, Estancia Canadon Largo, Santa Cruz Province, Argentina. Dinosaur egg of unknown affinities. x 1.4 (F. Coffa, courtesy of the Instituto Miguel Lillo and H. Powell)

376 A skull of the ornithosuchid thecodont *Riojasuchus tenuiceps* from the Late Triassic Los Colorado Formation of Quebrada De Los Jachaleros, La Rioja Province, Argentina. The carnivorous thecodonts were a major element in the South American Triassic faunas. x 0.8 (F. Coffa, courtesy of the Instituto Miguel Lillo and H. Powell)

377 A skull of *Massetognathus* from the Middle Triassic of Los Chanares, La Rioja Province, Argentina. x 0.9 (F. Coffa, courtesy of the Instituto Miguel Lillo and H. Powell)

378 The advanced mammal-like reptile *Cynognathus minor* from the late Early Triassic *Cynognathus* Zone of the Puesto Viejo Formation, near San Rafael, Mendoza Province, Argentina. Reptiles in this genus are also common in South African rocks of the same age. x 0.4 (F. Coffa, courtesy of the Instituto Miguel Lillo and H. Powell)

379 A mammal-like traversodont reptile, *Pascualgnathus polanskii*, from the late Early Triassic Puesto Viejo Formation of Argentina. x 0.7 (F. Coffa, courtesy of the Museo De La Plata and R. Pascual)

380 A skull of *Massetognathus* sp., in side view, from the Middle Triassic Los Chañares, La Rioja Province, Argentina. x. 0.6 (F. Coffa, courtesy of the Museo De La Plata and R. Pascual)

376

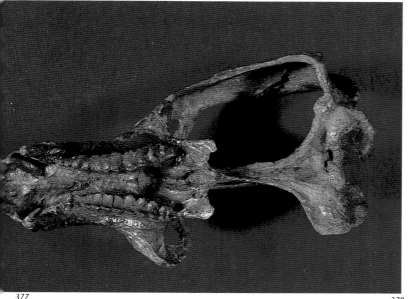

377

379

378

380

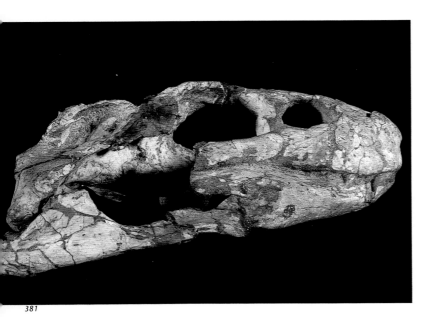

381

The Early Cretaceous tetrapod faunas are so scanty as to make comparisons between different geographic areas rather meaningless. By the Late Cretaceous, however, when a variety of terrestrial vertebrate communities are known on a world scale, they show strong continental differences, typical of the modern world. East Asian and North American faunas share many vertebrate groups in common. But the chronological overlap is not exact, for the Asian record is concentrated in the earlier part of the Late Cretaceous, whilst the North American record extends up to and across the Cretaceous–Palaeocene boundary.

381 Side view of the "mammal-like" terrestrial crocodilian Notosuchus terrestris from the Late Cretaceous of Paso Del Sapo, Neuquen Province, Argentina. x 0.7 (F. Coffa, courtesy of the Museo De La Plata and R. Pascual)

382 A skin impression of the carnivorous theropod dinosaur Carnotaurus sastrei from Early Cretaceous (probably Albian) sediments of Chubut Province, Argentina. This specimen is one of the rare skin impressions of a carnivorous dinosaur; most are from herbivores. x 0.5 (F. Coffa, courtesy of the Museo Argentino De Ciencias Naturales "Bernardino Rivadavia" and J. Bonaparte)

383 The skull of the carnivorous theropod dinosaur Carnotaurus sastrei from the Early Cretaceous (probably Albian) sediments of Chubut Province, Argentina. x 0.4 (F. Coffa, courtesy of the Museo Argentino De Ciencias Naturales "Bernardino Rivadavia" and J. Bonaparte)

384 The lower jaw of the carnivorous theropod dinosaur Carnotaurus sastrei from Early Cretaceous (probably Albian) sediments of the Chubut Province, Argentina. x 0.25 (F. Coffa, courtesy of the Museo Argentino De Ciencias Naturales "Bernardino Rivadavia" and J. Bonaparte)

385 Tracks of mammals from the Jurassic of Santa Cruz Province, Pantagonia, Argentina. x 1.0 (F. Coffa, courtesy of the Museo Argentino De Ciencias Naturales "Bernardino Rivadavia" and J. Bonaparte, and the Instituto Miguel Lillo and H. Powell)

Another difference between the two areas is that the published Asian record comes mostly from Mongolia, China and those parts of the Soviet Union which were then intermontane valleys and possibly quite dry. In contrast, the vast majority of North American Late Cretaceous sites were in areas of lush vegetation on the margins of the shallow inland sea which bisected the continent from north to south. Thus, it is not surprising to learn that although the two regions have much in common, ceratopsians (the horned dinosaurs) are represented in Asia mainly by the most primitive members of the group, such as *Protoceratops*, while in North America, not only is there a comparable form, *Leptoceratops*, but also a multitude of larger, more advanced forms including the well known *Triceratops*. Quite recently a more advanced form, *Turanoceratops*, has been described from central Asia, so perhaps there is still much to learn about the faunas from this region. Like the ceratopsians, hadrosaurs (or duckbilled dinosaurs) arose in the Late Cretaceous and are well known in North America. While Asiatic specimens are known, they are not nearly so diverse as those in North America. A possible Cretaceous marsupial may have been present in Asia, but the group is otherwise unknown from that continent until the Oligocene. In sharp contrast, three marsupial families were living in North America by the end of the Cretaceous.

Despite their differences, when compared with the Late Cretaceous terrestrial vertebrate faunas of South America, the Asian and North American ones do form a distinctive unit. In South America, Late Cretaceous sauropod dinosaurs are quite diverse, represented by several genera in the family Titanosauridae and characterized by the unique form of their vertebrae. *Alamosaurus* is a North American titanosaurid that may have entered from South America in the Late Cretaceous, marking a reinvasion of the continent by sauropods after the group became extinct there in the Early Cretaceous. In the Late Cretaceous of Asia, too, there was a diplodocid sauropod, *Nemegetosaurus*, implying that this more typically Late Jurassic group survived until near the end of the Mesozoic. While undoubtedly present in North America and Asia, Late Cretaceous sauropods were never the dominant element there that they were in contemporaneous South America.

Two genera of Late Cretaceous hadrosaurs are present alongside the titanosaurids in South America, but ornithischian dinosaurs are otherwise unknown there, with the exception of a toothless jaw fragment probably belonging to a ceratopsian. Also present in South America at that time were at least three theropods. They seem to be ceratosaurs, not carnosaurs, while ceratosaurs are unknown later than the Jurassic in North America and extra-Indian Asia. *In toto*, it appears that the Late Cretaceous dinosaur fauna of South America differed significantly from that of the Northern Hemisphere in proportional representation of the groups, rather than in the presence of groups unknown elsewhere. Nevertheless, the existing representation of ornithischians there is based on an extremely meagre amount of fossil material, and it is highly likely that further discoveries might change this picture drastically.

382

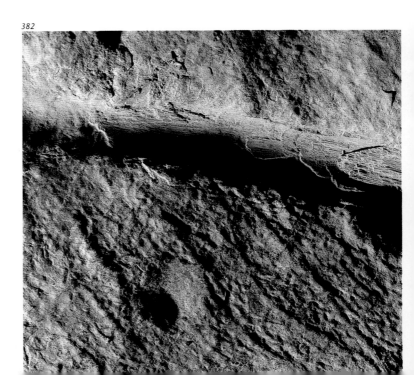

383

384

385

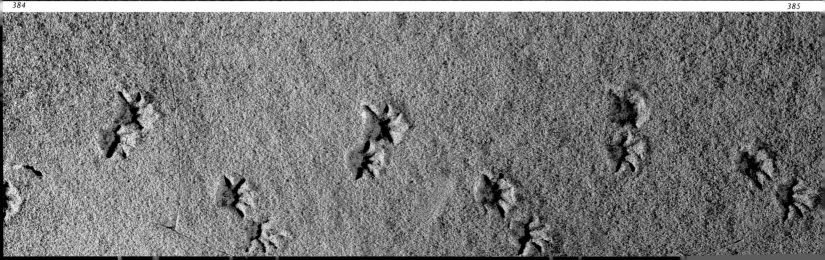

386 The skin impression and scutes from a late Cretaceous titanosaurid sauropod dinosaur from Argentina. x 0.3 (F. Coffa, courtesy of the Instituto Miguel Lillo and H. Powell)

387 A skeleton of the possibly plankton-feeding pterosaur, or flying reptile, *Pterodaustro* sp., from the Early Cretaceous of Argentina. x 0.6 (F. Coffa, courtesy of the Instituto Miguel Lillo and H. Powell)

388 A side view of a skull of the rodent *Pseudoneoreomys mesorhynchus* from the Santacruzian Miocene sediments of Corriguen-Kaik, Peru. x 0.8 (F. Coffa, courtesy of the Museo Argentino De Ciencias Naturales "Bernardino Rivadavia" and J. Bonaparte)

389-390 Ventral and dorsal views of the skull of a mesosuchian ziphodont crocodile, *Lomasuchus* sp., which was a peirosaurid, from the Late Cretaceous of Neuquén Province, Argentina. x 0.4 (F. Coffa, courtesy of the Museo De La Plata and R. Pascual)

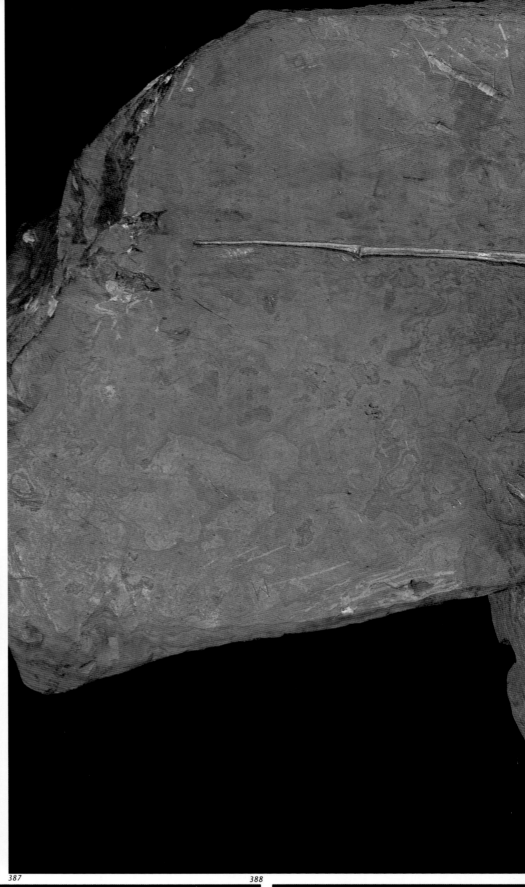

386

387

388

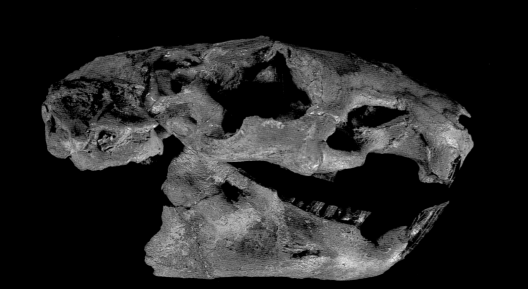

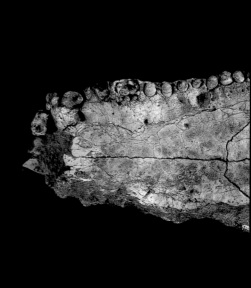

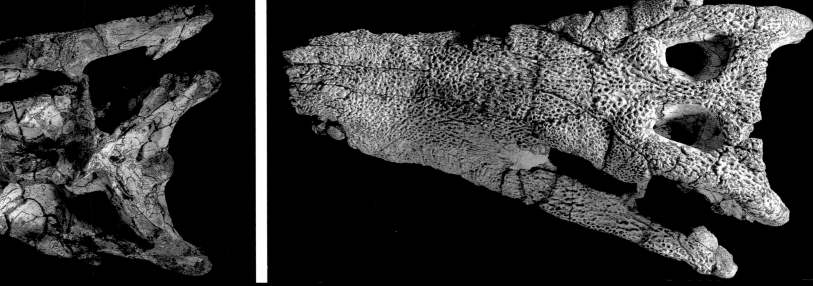

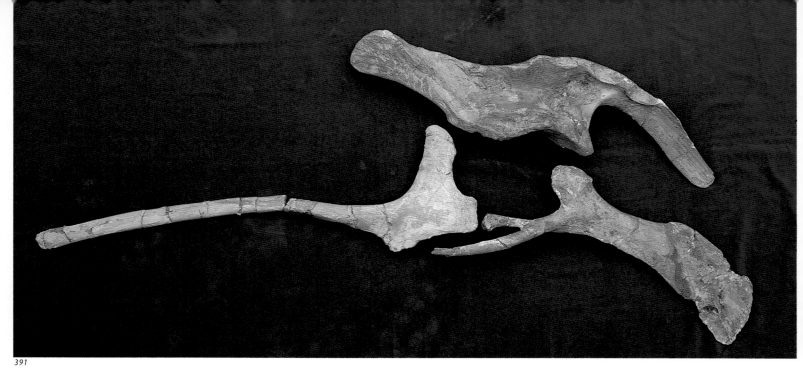

Like the dinosaurs, what little is known of the South American, and also Australian, avifauna at this time is distinctive. In South America and Australia the enantiornithine birds, clearly quite primitive members of the class Aves (which is somewhat intermediate between theropod dinosaurs and the more advanced ornithuran birds), are unique to these once seemingly isolated areas of Gondwana. The enantiornithines were flying birds, but a seemingly closely related, unnamed group from the Campanian (Late Cretaceous) Rio Colorado Formation in north-western Patagonia was apparently flightless and highly cursorial. About the size of a chicken, this unnamed bird had a very small fore limb, about 50 per cent smaller than the hind limb, no wishbone (furcula), a sternum lacking a keel (where normally the flight muscles attach), and lacked the special saddle-shaped articulations so characteristic of bird vertebrae. The structure of its shoulder girdle, however, with a broad sternal articulation of the coracoid and the low angle with which the scapula articulates, suggests that this unique bird was derived from flying ancestors. Thus, the birds, as well as strange, mammal-like crocodilians and many of the dinosaurs all indicate isolation of South America in the late Mesozoic.

The situation seems even clearer in the case of mammals which reflect a pronounced distinction between the faunas of South America and the Northern Hemisphere at this time. There are signs that the terrestrial vertebrate faunas of the various continents, or clusters of them, had taken on distinctive characters by the Late Cretaceous.

Many of the same titanosaurids found in South America have been reported from the Late Cretaceous of India. Titanosaurs are also known from Africa as well as from the northern continents, although they are assigned for the most part to different genera. Taken at face value, this distribution would suggest a stronger Late Cretaceous link among titanosaurs between India and South America than with those of Africa. However, given the position of Africa between them, this anomaly cries out for reanalysis before any such conclusion is warranted.

India is a particular biogeographic puzzle in another regard. All plate tectonic reconstructions since 1970 of the late Mesozoic and early Cainozoic history of the subcontinent show it well south of the Equator and separated from the other Gondwanan lands by perhaps 145 million years ago at the Jurassic–Cretaceous boundary, possibly even as recently as 120 million years ago, in the middle Early Cretaceous. Much as was the case with Australia during the Late Cretaceous and Cainozoic, India separated from Antarctica and became an island-continent. Between 50 and 100 million years ago

India drifted as a lonely ark northwards, eventually colliding with Asia in the Cainozoic, the impact creating the Himalayas.

The Cainozoic and modern terrestrial vertebrate fauna of Australia clearly reflects its isolation episode. But such may not be the case for India. In the Indian Late Cretaceous no dinosaurs have been reported which are as distinctive as the marsupial kangaroo is for the modern Australian fauna. Rather, the known Indian dinosaurs are thought to be cosmopolitan forms, differing little from those on other lands. Yet, when the known members of the Late Cretaceous Indian assemblage were living animals, the subcontinent had presumably been isolated for at least 50 million years, a period of time comparable to Australia's present isolation. This anomaly, if real, casts doubt upon the inferred period of isolation. Perhaps, as India moved from a position attached to Antarctica to its present one, the intervening route was not significantly isolated. For example, had there been an archipelago north of the subcontinent that was subsequently swept against or under the Asian landmass as the two regions collided, it might have served as a stepping stone for the continual interchange of faunal elements during the Cretaceous. Another possibility is that the amount of separation from Africa might have been much less than has been thought to date. Currently, however, there is debate concerning the identity of Indian dinosaurs. Ralph Molnar, Curator Of Reptiles at the Queensland Museum, sees no close relationships between the carnosaurs of India and North America and Asia. More work is essential on the Indian fossils in order to resolve, once and for all, this issue.

392

391 A pelvis of Kritosaurus australis, a Late Cretaceous hadrosaur, or duck-billed dinosaur, from Argentina. x 0.5 (F. Coffa, courtesy of the Museo Argentino De Ciencias Naturales "Bernadino Rivadavia" and J. Bonaparte)

392 A skeleton of the frog Saltenia ibanezi, a pipid, from the early Late Cretaceous (probably Cenomanian) Las Curtiembres Formation near Alemania, Salta Province, Agentina. x 4.0 (F. Coffa, courtesy of the Museo Argentino De Ciencias Naturales "Bernadino Rivadavia" and J. Bonaparte)

THE CAINOZOIC RECORD

By the Eocene, the fossil land mammals of India clearly resembled those of Asia and Europe. For the balance of the Cainozoic, this faunal affinity with the landmasses to the north is still evident, superimposed on the regional nature of the tetrapod fauna.

For the other Gondwanan continents, too, the biogeographic origins of the terrestrial vertebrates are much clearer in the Cainozoic than in previous times. This is due primarily to the much better fossil record for this era than for the earlier ones. The African fauna, for example, like that of India, had significant interchange with faunas of Asia and Europe, from time-to-time during the Cainozoic, and yet was isolated enough to maintain a distinctive flavour of its own.

Apart from Australia, South America was the most isolated of the Gondwanan continents during the Cainozoic and developed a highly endemic continental fauna. After interchange with North America during the Cretaceous and earliest Cainozoic, when both dinosaurs and mammals passed between the two continents, only two significant interchanges occurred prior to the final establishment of the Panamanian landbridge in Pliocene times. These two episodes of faunal mixing were in the Oligocene, which saw the arrival of rodents and primates from Africa or North America. In the Late Miocene and continuing through the Early Pleistocene several mammalian groups also crossed between North America and South America. More groups went south than north during this latter mix, but the exchange was nonetheless far from one way, with forms like porcupines, opossums, and ground sloths entering North America. Groups which had been highly diverse and successful in South America for much of the Cainozoic, such as the placental notoungulates, became extinct at this time, perhaps giving way to the invading perissodactyls (horses and tapirs) and artiodactyls (including llamas and deer). North America had no comparable extinction event, which supports the idea that the impact of the interchange was much greater on the South American fauna than vice versa.

In a few tens of millions of years Australia will make contact with Asia and the routes of interchange will become more readily accessible to terrestrial vertebrates. Had humans not disturbed the fauna already, presumably events analogous to those that occurred between the Americas during the late Cainozoic would have taken place between Australia and Asia. The arrival in Australia from Asia of rodents unassisted by humans in the Pliocene is in many ways analogous to the earliest phase of interchange that occurred between the Americas, with the appearance of racoons or procyonids during the Late Miocene in South America and two different families of ground sloth at the same time in North America, all having crossed significant water gaps.

Although Antarctica has long been cited as a route of interchange for mammals and birds between Australia and South America near the Mesozoic–Cainozoic boundary, no relevant fossil evidence was known prior to 1980. However, Eocene marsupials, notoungulates and paratheres have now been discovered on Seymour Island off the Antarctic Peninsula. They are so similar to their contemporaries in South America that it would be quite easy to believe they were found in Patagonia instead.

393 The chelid turtle Acanthochelys sp., from the Pliocene of Rio Cosquin, Cordoba Province, Argentina. x 0.9 (F. Coffa, courtesy of the Museo De La Plata and R. Pascual)

394 A pelomedusid turtle from the Late Palaeocene to Early Eocene Gordo Formation of Rio Casa Grande, Argentina. x 0.7 (F. Coffa, courtesy of the Museo De La Plata and R. Pascual)

393

394

395

396

397

395 A specimen that is probably Percichthys sp., from the Early Miocene of Las Baja, Rio Negro Province, Argentina. This genus still survives in South America today. x 2.3 (F. Coffa, courtesy of the Museo De La Plata and R. Pascual)

396-397 Small fragments of Gondwana, such as New Caledonia (Isle De Pins) and Norfolk Island (Cemetery Beach), have produced fossil remains in limestone sinkholes and in dunes. Remnants of an ancient fauna are preserved in New Caledonia, but the fossil fauna of Norfolk Island appear to represent relatively new invaders to this rather recent island that sits atop a tectonically active submarine ridge connecting New Zealand and New Caledonia. During the Cainozoic, islands must have been constantly appearing and disappearing along this ridge. (P. Vickers-Rich)

398 Partial jaws of Eocene polydolopids from South America and Antarctica. Shown are (right to left): an undetermined genus from Argentina; a Polydolops cf. thomasi maxilla from Argentina; and an undetermined genus from Seymour Island, Antarctica. x 4.0 (F. Coffa, courtesy of the Museo De La Plata and R. Pascual)

399 Foot of the phororhacoid Psilopterus colzecus from the Late Miocene Vivero Member of the Arroyo Chasico Formation, Buenos Aires Province, Argentina. x 1.0 (F. Coffa, courtesy of the Museo De La Plata and R. Pascual)

400 Humerus of the penguin Anthropornis sp., from the Late Eocene La Meseta Formation, Seymour Island, West Antarctica. x 0.7 (F. Coffa, courtesy of the Museo De La Plata and R. Pascual)

401 Armoured skin of the glyptodont parathere Panochtus sp., from the Pleistocene of Buenos Aires Province, Argentina. x 1.0 (F. Coffa, courtesy of the Museo De La Plata and R. Pascual)

To sum up, it would appear that although in the past fossil evidence has been interpreted to show separate faunal provinces for terrestrial vertebrates through much of the Phanerozoic, the evidence for this situation is equivocal. Considerations of the effects of *differences in the depositional environments* and *gaps* in the fossil record on various continents appear to be more than sufficient to explain the apparent regional differences of vertebrate faunas seen prior to the Cretaceous, at least back to the middle part of the Palaeozoic. The many discoveries of fossils outside their well known ranges after more than a century of collecting supports the view that caution is still very necessary when evaluating negative evidence. Therefore, analysis of the available information suggests that the degree of regional differentiation of present continental vertebrate faunas is something that developed no earlier than the Cretaceous, if even that early, and reached its present degree only within the Cainozoic. After the early part of the Palaeozoic, the Gondwanan continents were probably only a meaningful geographic entity for *terrestrial vertebrate evolution* in a fleeting geological moment in the Cretaceous when the division of Pangaea was passing through a geologically brief stage during which the landmasses of Earth were divided into northern and southern moieties.

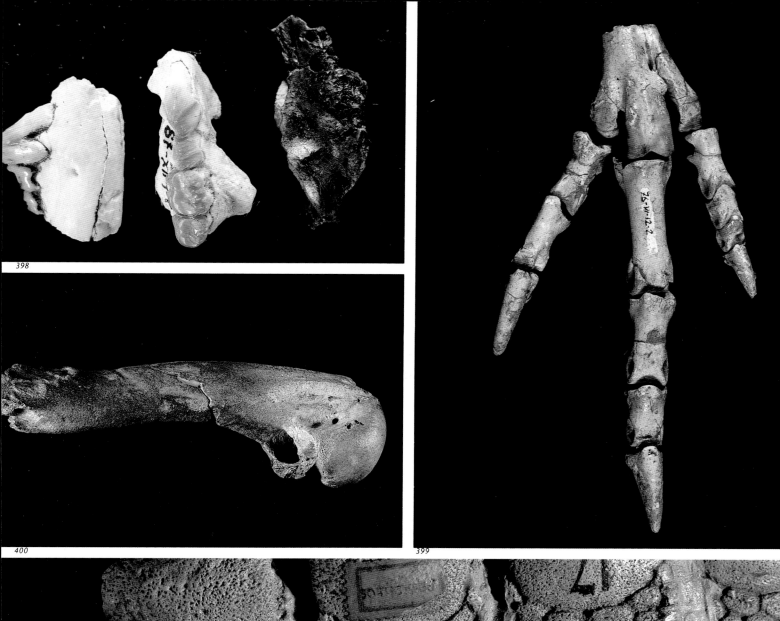

398

399

400

401

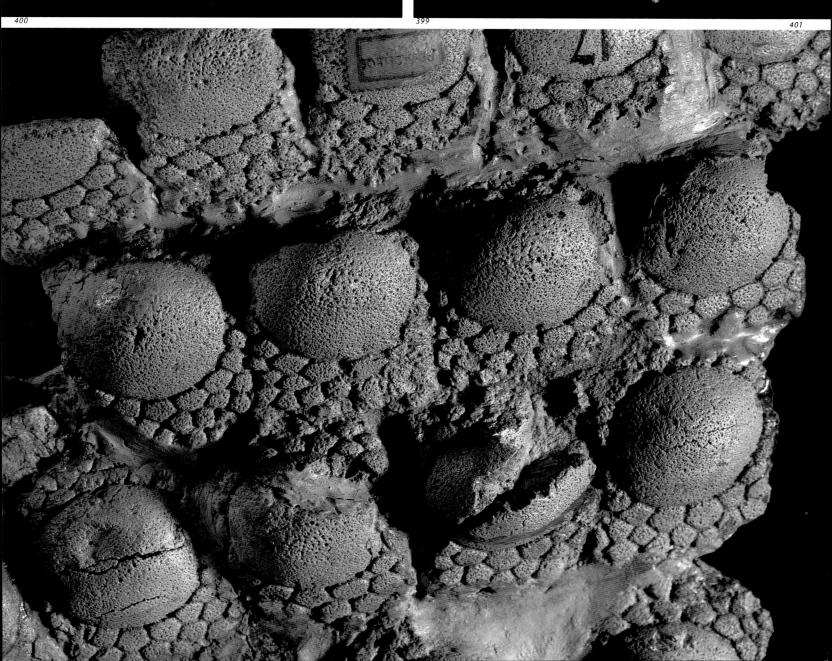

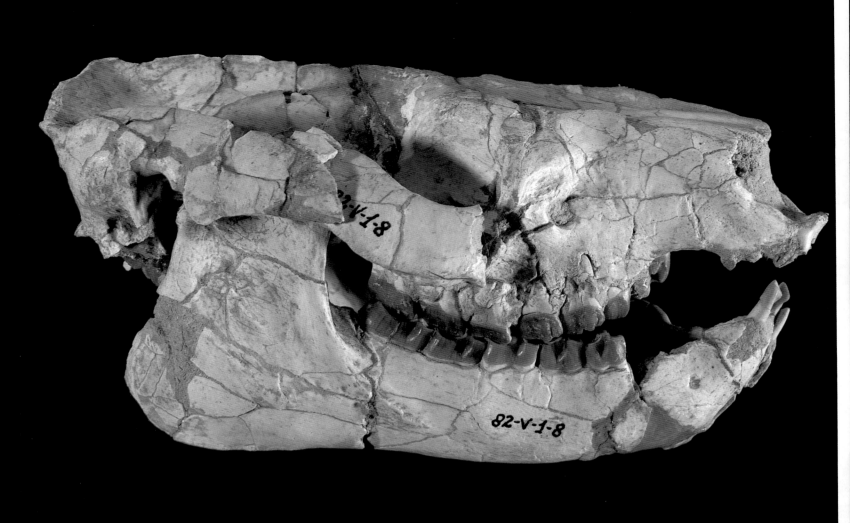

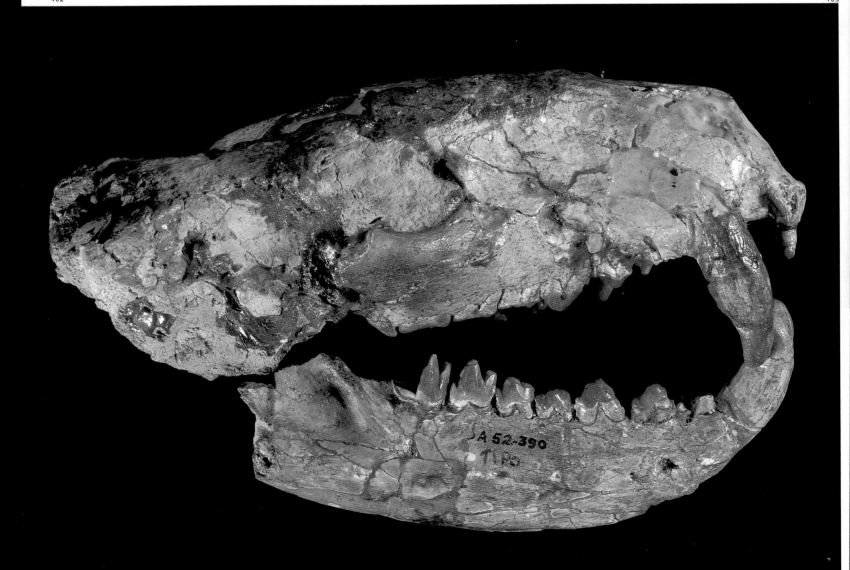

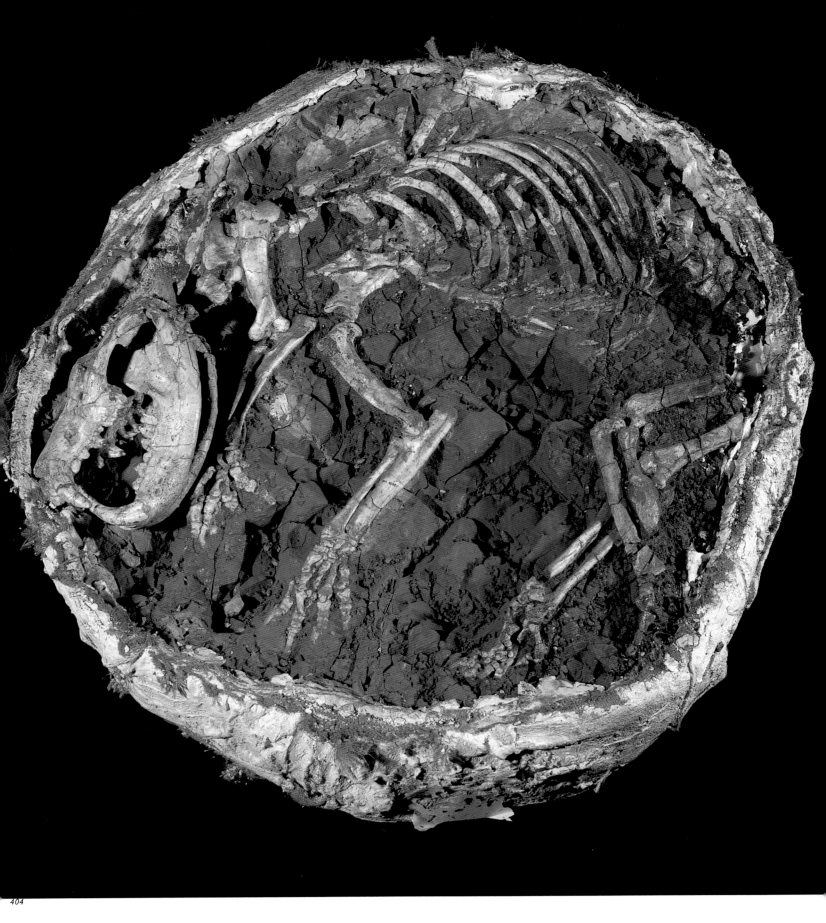

404

402 A skull of the primitive placental *Notostylops*
sp., from the Middle Eocene (Mustersan) of Valle
Hermoso, Chubut Province, Argentina. x 1.3 (F. Coffa,
courtesy of the Museo De La Plata and R. Pascual)

403 As no placental carnivores entered South America
until the end of the Miocene, other groups took up the
terrestrial carnivore niche. Prominent among them
were the borhyaenid marsupials of which there was a
great variety in the course of the Cainozoic, ranging in
size from domestic cats to bears. One of the larger of
these marsupials was *Pseudoborhyaena
macrodonta* from the Late Oligocene of Argentina.
x 1.0 (F. Coffa, courtesy of the Museo Argentino De
Ciencias Naturales "Bernadino Rivadavia" and
J. Bonaparte)

404 The skeleton of a highly predatory borhyaenid
marsupial from the Early Eocene Lumbrera Formation
of Pampa Grande, Salta Province, Argentina. x 0.3
(F. Coffa, courtesy of the Instituto Miguel Lillo and
H. Powell)

405-406 The armoured skull and carapace of the armadillo *Chorobates* sp. from the Pliocene (Chapadmalalan) of Chapadmalal, Buenos Aires Province, Argentina. Skull x 1.2 (F. Coffa, courtesy of the Museo De La Plata and R. Pascual)

407 A skull of the carnivorous sabre-toothed marsupial *Thylacosmilus lentis* from the Late Miocene to Pliocene ("Araucanian") of Catamarca Province, Argentina. x 0.5 (F. Coffa, courtesy of the Museo De La Plata and R. Pascual)

408-409 The fore foot and skull (in occlusal view) of the notoungulate *Periphragnis* sp., from the Middle Eocene (Mustersan) of Cerro Del Humo, Chubut Province, Argentina. Skull x 0.8 (F. Coffa, courtesy of the Museo De La Plata and R. Pascual)

405

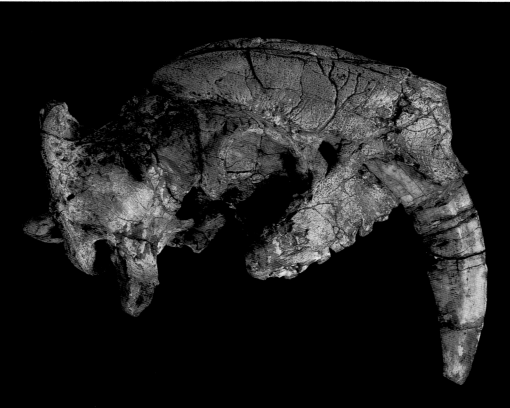

407

409

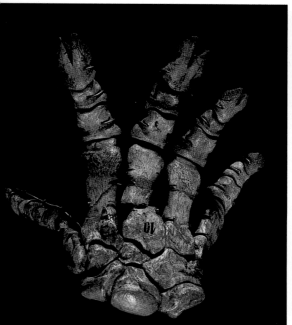

408

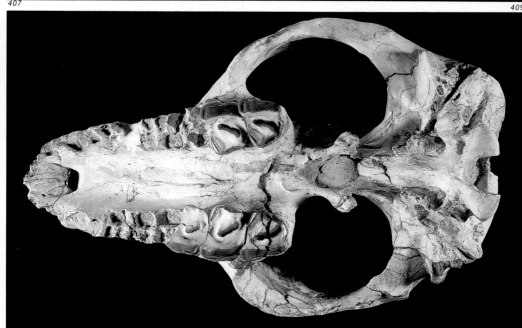

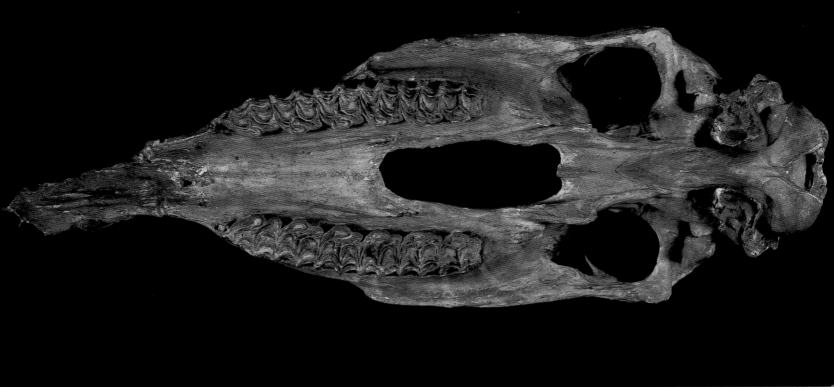

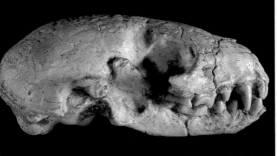

*410 True horses reached South America by the Plio–
Pleistocene. By then, some had differentiated from their
North American ancestors enough to be regarded as
distinct genera, such as Hippidion, the skull of which
is shown here in occlusal view. x 0.3 (F. Coffa,
courtesy of the Museo De La Plata and R. Pascual)*

*411-412 Skull, in occlusal and lateral views, of
Proterotherium mixtum, a horse-like litoptern
from the Miocene of Argentina. Litopterns were
just one element of the diversified order Noto-
ungulata which flourished in South America
during the Cainozoic only to become extinct in the
Pleistocene. Beside converging on horses of other
continents, some notoungulates emulate camels.
x 1.2 (F. Coffa, courtesy of the Museo Argentino
De Ciencias Naturales "Bernadino Rivadavia"
and J. Bonaparte)*

*413 A skull of the carnivorous skunk Conepatus sp.,
from the late Cainozoic of Argentina. This animal was
an invader from the north which in part brought about
the demise of many of the endemic South American
placentals. x 1.0 (F. Coffa, courtesy of the Museo De
La Plata and R. Pascual)*

*414 A skull of the cervid Morenelaphus sp., from the
Pleistocene (Lujanian) of Buenos Aires Province,
Argentina. This animal was one of the northern
invaders of the late Cainozoic Great American
Interchange that displaced some of the South
American endemic ungulates. x 0.3 (F. Coffa, courtesy
of the Museo De La Plata and R. Pascual)*

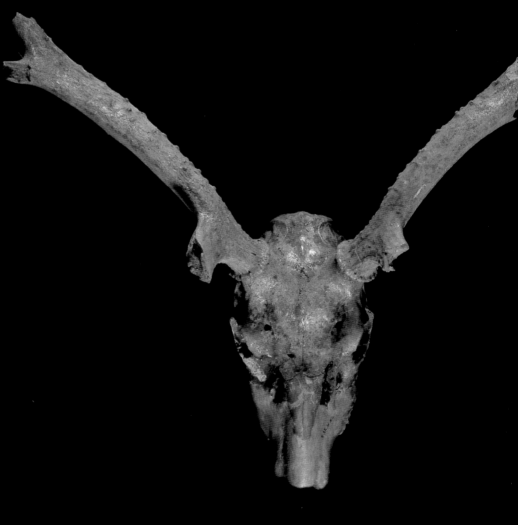

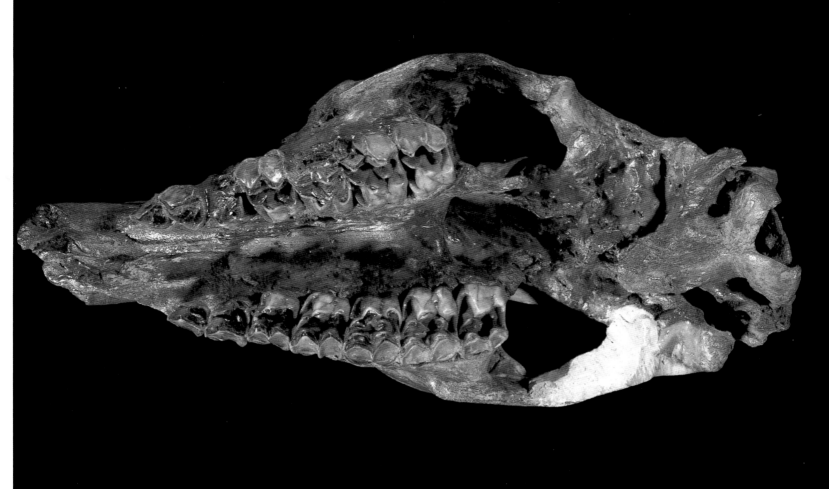

411

412

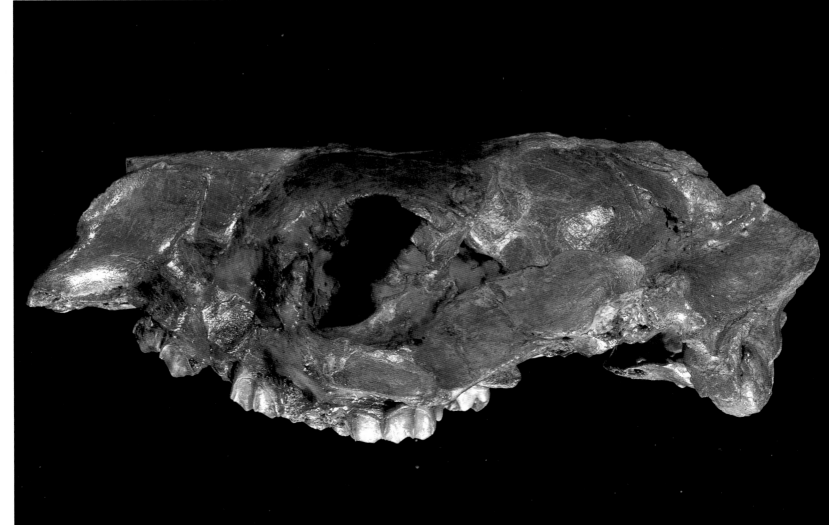

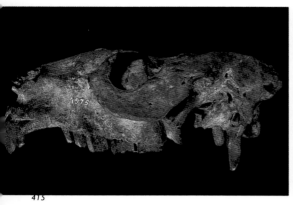

415

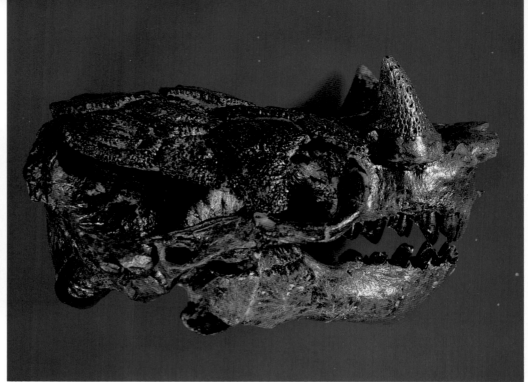

416

415 Skull of *Typotherium pseudopachygnathus*, a primitive ungulate placental endemic to South America, from the Pliocene of Monte Hermoso, Buenos Aires Province, Argentina. x 0.5 (F. Coffa, courtesy of the Museo Argentino De Ciencias Naturales "Bernardino Rivadavia" and J. Bonaparte)

416 Side view of the skull and lower jaw of the edentate glyptodont *Peltephilus* sp., from Miocene terrestrial sediments near Sehuen, Patagonia, Argentina. These primitive placental mammals were heavily armoured, both on the head and the body. x 1.0 (F. Coffa, courtesy of the Museo Argentino De Ciencias Naturales "Bernadino Rivadavia" and J. Bonaparte)

417-418 Side and occlusal views of a skull of *Padeotherium* sp. which belongs to the South American endemic placental group the hegetotheres, from the Pliocene (Chapadmalalan) of Chapadmalal, Buenos Aires Province, Argentina. Hegetotheres were herbivores, convergent on some rodents and rabbits from the Northern Hemisphere. x 1.6 (F. Coffa, courtesy of the Museo Argentino De Ciencias Naturales "Bernadino Rivadavia" and J. Bonaparte)

419 Tail armour of the glyptodont edentate *Hoplophorus ornatus* from the Pleistocene Pampean Formation near Buenos Aires, Argentina. Glyptodonts were one of the groups that developed and prospered in South America, and even moved into North America, but were finally swamped by the more advanced placental herbivores that flooded into South America in the late Cainozoic. x 1.2 (F. Coffa, courtesy of the Museo Argentino De Ciencias Naturales "Bernadino Rivadavia" and J. Bonaparte)

420 Upper and lower dentitions of the horse-like litoptern *Macraucheniopsis* sp., from the Pleistocene (Ensenadan) of Buenos Aires Province, Argentina. Many of the primitive placental mammals in South America, in the absence of northern placentals, developed lifestyles and tooth types similar to their northern counterparts, an example of convergent evolution. When North America and South America were connected by the Central American landbridge in the late Cainozoic, many of the southern forms succumbed to the northern invaders as they competed for limited resources. x 0.5 (F. Coffa, courtesy of the Museo De La Plata and R. Pascual)

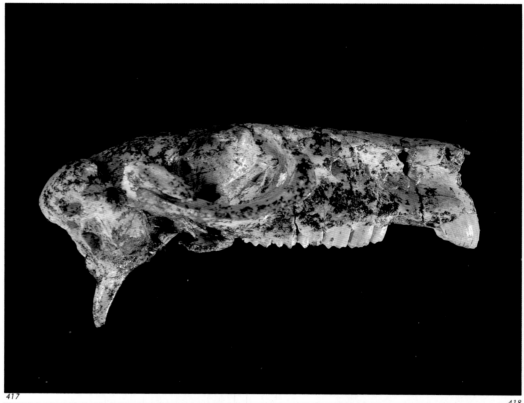

417

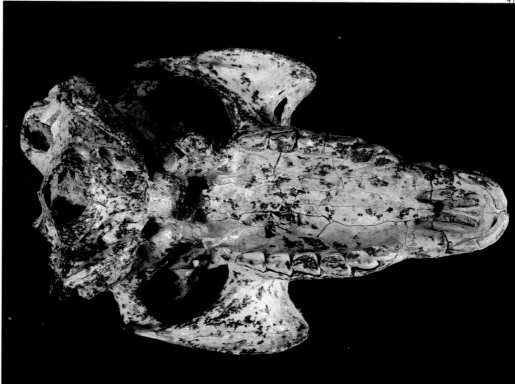

418

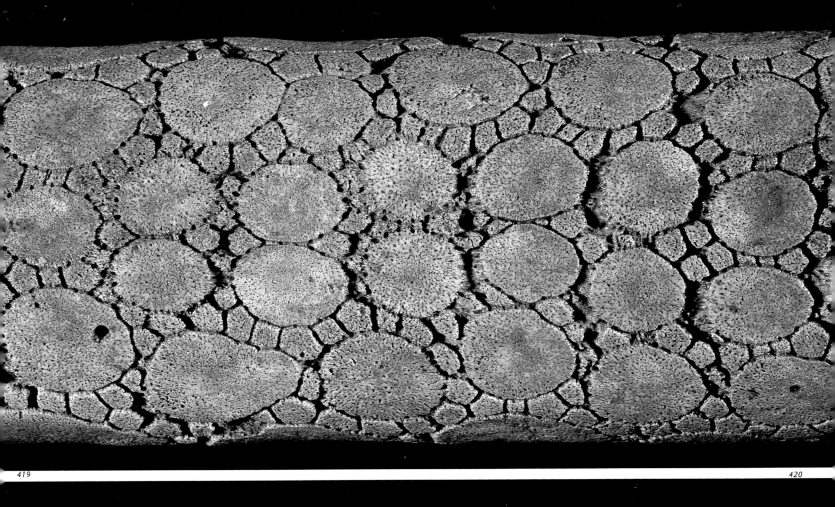

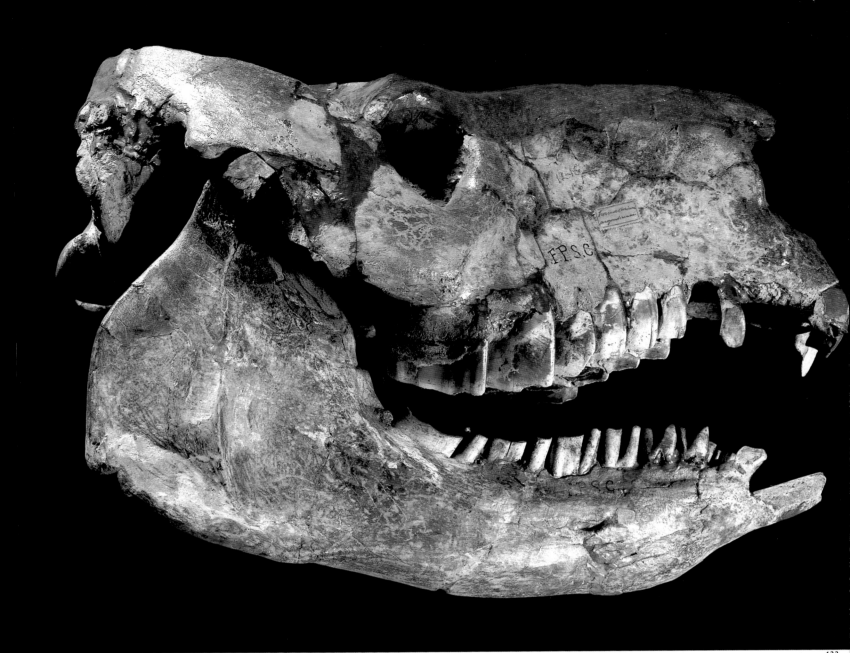

421

422

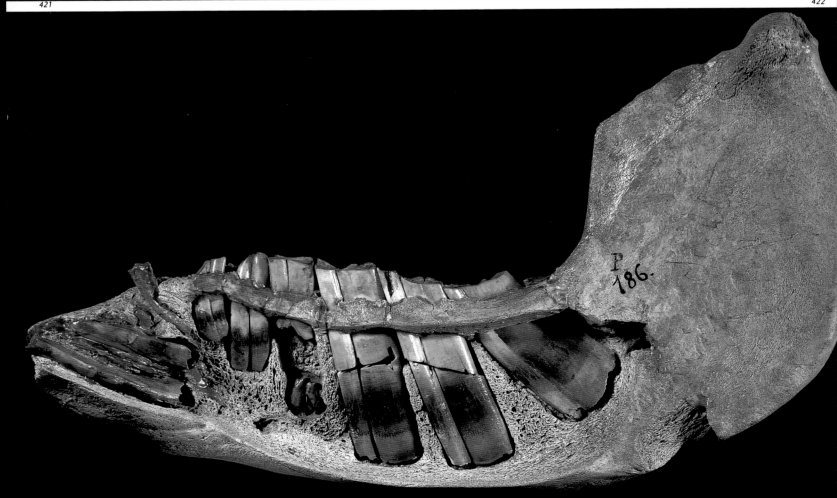

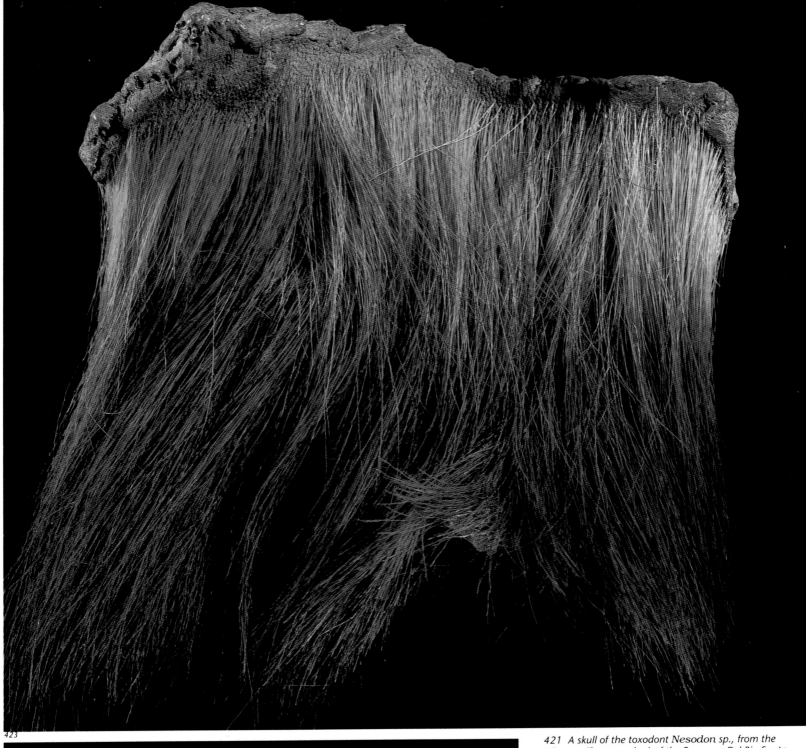

421 A skull of the toxodont Nesodon sp., from the Miocene (Santacruzian) of the Barranca Del Rio Santa Cruz, Santa Cruz Province, Argentina. x 0.5 (F. Coffa, courtesy of the Museo De La Plata and R. Pascual)

422 A lower jaw with bone cut away showing the roots of the ever growing teeth of the primitive ungulate Toxodon sp., from the Pleistocene (Lujanian) of Salto, Buenos Aires Province, Argentina. x 0.5 (F. Coffa, courtesy of the Museo De La Plata and R. Pascual)

423 Hair of a ground sloth, Glossotherium listai, from the Late Pleistocene Ultima Esperanza Cave in southern Chile. x 0.8 (F. Coffa, courtesy of the Museo Argentino De Ciencias Naturales "Bernardino Rivadavia" and J. Bonaparte)

424 Dermal bones that ossified in the skin of a ground sloth, Mylodon, from the Middle Pleistocene of Necochea, Buenos Aires Province, Argentina. x 1.0 (F. Coffa, courtesy of the Museo Argentino De Ciencias Naturales "Bernadino Rivadavia" and J. Bonaparte)

AFTERWORD: NEW DISCOVERIES IN GONDWANA

Since the publication of the first edition of *Wildlife of Gondwana*, a number of new and significant discoveries have been made on the Gondwana continents, and even on continental masses that until recently had been thought to have only a Laurasian (Europe plus Asia plus North America) history. Studies by Metcalfe (1998) and others (Hall & Holloway 1998; Polcyn, Tchernov & Jacobs in press; and Shishkin 1993) have indicated that pieces of Gondwana may well be "glued" firmly onto the now northern continents, and that these fragments may well bear the "Viking funeral goods" of dinosaurs and other vertebrates that lived when these continental fragments were part of Gondwana (see reconstruction in Metcalfe of the Early Cretaceous). Thus, fossils found even on the now northern continents may not belong there as once living animals—and this caution needs to be kept in mind.

Because of the rapidity of discovery in the last decade on the Gondwana continents, it would be impossible in such a short space to cover in any detail the great variety of fossil vertebrates recovered from these mostly now southern continents—and so only a selection will be mentioned, some of the more spectacular in this final chapter of *Wildlife of Gondwana*.

An interesting idea that is developing based on discoveries since the early 1990s is that a number of groups once thought to have originated in the north and then to have migrated to the south, or thought not to have a southern history at all (e.g., the discovery by Anne Warren [in press, *Trans. Roy. Soc. Edinburgh*] of a tupilakosaurid amphibian in South Africa, a group once thought confined to Russia and Greenland), are beginning to appear in quite ancient rocks on the Gondwana continents. Although the identity of many of

these new discoveries is still under review, such groups as ceratopsian (Rich & Vickers-Rich 1994) and oviraptorosaurian dinosaurs (Currie et al. 1996) and even possible placental mammals (T.H. Rich et al. 1997, 1999) have more ancient, or just as ancient (albeit much poorer), records on the southern continents than those on the northern continents. It is even possible that these animals reached Laurasia on the microcontinental fragments that broke off Gondwana from numerous sites between Africa and Australia and drifted northward. Sereno et al. (1996 and in press) have proposed the idea that many groups of dinosaurs were initially Pangaeic in their distribution, and only after the break-up of this supercontinent did the distinctive groups of Gondwanan and Laurasian nature appear—and they further point out that the record on many of the Gondwana fragments (such as India, Antarctica and Australia) is insufficient to adequately characterize their dinosaur faunas during the late Mesozoic. This is certainly true at other times during the Phanerozoic and for other vertebrate groups.

Examination of tissues from living mammals is lending support to the idea that Africa was home to one-third of the living placental orders, which had arisen there by the mid-Cretaceous, a time when the terrestrial micro-vertebrate fossil record for that continent is all but non-existent (Hedges et al. 1996; Stanhope et al. 1998). These placentals of African origin have been grouped together as the Afrotheria.

As the poor record of the Gondwana continents improves, the entire view of mammalian palaeobiogeography is being substantially modified from the vision proposed by A.R. Wallace, W.D. Matthew and G.G. Simpson. Wallace (1876) put forward the view that the

425 Dorsal view of skull of **Brachyops laticeps** *(BMNH R4414) from the Mangli Beds, Early Triassic, India. (Photo by Russell Baader, courtesy of Anne Warren and the British Museum [Natural History]) This is the first described member of the Brachyopidae from India; skull is approximately 11 cm long.*

425

marsupials and placentals originated in the Northern Hemisphere and later dispersed into the Southern. This idea has been refined during the past century, but is still the prevalent one. In addition, the long-accepted hypothesis has been that the differentiation of the majority of placental orders occurred just after the dinosaurs became extinct in the earliest Cainozoic.

Recently, however, analysis of genetic information, combined with an Australian fossil discovery, has suggested to some, but not all, scientists an alternative interpretation. The genetic studies place the differentiation of placental orders in the mid-Cretaceous (Hedgeset al. 1996; Kumar & Hedges 1998). There is not universal agreement about this (Foote et al. 1999). Furthermore, not all of the placental orders necessarily originated on the Laurasian continents.

In the Early Cretaceous of southeastern Australia, two jaws interpreted as those of placentals have been found in the last two years, and others are yet undescribed. Although the mandibles of some of these small mammals are extremely primitive for placentals, the teeth show remarkable similarities to the earliest erinaceids (which include the living European hedgehogs) in the Northern Hemisphere, which are just over half their age. If this relationship is not an artifact of convergence, it suggests that microcontinental plates, which are known to have separated from the Australian region of Gondwana and drifted northward to become parts of Southeast Asia, may have provided the method whereby these mammals entered the Northern Hemisphere (Rich et al.1999). The Erinaceidae are far removed from the Afrotheres (Stanhope et al. 1998), suggesting that there may have been separate places of origin within Gondwana for different placental groups.

426 Early Cretaceous (120 million years ago) palaeoreconstruction from Metcalfe (1998).

427 Mangahouanga Stream, Early Cretaceous of New Zealand has yielded much of the dinosaur material known from this country and collected through the efforts of Joan Wiffen and her team; the image on the left shows the general area, the one on the right, a close-up of a bone still in the rock. (Photos courtesy of J. Wiffen)

428 Moa art. a.) Mummified head and neck of the Upland Moa Megalapteryx didinus. (Museum of New Zealand Te Papa Tongarewa S400) (Scale: distance from crown of skull to base of neck, about 180 mm). b.) M. didinus feathers (MNZ S27950) are presumed to have been collected in 1949 by R. A. Falla from a cave in Takahe Valley, Fiordland (Scale: main shaft of upper left double-shafted feather, about 75 mm). c.)The leg of this same species from Waikaia, Old Man Range and in the collection of the Otago Museum, Dunedin (figured in Oliver 1949). (All illustrations by P. Trusler, from Vickers-Rich et al. 1995)

In Australia, continued work at the rich Cainozoic sites on Riversleigh Station in north Queensland has led to many new groups of marsupials being recognised and a substantial literature developed (major summary in volume 41, number 2 of the *Memoirs of the Queensland Museum,* Brisbane, 30 June 1997). Peter Murray's work on the diverse assemblage of dromornithid birds has led him to propose (1998) that this group with elusive ancestry is, in fact, derived from the Anseriformes (ducks and kin), and that the feeding style is indeed bizarre. The record of this group has now been extended into the Eocene with the discovery of the impression of a partial foot (Vickers-Rich & Molnar 1997). The occurrence of a brachyopid amphibian, *Koolasuchus,* in the Early Cretaceous of southeastern Australia has confirmed the existence of "labyrinthodonts" into the Cretaceous, yet another anachronistic vertebrate from this part of Gondwana—a group surviving long beyond its time of extinction elsewhere in the world (Warren et al. 1997).

Further back in time, the discovery of a diverse fauna of Early Carboniferous vertebrates in the Ducabrook Formation of Queensland is bound to lead to a much better understanding of early

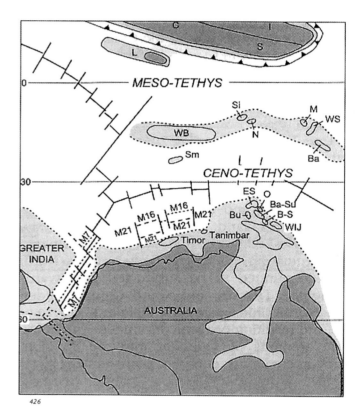

426

427

428b

428c

428a

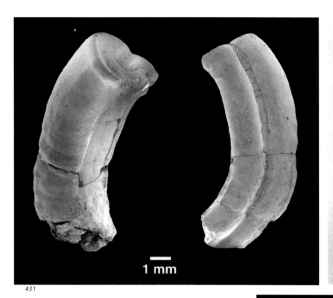

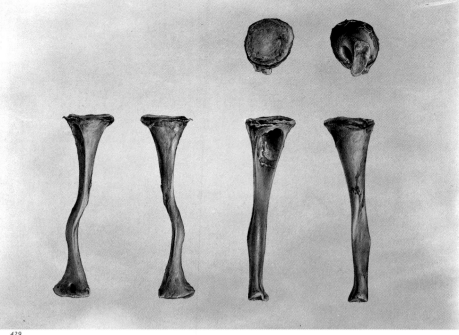

429

430

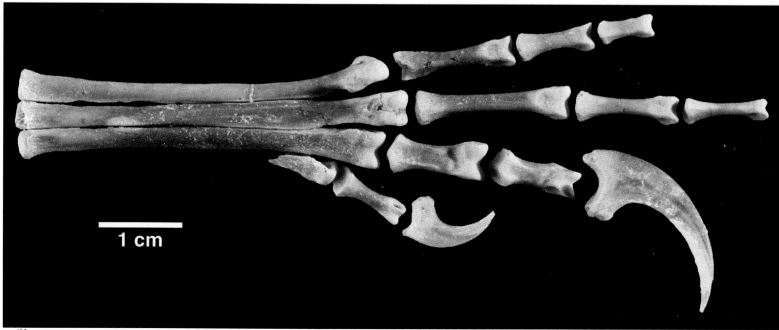

432

429 Stapes (middle ear bone) of the moa *Megalapteryx didinus* (MNZ S400) is distinct from the stapes of all other ratite birds. There are some similarities with those of the living Emu and Ostrich, but little with the kiwis or tinamous. This stapes was extracted from the skull of one of the few mummies of the moas. (Illustration by P. Trusler)

430 Reconstruction of the skull and lower jaws of *Majungatholus atopus* (FMNH PR 2100), an abelisaurid theropod, from the Late Cretaceous of Madagascar. (Courtesy of D.W. Krause and modified from image in Science)

431 *Lavanify miolaka* from the Upper Cretaceous Maevarano Formation, Mahajanga Basin, northwestern Madagascar (UA 8653), a sudamericid gondwanathere mammal. Gondwanatheres are multituberculate or multituberculate-like primitive mammals (UA = Universite d'Antananarivo). (Courtesy of D.W. Krause, modified from image in Nature, 1998 [390]: 505)

432 Hind limb of *Rahonavis ostromi*, possibly a primitive bird from the Late Cretaceous of Madagascar. (Courtesy of D. Krause & Cathy Foster; image modified from that in Science)

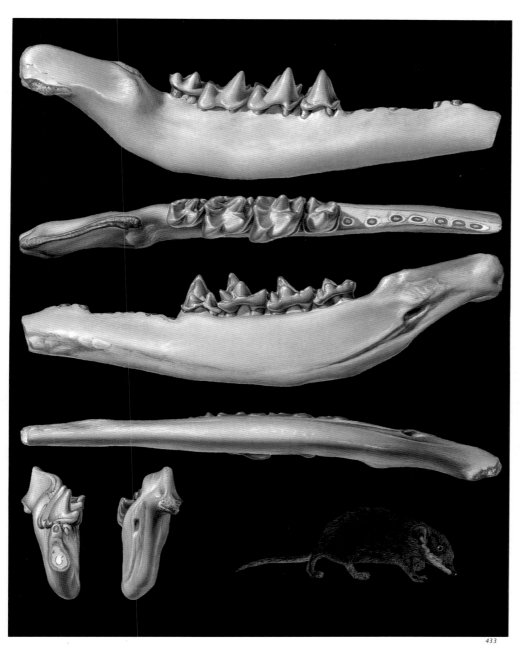

433 Ausktribosphenos nyktos (MSC007/NMV P208090), a possible placental mammal from the Early Cretaceous of southern Australia. (Specimen art by Peter Trusler; reconstruction by D. Gelt; computer graphics by S. Morton & Adrian Dyer; courtesy of the Monash Science Centre and Museum Victoria [NMV]) l.=16mm.

433

terrestrial vertebrates, almost unrepresented in this part of Gondwana before this find. The find also signals that "several major groups of tetrapods were distributed worldwide through equatorial regions during the Early Carboniferous" (Thulborn et al. 1996).

In Australia, also, sedimentological evidence discovered very recently has added strength to the suggestion that some dinosaurs were able to cope with fairly severe climatic conditions. Frozen ground structures (cryoturbated) in Early Cretaceous sediments from along the southern coast of Australia which had a palaeo-latitude of about 78° South are associated with rocks bearing both hypsilophodontid and ornithomimid dinosaurs (Constantine et al. 1998). Studies on the histology of the bones of the two groups of dinosaurs established that the hypsilophodonts grew continuously, while the ornithomimids had well-developed LAGs (lines of arrested growth, indicating that they almost ceased growing altogether at times). The hypothesis has been advanced that these dinosaurs coped with the low temperatures of southern polar winters with two different strategies—one remained active (and was very likely warm-blooded), while the other hibernated. Temperatures have been calculated using oxygen isotopes at a MAT (mean annual temperature) of –2°C +/–5°C. The presence of the sedimentary structures indicative of seasonally frozen ground suggest temperatures of –6°C to +3°C—which is in agreement with the geochemical evidence.

Work in Antarctica by several researchers, namely from the British Museum (Hooker and Milner), Argentina (Gasparini & Pereda-Suberbiola working with Australian R. Molnar), and the U.S. (Hammer & Hickerson), has led to the discovery of a number of new dinosaurs. These include a partial skeleton of an ankylosaur from James Ross Island in the Antarctic Peninsula (Gasparini et al. 1996), a fragment of a theropod (Molnar et al. 1996), some scraps of hadrosaurs (being studied by J. Case et al. in press and T. Rich et al.

in press), a partial skeleton of an ornithopod (Hooker et al. 1991)—all from West Antarctica and of Late Cretaceous age. Without a doubt, the most spectacular dinosaur skeleton from the Antarctic is that of a crested theropod, *Cryolophosaurus ellioti* (Hammer & Hickerson in press), found in Early Jurassic sediments in the Beardsmore Glacier area of mainland East Antarctica. Work on early Cenozoic terrestrial vertebrate faunas of West Antarctica show them to be very similar to faunas from southern South America (Marenssi et al. 1994). New fossil finds of birds include the presbyornithid remains from Cape Lamb on Vega Island reported on from the Late Cretaceous by Noriega & Tambussi (1995).

In Madagascar, finds of new Late Cretaceous dinosaurs by Sampson, Krause, Dodson and others (Sampson et al. 1998) include some wonderfully complete cranial material of abelisaurid theropods (*Majungatholus atopus*) as well as titanosaurs, crocodiles, snakes, turtles, fish, frogs, birds and mammals. Abelisaurids are restricted to three Gondwanan fragments (Madagascar, South America and India), except for a few possible fragments from England (which need further confirmation). This is a similar distribution shared with the gondwanathere mammals, originally discussed by Bonaparte. This sort of distribution can be explained in two different ways: abelisaurids and gondwanatheres originated prior to the main continental splitting in the Early Cretaceous and spread across most of Gondwana, and even, in the case of the abelisaurids, into Laurasia. Alternatively, abelisaurids and gondwanatheres originated sometime in the Early Cretaceous after the tectonic isolation of Africa. In the case of the abelisaurids, according to this theory, they may never have existed on Africa, but dispersed between South America and Indo-Madagascar via Antarctica, utilizing the land bridge across the Kerguelen Plateau. Either view could be correct (Sampson et al. 1998). A third hypothesis championed by Sereno (1996 and in press)

434 The fossil dolphin, Waipatia maerewhena, skull and jaws, from near "The Earthquakes," North Otago, Late Oligocene of New Zealand. (Courtesy of E. Fordyce, University of Otago) l.=556mm.

435 Teeth, petrosals and tympanic bullae (box in lower right) of a squalodontid dolphin from near "The Earthquakes," North Otago, Late Oligocene of New Zealand. (Courtesy of E. Fordyce, University of Otago) (length of tympanic bulla in lower left is 63.5 mm. point to point)

436 Skull and jaws of a cetothere (archaic baleen whale), Hakatarmea Valley, South Canterbury, Late Oligocene of New Zealand. Preparator Andrew Grebneff for scale. (Courtesy of E. Fordyce, University of Otago)

suggests that most dinosaur groups had a broad, cosmopolitan distribution that later became much more provincial as Pangaea dispersed.

On the African mainland, Paul Sereno's (Sereno et al. 1998) long-snouted, fish-eating spinosaurid from the Early Cretaceous of Niger is an outstanding surprise in the shape that dinosaurs assumed. The distribution of the spinosaurids is broad and suggests that there was at least one dispersal event across the Tethyan seaway during the Early Cretaceous.

In India, the discovery of burrows left by metazoans in the Chorhat Sandstone of the Vindhyan Basin of central north India pushes back the origin of this group to more than 1 billion years, more than 400 million years further back in time than previously thought (Seilacher, Bose & Pfluger 1998)—and even though they were not vertebrates, this again indicates how sparse the record is during this time on Gondwana, and how new discoveries can radically change ideas about evolutionary and biogeographic pathways. Much new information is now available on the late Mesozoic of India, summarized in a publication by Prasad et al. (1994) and Prasad & Khajuria (1995). The Early to Middle Jurassic Kota Formation has yielded a number of mammals, such as the symmetrodont *Kotatheirum.* Occurring in the significantly younger Late Cretaceous intertrappean rocks are what appear to be primitive palaeoryctids, such as *Deccanolestes* (Prasad et al. 1994), which seem to share many primitive (plesiomorphic) characters with Laurasian forms. However, since Gondwanan Late Cretaceous mammal faunas are so very poorly known, another hypothesis that cannot be ruled out is that these Gondwanan forms gave rise to those in the north. Certainly the finding of these palaeoryctids has led to a lively debate concerning why the Late Cretaceous India terrestrial vertebrate faunas show so little endemism: (Hypothesis 1) the Mascarene Plateau and the Chago-Laccadive ridges in the south and the Dras island-arc in the north acted as dispersal corridors allowing interchange between the north and the south; (Hypothesis 2) India itself was located between northeastern Africa (Somalia) and Asia and acted as a stepping stone in dispersal events; and (Hypothesis 3) the India/Asia collision took place at the Cretaceous–Tertiary boundary instead of during the Eocene as traditionally thought, and thus amalgamation of the faunas explains the lack of a unique, latest Cretaceous Indian biota (Prasad et al. 1995).

In South America, the finding of nests of sauropod eggs with embryos inside and the discovery of several complete or nearly complete skeletons of the Late Cretaceous *Gasparinisaura* (thought to be a basal iguanodontian) (Coria in press) are both spectacular. The eggs containing the embryos (discovered by Rudolfo Coria, Luis Chiappe, Lowell Dingus and their crews) have also preserved remains of some of the soft tissues—the impression of pieces of skin, with mosaics of tiny, lizard-like scales. The scale pattern of rosettes in parallel rows is similar to the pattern known on the backs of titanosaurs, which have developed into proper armour plates. The bones of the embryos also identify the young as titanosaur sauropods, a group very common on the Gondwana continent, but also with a Pangaeic distribution. The most spectacular aspect of this discovery is that for the first time there are examples of such large, spherical eggs that definitely contain sauropod embryos. This has never been established before, and at least proves that some sauropods (if not all) laid eggs and did not "give birth" to live young, which has been suggested in the past. Along with this, the partial skeleton of the largest known carnosaur, *Giganotosaurus*, adds to the list of new forms that keep turning up in an area that has great promise of many future discoveries, both because of the great expanse of exposures yet to be prospected and because of the increasing activity in the region. The exquisitely preserved material from the Early Cretaceous Santana Formation of northeastern Brazil, containing a wide array of both invertebrates and vertebrates (with an emphasis on fish), is beautifully illustrated and discussed in Maisey (1991). The fish fauna is certainly endemic at the generic level to Gondwana, but many of the insect taxa are more widespread.

The record of birds in South America continues to improve,

436

especially with the work of Chiappe (1996) and Novas (1996). Such forms as the hen-sized, flightless *Patagopteryx* from the Late Cretaceous of northern Patagonia appear to be related to forms that gave rise to modern birds, but because of the presence of LAGs (lines of arrested growth) in the bones of this form, it did not behave physiologically like all other modern birds (Ornithurae). One group, the Enantiornithes, stands out as a diverse array, thought to be a group more primitive than modern birds—a sister-group of the ornithureans and *Patagopteryx*. Like *Patagopteryx,* enantiornithines possess LAGs in their bones, suggesting times when bone was not being deposited in the life cycle of an individual bird—a time of growth shutdown. Enantiornithines are known from the Coniacian to the Maestrichtian and are quite diverse in South America, but as pointed out by Chiappe, they are widespread across the globe. Thus, earlier claims that this group originated and spread out from South America are yet untestable because of the sparseness of the record.

The record of the oldest and most primitive member (*Eutreptodactylus itaboraiensis*) of a modern avian family, the Cuculidae, has been reported from the Late Palaeocene karst deposits of eastern Brazil (Baird & Vickers-Rich 1997). This form shows the specialization for zygodactyly (two toes forward and two toes back) but may represent a bird that had a choice in the placement of one of those toes—perhaps three forward and one back was possible—thus a primitive state for this group, and not unexpected in the most ancient fossil for the family.

As the record continues to improve as a result of the increasing activity of palaeontologists on the southern continents, there are certain to be some startling and outstanding, unpredicted discoveries. The record is still so poor that only the surface has been scratched of this intriguing megacontinent, and there is still ample opportunity for the intrepid adventurer to make fundamental discoveries—there is still a great deal of pioneering work to be done.

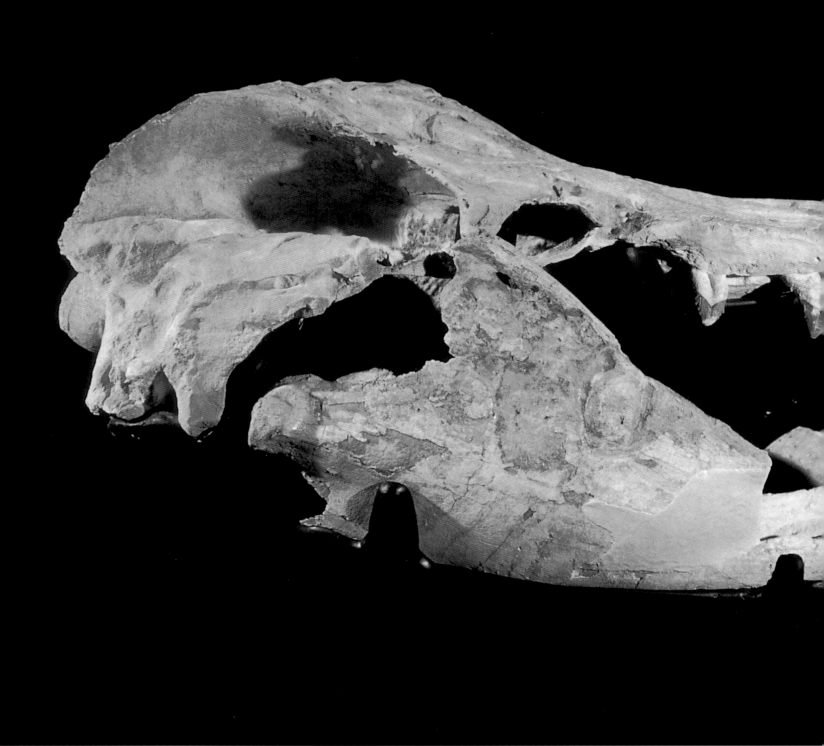

437 Large shark-toothed dolphin skull and jaws from near "The Earthquakes," North Otago, Late Oligocene of New Zealand. (Courtesy of E. Fordyce, University of Otago) l.=1m.

438 Penguin skull from Burnside, near Dunedin, New Zealand, Late Eocene. (Courtesy of E. Fordyce, University of Otago) l.=198mm.

439 Vertebrae of a mosasaur from Shag Point from sediments of Late Cretaceous age in North Otago, New Zealand. (Courtesy of E. Fordyce, University of Otago)

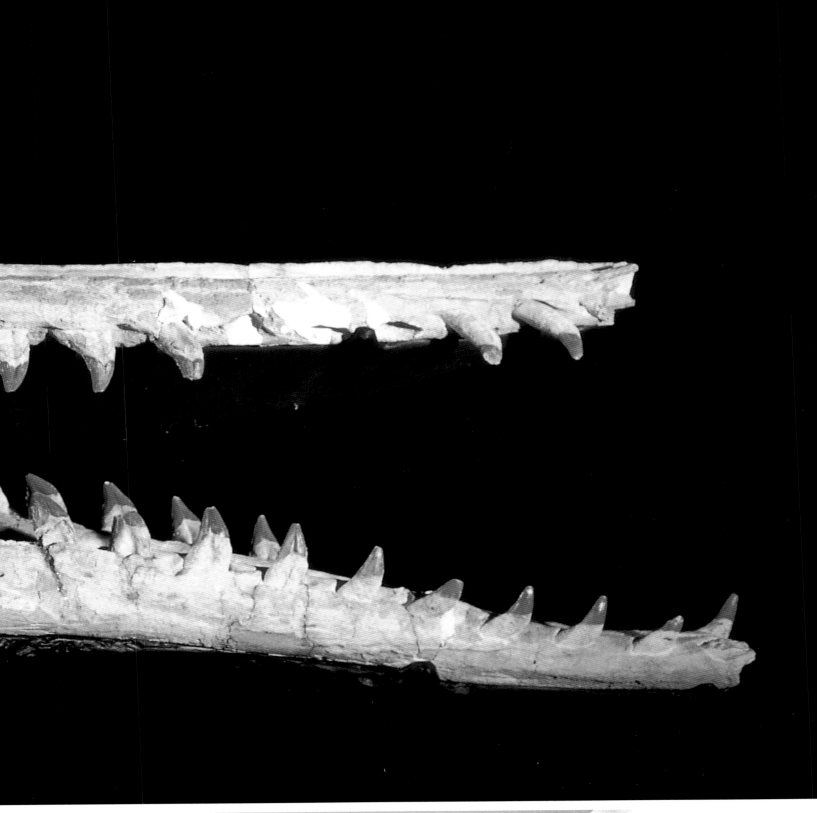

439

440 a,b Specimen and reconstruction of the Antarctic
dinosaur *Cryolophosaurus ellioti* from the Early Jurassic
Hanson Formation, unusual in possessing a bony crest.
(Courtesy of William R. Hammer, Augustana College,
Illinois. Reconstruction by Akio Itoh and Yaskuko
Okamoto.) Skull length=65cm.

440a

440b

441 A large complete specimen of *Lepidotes deccanensis* (ISI P12) collected from Boraigurdem locality of early Middle Jurassic Kota Formation of Pranhita-Godavari valley, India. (Figures 441 to 451 are courtesy of Saswati Bandyopadhyay. All of the specimens in these photos are housed in the Geology Museum of Indian Statistical Institute, Calcutta, India.) About 0.4x.

442 A complete specimen of *Paradapedium egertoni* (ISI P32) collected from Kota ledge of early Middle Jurassic Kota Formation of Pranhita-Godavari valley, India. About 0.45x.

443 Dorsal View of the skull and mandible of *Parotosuchus rajareddeyi* (ISI A18) collected from the Middle Triassic Yerrapalli Formation of the Pranhita-Godavari valley, Deccan India. About 0.15x.

441

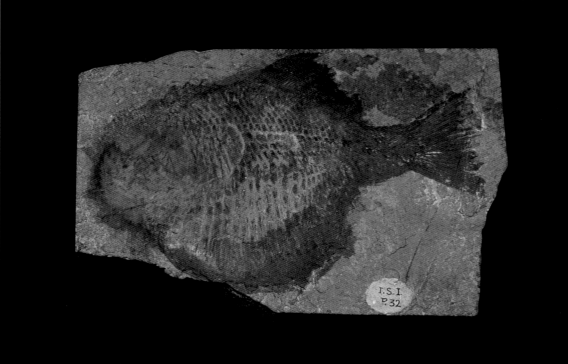

442

443

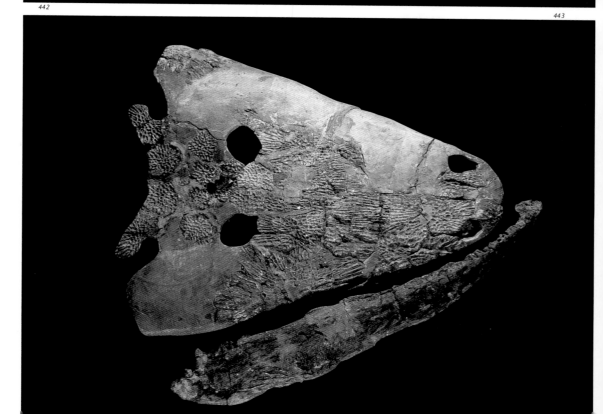

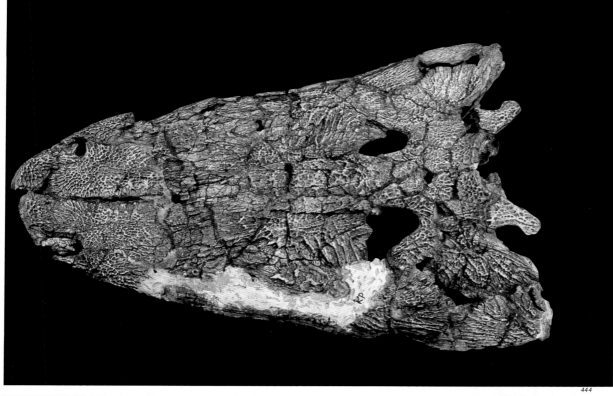

444

444 Dorsal View of the complete skull of Parotosuchus denwai (ISI A54) collected near Purtala village of the Middle Triassic Denwa Formation of the Satpura Gondwana basin, Central India. About 0.1x.

445 Dorsal View of the complete skull of Parotosuchus crook-shanki (ISI A55) collected from Sahavan village of Middle Triassic Denwa Formation of the Satpura Gondwana basin, Central India. About 0.4x.

446 Dorsal View of the complete skull of Parotosuchus maleriensis (ISI A56) collected near Aigerapalli of the Late Triassic Maleri Formation of the Pranhita-Godavari valley, Deccan, India. About 0.3x.

446

445

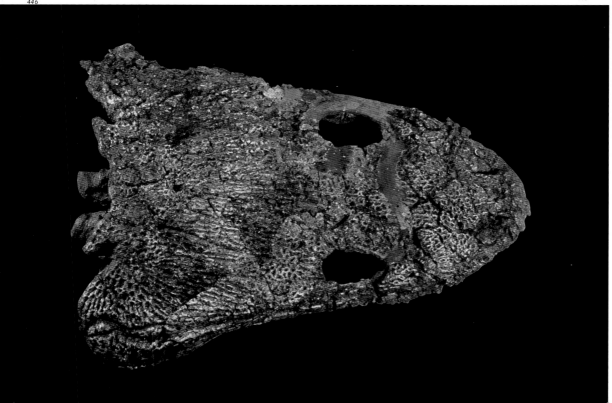

447

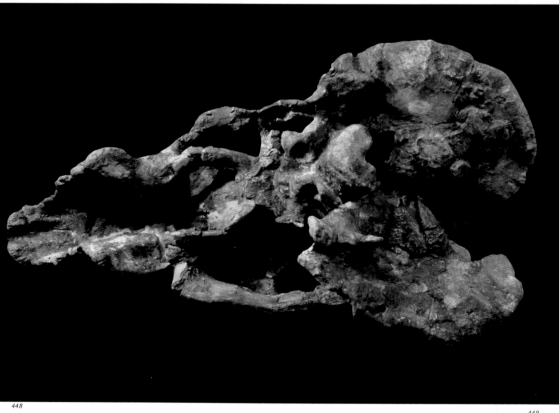

448

449

447 Dorsal View of the complete skull and mandibles of **Compsocerops cosgriffi** (ISI A 33) collected near Rechni village of the Late Triassic Maleri Formation of the Pranhita-Godavari valley, Deccan, India. About 0.25x.

448 Palatal view of the skull of **Wadiasaurus indicus** (ISI R38) collected near Yelkesram village of the Middle Triassic Yerrapalli Formation of Pranhita-Godavari valley, India. About 0.4x.

449 Lateral view of the skull of **Rechnisaurus cristarhynchus** (ISI R37) from Rechni village locality of the Middle Triassic Yerrapalli Formation of Pranhita-Godavari valley, India. About 0.35x.

451

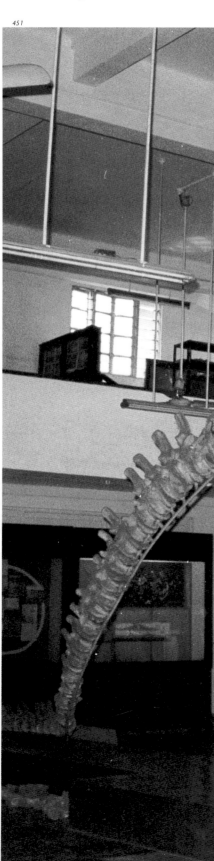

450 Lateral view of complete mounted skeleton of *Paradapedon huxleyi* (ISI R1) collected from Maleri village locality of Late Triassic Maleri Formation of Pranhita-Godavari valley, India. About 0.1x.

451 Complete mounted skeleton of *Bararpasaurus tagorei* (ISI R50-149) collected from Pochampalli village of early Middle Triassic Kota Formation of Pranhita-Godavari valley, India.

450

ACKNOWLEDGEMENTS

The authors acknowledge their gratitude to many people and institutions.

Frank Coffa and Steven Morton meticulously and untiringly photographed most of the specimens in this book, oft-times under difficult circumstances and on a short time scale. Their dedication to this project and unending good humour made this project work. Jim Frazier beautifully photographed several of the specimens in the Western Australian Museum collection.

Peter Trusler painstakingly and with marked sensitivity created the three centerpiece restorations. His combination of skills as a fine artist and trained zoologist has led to the extraordinary beauty and believability of these works. We also appreciate the use of four of Peter's pen and ink illustrations allowed by both his permission and that of Doubleday (New York). Information for the restorations was provided by many people including Andrew Wegener, Nicolas Day, Pierre Testart, Professor Ken Campbell, Dick Barwick, Dr Peter Pridmore, Robert Fencham, Jeff Davies, Erich Hartler, Craig Cleeland, Gael Trusler, Doris Bogie, Bill Trusler, Gary Backhouse, Greg Mayo, Roley Oldricks, Barbara Wagstaff, Dr Jennifer McEwen Mason, Dr Neil Archbold, Tess Kloot, Jack Harrison, Dr. Ralph Molnar, Dr. Alex Ritchie, Robert Jones, and Healesville Sanctuary.

Draga Gelt drafted and computer generated the graphics scattered throughout the book. Steven Morton and Adrian Dyer provided photographic processing needed in this regard.

The Earth Sciences Department of Monash University provided us with a variety of MacIntosh computers and software packages that allowed drafting and manuscript preparation to proceed. Colleagues in the Earth Sciences Department and at the Museum Of Victoria, especially Ray Cas, Lesley Kool, Jennifer Monaghan, Elizabeth Thompson, Mary Walters and Corrie Williams all provided information, proofreading assistance or data gathering. Francis d'Souza gave much assistance with computer software.

A number of scientists gave many hours of their time in providing data on specimens and allowed us unrestricted access to their collections, many specimens of which had never been described before. We are especially grateful to them, for without their enthusiastic co-operation, such a book as this would not have been possible: Max Banks (University Of Tasmania, Hobart); José Bonaparte, Luis Chiappe, Fernando Novas, Maximiliano Rocca, Guellermo Rougier and Graciela Esteban (Museo Argentino De Ciencas Naturales "Bernardino Rivadavia", Buenos Aires); Ken Campbell (the Australian National University, Canberra); Noel Kemp (Tasmanian Museum And Art Gallery, Hobart); John Long and Ken McNamara (Western Australian Museum, Perth); Ralph Molnar (Queensland Museum, Brisbane); Peter Murray (Spencer And Gillan Museum, Alice Springs); Rosendo Pascual, Zulma Gasparini, Eduardo Tonni and Marcello Fuente (Museo De La Plata, La Plata); Neville Pledge (South Australian Museum, Adelaide); Jaime Powell, Norma Nasif and Fernando Abdala (Instituto Miguel Lillo, Tucuman); Alex Ritchie (Australian Museum, Sydney); Chris Tassell (Queen Victoria Museum And Art Gallery, Launceston); Gavin Young (Bureau Of Mineral Resources, Canberra); and Rod Wells (Flinders University, Adelaide).

Several individuals and institutions provided information or actual specimens for photography and/or photographs of specimens or localities, which are essential to the broad coverage in this book: Michael Archer, Sue Hand and Henk Godthelp (University Of New South Wales, Sydney); Ewan Fordyce (University Of Otago, Dunedin); Brian Allison, Robert Green and Chris Tassell (Queen Victoria Museum And Art Gallery, Launceston); Cindy Hann (Wesley College, Melbourne); Leo Hickey (Yale University, New Haven); Miranda Hughes (Deakin University, Geelong); Robert Jones (Australian Museum, Sydney); Anne Kemp and Susan Turner (Queensland Museum, Brisbane); Larry Marshall (Institute For Human Origins, Berkeley); the Museum Of Victoria (Melbourne), especially with the help of Jim Bowler; Rosendo Pascual (Museo De La Plata, La Plata); Neville Pledge (South Australian Museum, Adelaide); Alex Ritchie, Ron Strahan, Hal Cogger and the Photographic Department of the Australian

Museum; Ramiro Soruco (Yacimientos Petroliferos Fiscales Bolivianos, Cochabamba); Richard Tedford (American Museum Of Natural History, New York); and Anne Warren (La Trobe University, Melbourne). Especially appreciated were the large number of slides on the Antarctic provided by Edwin Colbert (Museum Of Northern Arizona, Flagstaff) and a significant collection of large format transparencies of Chinese fossils taken especially for this project by the photography department of the Natural History Museum of Los Angeles County under the direction of Craig Black, all provided gratis. Peter Menzel, Mary Smith and the National Geographic Society allowed the use of photographs especially taken for a NGS project.

The Systematic, Geographic And Geologic Index was compiled by Anne Kemp, Noel Kemp, John Long, Alex Ritchie and Sue Turner (fish), Michael Tyler and Anne Warren (amphibians), Ralph Molnar (reptiles), Bob Baird, Pat Vickers-Rich and Corrie Williams (birds), and Timothy Flannery, Robert Bearlin, and Thomas Rich (mammals) with the assistance of Jennifer Monaghan.

Many others gave their support for this book in a number of ways. The Bureau Of Mineral Resources, Canberra, allowed use of its palaeogeographic map data, which were computer modified using Canvas 2.1 for this book. Marita and John Bradshaw and Trevor Powell of the BMR were especially helpful in this regard. Kay Jenkins and the Shire Of Dundas allowed the photography of the Glenisla trackway, which had been donated to the shire by Mr and Mrs Eric Barber, the owners of Glenisla. Dianne Logg, Denise Krake, Merril Healey and Brian Phillips of the Healesville Wildlife Sanctuary in Victoria gave much assistance in the photography of living Australian birds, mammals and reptiles. Many individuals provided photographs which are specifically acknowledged in the legends for the figures. Leaellyn Rich proofread several versions of the manuscript and together with André Coffa and Lisa Coffa organized much of the slide duplicate collections resulting from book preparation. Greg Taylor Productions of Melbourne allowed the use of their video production studio for the initial layout work with illustrations. Sally Webster of the Historic Houses Trust Of New South Wales provided information concerning and illustrations of Alexander Macleay.

Our thanks also go to scientists who have reviewed individual chapters for accuracy and completeness: Alex Ritchie for Parts One and Two and he and John Long for the chapters on fishes (1 and 2) in Part Three; Anne Warren, Chapter 3 in Part Three; Ralph Molnar, Chapter 4 in Part Three; Timothy Flannery, for chapters 5 and 6 in Part Three. Mary White graciously, but critically, commented on the floristic aspects of the book.

We both feel extremely lucky that Helen Grasswill edited the manuscript, sentence by sentence, thought by thought, in a way few editors manage. We appreciate her dedication and real interest in this book over the months. Equally appreciated is the designing effort by Bruno Grasswill, who successfully sought to understand and sort through more than 1000 illustrations provided for this book and place them in context with the words, and who also created most of the diagrams.

Bill Templeman, Publisher, Reed, Sydney, has our greatest respect for making this project work commercially. He provided the financial base for the book in monetarily difficult times, maintained both a determined schedule and an enduring interest in the project, and still kept his sanity and good humour. He has our sincere appreciation for all he has done.

Finally, our deepest appreciation to Mary White and Mary Vickers Macdonald. Mary White, author of *The Greening Of Gondwana*, gave us the inspiration to write this book, and together with Alex Ritchie suggested that we undertake the project. Mary provided much moral support and critical discussion, in addition to meals and a peaceful home where we could retreat occasionally. Mary Vickers Macdonald, Pat's mother, has given both of us incalculable support for the past few decades, both personally and with our scientific work. She was deeply involved in the technical aspects of manuscript and illustration preparation as well as proofreading for this book over the past two years. It is most appropriate, then, that this book is dedicated to these two Marys.

GLOSSARY

absolute dating A method which determines the time order in rock sequences, measured in years by radiometric techniques—that is, those techniques which depend on the regular and statistically predictable decay of radioactive atoms. This method is different from *relative dating*, which is based on fossil content or the order in which rock bodies are stacked one on top of the other.

acellular bone Bone lacking spaces to house bone secreting cells.

ammonites Extinct invertebrates in the phylum Mollusca, class Cephalopoda (which includes the squids and octupi), with shells divided into distinct chambers. The walls separating the chambers are complexly folded. The ammonites first appeared in the Palaeozoic Era and persisted until the end of the Mesozoic Era.

amniotes Reptiles, birds and mammals; all vertebrates that have an egg with special structures that allow development of the embryo independent of water. The amniote egg contains specialized membranes (yolk sac, amnion, chorion and allantois that respectively provide nutrients, protection from desiccation, protection from physical destruction and waste collection) for the developing embryo.

angiosperms Flowering plants, which have the seed enclosed in an ovary—making up the fruit. The group ranges from the Cretaceous, possibly Late Jurassic, to the present day. They are now the dominant plants on land.

ankylosaurs Armoured bird-hipped dinosaurs with the entire body covered by a mosaic of tiny, interlocking bony plates. Includes two groups: the ankylosaurids, which possess a bony club on the end of the tail; and the nodosaurids, which lack this structure.

annelids Worm-like invertebrates with a segmented body; they posses chitinous jaws, called scolecodonts, that are often the only fossil remains of this group other than burrows left behind in the sediments.

antorbital fenestra An opening in the skull of some vertebrates, lying in front of the orbit (eye socket). Along with other openings (temporal fenestrae) in the skull it is especially useful in determining the relationships of many reptile groups.

apatite A group of minerals with six-fold crystal symmetry. Composed of calcium, phosphate, carbonate, fluoride, and chloride, apatite has the general chemical formula $Ca_5 (PO_4, CO_3)_3 (F, OH, Cl)$. It is the basic hard constituent of vertebrate bone.

Apollo objects Extraterrestrial objects that from time to time cross the Earth's orbit, such as meteorites, comets, asteroids, etc.

Archean A period of time (an Eon) early in the Earth's history, from 4000 million to 2500 million years ago.

aridification Desiccation; increased aridity over a period of time due to a decrease in effective rainfall.

arthropods Invertebrates in the phylum Arthropoda ("joint footed") characterized by their jointed appendages (limbs, antennae and mouth parts) and segmented bodies. Examples include insects, trilobites and crabs. They range from the Cambrian to the present.

astragalus An ankle bone with which the lower limb (the tibia) articulates.

atlas vertebra The first vertebra in the vertebral column; articulates directly with the occipital condyle on the back of the skull.

Australasia The biogeographic region that includes Australia, New Guinea, New Zealand and the south-western Pacific area.

australopithecines A group of primates from which humans descended.

axis vertebra The second vertebra in the vertebral column; articulates with the atlas in front and the first cervical vertebra behind.

basalt A dark coloured rock with a fine-grained crystalline structure, generally composed of such minerals as olivine, quartz and hypersthene, but often with associated apatite and magnetite. Can be formed from the eruption of a volcano (extrusive formation) or underground (intrusive). Solidifies quickly from a melt, which accounts for the small size of the crystals.

Bering Land Bridge A terrestrial link between eastern Asia and Alaska which emerged several times during the Pleistocene (the last 1.64 million years) as sea levels fell in response to increased amounts of sea water being frozen into the huge ice caps that at times covered much of North America, Europe and Asia.

BIFs Banded iron formations deposited by microorganisms prior to about 2500 million years ago when atmospheric oxygen was decidedly lower than present levels.

Big Bang Hypothesis The idea that the expansion presently observed in the Universe can be extrapolated back to a time when it first began: the idea of a primaeval cosmic fireball.

biostratigraphy The ordering of rock sequences based on the assessment of their fossil content. Certain fossils occur only during specific time intervals, thus their first and last appearances, and joint occurrences, can be useful in determining the contemporaneity of widely separated sedimentary sequences.

biota The sum total of animals and plants in a certain area.

biotic diversity The number of different species of living organisms.

black coals The most highly altered (metamorphosed) of coals, which are derived from plant material that has been subjected to heat and pressure during burial.

brachiopods (articulate, inarticulate) Invertebrates in the phylum Brachiopoda ("arm footed"). They are solitary, bivalve marine organisms whose two shell parts are not mirror-images of each other, thus differing from clams. Generally they are attached to the sea floor or some other object by the pedicle, a muscular organ. Articulate brachiopods possess shells that have a complex hinge area where the two shell parts articulate; inarticulates have a smooth surface in this area, the shell parts being held together by musculature.

branchial skeleton (visceral skeleton) A series of cartilaginous or bony supports between the gills.

branchiostegals (gill bars) The cartilaginous or bony structures that make up the branchial skeleton, each with a dorsal (top) and a ventral (bottom) segment—the epibranchial and ceratobranchial respectively.

bryozoans Invertebrates in the phylum Bryozoa ("moss animals"), characterized by their small size, their habit of growth in colonies, a U-shaped alimentary canal with a mouth and an anus, and their feeding structure called a lophophore which waves about the mouth collecting food particles—this latter they share with the brachiopods.

Cainozoic Era ("Modern Life") The period of time between 65 million years ago and the present.

calcaneum A bony element of the ankle which articulates with the astragalus and forms the heel in many vertebrates.

Carbon-14 dating A radiometric dating technique which utilizes the known decay rate of the unstable Carbon 14 isotope into its stable daughter product, Nitrogen 14. This method gives an absolute age in years for once living organisms such as shells, plant material and bones which incorporate carbon into their cellular structure. Once the organism dies, no further carbon is incorporated, and thus the ratio of the radioactive Carbon 14 to the stable carbon isotopes, Carbon 12 and Carbon 13, will allow determination of the time of death.

cartilage A flexible, often translucent material that consists of a sulphated polysaccharide, which forms support structures in vertebrates. Through the gel-like matrix of cartilage are spread a network of connective tissue fibres and rounded spaces for cartilage-depositing cells. Several types of cartilage are known, including some that have calcium carbonate salts deposited in them, giving added strength.

caudal fin Tail fin.

choanae (internal nostrils) An opening or openings in the palate of some vertebrates, such as crossopterygian fishes and amphibians. In some crocodilians and all mammals such internal nostrils are absent, the air channel from the nose having been completely shut off from the mouth cavity by the development of a secondary palate. This structure allows eating and breathing to occur at the same time.

choanates Vertebrates that possess internal nostrils.

concretions Hard and compact, usually rounded structures most often formed by precipitation of minerals from an aqueous solution, often centred on some object such as a fossil. Their composition can vary greatly, but carbonate and silicate concretions are common. They are often, because of their unusual external form and frequently layered internal structure, misidentified as fossils.

Continental Drift The theory that continents on the Earth's surface have not always been where they are today, but instead have drifted about in times past.

conularids Invertebrates known from Cambrian to Triassic times. They have chitinous shells (cone-shaped, and square in cross-section) and are thought to be jellyfish relatives.

core The central mass of the Earth, about 7000 kilometres in diameter. It is composed of a solid Inner Core and a liquid Outer Core. The Earth's rotation most likely causes the liquid core to circulate, which in turn generates the magnetic field of this planet.

cranial (cranium) Skull or parts of the skull.

crustal plates (lithospheric plates) Major subdivisions of the Earth's outer skin, the Crust, which are in constant movement relative to each other.

Cruziana The trace fossils (trackways) thought to have been left behind in sediments by the movement of trilobites across the sea floor; known only in rocks of Palaeozoic age.

cryptodire Turtle in the suborder Cryptodira, characterized by having a neck that is retracted into the shell by vertical flexure. A second living group of turtles, the Pleurodira (side-necked turtles), retract their necks by lateral flexure. Both of these turtle groups had appeared by the end of the Jurassic.

cuboid A bone of the ankle in some terrestrial vertebrates.

Curie Point The temperature above which thermal agitation does not allow the magnetic minerals to be oriented in line with the prevailing magnetic field of the Earth.

cursorial Running, referring to organisms adapted to fast movement over open expanses.

cyclothem A series of sedimentary beds deposited during a single sedimentary cycle that represents the transgression (incoming) and regression (retreat) of the sea relative to land. Terrestrial sediments (such as coals, and river- and lake-laid sediments) characterize the lower half of the cyclothem, and marine rocks the top half.

Deep Sea Drilling Project (JOIDES) A long term project begun in the 1960s with the aim of drilling deep cores in the sea floors of the world in an attempt to understand the origin and evolution of the oceans.

dentary A bone of the lower jaw in vertebrates; the only bone in the lower jaw of most mammals.

dermal bone (membrane bone) Bones deposited in the skin of vertebrates, as opposed to endochondral bones which are formed as replacements to a cartilaginous precursor. In fish and amphibians, dermal bones form over most of the body, but in higher vertebrates such as birds and mammals they are normally found only in the skull, jaws and shoulder girdle.

dolerites A dark coloured intrusive rock formed mainly of the minerals labradorite and pyroxene, which are silicates rich in iron and magnesium.

dolomite A common mineral in sedimentary rocks, consisting of calcium, magnesium and carbonate $(CaMg(CO_3)_2$. It can be formed by direct precipitation or as an alteration product of limestone metamorphosis.

dorsal The top part of an object; upper.

echinoderms Solitary marine invertebrates in the phylum Echinodermata ("spiny skinned"), characterized by having radial symmetry, an internal skeleton formed of small calcareous plates and a water vascular system which is used in feeding and locomotion. Echinoderms range from the late Precambrian to the present.

echinoids Echinoderms in the class Echinoidea, characterized by a spherical shape, a skeleton of calcareous plates that interlock, and movable appendages; includes such forms as sea urchins and sand-dollars.

ecomorphic types Organisms that despite their ancestry take on certain kinds of morphology when affected by similar environmental conditions.

ecospace Space available in which the living animals and plants—the biota—can exist at any one time.

electrosensory system A series of soft tissues, occupying the canals incising the bones of many fishes and amphibians, which were sensitive to electric fields in the aqueous environments that these vertebrates inhabited.

endemic Native to and restricted to a particular place or environment.

endochondral bone Bone formed as a replacement of an embryonic cartilage precursor. Examples are the long bones of the limb such as the humerus and the femur.

eucaryotes Organisms that have a nuclear membrane separating the DNA from the cytoplasm of the rest of th cell.

euryhaline Having a wide salt concentration tolerance; in contrast to stenohaline, or narrow salt concentration tolerance. Brachiopods and trilobites, for example, are and were stenohaline organisms, being able to survive only in marine environments; whereas bivalves are often more euryhaline, having the ability to tolerate a wider range of salt concentrations.

evaporites Salts that are precipitated and left behind as a water body evaporates; examples are the deposits in salt or playa lakes (Lake Eyre) and in coastal lagoons.

evolutionary explosion (adaptive radiation) Rapid development of new kinds of organisms as a result of some significant change in genetic make up or an environmental change.

evolutionary rates The speed at which genetic change, reflected by the development of new structures and new species, occurs.

exoskeleton An external skeleton such as that of arthropods and the bony armour of placoderm fishes, in contrast to an internal skeleton such as that supporting the internal organs and muscles (vertebrae, limb bones, etc) in vertebrates.

facies The sum total of all the lithologic (that is, rock types, such as sandstone, siltstone, etc) and palaeontologic (fossil content) characteristics of a sedimentary rock body.

filter feeding A feeding style which involves the removal of particles from the water by means of some specialized structure, such as the lophophore in bryozoans and brachiopods or the baleen in whales.

fission tracks Paths of radiation damage made by the passage of nuclear particles in mineral structures or natural glass, caused by the spontaneous splitting of Uranium 238 impurities. Fission track damage can be assessed only after the crystals have been acid etched and observed under a microscope. The density of the tracks in such crystals gives a way of determining the age of the containing rocks.

furcula The fused clavicles, or "wishbone", forming part of the shoulder girdle in birds and some reptiles.

Ga (Giga years) 1000 million years.

ganoine A shiny enamel-like material that forms the outer layers of scales in many primitive fishes.

geographic barriers Obstacles such as mountain ranges, water bodies, etc., that can prevent organisms from moving from one part of the globe to another.

gill bars See **branchiostegals.**

glacial cycle The cycle that develops during glacial times, whereby glaciers build up due to many factors including increased precipitation with concomitant lowering of sea level and temperature, which is followed by a warmer period, an interglacial, when glaciers retreat. During any glacial period these cycles repeat themselves many times.

glacial dropstones Rocks that have been carried in icebergs and dropped as unusual (erratic) elements onto the lake floor or sea floor.

Glossopteris A seed-fern that is characteristic of the sediments being deposited on Gondwana during the Permian Period.

Gondwana (formerly Gondwanaland) A massive supercontinent of the past that included South America, Africa, Antarctica, Australia, New Zealand, India and parts of China. It existed during the late Palaeozoic but began to fragment during the Mesozoic.

gnathostomes Vertebrates with jaws, including some fishes and all higher vertebrates.

granite An intrusive igneous or metamorphic rock (that is, rock formed underground) that results from either the cooling from a molten condition or the alteration of some previously existing rock through metamorphism (application of heat and pressure); granite is composed mainly of the light coloured minerals, quartz and potash (high content of potassium), feldspar (orthoclase) and some biotite mica.

graptolites Extinct invertebrates belonging to the phylum Graptolithina that lived in colonies. Graptolites had exoskeletons composed of a hard tissue resembling that of vertebrates. They often grew in branching colonies; some species were attached bottom dwellers while others were planktonic. They range from the Cambrian to the Carboniferous, and planktonic forms are useful biostratigraphic tools.

greenhouse effect An increase in temperature due to the build-up of gases which prevent heat loss into space. The Earth has experienced greenhouse conditions several times in the past, the most notable being during the Mesozoic when dinosaurs dominated terrestrial vertebrate faunas. It appears that the Earth is now headed towards another greenhouse period, a circumstance accelerated by the activities of humans in their burning of fossil fuels.

half-life The time it takes for one half of the atoms in a sample of a radioactive isotope, such as Carbon 14, to decay.

hematite A common iron mineral with a composition of Fe_2O_3; a principle ore of iron, it gives color to ochre and many weathered soil profiles.

hexacoral (scleractinian coral) Corals belonging to the order Scleractinia, which contains both solitary and colony-forming individuals that possess a six-fold symmetry in both hard and soft parts. They range in age from the Triassic to the present and have dominated reef environments through much of the Mesozoic and the Cainozoic.

hipposiderid bats A group of cave dwelling, generally tropical bats (the leaf-nosed bats) almost entirely restricted to the Old World and with a long fossil record reaching back into the early Cainozoic. Six species in two genera live in Australia today.

holotype A specimen designated to characterise a species or subspecies of organism at the time it is originally described.

horse latitudes Oceanic areas between 30° and 35° North and South that are typically characterized by calms or only light winds, heat and aridity. They are the areas of descending airmasses with little contained moisture, and they move north and south (about 5°) each year with the Sun.

humerus The upper arm bone in vertebrates, which articulates proximally (bodyward) with the shoulder girdle and distally (handward) with the radius and ulna.

humic acids Black, highly acidic organic matter that is found in soils and decaying plant matter including some coals (for example, brown coals that have not been highly metamorphosed).

hydrocarbons Compounds composed mainly of hydrogen and carbon, examples being the liquid hydrocarbon oil and the solid hydrocarbon coal.

hydrosphere The water bodies on the Earth's surface—the oceans, lakes and rivers.

hypobranchial musculature The musculature that attaches and controls the gill bars (or branchiostegals).

hypsilophodont dinosaurs ("high lophed teeth") Ornithischian or bird-hipped dinosaurs that lived during the Jurassic and the Cretaceous nearly worldwide. Generally small dinosaurs, light bodied and cursorial; bipedal reptiles with four hind toes and five fingers in the hand. They were herbivores with teeth reminiscent of the plant-eating iguanid lizards. In Australian polar faunas of the Early Cretaceous this group is unusually diverse compared to its occurrence in other parts of the world.

hystricomorph rodents Rodents in the order Hystricomorpha which are unique in the structure of their tooth enamel and in that the skull has a foramen or opening below the orbit (eye socket) which is quite large and through which passes a part of the masseteric jaw musculature that originates on the snout and inserts on the lower jaw. Both Old and New World porcupines are in this group as are the guinea pigs. Hystricomorphs first appeared in the Late Oligocene of both Africa and South America.

icehouse conditions Times in the Earth's history when glacial conditions existed, for example during the late Precambrian, the late Palaeozoic, and the Pleistocene.

indigenous Native to an area.

Inner Core Innermost part of the Earth's Core. Extending from a depth of about 5100 kilometres to the centre at 6371 kilometres, its radius is about a third of the entire Core. Based on the study of earthquake waves and their behaviour as they pass through this part of the Earth, the Inner Core is thought to be solid; it is probably composed of pure iron.

intrusive bodies Bodies of rock that have been injected or emplaced within pre-existing rocks. An example is a granite body such as those that often form the cores of mountain ranges.

invertebrates Animals lacking a true backbone.

kerogen Fossilized and insoluble organic material found in sedimentary rocks, most usually in shales. It can be distilled to form petroleum products. The colour of the kerogen impurities in rocks can give some idea of how much heat and pressure have affected them, and thus give some idea of how deeply the sediments have been buried underground.

kimberlite A rock with a wide range of crystal sizes (porphyritic). It is composed of dark minerals such as olivine, which is an iron-magnesium silicate $[(Mg, Fe)_2SiO_4]$ that forms under high pressure conditions associated with diamond formation. Named for the well-known kimberlites in the Kimberley region of southern Africa.

lacustrine Of lake origin.

lag deposit Deposits left behind after the processes of weathering and erosion. An example is the gibber plains, where water and wind have worn away all fine-grained and soluble materials and left behind a pavement of silica-rich rocks.

lampreys A group containing one of the few living jawless fishes (class Agnatha). Eel-like in appearance, they are scaleless and lack bones, having only a cartilage skeleton. Adult lampreys are predators that feed on other fishes by attaching to them and using their rasping tongue-like structure within the mouth to slowly eat their prey.

lateral line system A system of interconnecting canals, reflected in the channels preserved on the bones of fossil fishes and amphibians, that must have had some sensory function, perhaps picking up changes in the electric field in the aqueous environment within which these vertebrates lived. The pattern of the lateral line canals on the bone surface can be very characteristic of individual species of fossil organisms and thus useful in classification.

Laurasia A great Northern Hemisphere supercontinent of the past, comprising most of North America, Greenland, Europe and Asia (excluding India); a geographic entity distinct from Gondwana.

Law Of Superposition The practical observation that the oldest rocks lie on the bottom of a rock sequence and the youngest on the top, provided there is no structural reason to think that the sequence has been overturned. In order to decide if there has been "post-mortem" alteration it is important to have a good idea of the geology of the region being studied.

limestones Sedimentary rocks with a composition dominated (more than 50 per cent) by calcium carbonate. Deposited in both marine and freshwater conditions, though the majority are marine. Limestones include chalk and coquina (made up of a large proportion of fragmental limestone and fossil shells). If a drop of acid is put on limestone, it effervesces.

lithospheric plates See **crustal plates.**

lumpy jaw A diseased condition in vertebrates that leads to the deposition of excess bone in the jaws, thus leaving behind a lumpy appearance. Very often a condition associated with highly stressed populations of animals such as might be found in times of excessive drought.

lycopsids (lycopods, club mosses) Primitive vascular, spore-bearing plants (those with vessels that carry water and nutrients from the roots upwards into the plant) characterized by dense, spirally arranged leaves. They were among the first plants to colonize the terrestrial environment and by some time in the Devonian were forest formers. Although diverse in the Palaeozoic, they declined in diversity in Mesozoic and Cainozoic times and are now represented by only a few herbaceous forms such as *Lycopodium.*

macrofossils Fossils that can be discerned with the naked eye, without the aid of a microscope.

magnetic anomalies Differences in the magnetic fields recorded over the surface of the Earth which reflect, in many cases, changes in the prevailing direction of the magnetic field in times past. In ocean basins these anomalies are seen as linear patterns in ocean floor basalts, their magnetic minerals having aligned themselves with the prevailing magnetic field of the Earth at the time the basalts solidified.

magnetite A black, highly magnetic mineral containing iron, magnesium and oxygen $[(Fe, Mg)Fe_2O_4]$.

Mantle The part of the Earth below the Crust and above the Core, extending to a depth of 3480 kilometres.

marsupials Mammals that give birth to their young at an early stage of development and then suckle them generally in a pouch until complete development has occurred. In addition to reproductive differences, they have skeletal and dental characteristics that for the most part clearly separate them from the placentals.

maxilla The upper jaw.

median dorsal plate One of many bony plates, made up of dermal bone, that form the head and thoracic armour of some fishes, in this case of placoderms.

mesosaurs The oldest amniotes to become fully aquatic. Reptiles with a primitive skull structure lacking any skull fenestration, from the Permian of southern Africa and eastern South America. The skull of these moderate-sized reptiles (about 1 metre in total length) was elongated and the jaws were filled with a large number of long slender teeth. Both the neck and the tail were elongated. The freshwater mesosaurs have been used as evidence suggesting that Africa and South America were closely apposed at the time these little reptiles existed.

Mesozoic ("Middle Life") The period of time between 251 million and 65 million years ago.

microfossils Fossils that are so small they must be studied with the aid of a microscope.

moas Moderate sized to large ground-dwelling birds that lived during the Pliocene and the Pleistocene in New Zealand. Their exact relationships to other birds are in debate, even though traditional classifications ally them to the ratites—kiwis, ostriches, rheas, cassowaries and emus.

molluscs Solitary, mobile invertebrates belonging to the phylum Mollusca. They have non-segmented, bilaterally symmetrical bodies. Included in this group are clams, snails, octopuses, squid and several now-extinct groups like the belemnites and ammonites.

multituberculates Primitive mammals that lived during the Mesozoic and early Cainozoic (until the Oligocene), which had distinctive molars made up of two or three rows of cusps and a premolar commonly with a side to side flattened blade. They were the most varied of the Mesozoic mammals and may have been displaced in the middle Tertiary by the rodents, which moved into the fruit and seed-eating niches that must have been occupied earlier by the multituberculates. They ranged from mouse-sized to possum-sized and were probably tree-dwellers.

Natural Selection The process whereby the offspring of certain individuals ensure the perpetuation of a species because they possess some characteristic that gives them a slight advantage of survival. It works in much the same way as the artifical selection that humans impose on domestic animals, only the natural environment is the controlling agent. A theory originally articulated by Charles Darwin and Alfred Russel Wallace in the mid-nineteenth century.

neural crest Embryonic tissue that for the most part gives rise to the nervous system.

niche The "job" an organism performs in a community (predator or prey, carnivore or herbivore), what types of locomotion styles it exhibits, and the like).

notochord ("back chord") A long flexible rod that is the main structural support of the trunk in embryonic and many fish groups. It is the basis of the name of the phylum to which vertebrates belong, the Chordata, which includes not only the vertebrates but some forms intermediate between these "backboned" animals and invertebrates.

nuclear membrane A membrane that separates the nucleus and thus the genetic material of the cell from the cytoplasm of the rest of the cell. Eucaryote organisms, but not procaryotes, possess this membrane.

Oxygen 18 to Oxygen 16 ratio The ratio of two isotopes of oxygen determined by analysis of carbonate concretions and fossils. It is important in determining the water temperature at the time the sedimentary rocks containing the concretions or fossils were deposited. Based on studies of these ratios in today's oceans, it is clear that they are temperature dependent.

occipital condyles Articular structures, on the back of the skull in vertebrates, that connect with the atlas vertebra of the backbone.

occlusal The chewing surface of a tooth.

olfactory epithelia A regular and compact arrangement of cells in a sheet that lines the part of the body associated with detecting smell; in the nasal (nose) region.

ophiolites Rocks made up of dark minerals such as basalt and gabbro (rich in serpentine, chlorite, epidote), clearly metamorphosed (altered from their original composition by the application of heat and/or pressure). They are associated with tectonic trenches where one great lithospheric plate of the Earth's Crust is sliding beneath another.

optic lobes That part of the brain whose function is to process signals picked up by the eyes.

ostracod An aquatic crustacean (an arthropod), belonging to the subclass Ostracoda, which has two shells and is made up of for the most part of calcium carbonate. Most ostracods are microscopic and occupy marine environments, but some freshwater forms may reach up to 30 millimetres in length. They range from the Cambrian to the present.

otic notch A notch in the skull of some primitive tetrapod vertebrates, located behind the orbit (eye socket); the site of attachment of the tympanic membrane, the eardrum.

outcrops The exposed areas of rock bodies, not covered by vegetation and soils.

Outer Core The outer part of the Core of the Earth, extending from the base of the Mantle to a depth of 5100 kilometres. It is presumed to be liquid, based on the characteristic behaviour of earthquake waves, some of which (shear waves) will not pass through it.

paedomorphosis The retention of juvenile characteristics by adult breeding individuals, such as the retention of external gills in some forms of salamander.

palaeopoles Fossil polar positions. Because continents have moved position relative to one another and relative to the magnetic field of the Earth, the positions of the North Pole and the South Pole over the past 4 billion years have changed relative to the continents; these fossil polar positions are called palaeopoles.

Palaeozoic Era ("Ancient Life") A period of time between 545 million and 251 million years ago.

paludal Of swamp origin.

Pangaea A megacontinent of the past, comprising all the continents we know today.

pantotheres Small, mouse- and rat-sized Mesozoic mammals which were probably the ancestral stock that gave rise to most living mammals.

paratheres Edentates or primitive placentals that lack enamel on their teeth. They include sloths, armadillos and a variety of now extinct animals.

pectoral fins The paired fins that attach to the shoulder girdle. The front-most pair of fins.

phalangeal count The number of individual bones in each digit, beginning with those in the first digit. For example, the count 2-3-3-3-3 would indicate two bones, or phalanges, in the first finger or toe, and three in each of digits 2 to 5.

Phanerozoic Eon The period of time between 545 million years ago and the present, when fossils have been abundant and organisms have had well developed hard parts.

pharangeal cavity The space inside the pharynx (which is a segment of the digestive tract between the mouth and the esophagus), where the glottis opens to the lungs and dorsally to the eustachian tubes which lead to the middle ear cavities. It is of minor importance in higher vertebrates such as mammals and birds, but in the lower vertebrates such as fishes it is very important because it is the area in which the gills developed, structures essential for respiration in these forms.

phytosaurs (parasuchians) Crocodile-like, highly carnivorous reptiles whose nasal openings are situated far back near the orbits instead of opening at the end of the snout. They were abundant in the Late Triassic of western North America, Europe, Madagascar and India. So far the group has not been found on most of the Gondwanan continents.

pillow lavas Lavas erupted in aqueous environments (seas, lakes) which take on a pillow-like shape and have a distinctive internal structure produced by the hot magma collapsing inwards, as it cooled very rapidly upon coming into contact with water. A common kind of lava found along mid-ocean ridges.

pineal eye A single sensory organ located in the middle of the dorsal surface of the skull in many lower vertebrates, including reptiles. In higher vertebrates this structure becomes the pineal organ which has no external opening, although in lower vertebrates it is light sensitive.

placentals Mammals that develop a specialized tissue, the placenta, which nourishes the developing embryo. In contrast to marsupials, placental females carry their young inside the womb for much longer periods before birth.

Plate Tectonics The theory that the Earth's Crust is divided into a series of large lithospheric plates which are in constant motion with respect to one another. An explanation for the apparent movement of continents relative to one another, for the formation of ocean basins and mountain ranges, and for the distribution of earthquakes, volcanoes and uneven heat flow on the Earth's surface.

playa lakes Dry lakes that are the result of total evaporation of the water that may periodically fill them. During times of desiccation such lakes concentrate a number of salts; an example is Lake Eyre in Central Australia.

polar wander path A path constructed by plotting through time the apparent position of the North Pole or the South Pole relative to any one continental mass.

polar wandering If the continents are retained in their present positions and the position of the North Pole or the South Pole is determined for different times in the past, it appears that the Pole has wandered about. In fact, what appears to have happened is that the continents, and not the Poles, have moved.

polarity The direction of the prevailing magnetic field of the Earth which is generated by the electrical current set up as a result of the rotation of the planet's solid iron-rich Core. For some not quite understood reason, periodically this electric current and thus the sense of the magnetic field that it creates reverse direction; that is, reverse polarity.

potassium-argon dating A method of radiometric dating that depends on the radioactive decay rate of Potassium 40 to the gas Argon 40. A method applied to materials generally more than a million years old.

premaxilla A bone of the vertebrate upper jaw that lies in front of the maxilla, often forming the front end of the upper jaw and often bearing teeth.

procaryotes Organisms which lack a nuclear membrane to separate the genetic material (DNA) from the rest of the cell. Examples are bacteria and blue-green algae.

psychrosphere Cold, descending watermasses that develop around Antarctica and spill northwards into the ocean basins, causing upwelling of watermasses further north.

quadrate A bone of the skull of vertebrates that is involved in jaw articulation in lower vertebrates and birds but becomes a part of the middle ear (incus) in mammals.

relative dating The use of fossils or sequences of rocks to determine the age of one rock body relative to another, in contrast to absolute dating where the age is calculated in years.

retroarticular process A process on the back part of the lower jaw in vertebrates for the attachment of muscles which open the jaw upon contracting.

rift valley An area where the Crust of the Earth is being stretched and thinned, due to two lithospheric plates spreading apart from one another. Examples are the Rift Valley of East Africa and the Red Sea. These are areas of high heat flow as new molten material is being added from below, and of earthquakes and tensional faulting.

rudists A group of bivalve molluscs, belonging to the clams, that have two very unequal shells and take on the appearance of solitary corals. They were major reef formers and lived from the Late Jurassic to the Late Cretaceous, perhaps into the Palaeocene.

sandstone A sedimentary rock that is made up primarily of sand-sized particles and cemented in most cases by silica, iron oxide or calcite.

scapulocoracoid A fusion of two bones of the shoulder girdle, the scapula and the coracoid. This fusion often occurs in birds that have lost the power of flight.

sclerotic capsule A capsule surrounding the eyeball that is often reinforced with bony plates for the protection of the eye.

Sea Floor Spreading The theory that the addition of new molten volcanic material along mid-ocean ridges causes the growth of the ocean basins over time. This idea forms a part of the more wide-ranging Theory Of Plate Tectonics.

seed-ferns (pteridosperms) Gymnospermous plants (those with naked seeds) with fern-like foliage and true seeds borne on the leaves, not in cones. They ranged in time from the Devonian to the Mesozoic.

seismology The study of earthquakes.

shocked quartz Quartz that has been affected by high pressure which has left its effects in the microstructure of the crystalline make-up of this mineral. Attributed by many to the effects of

the crash of an Apollo object on the Earth's surface, but explained by others as due to high pressure vulcanism.

shoulder girdle The bones that make up the shoulder region of vertebrates, including such bones as the scapula (shoulder blade), clavicles (collar bones) and coracoids.

stabilist theory The idea that continents have never moved, but have always remained in their present positions. This was a prevailing idea in science until the mid-twentieth century when it was superseded by the Theory Of Continental Drift.

stapes One of the bones of the middle ear in mammals; involved in the transmission of sound in most vertebrates.

sternum The breastbone.

streptostyly Due to the loss of the lower temporal bar in some reptiles, such as lizards and snakes, the quadrate which attaches the upper and lower jaws has a great freedom of movement, a condition called streptostyly. In lizards and even more so in snakes, this condition allows a wide opening of the mouth thus enabling ingestion of very large prey relative to the actual size of the predator—a major factor in the success of these groups.

stromatolites Layered, frequently calcium carbonate-rich structures deposited by mats of algae and bacteria that occur most frequently in marine environments but can form in fresh waters. Were particularly abundant during the Precambrian, but still form today in places like Shark Bay in Western Australian and in the Persian Gulf.

subduction zones Parts of the lithospheric plates of the Earth's surface that are plunging beneath one another. Areas where subduction zones occur are characterized by oceanic trenches, deep earthquakes and vulcanism, such as the area around Japan and the Tonga Trench.

swim bladder An elongate sac that develops as a dorsal outgrowth from the front part of the digestive tract in some fishes. This sac is inflatable and is filled with some type of gas, sometimes air. It serves as a hydrostatic organ; that is, by filling or emptying the sac, the specific gravity of the fish is changed and thus it aids in regulating the depth at which the fish operates.

symmetrodonts A primitive group of Mesozoic mammals which range from the Late Triassic to the end of the Cretaceous but are very rare as fossils. The name of this group arises because of the symmetrical appearance of the upper and lower molar teeth which have a triangular arrangement of the cusps but are compressed from side to side.

tarsometatarsus The fusion of some tarsal bones with the metatarsals, as occurs in birds, producing the most distal of the leg bones. A bone that is very useful in identifying fossil bird groups because it is so complex and because it is so densely ossified that it is commonly preserved as a fossil.

tectonic trench See **subduction zones.**

tensional features Features created when material is being stretched, such as in the case of two

crustal plates pulling apart from one another. Rift valleys are examples of tensional features.

Tethys A sea, similar to the modern Mediterranean Sea, that was present for long periods of geologic time between the northern and southern continents in the Eastern Hemisphere. It existed primarily from the Permian to the early Cainozoic.

thoracic armour Heavy bony plates that cover the trunk region of many early fish groups. This armour would have protected most of the essential organs of these animals.

thoracic vertebrae Vertebrae of the trunk of an organism, situated between the neck vertebrae (cervicals) and those that connect with the pelvis, the sacral vertebrae.

tibiotarsus A fusion of the tibia and some tarsal bones characteristic of birds.

tillite A sedimentary rock formed by unsorted and unstratified material picked up by a glacier as it moved across a land surface. When the glacier melted it deposited all the sedimentary particles, varying from clay to boulders, that it once contained.

titanosaurs A group of large herbivorous, quadrapedal sauropod dinosaurs that are primarily of Cretaceous age and mainly restricted to the southern continents. A few specimens are known in North America and Europe.

transform faults Faults that offset tectonic ridges (areas of divergence of crustal plates where new molten material is being added from below) or tectonic trenches (places where crustal plates converge and one dives beneath the other).

triconodontans Primitive Mesozoic mammals which existed from the Late Jurassic to the Late Cretaceous. The teeth of these mammals comprise three main cusps arranged in a linear array, the teeth being compressed laterally (from inside to outside). They had one or two pairs of incisors, a pair of canines, as many as four pairs of premolars and up to five pairs of molars.

trilobites Marine arthropod invertebrates belonging in the class Trilobita. They had a segmented external skeleton divided into three distinct lobes a central axial lobe and two pleural lobes, one to either side of the central—and the body was divided into a head, thorax and tail. They ranged from the Early Cambrian to the Permian and were responsible for leaving behind trackways called *Cruziana*.

ventral The bottom side.

weathering Destructive processes where earthy and rocky materials upon exposure to the atmosphere at or near the Earth's surface are altered in colour, texture, composition, firmness or form with little or no transport of material involved.

zebra stripes Magnetic anomalies on the sea floor.

zircon A mineral that has a tetragonal crystal structure and comes in a variety of colours (from brown to colourless). Common in igneous and metamorphic rocks, it has the chemical composition $ZrSiO_4$. Often used in dating rock sequences.

SYSTEMATIC, GEOGRAPHIC AND GEOLOGIC INDEX

Included in this index are all those genera of vertebrates that have a pre-Holocene record in Australia, New Guinea, New Zealand and the southwest Pacific. If a Holocene history is known for those taxa with a longer history, that is included, but forms with only a Holocene record are not incorporated. Families and genera yet unnamed are mentioned only where they significantly extend the range of the group. Due to insufficient time we have been unable to check for recent occurrence of many New Zealand and Southwest Pacific fish genera. This index was compiled by Anne Kemp, Noel Kemp, John Long, Alex Ritchie and Sue Turner (fish), Michael Tyler and Anne Warren (amphibians), Ralph Molnar (reptiles), Bob Baird, Pat Vickers-Rich and Corrie Williams (birds), and Timothy Flannery, Robert Bearlin, and Thomas Rich (mammals) with the assistance of Jennifer Monaghan and Pat Vickers-Rich.

SUPERCLASS: AGNATHA
CLASS: PTERASPIDOMORPHI
SUBCLASS: HETEROSTRACI
ORDER: ARANDASPIDIFORMES
Arandaspididae. *Arandaspis*, M. Ord. N.T.; *Porophoraspis*, M. Ord. N.T.; *Sacabambaspis*, M. Ord. N. T.

ORDER: UNCERTAIN
Family Uncertain. *Areyongia*, M. Ord. N. T.; *Apedolepis*, M. Ord. N. T.; Genus Unnamed, L. Camb. Qd.

SUBCLASS: THELODONTI
ORDER: THELODONTIDA
Turiniidae. *Turinia*, E.-M. Dev. N.S.W., N.T., Qd., S.A., Vic., W.A.; *Australolepis*, L. Dev. W.A. **Nikoliviidae.** *?Gampsolepis*, E.-M. Dev. Qd.; *Nikolivia*, E. Dev. N.S.W., L. Dev. W.A.

SUPERCLASS: GNATHOSTOMATA
CLASS: PLACODERMI
SUPERORDER: PTYCTODONTOMORPHA
ORDER: PTYCTODONTIDA
Ptyctodontidae. *Austroptyctodus*, *Campbellodus*, L. Dev. W.A.; Genus indet., E. Dev. N.S.W.

SUPERORDER: PETALICHTHYOMORPHA
ORDER: ACANTHOTHORACI
Weejasperaspidae. *Weejasperaspis*, E. Dev. N.S.W.; *Murrindalaspis*, E. Dev. N.S.W., Vic. **Family Uncertain.** *Brindabellaspis*, E. Dev. N.S.W.; "*Ohioaspis*", body scales, E. Dev. N.S.W., Qd., Vic.

ORDER: PETALICHTHYIDA
Macropetalichthyidae. *Notopetalichthys*, *Shearsbyaspis*, *Lunaspis*, E. Dev. N.S.W.; *Wijdeaspis*, E. Dev. N.S.W., vic.

SUPERORDER: DOLICHOTHORACOMORPHA
ORDER: ARTHRODIRA
SUBORDER: ACTINOLEPIDOIDEI
INFRAORDER: ACTINOLEPIDI
Actinolepididae? Genus indet., E.-M. Dev. N.S.W., Qd.

INFRAORDER: WUTTAGOONASPIDI
Wuttagoonaspididae. *Wuttagoonaspis*, E.-M. Dev. N.S.W., Qd.

INFRAORDER: PHYLLOLEPIDI
Phyllolepididae. *Austrophyllolepis*, L. Dev. N.T., Vic.; *Placolepis*, L. Dev. N.S.W., N.T.; *?Phyllolepis*, L. Dev. N.S.W.

SUBORDER: PHLYCTAENIOIDEI
INFRAORDER: PHLYCTAENII
Phlyctaeniidae. *Denisonosteus*, M. Dev. N.S.W. **Groenlandaspididae.** New Genus. M. Dev. N.S.W., Qd.; *Groenlandaspis*, L.Dev. N.S.W., Vic. **Holonematidae.** *Holonema*, L. Dev. W.A.**Williamsaspididae.** *Williamsaspis*, E. Dev. N.S.W.

INFRAORDER: BRACHYTHORACI
Goodradigbeeonidae. *Goodradigbeeon*, E. Dev. N.S.W. **Burrinjucosteidae.** *Burrinjucosteus*, E. Dev. N.S.W. **Buchanosteidae.** *Buchanosteus*, E. Dev. N.S.W., Vic. **Taemasosteidae.** *Taemasosteus*, E. Dev. N.S.W., Vic.; **Family uncertain.** *Arenipiscis*, *Errolosteus*, E. Dev. N.S.W., Vic. **Homosteidae.** *cf. Homosteus*, *Atlantidosteus*, M. Dev. Qd. **cf. Plourdosteidae.** *Harrytoombsia*, *Kimberleyichthys*, *Goujetosteus* (=*Torosteus*), *Mcnamaraspis*,, L. Dev. W.A. **Incisoscutidae.** *Incisoscutum*, L. Dev. W.A. **Camuropiscidae.** *Latocamurus*, *Camuropiscis*, *Rolfosteus*, *Tubonasus*, *Fallacosteus*, L. Dev. W. A. **Mylostomatidae.** *Kendrickichthys*, *Bruntonichthys*, *Bullerichthys*, L.Dev. W.A. **Family uncertain.** *Simosteus*, *Pinguosteus*, L. Dev. W.A. **Dinichthyidae.** *Eastmanosteus*, *Westralichthys*, L. Dev. W.A.

INFRAORDER: ACANTHOTHORACI
Family Uncertain. *Kadunglelepis*, *Jerulepis*, E. Dv. N.S.W.

ORDER: ANTIARCHI
SUBORDER: SINOLEPIDOIDEI
Sinolepididae. *Grenfellaspis*, L. Dev. N.S.W.

SUBORDER: BOTHRIOLEPIDOIDEI
Bothriolepididae. *Monarolepis*, M. Dev. N.S.W.; *Briagalepis*, M.-L. Dev. Vic.; *Bothriolepis*, M.-L. Dev. N.S.W., N.T., Qd., Vic., W.A.; *Nawagiaspis*, M. Dev. Qd.

SUBORDER: ASTEROLEPIDOIDEI
Pterichthyodidae. *Sherbonaspis*, M. Dev. N.S.W.; *Wurungulepis*, M. Dev. Qd. **Asterolepididae.** *Pambulaspis*, *Remigolepis*, L. Dev. N.S.W.

CLASS: CHONDRICHTHYES
SUBCLASS: ELASMOBRANCHII

ORDER: XENACANTHIFORMES
Family Uncertain. *Antarctilamna*, M. Dev. N.S.W.; **Phoebodontidae.** *Phoebodus*, M. Dev.-E. Carb. N.S.W., Qd.; *Thrinacodus* (*Harpagodens*), L. Dev.-E. Carb. Qd., W.A. **Family Uncertain.** cf. *Diploselache*, E. Carb. Qd.; *Pleuracanthus*, E. Trias. N.S.W.

ORDER: SYMMORIIFORMES
Symmoriidae. *Symmorium*, L. Dev.- E. Carb. N.S.W. **Stethacanthidae.** *Stethacanthus*, E. Carb. N.S.W., W.A.

ORDER: EUGENEODONTIFORMES
SUPERFAMILY: EDESTOIDEA
Agassizodontidae. *Helicoprion*, M. Perm. W.A.

ORDER: ORODONTIFORMES
Orodontidae. *Orodus*, E. Carb. W.A.

ORDER: PETALODONTIFORMES
Family uncertain. *Ageleodus*, E. Carb. Qd.

ORDER: UNCERTAIN
Family uncertain. *Gondwanalepis*, *Notiolepis*, *Aussilepis*, *Ohiolepis*, *Cladolepis*, M. Dev. Qd.; *Poecilodus*, Perm. W.A.; *Holmesella*, L. Carb. Qd.; *Tomodus*, Carb.-Perm. Qd.; *Deltoptychius*, Carb. Qd; *Crassidonta*, Perm. W.A.; *Helodus*, E. Carb. Qd., W.A.; *Ohiolepis*, E. Dev. N.S.W. **Psammodontidae.** *Psammodus*, E. Carb. Qd.

COHORT: EUSELACHII
SUPERFAMILY: PROTACODONTOIDEA
Protacodontidae. *Protacrodus*, L. Dev.-E. Carb. Qd.

SUPERFAMILY: CTENACANTHOIDEA
Ctenacanthidae. *Ctenacanthus*, E. Carb. Qd., W.A.

SUPERFAMILY: HYBODONTOIDEI
Family Uncertain. *Geus* undetermined, M. Jar. W.A. **Ptychodontidae.** *Tristychius*, E. Carb. Qd.; *Ptychodus*, L. Cret. W.A.

SUBCOHORT: NEOSELACHII
SUPERORDER: SQUALIMORPHII
ORDER: HEXANCHIFORMES
Family uncertain. *Mcmurdodus*, M. Dev. Qd. **Hexanchidae.** *Hexanchus*, L. Cret. N.Z., L. Palaeo. W.A., L. Palaeo.-Eoc. Vic., M.-L. Eoc., L. Olig. S.A., Rec.; *Notorynchus*, L. Cret.-L. Eo. N.Z., L. Cret. W.A., E. Mio.-E. Plio. Vic., Rec. **Heptranchidae.** *Heptranchias*, L. Eoc., S.A., Rec.

ORDER: SQUALIFORMES
Squalidae. *Centrophoroides*, *Protosqualus*, L. Cret. W.A., L. Cret.-E. Mio. N.Z.; *Scymnorhinus*, M. Eo.-E. Mio. N.Z.; *Squalus*, L. Cret.-Palaeo. N.Z. **Echinorhinidae.** *Pseudoechinorhinus*, Palaeo. N.Z.

ORDER: PRISTIOPHORIFORMES
Pristiophoridae. *Ikamauius*, L. Eo.-L. Plio. N.Z.; *Pristiophorus*, E. Cret. Qd., E. Eo.-E. Mio. N.Z., E. Mio.-E. Plio. Vic., E. Plio. S.A., Rec.

SUPERORDER: SQUATINOMORPHII
ORDER: SQUATINIFORMES
Squatinidae. *Squatina*, L. Cret. W.A., Palaeo. N.Z., Rec.

SUPERORDER: GALEOMORPHII
ORDER: HETERODONTIFORMES
Heterodontidae. *Heterodontus*, M.-L. Eo. N.Z.,

L. Olig.-E. Plio. Vic., E. Mio. Tas., E. Plio. S.A.,
L. Plio. Fl. Is., Rec.; *Synechodus*, L. Cret.-Palaeo.
N.Z.

ORDER: ORECTOLOBIFORMES
Orectolobidae. *Orectolobus*, E.-L. Mio. Vic.,
E. Plio. S.A., Rec.

ORDER: LAMNIFORMES
Family Uncertain. *Paranomotodon*,
Anomotodon, E. Cret. Qd., L. Cret. W.A.
Odontaspididae. *Carcharias*, L. Cret. W.A.; L.
Palaeo.-E. Eoc., L. Olig.-E. Plio. Vic., L. Palaeo.
W.A., M. Eoc., E.-M. Mio., Plio. S.A., E. Mio. Tas.,
?L. Mio. & "Tertiary" N.S.W., Rec.; *Eugomphodus*,
L. Cret.-E. Plio. N.Z.; *Hispidaspis*, L. Cret. W.A.;
Johnlongia, L. Cret. W.A.; *?Odontaspis*, E. Mio.
Tas., Rec. **Mitsukurinidae.** *Mitsukurina*, M.-L.
Eoc. S.A., Rec.; *Scapanorhynchus*, E. Cret. Qd., L.
Cret. W.A. **Lamnidae.** *Carcharodon*, M. Eo.-L.
Pleist. N.Z., L. Olig.-E. Mio., E. Plio. S.A., L. Olig.-
E. Plio., Pleist. Vic., E. Mio. Tas., L. Plio. Fl. Is., E.-
M. Mio. W.A., Rec.; *Carcharhoides*, L. Olig.-E. Mio.
Vic.; *Isurus*, Cret.-M. Mio. N.Z., L. Olig.-E. Plio.
Vic., L. Olig.-M. Mio., E. Plio. S.A., E. Mio. Tas.,
Rec.; *Lamna*, (including *"Lamna"*), E. Cret. Qd.,
L. Cret. W.A., Rec. **Cretoxyrhinidae.**
Archaeolamna, L. Cret. W.A.; *Cretolamna*,
Cretoxyrhina, E. Cret. Qd., L. Cret. W.A.;
Leptostyrax, L. Cret. W.A.; *Paraisurus*, E. Cret.
Qd.; *?Protolamna*, L. Cret. W.A. *Pseudoisurus*, L.
Cret. W.A. **Otodontidae.** *Otodus*, E.-M. Palaeo., L.
Eoc.-E. Olig. Vic., M. Eoc. S.A. **Anacoracidae.**
Pseudocorax, *Microcorax*, E. Cret. Qd.;
Squalicorax, L. Cret. W. A.

ORDER: CARCHARHINIFORMES
Scyliorhinidae. *Megascyliorhinus*, L. Olig.-E. Mio.
N.Z., L. Mio.-E.Plio. Vic.; *Galeorhinus*, M. Mio.
Vic., E. Plio. S.A., Rec. **Hemigaleidae.**
Hemipristis, E. Mio. S.A., Rec. **Carcharhinidae.**
Carcharhinus, L. Olig.-E. Mio. Vic., E. Mio. Tas.,
L. Plio. Fl. Is., E.-M. Mio. E. Plio. S.A., ?L. Mio.
N.S.W., Plio.-Pleist. W.A., Rec.; *Galeocerdo*, L.
Olig.-E.Plio. N.Z., Vic., E.-M. Mio., E. Plio. S.A.,
Rec. **Sphyrnidae.** *Scapanorhynchyus*, L. Cret.
N.Z.; *Sphyrna*, E. Mio. Tas., E.-M. Mio. Vic.,
E. Plio. S.A., Rec.

SUPERORDER: BATOMORPHII
ORDER: MYLIOBATIFORMES
Dasyatidae. *Dasyatis*, L. Olig.-E. Mio. N.Z., ?L.
Mio. N.S.W., E. Plio. S.A., Rec. **Myliobatidae.**
Myliobatis, M. Eo.-E. Mio. N.Z., M.-L. Eoc., Plio.
S.A., L. Olig.-E. Mio. Pleist. Vic., E. Mio. Tas., ?L.
Mio. N.S.W., E. Plio. S.A., Rec. **Sclerorhynchidae.**
Onchopristis, L. Cret. N.Z.

SUPERORDER : GALEOMORPHI
Palaeospinacidae. *Paraorthacodus*, L. Cret. W.A.;
Synechodus, E. Cret. Qd., L. Cret. W.A.Dev. N.S.W.;
MU

SUBCLASS: HOLOCEPHALI
ORDER: CHIMAERIDA
Callorhinchidae. *Ptyktoptychion*, E. Cret. Qd.
Chimaeridae. *Chimaera*, Tert. N.Z.; *Edaphodon*,
E. Cret. S.A., L. Cret. W.A., L. Mio.-E. Plio. Vic.;
Ischyodus, L. Cret. N.Z., E. Mio. Tas., L. Mio.-E.
Plio. Vic.

CLASS: TELEOSTOMI
ORDER: UNCERTAIN
Family Uncertain. *Yealepis*, L. Sil. Vic.

SUBCLASS: ACANTHODII
ORDER: ISCHNACANTHIDA
Ischnacanthidae. *Rockycampacanthus*, E. Dev.

Vic.; *Taemasacanthus*, E. Dev. N.S.W.;
Cheiracanthoides, E.-M. Dev. N.S.W., Qd.;
Gomphonchus, L. Sil. Qd., E.-M. Dev. Qd., N.S.W.,
S.A., Vic.; *Poracanthodes*, L. Sil. Qd.

ORDER: CLIMATIIDA
SUBORDER: CLIMATOIDEI
Family uncertain. *Nostolepis*, L. Sil. Qd., E. Dev.
N.S.W., Qd., Vic., M. Dev. N.S.W.

SUBORDER: DIPLACANTHOIDEI
Culmacanthidae. *Culmacanthus*, *Howittacanthus*,
M. Dev. N.S.W., Vic. **Family uncertain.**
Striacanthus, L. Dev. Vic. **Gyracanthidae.**
Gyracanthides, E. Carb. Qd., Vic.

ORDER: ACANTHODIDA
Acanthodidae. *Howittacanthus*, L. Dev. Vic.;
?Acanthodes, M. Dev. N.T., Qd., E.
Carb. Vic.

SUBCLASS: OSTEICHTHYES
ORDER: LOPHOSTEIFORMES
Lophosteidae. *Lophosteus*, E. Dev. N.S.W.

INFRACLASS: ACTINOPTERYGII
ORDER: PALAEONISCIFORMES
Family uncertain. *Ligulalepis*, E.-M. Dev. N.S.W.,
Qd.; Terenolepis, E. Dev. N.S.W. **Mimiidae.**
Howqualepis, *Mimia*, L. Dev. Vic. *Terenolepis*,
E. Dve. N.S.W. **Stegotrachelidae.** *Mimia*,
Moythomasia, L. Dev. W.A. **Gonatodidae.**
Novogonatodus, E. Carb. Vic. **Family uncertain.**
Mansfieldiscus, E. Carb. Qd., Vic.
Cryphiolepidae. *Cryphiolepis*, E. Carb. Qd.,
Perm. W.A. **Elonichthyidae.** *?Elonichthys*, Perm.-
E. Trias. N.S.W. **Bobasatraniidae.** *Ebenaqua*,
L. Perm. Qd. **Urosthenidae.** *Urosthenes*, Perm.
N.S.W. **Tegeolepidae.** *Apateolepis*, E. Trias.
N.S.W. **Acrolepidae.** *Acrolepis*, E. Trias. N.S.W.,
Tas., ?Perm. W.A.; *Leptogenichthys*, M. Trias.
N.S.W. **Palaeoniscidae.** *Agecephalichthys*,
M. Trias. N.S.W. **Coccolepidae.** *Coccolepis*, L. Jur.
N.S.W., E. Cret. Vic.; *Psilichthys*, E. Cret. Vic.
Platysomidae. *Platysomus*, L. Trias. N.S.W.
Family uncertain. *Atherstonia*, *Belichthys*, Trias.
N.S.W.; *Elpisipholis*, L. Trias. N.S.W.; *Leighiscus*,
Trias. S.A.; *Megapteriscus*, *Mesembroniscus*,
Myriolepis, Trias. N.S.W.

SUBORDER: PHOLIDOPLEUROIDEI
Pholidopleuridae. *Macroaethes*, M. Trias. N.S.W.

SUBORDER: REDFIELDOIDEI
Redfieldiidae. *Beaconia*, *Brookvalia*,
Dictyopleurichthys, *Dictyopyge*,
Geitonichthys, *Molybdichthys*, *Phlyctaenichthys*,
Schizurichthys, Trias. N.S.W.; Genus indet.,
L. Perm. Qd.

SUBORDER: PERLOIDEI
Perleididae. *Chrotichthys*, *Manlietta*, *Pristisomus*,
Procheirichthys, *Tripelta*, *Zeuchthiscus*, Trias.
N.S.W. **Cleithrolepidae.** *Cleithrolepis*, Trias.
N.S.W., Tas. **Pholidophoridae.** *?Thoracopterus*,
Trias. N.S.W.

ORDER: ?ACIPENSERIFORMES
Saurichthyidae. *Saurichthys*, E.-M. Trias. N.S.W.,
Qd., Tas., W.A.

ORDER: ASPIDORHYNCHIFORMES
Aspidorhynchidae. *Aspidorhynchus*,
Belonostomus, E. Cret. Qd.

ORDER: SEMIONOTIFORMES
SUBORDER: SEMIONOTOIDEI
Semionotidae. *Aetholepis*, *Aphnelepis*,

Corungenys, *Enigmatichthys*, Trias. N.S.W.
Parasemionotidae. *Promecosomina*, M.-L. Trias.
N.S.W. **Macrosemionotidae.** *Uarbryichthys*,
L. Jur. N.S.W.

ORDER: PHOLIDOPHORIFORMES
Archaeomaenidae. *Archaeomaene*, *Madariscus*,
L. Jur. N.S.W.; *Wadeichthys*, E. Cret. Vic.

COHORT: TELEOSTEI
SUPERORDER: LEPTOLEPOMORPHA
ORDER: LEPTOLEPIFORMES
Leptolepidae. *Leptolepis*, L. Jur. N.S.W., E. Cret.
Vic.

SUPERORDER: ELOPOMORPHA
ORDER: ELOPIFORMES
SUBORDER: ELOPOIDEI
Elopidae. *Elops*, E. Eo.-?Mio. N.Z.;
Flindersichthys, E. Cret. Qd. **Megalopidae.**
Megalops, L. Olig.-M. Mio. Vic., Rec.

SUPERORDER: ALBULOIDEI
Albulidae. *Pterothrissus*, L. Olig. Vic.
ORDER: ANGUILLIFORMES
SUBORDER: ANGUILLOIDEI
Congridae. Genus indet. L. Eo.-M. Mio. N.Z.;
Astroconger, L. Olig. Vic., Rec.; *Conger*, Mio. N.Z.;
Congridarum, E. Mio. N.Z.; *Gnathophis*, E. Eo.-
Plio. N.Z.; *Maxwelliella*, E. Eo. N.Z.; *Scalanago*, E.
Eo. N.Z.; *Uroconger*, L. Olig., Plio. Vic., Rec.
Heterenchelyidae. *Gnathophis*, E. Eo.-Plio. N.Z.;
Heterenchelyidarum, E. Mio. N.Z.; *Heterenchelys*,
L. Olig.-M. Mio. Vic., Rec. **Muraenesocidae.**
Muraenesox, M. Mio. Vic., Rec. **Ophichthidae.**
?Mystriophis, M. Mio. N.Z.

SUPERORDER: CLUPEOMORPHA
ORDER: CLUPEIFORMES
Clupeidae. Genus indet., E. Mio. N.Z.; *Anchoa*,
E. Mio. N.Z.; *Diplomystus*, Cret. N.Z.;
Scombroclupea, Eo. N.Z. **Ichthyodectidae.**
Cooyoo, *?Cladocyclus*, E. Cret. Qd.; *Xiphactinus*,
L. Cret. N.Z. **Pachyrhizodontidae.**
Pachyrhizodus, E. Cret. Qd., L. Cret. N.Z.;
?Thrissopater, Cret. N.Z.

SUPERORDER: CLUPEODEI
Koonwarriidae. *Koonwarria*, E. Cret. Vic.

SUPERORDER: OSTEOGLOSSOMORPHA
ORDER: OSTEOGLOSSIFORMES
SUBORDER: OSTEOGLOSSOIDEI
Osteoglossidae. *Phaeroides*, Eoc.-Olig. Qd.;
Scleropages, Olig.-?Mio. Qd., Rec.

SUPERORDER: PROTACANTHOPTERYGII
ORDER: GONORHYNCHIFORMES
SUBORDER: GONORHYNCHOIDEI
Gonorhynchidae. *Notogoneus*, Eoc.-Olig. Qd.

SUPERORDER: OSTARIOPHYSI
ORDER: SILURIFORMES
Plotosidae. *Tandanus*, ?Pleist. Qd., Rec. **Ariidae.**
Genus indet., L. Olig./M. Mio. S.A.; *Tachysurus*,
Plio. N.Z.

SUPERORDER: ?
ORDER; SALMONIFORMES
Argentinidae. *Argentina*, E. Mio. N.Z.;
Galaxiidae. Galaxis, Mio. N.Z.
Photichthyidae. *Polymetme*, E. Mio. N.Z.

SUPERORDER: STENOPTERYGII
ORDER STOMIIFORMES
Gonostomatidae. *Gonostoma*, E. Mio. N.Z.
Sternoptychidae. *Maurolicus*, E. Mio. N.Z.;
Polyipnus, E. Mio. N.Z., Rec.

SUPERORDER: SCLEROMORPHA
ORDER: AULOPIFORMES
Aulopiidae. *Aulopus*, E. Eo. N.Z.
Chlorophthalmidae. *Chloropht-halmus*, L. Eo.-E. Mio. N.Z., Rec. **Scopelarchidae.** *Scopelarchus*, L. Eo.-E. Mio. N.Z., Rec.

ORDER: MYSTOPHIFORMES
Myctophidae. *Benthosema*, Mio. N.Z.; *Bolinichthys*, M. Mio. N.Z.; *Diaphus*, E. Olig.-L. Mio. N.Z., Rec.; *Diogenichthys*, E. Mio. N.Z.; *Hygophum*, Mio. N.Z.; *Lampanyctodes*, E. Mio.-Plio. N.Z.; *Lampanyctus*, M.-L. Mio. N.Z.; *Lampichthys*, L. Mio. N.Z.; *Myctophidarum*, L. Mio. N.Z.; *Myctophum*, M. Mio. N.Z., Rec.; *Notoscopelus*, E. Mio.-Plio. N.Z.; *Symbolophorus*, E. Mio.-Plio. N.Z.

SUPERORDER: ?
ORDER: LOPHIIFORMES
Lophiidae. *?Lophius*, E. Mio. N.Z.
Ogcocephalidae. *?Dibranchus*, L. Eo. N.Z.

SUPERORDER: PARACANTHOPTERYGII
ORDER: GADIFORMES
Moridae. *Actuariolum*, L. Olig.-Mio. N.Z.; *Austrophycis*, E. Mio. N.Z.; *Lemonema*, M. Mio. N.Z.; *Lotella*, E. Mio. N.Z.; *Physiculus*, E. Mio. N.Z.; *Raniceps*, Mio. N.Z.; *Tripterophycis*, L. Eo. N.Z. **Melanonidae.** *Karrerichthys*, Mio. N.Z.
Bythididae. Genus indet., M. Mio. N. Z.; *Oligopus*, E. Mio. N.Z.; *Bythitidarum*, E. Mio. N.Z.; *Saccogaster*, Plio. N. Z. **Euclichthyidae.** *Euclichthys*, L. Eo.-E. Mio. N.Z.

SUBORDER: GADOIDEI
Gadidae. *Bregmaceros*, Eo. N.Z., M. Mio. Vic., Rec.; *Gadiculus*, E. Mio. N.Z., Rec.; *Gadus*, Mio. N.Z.; *Macruronus*, L. Olig.-Plio. N.Z., Rec.; *Merluccius*, Mio. N.Z., Rec.; *Micromesistius*, Plio. N.Z. **Merlucciidae.** *Merluccius*, L. Olig.-M. Mio. Vic., N.Z., Rec.

SUBORDER: OPHIDOIDEI
Ophidiidae. Genus indet., M. Mio. N.Z.; *Ampheristus*, E. Eo. N.Z.; *Genypterus*, Plio. N.Z.; *Monomitopus*, L. Eo.-L. Mio. N.Z.; *Neobythites*, M. Mio. N.Z., Rec.; *Nolfophidion*, L. Eo. N.Z.; *Ophidion*, L. Olig.-M. Mio. Vic., Mio.-Plio. N.Z., Rec.; *Ophidiidarum*, Nio. N.Z.; *Sirembinorum*, E. Mio. N.Z. **Carapidae.** *Carapus*, Olig.-M. Mio. N.Z., M. Mio. Vic., Rec.

SUBORDER: MACROUROIDEI
Macrouridae. *Bathygadus*, E. Mio. N.Z.; *Coelorinchus*, L. Olig.-M. Mio. Vic., L. Olig.-Plio. N.Z., Rec.; *Lepidorhynchus*, E. Mio.-Plio. N.Z.; *Macrouridarum*, E. Mio. N.Z.; *Macrourus*, Mio. N.Z., Rec.; *Macrurulus*, Eo. N.Z.; *Maorigadus*, E. Eo. N.Z.; *Nezumia*, L. Mio. N.Z.; *Trachyrincus*, Mio. N.Z.; *Ventrifossa*, Mio. N.Z.

SUPERORDER: ACANTHOPTERYGII
ORDER: BERYCIFORMES
SUBORDER: BERYCOIDEI
Berycidae. *Centroberyx*, E. Mio. N.Z.
Trachichthyidae. *Egregioberyx*, L. Eo. N.Z.; *Hoplostethus*, E. Mio.-Plio. N.Z., REC.; *Paratrachichthys*, E.-M. Mio. N.Z.; *Trachichthodes*, Tert. Aust., L. Eo. N.Z.; **Holocentridae.** Genus indet., L. Eo. N.Z.; *Adioryx*, E. Eo. N.Z.
Melamphaeidae. *Melamphaes*, Mio. N.Z.
Monocentridae. *Cleidopus*, L. Olig.-M Mio. Vic., Rec.; *Monocentris*, Plio. S.A., Rec.

ORDER: ZEIFORMES
Caproidae. *Antigonia*, M. Eo. N.Z., M. Mio. Vic., Rec.

ORDER: SCORPAENIFORMES
Cottidae. Genus indet., L. Eo. E.Mio. N.Z.; *Cottus*, Olig./Mio. N.Z., Rec.; *?Cottoideorum*, E. Mio. N.Z.
Cyclopteridae. *Cyclopterus*, Mio. N.Z., Rec.
Hoplichthyidae. *Hoplichthys*, E. Mio. N.Z.; *Praehoplichthys*, M.-L. Eo. N.Z. **Scorpaenidae.** *Sebastodes*, M. Mio. Vic.

SUBORDER: PLATYCEPHALOIDEI
Platycephalidae. *Platycephalis*, M. Eo.-E. Mio. N.Z., E. Mio. Tas., M. Mio. Vic., Rec.

ORDER: PERCIFORMES
SUBORDER: PERCOIDEI
Apogonidae. *Apogonidarum*, L. Olig.-M. Mio. N.Z. **Centropomidae.** *Muccullochella*, Mio. N.S.W., Pleist. Qd.; *Percaletes*, Eoc.-?Mio. Qd., Mio. N.Z., Rec. **Lutjanidae.** *Lutjanus*, ?Mio. Qd., Rec. **Lactariidae.** *Lactarius*, E. Mio. Vic., Rec.
Pomadasyidae. Genus indet., E. Eo.-E. Mio. N.Z.; *Pomadasyidarum*, E. Mio. N.Z. **Cepolidae.** *Cepola*, E. Mio. N.Z., Rec. **Theraponidae.** Genus indet., ?Olig. Qd. **Sillaginidae.** *Sillago*, L. Olig.-Plio. Vic., E. Mio. N.Z., Rec. **Ambassidae.** *Dapalis*, E.-M. Eo. N.Z. **Sparidae.** *Dentex*, E. Eo.-Mio. N.Z., Rec.; *Pagellus*, Mio. N.Z.; *Pagrosomus*, Pleist. Vic., Rec.; *Sargus*, E. Mio. Vic., Mio. N.Z., Rec.; *?Chrysophrys*, E. Mio. S.A. **Oplegnathidae.** *Oplegnathus*, E. Mio.-E. Plio. Vic.; *?Percidarum*, M. Mio. Vic., L. Mio. N.Z. **Percichthyidae.** Genus indet., L. Olig./M. Mio. S.A. **Monacanthidae.** E. Plio. S.A. **Serranidae.** Genus indet., M. Eo.-E. Mio. N.Z.; *Serranus*, Mio. N.Z.

SUBORDER: MUGILOIDEI
Sphyraenidae. *Sphyraena*, Pleist. Vic., Rec.

SUBORDER: LABROIDEI
Labridae. *Labridarum*, E. Mio. N.Z.; *Labrodon*, E. Mio.-E. Plio. Vic., Mio. N.Z.; *Nummopalatus*, E. Mio-E. Plio. Vic., Rec.

SUBORDER: TRACHINOIDEI
Trichodontidae. *Trichodon*, E. Mio. N.Z., Rec.

SUBORDER: SCOMBROIDEI
Gampeilidae. Genus indet., L. Mio. Vic.
Trichiuridae. Genus indet., L. Mio. Vic.

SUBORDER: NOTOTHENOIDEI
Nototheniidae. *Notothenia*, M.-L. Mio. N.Z., Rec.
Epigonidae. *Epigonus*, E.-M. Mio. N.Z.
Lactariidae. *Lactarius*, L. Olig.-Plio. N.Z.; *Neoscombrops*, E. Eo. N.Z.; *Paralactarius*, E. Eo.-E. Mio. N.Z.

SUBORDER: ?
Acropomatidae. *Acropoma*, L. Eo. N.Z., L. Olig.-M. Mio. N.Z., Rec. **Gerreidae.** *?Gerreidarum*, E. Mio. N.Z.

SUBORDER: SCOMBROIDEI
Gempylidae. *Eothyrsites*, Olig. N.Z. **Scombridae.** Genus indet., E. Mio. N.Z.

SUBORDER: GOBIODEI
Gobiidae. *Gobiidarum*, E. Mio. N.Z. **Eleotridae.** *Gobiomorphus*, L. Pleist. N.Z.; *Paradiplospinus*, E. Mio. N.Z.

ORDER: PLEURONECTIFORMES
SUBORDER: PLEURONECTOIDEI
Bothidae. Genus indet., E. Mio. N.Z.; *Arnoglossus*, L. Olig.-E. Mio. N.Z. **Pleuonectidae.** *Pleuronectes*, M. Mio. Vic., Rec. **Soleidae.** *Achirus*, M. Mio. N.Z.; *Solea*, Mio. N.Z. **Eucitharidae.** *Citharus*, Mio. N.Z.; *Pleuronectidarum*, Mio. N.Z.

ORDER: TETRADONTIFORMES
SUBORDER: TETRADONTOIDEI
Family Uncertain. *Sparidarum*, Mio. N.Z.
Diodontidae. *Diodon*, E.-L. Mio. Vic., E. Plio. S.A., Rec.; *Aracana*, E. Mio. Tas., Rec.
Trigonodontidae. *Trigonodon*, Mio. N.Z.

SUPERORDER: MIGILOIDEI
Mugiloidae. Genus indet., M. Eo. N.Z.; *Parapercis*, L. Olig.-E. Mio. N.Z.

SUBORDER: TRACHINOIDEI
Trachinidae. *Trachinus*, Mio. N.Z.
Uranoscopidae. *Uranoscopus*, M. Eo. N.Z.
Leptoscopidae. *Leptoscopus*, L. Olig.-L. Mio. N.Z., Rec.

SUBORDER AND ORDER: ?
Hemerocoetidae. *Hemerocoetus*, E. Mio.-Plio. N.Z.; *Krebsiella*, L. Eo.-E. Mio. N.Z.; *Waitakia*, L. Olig.-E. Mio. N.Z.

INFRACLASS: DIPNOI
Dipnorhynchidae. *Dipnorhynchus*, E. Dev. N.S.W., Vic. **Speonesydrionidae.** *Speonesydrion*, E. Dev. N.S.W.; *Ichnomylax*, E. Dev. Vic.
Chirodipteridae. *Adololopas*, L. Dev. W.A.; *Chirodipterus*, L. Dev. W.A.; *Gogodipterus*, L. Dev. W.A.; *Pillararhynchus*, L. Dev. W.A. **Dipteridae.** Genus indet., L. Dev. N.T. **Fleurantiidae.** *Barwickia*, M. Dev Vic.
Rhynchodipteridae. *Griphognathus*, L. Dev. W.A.; *Soederberghia*, L. Dev. N.S.W.
Holodipteridae. *Holodipterus*, L. Dev. W.A. ?
Ctenodontidae. *Eoctenodus*, L. Dev. Vic.; *Delatitia*, E. Carb. Vic. **Ceratodontidae.** *Ceratodus*, Trias. N.S.W., Qd., Tas., Vic., W.A., E. Cret. N.S.W., E.-M. Plio. Qd., L. Plio. N.S.W.; *Gosfordia*, Trias. N.S.W. **Ceratodontidae.** *Metaceratodus*, E. Cret. N.S.W.; Olig.-Pleist., S. Aust. **Neoceratodontidae.** *Mioceratodus*, Olig.-Mio. Qd.;*Neoceratodus*, E. Cret.-Pleist., N.S.W., N.T., Qd., S.A., Rec.

INFRACLASS: CROSSOPTERYGII
ORDER: ONYCHODONTIDA
Onychodontidae. *Onychodus*, Genus indet., E. Dev. N.S.W., Qd., M. Dev. Qd., L. Dev. W.A.

ORDER: ACTINISTIA
Miguashiidae. *Gavinia*, M. Dev. Vic.
Coelacanthidae. Genus indet., Trias. Tas., L. Jur. N.S.W. **Miguashiidae.** *Gavinia*, M. Dev. Vic., L. Dev. N.S.W.

ORDER: POROLEPIFORMES
SUPERFAMILY: HOLOPTYCHOIDEA
Holoptychiidae. Genera indet., E. Dev. N.S.W.; *? Glyptolepis*, L. Dev. Vic.; *Holoptychius*, L. Dev. N.S.W.

ORDER: RHIZODONTIDA
Family Uncertain. *Gooloogongia*, L. dev. N.S.W.
Rhizodontidae. *Barameda*, E. Carb. Vic.; Genus indet., E. Carb. Qd., W.A.

ORDER: OSTEOLEPIFORMES
Canowindridae. *Beelarongia*, M. Dev. Vic.; *Canowindra*, L. Dev. N.S.W. **Osteolepididae.** *Gogonasus*, L. Dev. W.A.; *? Gyroptychius*, M. Dev. N.S.W. **Megalichthyidae.** *Cladarosymblema*, E. Carb. N.S.W.; *Megalichthys*, E. Carb. Qd. **Tristichopteridae (=Eusthenopteridae).** *Marsdenichthys*, L. Dev. Vic.; *Mandageria*, *Cabonnichthys*, L. Dev. N.S.W. **Rhizodopsidae.** *Marsdenichthys*, M. Dev. Vic.

CLASS: AMPHIBIA

SUBCLASS: LABYRINTHODONTIA
ORDER: TEMNOSPONDYLI
Family uncertain. *Metaxygnathus*, L. Dev. N.S.W.
New Family (To Be Named): *Lapillopsis*, E. Trias.
Qd.; New Genus (To Be Named), E. Trias. Tas.
Lydekkerinidae. *Chomatobatrachus*, E. Trias.
Tas. **Capitosauridae.** *Parotosuchus*, E. Trias.
N.S.W., Qd., Tas., W.A.; *Paracyclotosaurus*, M.
Trias. N.S.W. **Rhytidosteidae.** *Rewana, Arcadia,
Acerastea*, New Genus (To Be Named), E. Trias.
Qd.; *Trucheosaurus*, L. Perm. N.S.W.; *Deltasaurus*,
E.Trias. Tas., W.A.; *Derwentia*, E. Trias. Tas.
Trematosauridae. *Erythrobatrachus*, E. Trias.
W.A.; Genus Uncertain, E. Trias. Qd.
Brachyopidae. *Bothriceps*, E. Trias. N.S.W.;
Platycepsion, E. Trias. N.S.W.; *Banksia*, E. Trias.
Tas.; *Batrachosuchus*, E. Trias. W.A.;
Xenobrachyops, E. Trias. Qd.; *Notobrachyops*,
M. Trias. N.S.W.; *Austropelor*, E. Jur. Qd.
Chigutisauridae. *Keratobrachyops*, E. Trias. Qd.;
Siderops, E. Jur. Qd.; *Koolasuchus*, E. Cret. Vic.
Family Uncertain. *Plagiobatrachus*, E. Trias. Qd.

SUBCLASS: LISSAMPHIBIA
ORDER: ANURA
Hylidae. *Australobatrachus*, L. Olig./M. Mio. S.A.;
Litoria, L. Olig./M. Mio. S.A. M. Mio. Qd. Pleist.
S.A., W.A., Rec. **Leptodactylidae.** *Limnodynastes*,
L. Olig./M. Mio. Qd., S.A., M. Mio. Qd., Pleist. S.A.,
Tas., W.A., Rec.; *Crinia*, L. Olig./M. Mio. Qd.,
Pleist. S.A., W.A., Rec.; *Heleioporus/Neobatrachus*,
Geocrinia, Pleist. S.A., W.A.; *Lechriodus*,
Kyarranus, L. Olig./M. Mio. Qd.; *Neobatrachus*,
Plio.-Pleist. S.A., Pleist. W.A.

SUPERCLASS: AMNIOTA
CLASS: ANAPSIDA
Procolophonidae. E. Trias. Qd.

SUBCLASS: TESTUDINES
SUBORDER: INCERTAE SEDIS
Families uncertain. *Chelycarapookus*, E. Cret.
Vic.; *Cratochelone*, E. Cret. Qd. **Protostegidae.**
Genus indet., L. Cret. N.Z.

SUBORDER: PLEURODIRA
Chelidae. *Chelodina*, Mio. ?Plio. Qd., Pleist. Qd.,
Rec.; *Emydura*, Olig./Mio. Tas., L. Olig./M. Mio.,
S.A., Plio. N.S.W., Qd., Rec.; *Pseudemydura*, Mio.
Qd.

SUBORDER: CRYPTODIRA
PARVORDER: EUCRYPTODIRA
Family uncertain: *Otwayemys*, E. Cretaceous,
Vic.

SUPERFAMILY: BAENOIDEA
Meiolaniidae. Genus indet., L. Olig./M. Mio. N.T.,
S.A.; Genus indet., Pleist. N. Caledonia (Tiga Is.,
Mainland); *Meiolania*, L.Olig./M. Mio. N.T., Qd.,
S.A., Plio. N.S.W., Qd., Pleist. Lord Howe Is., N.
Caledonia (Walpole Is.), Qd. *Warkalania*, Mio. Qd.

SUPERFAMILY: TRIONYCHOIDEA
Carettochelyidae. *?Carettochelys*, Pleist. or Rec.
W.A. **Trionychidae.** Genus indet. E. Tert.-Pleist.
Qd., Mio. N.T.

SUPERFAMILY: CHELONIOIDEA
Family uncertain. *Notochelone*, E. Cret. Qd.

CLASS: DIAPSIDA
SUBCLASS: LEPIDOSAUROMORPHA
SUPERORDER: LEPIDOSAURIA
ORDER: SPHENODONTA
Sphenodontidae. *Sphenodon*, L. Pleist. N.Z., Rec.

ORDER: SQUAMATA
Family uncertain. *Kudnu*, E. Trias. Qd.

INFRAORDER: IGUANIA
Agamidae. *Pogona*, Pleist. S.A.; Genus indet.,
?Qd., Rec.; *Chlamydosaurus*, Pleist. Qd., Rec.;
Physignathus, M. Mio. Qd., Rec.; *Sulcatidens*, M.
Mio. Qd., Rec.

INFRAORDER: GEKKOTA (NYCTISAURIA)
Gekkonidae. Genera indet., Mio. Qd.
Pygopodidae. *Pygopus*, Mio. Qd.

INFRAORDER: SCINCOMORPHA
(LEPTOGLOSSA)
Scincidae. Genera indet., L. Olig./M Plio. Qd.;
Egernia, L. Olig./M. Mio., Pleist. S.A., Rec.;
?Sphenomorphus, Pleist. S.A., Rec.; *Tiliqua*, Mio.
S.A., M. Plio.-Pleist. Qd., S.A., Rec.;
Trachydosaurus, Pleist. S.A., Rec.

INFRAORDER: ANGUIMORPHA
SUPERFAMILY: VARANOIDEA
Varanidae. Genus indet., M. Mio. Qd., M. Plio.
S.A., Plio.-Pleist., S.A. *Megalania*, Plio. Qd.;
Pleist. N.S.W., Qd., S.A.; *Varanus*, L. Plio.-E.
Pleist. Qd., Pleist. N. Caledonia, Rec.
Mosasauridae. Genus indet., L. Cret. W.A.;
Moanasaurus, L. Cret. N.Z.; *Mosasaurus*, L. Cret.
N.Z.; *Prognathodon*, L. Cret. N.Z.; *Rikisaurus*, L.
Cret. N.Z.; *Taniwhasaurus*, L. Cret. N.Z.;
Tylosaurus, L. Cret. N.Z.

INFRAORDER: SERPENTES
Boidae . *Morelia*, L. Olig./Mio. N.T., Plio. Qd.,
Rec.; *Montypythonoides*, L. Olig./M. Mio. Qd.
Madtsoiidae. *Nanowana*, Mio. Qd.; *Wonambi*,
Mio. Qd., Mio.-Pleist. S.A.; *Yurlunggur*, Mio. Qd.
Acrochordidae. Genus indet., Plio. Qd.
Elapidae. Genus indet., M. Mio. Qd.; Genus
indet., Mio.-Plio. S.A.; *Notechis, Pseudechis,
Pseudonaja*, Pleist. S.A., Rec. **Typhlopidae.**
?Rhamphotyphlops, M. Mio. Qd.

SUPERORDER: SAUROPTERYGIA
SUPERFAMILY: PLESIOSAUROIDEA
Elasmosauridae. Genus indet., L. Cret. N.Z.;
Genus indet., M. Jur. W.A.; *Mauisaurus*, L. Cret.
N.Z.; *Tuarangisaurus*, L. Cret. N.Z.;
Woolungasaurus, E. Cret. Qd., ?S.A.
Cimoliasauridae. *Cimoliasaurus*, E. Cret. N.S.W.

SUPERFAMILY: PLIOSAUROIDEA
Family uncertain. Genus indet., L. Cret. N.Z.;
Genus indet., Jur. Qd. **Cryptoclididae.** Genus
indet., L. Cret. N. Z. **Dolichorhynchopidae.**
?Dolichorhynchops, E. Cret. N.S.W. **Pliosauridae.**
Kronosaurus, E. Cret. Qd.; *Leptocleidus*, E. Cret.
N.S.W., W.A.

SUBCLASS: ICHTHYOPTERYGIA
Family uncertain. Genera indet., M. Trias.-E.
Cret. N.Z. **Leptopterygiidae or
Stenopterygiidae.** Genus Indet., E. Cret. N.Z.;
Platypterygius, E. Cret. N.T., Qd., S.A., W.A.; L.
Cret. W.A.

SUBCLASS: ARCHOSAUROMORPHA
SUPERORDER: PROTOROSAURIA
ORDER: EOSUCHIA
Prolacertidae. *Kadimakara*, E. Trias. Qd.

Superorder: Archosauria
ORDER: PROTEROSUCHIA
Chasmatosauridae. *Kalisuchus*, E. Trias. Qd.;
Tasmaniosaurus, E. Trias. Tas.

ORDER: CROCODYLOTARSI
SUBORDER: CROCODYLOMORPHA
INFRAORDER: NEOSUCHIA
Family uncertain. "*Crocodylus*", E. Cret. N.S.W.
Crocodylidae. Genus indet., E. Cret. N.S.W.; L.
Cret. Qd.; *Australosuchus*, Olig.-Pleist., S.A.; *Baru*,
Olig./Mio. Qd., S.A.; *Crocodylus*, Plio.-Quat. Qd.,
Rec.; *Harpacochampsa*, Mio., N.T.; *Pallimnarchus*,
L. Mio. N.T., Plio. Pleist. Qd.; *Quinkana*, Plio.,
Pleist. Qd. **Mekosuchidae.** *Mekosuchus*, Quat. N.
Caledonia. **Sebecosuchia or Pristichampsinae.**
Plio. S.A.

ORDER: ORNITHODIRA
SUBORDER: PTEROSAURIA
Ornithocheiridae. Genus indet., E. Cret. Qd., ?Vic.
?Anhangueridae. *?Anhanguera*, Cret. Qd.; E. Cret.
N. Z. **?Azhdarchiidae.** Genus indet., L. Cret. W.A.
SUBORDER: SAURISCHIA
INFRAORDER: THEROPODA
Family uncertain. Genera indet., E. Cret. Vic., S.A.,
W.A.; L. Cret. N.Z., W.A.;*Changpeipus* (footprint),
M. Jur. Qd.; *Kakuru*, E. Cret. S.A.; *Megalosauropus*
(footprint), E. Cret. W.A.; *Ozraptor*, M. Jur. W.A.;
Rapator, E. Cret. N.S.W.; *Tyrannosauropus* (foot-
print), L. Cret Qd.*Walgettosuchus*, E. Cret. N.S.W.
Allosauridae. *Allosaurus*, E. Cret. Vic.;
Oviraptorosauridae. Genus indet., E. Cret. Vic.
Ornithomimosauria. *Timimus*, E. Cret. Vic.
Caenagnathidae. Unnamed Genus, E. Cret. Vic.

INFRAORDER: SAUROPODOMORPHA
SUPERFAMILY: SAUROPODA
Family uncertain. Genus indet., L. Cret. N. Z.;
Footprint Genus indet., E. Cret. W.A.; *Austrosaurus*,
E.-L. Cret. Qd. **Brachiosauridae.** Genus indet., E.
Cret. Qd. **Cetiosauridae.** *Rhoetosaurus*, M. Jur. Qd.;
Genus indet., M. Jur. W.A.; **?Titanosauridae.**
Austrosaurus, E. or L. Cret. Qd.

SUBORDER: ORNITHISCHIA
INFRAORDER: ORNITHOPODA
Family uncertain. Genera indet., E. Cret. N.S.W.,
Qd., Vic., L. Cret. N.Z.; Footprint Genus indet., E.
Cret. W.A.; *Wintonopus* (footprint), L. Cret. Qd.
Anomoepodidae. *Anomoepus* (footprint), E. Jur.
Qd. **Hypsilophodontidae.** *Atlascopcosaurus*, E.
Cret. Vic.; *Fulgurotherium*, E. Cret. N.S.W., Vic.;
Leaellynasaura, E. Cret. Vic.; *Qantassaurus*, E. Cret.
Vic. **Iguanodontia or New Undescribed Family.**
Muttaburrasaurus, E. Cret. N.S.W., Qd.

INFRAORDER: STEGOSAURIA
Stegosauridae. Footprint Genus indet., E. Cret
W.A.

INFRAORDER: CERATOPSIA
SUPERFAMILY; NEOCERATOPSIA
?Protoceratopsidae. Unnamed Genus, E. Cret.,Vic.

INFRAORDER: ANKYLOSAURIA
Ankylosauridae. Genus indet., E. Cret. Vict.;
L. Cret. N. Z. *Minmi*, E. Cret. Qd.

SUBCLASS: SYNAPSIDA
SUBORDER: DICYNODONTIA
INFRAORDER: PRISTERODONTIA
?Kannemeyeriidae. E. Trias. Qd.

CLASS: MAMMALIA
ORDER: MONOTREMATA
Family uncertain. Genus indet., E. Cret. Vic.
Steropodontidae. *Steropodon*, E. Cret., N.S.W.
Kollikodontidae. *Kollikodon*, E. Cret. N.S.W.
Ornithorhynchidae. *Obdurodon*, L. Olig./M. Mio.
S.A.; *Ornithorhynchus*, E. Plio. N.S.W., Pleist. Aust.;

Rec. **Tachyglossidae.** *Tachyglossus*, Pleist. Aust., Rec.; *Zaglossus*, E.-M. Mio. ?Plio-Pleist. N.S.W., Qd., Pleist. N.G., Rec.

INFRACLASS: PANTOTHERIA
ORDER; EUPANTOTHERIA
Teinolophidae. *Teinolophos*, E. Cret. Vic.

SUBCLASS: THERIA
INFRACLASS: MARSUPIALIA
ORDER: UNCERTAIN
Family Uncertain. *Thylacotinga*, Eo Qd.
Yingabalanaridae. *Yingabalanara*, L. Olig./M. Mio. Qd.

ORDER: DASYUROMORPHIA
Family Uncertain. *Mayigriphus*, L. Olig./M. Mio. Qd. **Dasyuridae.** Genus indet., L. Olig. Tas.; *Ankotarinja*, L. Olig./M. Mio. S.A.; *Antechinomys*, Pleist. s.cent. Aust., Rec.; *Antechinus*, E. Plio., Pleist. Aust., Rec.; *Apoktesis*, L. Olig./M. Mio. S.A.; *Dasycercus*, Pleist. s.w. Aust., Rec.; *Dasykaluta*, ?Pleist. W.A., Rec.; *Dasylurinja*, L. Olig./M. Mio. S.A.; *Dasyuroides*, Plio., S.A., ?E. Pleist. N.S.W., W.A., Rec.; *Dasyurus*, Plio. N.S.W., Qd., S.A., Vic., Quat. Aust., Rec.; *Ganbulanyi*, L. Olig./M. Mio. Qd.; *Glaucodon*, L. Plio. Vic., M. Plio. S.A.; *Keeuna*, L. Olig./M. Mio. S.A.; *Myoictis*, Plio. N.G., Rec.; *Parantechinus*, Pleist. s.w. Aust., Rec.; *Phascogale*, ?M. Plio. S.A. Pleist. e.s.w. Aust., Rec.; *Planigale*, E. Plio. Qd., Pleist. Aust. except Tas., Rec.; *Pseudantechinus*, ?Pleist. W.A., Rec.; *Sarcophilus*, ?L. Plio., N.S.W., Pleist. Aust., Rec.; *Satanellus*, ?E. Plio. e.?w. Aust., N.G., Rec.; *Sminthopsis*, L. Plio. N.S.W., Qd, ?E. Pleist. e.s.w. Aust., Rec.; *Wakamatha*, ?M. Mio. S.A. **Myrmecobiidae** *Myrmecobius*, Pleist. s.cent. Aust., Rec. **Thylacinidae.** Genera indet., L. Olig./ M. Mio. Qd.; *Thylacinus*, L.Olig./M. Mio. Qd., L. Mio. N.T., Plio. N.G., N.S.W., Qd., S.A.; Pleist. Aust., Rec.; *Muribacinus*, L. Olig./M. Mio. Qd.; *Nimbacinus*, L. Olig./M. Mio. N.T., Qd.; *Ngamalacius*, L. Olig./M. Mio. Qd.

ORDER: PERAMELEMORPHIA
Family Uncertain. *Yarla*, L. Olig..M. Mio. Qd. **Peramelidae.** *Chaeropus*, E. Pleist. s.cent. Aust., Rec.; *Isoodon*, L. Plio.-Pleist. Vic., Rec.; *Perameles*, E. Plio. N.S.W., Qd., Vic., Pleist. Aust., Rec. **Thylacomyidae.** *Ischnodon*, Plio./E. Pleist. S.A.; *Macrotis*, Pleist. cent.e.s.w. Aust., Rec.

ORDER: NOTORYCTEMORPHIA
Notoryctidae. Genus indet., L. Olig./M. Mio. Qd.

ORDER: DIPROTODONTIA
SUBORDER: VOMBATIFORMES
INFRAORDER: PHASCOLARCTOMORPHIA
Phascolarctidae. *Koobor*, Plio. Qd.; *Nimiokoala*, L. Olig./M. Mio. Qd.; *Madakoala*, L. Olig./M. Mio. S.A; *Litokoala*, L. Olig./M. Mio. Qd., S.A.; *Perikoala*, L. Olig./M. Mio. S.A.; *Phascolarctos*, E. Plio. S.A., Pleist. Aust., Rec.

INFRAORDER: VOMBATOMORPHIA
Diprotodontidae. Subfamily Diprotodontinae. *Bematherium*, M. Mio. Qd.; *Diprotodon*, Plio. N.S.W., S.A., Pleist. Aust.; *Euowenia*, E. Plio.-Pleist. Qd., S.A., L. Plio. N.S.W., Plio.-Pleist. Vic.; *Euryzygoma*, E.-M. Plio.-Pleist. Qd.; *Meniscolophus*, L. Plio./E. Pleist. S.A.; *Nototherium*, Plio.-Pleist. N.G., N.S.W.; *Pyramios*, L. Mio. N.T.; Subfamily Zygomaturinae. *Alkwertatherium*, L. Mio. N.T.; *Hulitherium*, Pleist. N.G.; *Kolopsis*, L. Mio. N.T., Vic., Plio. N.G.; *Kolopsoides*, Plio. N.G.; *Nimbadon*, L. Olig./M.

Mio. N.T., Qd.; *Neohelos*, L. Olig./M. Mio. N.T., Qd., S.A.; *Plaisiodon*, L. Mio. N.T.; *Raemeotherium*, L. Olig./M. Mio. S.A.; *Silvabestius*, L. Olig./M. Mio. Qd.; *Zygomaturus*, L. Mio.-Pleist. N.S.W., Qd., S.A., Tas., Vic., W.A., ?E. Plio. N.T., Pleist, N.G. Subfamily indet. *Brachalletes*, Plio. Qd.; *Koalemus*, Plio. Qd..; *Sthenomerus*, ?Pleist. Qd.. **Palorchestidae.** *Ngapakaldia*, L. Olig./M. Mio. Qd., S.A.; *Pitikantia*, L. Olig./M. Mio. S.A.; *Palorchestes*, L. Olig./M. Mio. Qd., L. Mio. N.T., Vic., Plio. N.S.W., Qd., S.A., Pleist. Qd., N.S.W., Vic., S.A., Tas. **Wynyardiidae.** *Namilamadeta*, L. Olig./M Mio. Qd., S.A.; *Wynyardia*, L. Olig./M. Mio. Tas.; *Muramura*, L. Olig./M. Mio. S.A. **Ilariidae.** *Ilaria*, L. Olig./M. Mio. S.A.; *Kuterintja*, L. Olig./M. Mio. S.A. **Vombatidae.** *Lasiorhinus*, L. Plio. N.S.W., S.A., Rec.; *Phascolomys*, L. Plio. S.A., Pleist. e., s. Aust.; *Phascolonus*, Plio. N.S.W., S.A., Vic., W.A., Pleist. Aust.; *Ramsayia*, E. Plio. N.S.W., Pleist. N.S.W., Qd.; *Rhizophascolonus*, L. Olig./M. Mio. Qd.; *Vombatus*, Plio. N.S.W., Qd., S.A., Vic., Pleist. Aust., Rec.; *Warendja*, Pleist. S.A., Vic. **Thylacoleonidae.** *Thylacoleo*, Plio. N.S.W., Qd., S.A., Vic., Pleist. Aust.; *Wakaleo*, L. Olig./M. Mio. N.T., Qd., S.A.; *Priscileo*, L. Olig./M. Mio. S.A.

INFRAORDER: PHALANGERIDA
SUPERFAMILY: PHALANGEROIDEA
Phalangeridae. Subfamily Ailuropinae. *Ailurops*, Pleist. Sulawesi, Rec.. Subfamily Phalangerinae, Tribe Trichosurini. *Trichosurus*, L. Olig./M. Mio. Qd., Plio. Vic., Rec.; *Strigocuscus*, L.Olig./M. Mio. Qd., E. Plio. Vic., Rec. Tribe Phalangerini. *Phalanger*, E. Plio. Vic., Rec. **Miralinidae.** *Miralina*, L. Olig./M. Mio. S.A **Ektopodontidae.** *Ektopodon*, L. Olig./M. Mio. S.A.; *Chunia*, L. Olig./ M. Mio. S.A.; *Darcius*, E. Plio., E. Pleist. Vic.

SUPERFAMILY: MACROPODOIDEA
Potoroidae. Subfamily Bulungamayinae. *Bulungamaya*, L. Olig./M. Mio. Qd., S.A.; *Gunguroo*, L. Olig./M. Mio. Qd.; *Nowidgee*, L. Olig./M. Mio. Qd.; *Wabularoo*, L. Olig./M. Mio. Qd.; *Galanarla*, L. Olig./M. Mio. Qd. Subfamily Hypsiprymnodontinae. *Hypsiprymnodon*, L. Olig./M. Mio. Qd., Plio. Vic., Rec. Subfamily Propleopinae. *Ekaltadeta*, L. Olig./ M. Mio. Qd.; *Propleopus*, E. Plio.-Pleist. N.S.W., Qd., S.A., Vic. Subfamily Potoroinae. *Aepyprymnus*, Pleist. Qd., Rec.; *Bettongia*, Plio. N.S.W., S.A., Vic., Tas., W.A., N.S.W., Rec.; *Milliyowi*, E. Plio. Vic.; *Caloprymnus*, Pleist. N.S.W., W.A., Rec.; *Potorous*, M. Plio. S.A., L. Plio.-Pleist. Vic., Pleist. S.A., W.A., Tas., Rec.; *Purtia*, L. Olig./M. Mio. S.A.; *Wakiewakie*, L. Olig./M. Mio. N.T., Qd., S.A.; *Gumardee*, L. Olig./M. Mio. Qd., S.A. Subfamily Palaeopotoroinae. *Palaeopotorous*, L. Olig./M. Mio. S.A.
Macropodidae. Subfamily Balbarinae. *Balbaroo*, L. Olig./M. Mio. N.T., Qd., S.A.; *Wururoo*, L. Olig./ M. Mio. Qd. Subfamily Sthenurinae. *Lagostrophus*, L. Plio. W.A., N.S.W., Rec.; *Procoptodon*, Pleist. N.S.W., Qd., S.A., Vic.; *Sthenurus*, Plio. N.S.W., Qd. S.A., Vic., Pleist. Aust.; *Troposodon*, Plio. Qd., Vic., S.A., Pleist. Aust.; *Simosthenurus*, L. Plio. N.S.W., Vic., Pleist. Aust. Subfamily Macropodinae. *Bohra*, Pleist. N.S.W.; *Congruus*, Pleist S. A.; *Dendrolagus*, E.-M. Plio., Pleist. N.G., N.S.W., Vic., L. Plio. S.A., Rec.; *Dorcopsis*, L. Olig./ M. Mio. N.T., E. Plio. Vic., S.A., L. Plio. N.G.; *Dorcopsulus*, L. Mio. N.T., E. Plio. N.T.; *Dorcopsoides*, Pleist. N.G., Rec.; *Hadronomas*, L. Mio. N.T.; *Kurrabi*, Plio. N.S.W., S.A., Vic.; *Lagorchestes*, Pleist. N.S.W., S.A., W.A., Rec.; *Macropus*, Plio. N.S.W., Qd., Vic., Rec.; *Nambaroo*,

L. Olig./M. Mio. Qd., S.A.; *Onychogalea*, Pleist. N.S.W., W.A., Rec.; *Petrogale*, Plio. N.S.W., Qd., Rec. Aust. except Tas.; *Prionotemnus*, E.-M. Plio. N.S.W., Qd., L. Plio./E. Pleist. Qd., S.A.; *Protemnodon*, E. Plio. Vic., E. Plio.-Pleist. N.G., N.S.W., Qd., S.A., Tas., Vic., W.A.; *Setonix*, Pleist. W.A., Rec.; *Thylogale*, Plio. Vic., Rec.; *Wallabia*, Plio. N.S.W., Qd., Vic., Pleist. Aust., Rec.; *Watutia*, Plio. N.G.

SUPERFAMILY: BURRAMYOIDEA
Burramyidae. Genus indet., L. Olig.-E. Mio., Tas.; *Burramys*, L. Olig./M. Mio. Qd., S.A., Plio. Vic., Pleist.-Rec. Aust.; *Cercartetus*, L. Olig./M. Mio., Pleist. Qd., Pleist.-Rec. N.S.W., S.A., W.A.

SUPERFAMILY: PETAUROIDEA
Pseudocheiridae. *Marlu*, L. Olig./M. Mio. S.A.; *Paljara*, L. Olig./M. Mio. Qd. S.A.; *Pildra*, L. Olig./ M. Mio. Qd., S.A.; *Pseudocheirops*, L. Olig./M. Mio. Qd., L. Mio. N.T., E.-M. Plio. N.S.W., Pleist. N.G., Rec.; *Petauroides*, Pleist. e. Aust., Rec.; *Pseudocheirus*, Plio. Vic., Pleist. Aust. N.G., Rec.; *Pseudokoala*, E. Plio. Vic. **Petauridae.** Subfamily Petaurinae, *Gymnobelideus*, Pleist.-Rec. & Aust. N.G.; *Petaurus*, E. Plio. S.A., Vic., Rec. Dactylopsilinae. Genus indet., L. Olig./M. Mio. Qd. **Pilkipildridae.** *Pilkapildra*, L. Olig./M. Mio. S.A.; *Djilgaringa*, L. Olig./M. Mio. Qd., S.A.

SUPERFAMILY: TARSIPEDOIDEA
Tarsipedidae. *Tarsipes*, Pleist. W.A., Rec. **Acrobatidae.** *Acrobates*, L. Olig./M. Mio. Qd., Pleist. Aust., Rec.

ORDER: YALKAPARIDONTIA
Yalkaparidontidae. *Yalkaparidon*, L. Olig.-M. Mio. Qd.

INFRACLASS: PLACENTALIA
ORDER: AUSKTRIBOSPHENIDA
Ausktribosphenidae. *Ausktribosphenos*, E. Cret. Vic.

ORDER: ?CONDYLARTHRA
Family Uncertain. *Tingamarra*, Eo. Qd.

ORDER: CHIROPTERA
SUBORDER: MEGACHIROPTERA
Family Uncertain. *Australonycteris*, Eo. Qd. **Pteropodidae.** *Aproteles*, Pleist.-Rec. N.G.; *Dobsonia*, Pleist. N.G., Rec.; *Pteropus*, Pleist. N.G., Rec.

SUBORDER: MICROCHIROPTERA.
Family uncertain. Genus indet., E. Plio. Vic.; *Australonycteris*, Eo. Qd. **Rhinolophidae.** Genus indet., M. Mio., S.A.; *Rhinolopus*, Pleist. Qd., Vic., Rec. **Hipposideridae.** *Hipposideros*, L. Olig./M. Mio.-Pleist., Qd., Rec.; *Miophyllorhina*, L. Olig./M. Mio. Qd.; *Rhinonicteris*, L. Olig./M. Mio. Qd., Rec.; *Riversleigha*, L. Olig./M. Mio. Qd.; *Xenorhinos*, L. Olig./M. Mios. Qd. **Megadermatidae.** Genus indet., M. Mio. Qd.; *Macroderma*, L. Olig./M. Mio., Plio. Qd., E.-M. Plio. N.S.W., Pleist. N.S.W., N.T., Qd., S.A., W.A., Rec. **Molossidae.** Genus indet., L. Olig./M. Mio. Qd.; *Mormopterus*, Pleist. Vic., Rec.; *Tadarida*, Pleist. e., w. Aust., Rec. **Mystacinidae.** *Icarops*, L. Olig./M. Mio. Qd., M. Mio. N. T. **Vespertilionidae.** *Chalinolobus*, Pleist. Vic., W.A., Rec.; *Eptesicus*, Pleist. W.A., Rec.; *Miniopterus*, Pleist. Qd., Vic., Rec.; *Nyctophilus*, Pleist. Qd., Vic., W.A., Rec.; *Pipistrellus*, Pleist. Vic., W.A., Rec.

ORDER PRIMATES
Hominidae. *Homo,* L. Pleist. Aust., N.G., Rec.

ORDER: CARNIVORA
Otariidae. *Arctocephalus,* Pleist. N.Z., Rec.; *Neophoca,* M. Pleist. Vic., N.Z., Rec. **Phocidae.** *Mirounga,* Pleist. N.Z., Rec.; Monarchinae indet., L. Mio-E. Pleist.-E. Plio. Vic.; *Ommatophocca,* L. Plio. N.Z., Rec.

ORDER: RODENTIA
Muridae. Subfamily Hydromyinae. *Anisomys,* Pleist. N.G., Rec.; *Conilurus,* Pleist. Qd., S.A., Vic., Rec.; *Hydromys,* Pleist. N.S.W., Tas., Vic., W.A., Rec.; *Hyomys,* Pleist. N.G., Rec.; *Leggadina,* Pleist. Qd., Rec.; *Leporillus,* Pleist. N.S.W., S.A., W.A., Rec.; *Mallomys,* Pleist. N.G., Rec.; *Mastacomys,* Pleist. N.S.W., Vic., Tas., S.A., Rec.; *Melomys,* Pleist. N.G. Vic., Rec.; *Mesembriomys,* ?Pleist. W.A., Rec.; *Notomys,* Pleist. W.A., N.S.W., S.A., Rec.; *Pseudomys,* Plio. Qd., Vic., Pleist. Aust., Rec.; *Solomys,* Pleist. Sol. Is., Rec.; *Uromys,* Pleist. N.G., Rec.; *Zyzomys,* Plio. Qd., ?Pleist. N.T., Qd., W.A., Rec. Subfamily Murinae. *Rattus,* L. Plio.-E. Pleist. Qd., Pleist. Qd., N.S.W., Vic., S.A., W.A., Rec.

ORDER: CETACEA
SUBORDER: INCERTAE SEDIS.
Cetotilites, E. Olig. Vic.; *"Squalodon",* Olig. S.A.

SUBORDER: ARCHAEOCETI
Basilosauridae. *Kekenodon,* L. Olig. N.Z.

SUBORDER: MYSTICETI.
Family uncertain. Genera indet., Eo.-Olig., N.Z.; *Mammalodon,* L. Olig. Vic. **Cetotheriidae.** Genera indet., E. Mio.-E. Plio. Vic., L. Mio. S.A.; *Mauicetus,* L. Olig.-E. Mio. N.Z.; *cf. Parietobalaena.* (="Aglaocetus"), E. Mio. S.A.; *Pelocetus,* E. Mio. Vic. **Balaenopteridae.** Genera indet., M. Mio.-E. Plio. Vic., L. Mio. N.Z.; *Balaenoptera,* Plio. N.Z., Rec.; *Megaptera.,* ?L. Plio. Tas., L. Mio.-E. Plio. Vic., Rec. Aust. **Balaenidae.** Genera indet., L. Plio. N.Z., ?Mio.-L. Plio. S.A.; *Eubalaena,* Rec. Aust.; *cf. Eubalaena,* L. Plio. S.A., L. Mio.-E. Plio. Vic., Rec. Aust.; *cf. Morenocetus,* E.-?M. Mio. S.A.

SUBORDER: ODONTOCETI
Squalodontidae. Genera indet., L. Olig.-E. Mio. N.Z., E. Mio. Vic.; *Austrosqualodon,* L. Olig. N.Z.; *Metasqualodon,* L. Olig. e.S.A.; *Microcetus,* L. Olig.-E. Mio. N.Z.; *Parasqualodon,* L. Olig. Vic.; *Prosqualodon.* L. Olig.-E. Mio. N.Z., E. Mio. Tas.; *Tangaroasaurus,* E. Mio. N.Z. **Waipatiidae.** *Waipatia,* L. Olig. N. Z. **Eurhinodelphidae.** Genera indet., L. Olig.-L. Mio. N.Z.; *Phocaenopsis,* E. Mio. N.Z. **Ziphiidae.** Genera indet., ?E Plio. Tas., M.-L. Mio. Chatham Rise; *?Hyperoodon,* M.-L. Mio. N.Z.; *Mesoplodon,* L. Mio.-E. Plio. Vic., Rec.; *"Ziphius",* ?E Plio. Tas. **Physeteridae.** Genera indet., L. Olig.-E. Mio. N.Z., E. Mio.-E. Plio. Vic., ?E. Plio. Tas., L. Plio. N.Z.; *Physetodon.,* L. Mio-E. Plio. Vic.; *Scaldicetus,* M.-L. Mio. Chatham Rise, L. Mio.-E. Plio. Vic.; *Scaptodon,* ?E. Plio. Tas. **Kogiidae.** *?Kogia,* M.-L. Mio. Chatham Rise. **Kentriodontidae.** Genera indet., L. Olig.-E. Mio. N.Z. **Rhabdosteidae.** Genus indet., L. Olig./M. Mio. S.A **Delphinidae.** Genera indet., Neogene N.Z., L. Mio.-E. Plio. Vic., L. Plio. N.Z.; *Delphinus,* L. Plio. N.Z., Quat. Vic., Rec.;

Globicephala, Neogene N.Z., Rec.; *?Pseudorca,* L. Plio. N.Z.; *?Orcinus,* E. Pleist. N.Z.; *"Steno",* L. Mio.-E. Plio. Vic., E. Plio. N.Z.

ORDER: PERISSODACTYLA
Rhinoceratidae. *cf. Rhinoceros,* (not *Zygomaturus*) Quat. N. Caledonianot *in situ.*

ORDER: PROBOSCIDEA
Mastodon, N.S.W.; *Notoelephas,* Qd. (transported into Australia by trade, not found *in situ*).

ORDER: SIRENIA
Halicore, Age uncertain N.G.

CLASS: AVES
ORDER: ENANTIORNITHIFORMES
Enantiornithidae. *Nanantius,* E. Cret. Aust.

ORDER: SPHENISCIFORMES
Spheniscidae. *Anthropodytes,* E. Mio. Vic.; *Anthropornis,* L. Eo. Ant. Peninsula., S.A.; *Aptenodytes,* Mio.-Plio. N.Z., Rec.; *Archaeospheniscus,* L. Olig. Ant. Peninsula., N.Z.; *Duntroonornis,* L. Olig. N.Z.; *Eudyptes,* Quat. N.Z., Rec.; *Eudyptula,* Quat. N.Z., Rec.; *Korora,* L. Olig.-E. Mio. N.Z.; *Marplesornis,* Mio.-Plio. N.Z.; *Megadyptes,* Quat. Chatham Is. N.Z., Rec.; *Pachydyptes,* L. Eoc. N.Z.; *Palaeeudyptes,* L. Eoc.-e. Tert. Ant. Peninsula, S.A., Vic., L. Eo.-E. Mio. N.Z.; *Platydyptes,* L. Olig.-E. Mio. N.Z.; *Pseudoaptenodytes,* L. Mio. Vic.; *Pygoscelis,* Plio. N.Z., Rec.; *Tasidyptes,* Quat. Tas.; *Tereingaornis,* M. Plio. N.Z.

ORDER: CASUARIIFORMES
Casuariidae. *Casuarius,* Plio. N.G., Pleist. ?n. Aust., Rec.; *Emuarius,* L. Olig./M. Mio. Qd., N.T., S.A.; *Dromaius,* E.-M. Plio. Qd., Plio. S.A., Quat. Aust., Rec.

ORDER: DINORNITHIFORMES
Anomalopterygidae. *Anomalopteryx,* Quat. N.Z.; *Emeus,* Quat. N.Z.; *Euryapteryx,* Quat. N.Z.; *Megalapteryx,* Quat. N.Z.; *Pachyornis,* Quat. N.Z. **Dinornithidae.** *Dinornis,* Quat. N.Z.

ORDER: APTERYGIFORMES
Apterygidae. *Apteryx,* Quat. N.Z., Rec.

ORDER: PODICIPEDIFORMES
Podicipedidae. *Podiceps,* Quat. N.Z., Rec.; Genera Indet., L. Olig./M. Mio.-Quat. Aust.

ORDER: PROCELLARIIFORMES
Diomedeidae. *Diomedea,* L. Mio. Vic., Rec. **Pelecanoididae.** *Pelecanoides,* Quat. Aust. N.Z., Rec. **Procellariidae.** *Daption,* Quat. N.Z., Rec.; *Macronectes,* Quat. N.Z., Rec.; *Fulmarus,* Quat. N.Z., Rec.; *?Halobaena,* Quat. N.Z., Rec.; *Pachyptila,* Quat. N.Z., Rec.; *Pterodroma,* Quat. N. Caledonia, N.Z., Rec.; *Procellaria,* Quat. N.Z., Rec.; *Puffinus,* Pleist. Aust., N.Z., L. Howe Is., Rec. **Oceanitidae.** *Fregetta,* Quat. N.Z., Rec.; *Garrodia,* Quat. N.Z., Rec.; *Oceanites,* Quat. N.Z., Rec.; *Pelagodroma,* Quat. N.Z., Rec. **Pelecanoididae.** *Pelecanoides,* Quat. N.Z., Rec.

ORDER: PELECANIFORMES
SUBORDER: PELECANI
SUPERFAMILY: PELECANOIDEA
Pelecanidae. *Pelecanus,* L. Olig./M. Mio., Plio.-Pleist. S.A., E.-M. Plio. Qd., Quat. N.Z., Rec.

Pelagornithidae. *Neodontornis,* Plio. N.Z.; *?Odontopteryx, Pelagornis,* Mio. N.Z.; *Pseudodontornis,* Tert. N.Z.

SUPERFAMILY: SULOIDEA
Sulidae. *Sula,* Quat. N.Z., Rec. **Anhingidae.** *Anhinga,* L. Plio. S.A., Quat. S.A., Qd., Rec. **Phalacrocoracidae.** Genus indet., Mio. Qd.; *Microcarbo,* Plio. Qd.; *Leucocarbo,* Quat. N.Z., Rec.; *Phalacrocorax,* L. Olig./M. Mio.-Quat. Aust., Quat. N.Z., Rec.; *Stictocarbo,* Quat. N.Z., Rec.
ORDER: CICONIIFORMES
SUBORDER: ARDEAE
Ardeidae. Genus indet., L. Plio. S.A.; *Egretta,* Quat. N.Z., Rec.; *Botaurus,* Quat. N.Z., Rec.; *Ixobrychus,* Quat. N.Z., Rec.; *Platalea,* Quat. N.Z., Rec.

SUBORDER: CICONIAE
Ciconiidae. Genus indet., Mio. Qd., L. Plio. S.A.; *Ephippiorhychus,* Pleist. Aust., Rec.; *Ciconia,* Plio.-Quat. Qd.; *Xenorhychus,* Plio. n. Aust., Rec.; *Palaeopelargus,* Pleist. Qd.; **Threskiornithidae.** *Threskiornis,* Plio. Qd., Quat. Aust., Rec.

ORDER: PHOENICOPTERIFORMES
Phoenicopteridae. Genus indet., L. Mio. N.T.; *Ocyplanus,* L. Plio.-Quat. S.A.; *Phoeniconotius,* L. Olig./M. Mio. S.A.; *Phoenicopterus,* L. Olig./ M. Mio. S.A., Pleist. S.A., Rec.; *Xenorhynchopsis,* L. Plio.-Quat. S.A. **Palaelodidae.** *Palaelodus,* M. Mio.-M. Pleist. S.A.

ORDER: DROMORNITHIFORMES
Dromornithidae. *Barawertornis,* L. Olig./M. Mio. Qd.; *Bullockornis,* L. Olig./M. Mio. N.T., Qd.; *Dromornis,* L. Olig./Mio. Qd., L. Mio. N.T., Plio. N.S.W.; *Genyornis,* Pleist. Aust. except N.T.; *Ilbandornis,* L. Mio. N.T.

ORDER: ANSERIFORMES
SUBORDER: ANSERES
Anatidae. Genus indet., Mio. Qd.; *Anas,* Plio.-Quat. Aust., Quat. N. Caledonia, N.Z., Rec.; *Anseranas,* E.-M. Plio. Qd., Rec.; *Aytha,* E.-M. Plio. Qd., Quat. Aust. N.Z., Rec.; *Archeocygnus,* Quat. S.A.; *Biziura,* Plio.-Quat. Aust. N.Z., Rec.; *?Cereopsis,* E.-M. Plio. Qd., Rec.; *Chenopsis,* Quat. S.A.; *Cnemiornis,* Quat. N.Z., Rec.; *Cygnus,* E.-M. Plio. Qd., Quat. Aust. N.Z., Rec.; *Dendrocygna,* E.-M. Plio. Qd., Rec.; *Euryanas,* Quat. N.Z.; *Hymenolaimus,* Quat. N.Z., Rec.; *Malacorhynchus,* Quat. Aust. N.Z., Rec.; *Mergus,* Quat. N.Z., Rec.; *Oxyura,* Quat. N.Z., Rec.; *Pachyanas,* Quat. Chatham Is.; *Tadorna,* E.-M. Plio. Qd., Quat. Aust. N.Z., Rec.

ORDER: FALCONIFORMES
SUBORDER: FALCONES
SUPERFAMILY: FALCONOIDEA
Accipitridae. *Aquila,* Quat. Aust., Rec.; *Hieraeetus,* Quat. N.Z., Rec.; *Accipiter,* Quat. N. Caledonia, N.Z., Rec.; *Circus,* Quat. N.Z., Rec.; *Aviceda,* Quat. S.A., Rec.; *Harpagornis,* Quat. N.Z.; *Hieraaetus,* Quat. Aust., Rec.; *Haliaeetus,* Pleist. Kangaroo Is., Aust. N.Z., Rec.; *Ichthyophaga,* Quat. Chatham Is., Rec.; *Necrastur,* Plio. Qd.; *Pengana,* Mio. Qd.; *Taphaetus,* Quat. S.A., Qd. **Falconidae.** *Plioaetus,* Pleist. Aust.; *Falco,* Quat. Aust. N. Caledonia. N.Z., Rec.

ORDER: GALLIFORMES
SUBORDER: GALLI

Megapodiidae. *Megapodius,* Quat. N. Caledonia, Rec.; *Palaeopelargus,* Pleist. Aust.; *Progura,* Quat. Aust.; *Leipoa,* L. Pleist. Aust., Rec.; *Sylviornis,* Quat. N. Caledonia. **Phasianidae.** *Coturnix,* Quat. Aust. N.Z., Rec.

ORDER: GRUIFORMES
SUBORDER: TURNICES
Turnicidae. *Turnix,* Pleist. Aust., N. Caledonia,Rec.

SUBORDER: GRUES
Gruidae. Genus indet., L. Olig./M. Mio. S.A.; *Grus,* L. Plio.-Quat. S.A., ne. Aust., Rec. **Rallidae.** Genera indet., L. Olig./M. Mio.-Quat. S.A., Mio. Qd.; *Capellirallus,* Quat. N.Z.; *Porphyrio,* Quat. N. Caledonia, N.Z., Rec.; *Diaphorapteryx,* Quat. Chatham Is.; *Fulica,* Plio. Aust., Quat. Aust. Chatham Is., N.Z., Rec.; *Gallirallus,* Quat. Aust. Chatham Is., N. Caledonia, N.Z., Rec.; *Gallinula,* Quat. Aust. N.Z., Rec.; *Rallus,* Quat. Aust. N.Z., Rec.; *Porzana,* Quat. Aust., N. Caledonia, Rec.; *Tricholimnas,* Quat. N. Caledonia. **Aptornithidae.** *Aptornis,* Quat. N.Z. **Rhinochetidae.** *Rhinochetos,* Quat. N. Caledonia, Rec.

SUBORDER: OTIDES
Otididae. Genus indet., L. Plio. S.A.; *Ardeotis,* Pleist. S.A., Rec.

ORDER: CHARADRIIFORMES
SUBORDER: CHARADRII
Charadriidae. *Anarhynchus,* Quat. N.Z., Rec.; *Charadrius,* Quat. N. Caledonia, N.Z., Rec.; *Calidris,* Quat.-Pleist. W.A., S.A., Rec.; *Erythrogonys,* Quat. Vic., Rec.; *Peltohyas,* Quat. S.A., Rec.; *Pluvius,* Quat. N. Caledonia, Rec.; *Thinornis,* Quat. N.Z., Rec.; *Vanellus,* Quat. Aust., Rec. **Haematopodidae.** *Haematopus,* Quat. N.Z., Rec. **Scolopacidae.** *Arenaria,* Quat. N.Z., Rec.; *Coenocorypha,* Quat. Chatham Is., N. Caledonia, Rec.; *Calidris,* Quat. N.Z. S.A., Rec.; *Gallinago,* Quat. S.A., Kangaroo Is., Rec.; *Limosa,* Quat. N.Z., Rec.; *Numenius,* E.-M. Plio. Qd., Quat. N.Z., Rec.; *Tringa,* Quat. S.A., Rec. **Burhinidae.** Genus indet., L. Olig./M. Mio. S.A.; *Burhinus,* Quat. S.A., Rec. **Pedionomidae.** *Pedionomus,* L. Plio. or Quat., Vic., Rec. **Recurvirostridae.** *Himantopus,* Quat. N.Z., Rec.

SUBORDER: LARI
Laridae. *Larus,* Quat. Aust. N.Z., Rec.; *Hydroprogne,* Quat. N.Z., Rec.; *Sterna,* Quat. Aust. N.Z., Rec. **Stercorariidae.** *Stercorarius,* Quat. N.Z., Rec.

ORDER: COLUMBIDIFORMES
SUBORDER: COLUMBAE
Columbidae. Genus indet., L. Olig./M. Mio. S.A.; *Caloenas,* Quat. N. Caledonia, Rec.; *Chalcophaps,* Quat. N. Caledonia, Rec.; *Columba,* Quat. N. Caledonia, Rec.; *Drepanoptila,* Quat. N. Caledonia, Rec.; *Ducula,* Quat. N. Caledonia, Rec.; *Gallicolumba,* Quat. N. Caledonia, Rec.; *Hemiphaga,* Quat. N.Z., Rec.; *Leucosarcia,* Quat. Aust., Rec.; *Ocyphaps,* Quat. Aust., Rec.; *Phaps,* Pleist. Qd., S.A., Rec.; *Ptilinopus,* Quat. Aust., Rec.; *?Cyanorhamphus,* Quat. N.Z., Rec.

ORDER: PSITTACIFORMES
Cacatuidae. *Cacatua,* L. Olig./M. Mio. Qd.; *Calyptorhynchus,* Quat. Aust., Rec.; *Callocephalon,* Quat. Aust., Rec.; *Cacatua,* Pleist. Aust., Rec. **Loriidae.** *Glossopsitta,* Quat. Aust., Rec.; *Trichoglossus,* Quat. Aust., Rec. **Platyceridae.** *Cyanorhamphus,* Quat. N.Z., Rec.; *Psephotus,* Quat. Aust., Rec.; *Melopsittacus,* Plio. Qd., Quat. Aust., Rec.; *Geopsittacus,* Quat. Aust., Rec.; *Pezoporus,* Quat. Aust., Rec.; *Alisterus,* Quat. Aust., Rec.; *Polytelis,* Quat. Aust., Rec.; *Lathamus,* Quat. Aust., Rec.; *Purpureicephalus,* Quat. Aust., Rec.; *Platycercus,* Quat. Aust., Rec.; *Barnardius,* Quat. Aust., Rec.; *Northiella,* Quat. Aust., Rec.; *Neophema,* Quat. Aust., Rec. **Nestoridae.** *Nestor,* Quat. N.Z., Rec.; *Strigops,* Quat. N.Z., Rec.

ORDER: CUCULIFORMES
SUBORDER: CUCULI
Cuculidae. *Centropus,* Quat. Aust., Rec.; *Cuculus,* Quat. Aust., Rec.; *Chrysococcyx,* Quat. Aust., Rec., Chatham Is., *Eudynamys,* Quat. N.Z., Rec.; *Urodynamis,* Quat. N. Caledonia, Rec.

ORDER: STRIGIFORMES
Tytonidae. *Tyto,* Quat. Aust. N.Z., N. Caledonia, Rec. **Strigidae.** *Ninox,* Quat. Aust., N. Caledonia, N.Z., Rec. **Family uncertain.** *Sceloglaux,* Quat. N.Z., Rec.

ORDER: CAPRIMULGIFORMES
SUBORDER: CAPRIMULGI
Caprimulgidae. *Caprimulgus,* Quat. Aust., N. Caledonia, Rec. **Podargidae.** *Podargus,* Quat. W.A., Vic., Rec. **Aegothelidae.** *Aegotheles,* Quat. Aust., N. Caledonia, Rec.; *Megaegotheles,* Quat. N.Z.; *Quipollornis,* Mio. N.S.W.

ORDER: APODIFORMES
SUBORDER: APODI

Apodidae. Genus indet., Mio. Qd.; *Hirundapus,* Quat. Aust., Rec.; *Collocalia,* Quat. Aust., N. Caledonia, Rec.

ORDER: CORACIIFORMES
SUBORDER: ALCEDINES
Alcedinidae. Genus indet., Mio., Qd.; *Ceyx,* Quat. Aust., Rec.; *Halcyon,* Quat. Aust., N. Caledonia, N.Z., Rec.; *Dacelo,* Quat. Aust., Rec.

ORDER: PASSERIFORMES
Family uncertain. Genera indet., L. Olig./ M. Mio. S.A., Qd., L. Plio. S.A. **Menuridae.** *Menura,* L. Olig./M. Mio. Qd. **Atrichornithidae.** *Atrichornis,* Quat. Aust., Rec. **Hirundinidae.** *Hirundo,* Quat. Aust., Rec.; *Cecropis,* Quat. Aust., Rec. **Motacillidae.** *Anthus,* Quat. Aust., Rec. **Muscicapidae.** *Bowdleria,* Quat. N.Z., Rec; *Gerygone,* Quat. N.Z., Rec.; *Mohoua,* Quat. N.Z., Rec.; *Oreoica.* Quat. Aust., Rec.; *Pachycephala,* Quat. Aust., Rec.; *Petroica,* Quat. Aust. N.Z., Rec.; *Rhipidura,* Quat. N.Z., Rec. **Orthonychidae.** *Orthonyx,* L. Olig./M. Mio. Qd., Quat. Aust., Rec.; *Cinclosoma,* Quat. Aust., Rec.; *Psophodes,* Quat. Aust., Rec.; *Sphenostoma,* Quat. Aust., Rec. **Timaliidae.** *Pomatostomus,* Quat. Aust., Rec. **Sylviidae.** *Cincloramphus,* Quat. Aust., Rec.; *Megalurus,* Quat. Aust., Rec. **Acanthisittidae.** *Acanthisitta,* Quat. N.Z., Rec.; *Pachyplichas,* Quat. N.Z., Rec.; *Xenicus,* Quat. N.Z., Rec.; *Traversia,* Quat. N.Z., Rec. **Maluridae.** *Amytornis,* Quat. Aust., Rec.; *Anthus,* Quat. N.Z., Rec.; *Malurus,* Quat. Aust., Rec.; *Stipiturus,* Quat. Aust., Rec. **Acanthizidae.** *Acanthiza,* Quat. Aust., Rec.; *Daphoenositta,* Quat. Aust., Rec.; *Pycnoptilus,* Quat. Aust., Rec.; *Dasyornis,* Quat. Aust., Rec. **Climacteridae.** *Climacteris,* Quat. Aust., Rec. **Pardalotidae.** *Pardalotus,* Quat. Aust., Rec. **Neosittidae.** *Neositta,* Quat. Aust., Rec. **Meliphagidae.** *Anthornis,* Quat. N.Z., Rec.; *Melithreptus,* Quat. Aust., Rec.; *Notiomystis,* Quat. N.Z., Rec.; *Prosthemadera,* Quat. N.Z., Rec. **Ploceidae.** *Poephila,* Quat. Aust., Rec. **Oriolidae.** Genus indet., Mio. Qd. **Ptilonorhynchidae.** *Ptilonorhynchus,* Quat. Aust., Rec. **Paradisaediae.** *Turnagra,* Quat. N.Z., Rec. **Callaeidae.** *Callaeas,* Quat. N.Z., Rec.; *Heterolocha,* Quat. N.Z.; *Philesturnus,* Quat. N.Z., Rec. **Artamidae.** *Artamus,* Quat. Aust., Rec. **Grallinidae.** *Grallina,* Quat. Aust., Rec. **Cracticidae.** *Strepera,* Quat. Aust., Rec.; *Gymnorhina,* Quat. Aust., Rec. **Corvidae.** *Corvus,* Quat. Kangaroo Is. Aust., Rec.; *Palaeocorax,* Quat. N.Z.

BIBLIOGRAPHY

AGER, D.V., 1963. *Principles of Palaeoecology*. McGraw-Hill Book.

ALDERMAN, A.R., 1967. The development of geology in South Australia: a personal view. *Records of the Australian Academy of Science*, **1** (2): 30–52.

AN, Z., BOWLER, J.M., OPDYKE, N.D., MACUMBER, P.G. & FIRMAN, J.B., 1986. Palaeomagnetic stratigraphy of Lake Bungunnia: Plio-Pleistocene precursor of aridity in the Murray Basin, southeastern Australia. *Palaeogeography, Palaeoclimatology, Palaeoecology*, **54**: 219–239.

ANDERSON, C., 1926. The Wellington Caves. *Australian Museum Magazine*, **2**: 367–374.

ANDERSON, J. M. & A. R. I. CRUICKSHANK, 1978. The biostratigraphy of the Permian and the Triassic. Part 5. A review of the classification and distribution of Permo-Triassic tetrapods. *Palaeontologia Africana*, **21**: 15–44.

ANDERSON, W., 1914. Note on the occurrence of the sand-rock deposits of extinct species of marsupials on King Island, Bass Strait, Tasmania. *Records of the Australian Museum*, **10**: 275–283.

ANDREWS, S.M., MILES, R.S. & WALKER, A.D., eds., 1977. *Problems in Vertebrate Evolution*. Academic Press, London.

ARCHER, M., 1977. Origins and subfamilial relationships of *Diprotodon* (Diprotodontidae, Marsupialia). *Memoirs of the Queensland Museum*, **18**: 37–41.

ARCHER, M., 1978a. Australia's oldest bat, a possible rhinolophid. *Proceedings of the Royal Society of Queensland*, **89**: 23.

ARCHER, M., 1981. A review of the origins and radiations of Australian mammals. In: A. Keast, ed., *Ecological Biogeography in Australia*, W. Junk, The Hague: 1437–1488.

ARCHER, M., ed., 1982. *Carnivorous Marsupials*. Surrey Beatty & Sons Pty Ltd and the Royal Zoological Society of New South Wales, Sydney.

ARCHER, M., 1984. Effects of humans on the Australian vertebrate fauna. In: M. Archer & G. Clayton, eds., *Vertebrate Zoogeography & Evolution in Australasia*. Hesperian Press, Carlisle: 151–161.

ARCHER, M., ed., 1987. *Possums and Opposums: Studies in Evolution*. Surrey Beatty & Sons Pty Ltd and the Royal Zoological Society of New South Wales, Sydney.

ARCHER, M., 1988. Riversleigh. Window into Our Ancient Past. *Australian Geographic*, **9**: 40–57.

ARCHER, M. & CLAYTON, G., eds., 1984. *Vertebrate Zoogeography & Evolution in Australasia. Animals in Space & Time*. Hesperian Press, Carlisle.

ARCHER, M., FLANNERY, T.F., RITCHIE, A. & MOLNAR, R.E., 1985. First Mesozoic mammal from Australia—An early Cretaceous monotreme. *Nature*, **318**: 363–366.

ARCHER, M., GODTHELP, H., HAND, S. & MEGIRIAN, D., 1989. Fossil mammals of Riversleigh, northwestern Queensland: Preliminary overview of biostratigraphy, correlation and environmental change. *Australian Zoologist*, **25**: 29–65.

ARCHER, M., HAND, S. & GODTHELP, H., 1988. A new order of Tertiary zalambdodont marsupials. *Science*, **239**: 1528–1531.

AUGEE, M.L., 1988. *Marine Mammals of Australia*. Royal Zoological Society of New South Wales, Sydney.

BAIRD, R.F., 1989. Fossil bird assemblages from Australian caves: Precise indicators of Late Quaternary palaeoenvironments? *Palaeogeography, Palaeoclimatology, Palaeoecology*, **69**: 241–244.

BAIRD, R.F. & VICKERS-RICH, P., 1997. *Eutreptodactylus itaboraiensis* gen. et sp. nov., an early cuckoo (Aves: Cuculidae) from the Late Paleocene of Brazil. *Alcheringa*, **21**: 123–127.

BALKERS, M., DAVIES, S.J.J.F. & REILLY, P.N., 1984. *The Atlas of Australian Birds*. Melbourne University Press–Royal Australasian Ornithologists Union, Melbourne.

BANDYOPADHYAY, S., 1988. A kannemeyeriid dicynodont from the Middle Triassic Yerrapalli Formation. *Philosophical Transactions of the Royal Society of London*, **B 320** (1198): 185–233.

BANDYOPADHYAY, S., 1989. The mammal-like reptile *Rechnisaurus* from the Triassic of India. *Palaeontology*, **32**: 305–312.

BANKS, M.R., COSGRIFF, J.W. & KEMP, N.R., 1984. A Tasmanian Triassic stream community. In: M. Archer & G. Clayton, eds., *Vertebrate Zoogeography and Evolution in Australasia*, Hesperian Press, Carlisle: 291–297.

BARKER, W.R. & GREENSLADE, P.J.M., 1982. *Evolution of the Flora and Fauna of Arid Australia*. Peacock Publications, Frewville.

BARRETT, C., 1946. *The Bunyip and Other Mythical Monsters and Legends*. Reed & Harris, Melbourne.

BARRON, E.J., HAY, W.W. & THOMPSON, S., 1989. The hydrologic cycle: A major variable during earth history. *Palaeogeography, Palaeoclimatology, Palaeoecology*, **75**: 157–174.

BARTHOLOMAI, A., 1979. New lizard-like reptiles from the Early Triassic of Queensland. *Alcheringa*, **3**: 225–234.

BARTHOLOMAI, A. & HOWIE, A., 1970. Vertebrate fauna from the Lower Triassic of Australia. *Nature*, **225**: 1063.

BARTHOLOMAI, A. & MOLNAR, R.E., 1981. *Muttaburrasaurus*, a new iguanodontid (Ornithischia: Ornithopoda) dinosaur from the Lower Cretaceous of Queensland. *Memoirs of the Queensland Museum*, **20**: 319–349.

BARTHOLOMAI, A. & WOODS, J.T., 1976. Notes on the vertebrate fauna of the Chinchilla Sand. *Bulletin of the Bureau of Mineral Resources, Geology and Geophysics of Australia*, **166**: 151–152.

BASALLA, G., 1967. The spread of western science. *Science*, **156**: 611–622.

BAVERSTOCK, P.R., WATTS, C.H.S., HIGARTH, J.T. & ROBINSON, A.C., 1977. Chromosome evolution in Australian rodents. II. The *Rattus* group. *Chromosoma* **61**: 227–241.

BAYNES, A. & LONG, J., eds., 1999. Papers in vertebrate palaeontology. *Records of the Western Australian Museum, Supplement*, **57**: 1–424.

BEHRENSMEYER, A.K. & HILL, A.P., 1980. *Fossils in the Making*. University of Chicago Press, Chicago.

BEHRENSMEYER, A.K. & KIDWELL, S.M., 1985. Taphonomy's contributions to paleobiology. *Paleobiology*, **11** (1): 105–119.

BEMIS, W.E., 1984. Paedomorphosis and the evolution of the Dipnoa. *Paleobiology*, **10** (3): 293–307.

BENSLEY, B.A., 1901. A theory of the origin and evolution of the Australian Marsupialia. *American Naturalist*, **35** (415): 245–269.

BENTON, M.J., 1983. Dinosaur success in the Triassic: A non-competitive ecological model. *The Quarterly Review of Biology*, **58** (1): 29–35.

BENTON, M.J., 1986a. More than one event in the late Triassic mass extinction. *Nature*, **321** (6073): 857–861.

BENTON, M.J., 1986b. The Late Triassic tetrapod extinction events. In: K. Padian, ed., *The Beginning of the Age of Dinosaurs; Faunal Change across the Triassic–Jurassic Boundary*, Cambridge University Press, Cambridge: 303–320.

BENTON, M.J., 1987. Mass extinctions among families of non-marine tetrapods: The data. *Mémoires Societe géologie France*, n.s., **150**: 21–32.

BENTON, M.J., 1988a. Mass extinctions in the fossil record of reptiles: Paraphyly, patchiness, and periodicity(?). In: G.P. Larwood, ed., *Extinction and Survival in the Fossil Record*, Clarendon Press, Oxford: 269–294.

BENTON, M.J., ed., 1988b. *The Phylogeny and Classification of the Tetrapods*. Volumes 1, 2. Clarendon Press, Oxford.

BERGER, W.H. & LABEYRIE, L.D., eds., 1987. *Abrupt Climatic Change: Evidence and Implications*. D. Reidel Publishing Company, Tokyo.

BERGGREN, W.A., KENT, D.V., FLYNN, J.J. & VAN COUVERING, J.A., 1985. Cenozoic geochronology. *Geological Society of America, Bulletin*, **96**: 1407–1418.

BERNER, R.A. & LASAGA, A.C., 1989. Modeling the geochemical carbon cycle. *Scientific American*, **260** (3): 54–61.

BERRY, W.B.N., 1968. *Growth of a Prehistoric Time Scale*. W.H. Freeman, San Francisco.

BLAINEY, G., 1975. *Triumph of the Nomads: A History of Ancient Australia*. Sun Books, Melbourne.

BLOXHAM, J. & GUBBINS, D., 1989. The evolution of the Earth's magnetic field. *Scientific American*, **261** (6): 30–37.

BONAPARTE, J.F., 1982. Faunal replacement in the Triassic of South America. *Journal of Vertebrate Paleontology*, **2** (3): 362–371.

BONAPARTE, J.F., 1990. New Late Cretaceous Mammals from the Los Alamitos Formation, Northern Patagonia. *National Geographic Research*, **6**: 63–93.

BONAPARTE, J.F., 1996. *Dinosaurios de America del Sur*. Museo Argentino de Ciencias Naturales "Bernardino Rivadavia," Buenos Aires.

BONAPARTE, J.F. & KIELAN-JAWOROWSKI, Z., 1987. Late Cretaceous dinosaur and mammal faunas of Laurasia and Gondwana. *Occasional Papers of the Tyrrell Museum of Paleontology 3, Fourth Symposium on Mesozoic Terrestrial Ecosystems*: 24–29.

BONNEMAINS, J., FORSYTH, E. & SMITH, B., 1988. *Baudin in Australian Waters: The Artwork of the French Voyage of Discovery to the Southern Lands 1800–1804.* Oxford University Press, Melbourne.

BOTKIN, D.B., CASWELL, M.F., ESTES, J.E. & ORIO, A.A., 1989. *Changing the Global Environment: Perspectives on Human Involvement.* Academic Press, Boston.

BOWLER, J.M., 1976. Aridity in Australia: Age, origins and expression in aeolian landforms and sediments. *Earth Science Reviews,* **12**: 279–310.

BOWLER, J.M., 1980. Quaternary chronology and palaeohydrology in the evolution of Mallee landscapes. In: Storrier & M.E. Stannard, eds., *Aeolian Landscapes in the Semi-arid Zones of South Eastern Australia,* R.R. Australian Society of Soil Science (Riverina Branch), Wagga Wagga, N.S.W.: 17–36.

BOWLER, J.M. & THORNE, A.G., 1976. Human remains from Lake Mungo: Discovery and excavation of Lake Mungo III. In: R.L. Kirk & A.G. Thorne, eds., *The Origin of the Australians,* Australian Institute of Aboriginal Studies, Canberra: 127–138.

BOWLER, J.M., THORNE, A.G. & POLACH, H.A., 1972. Pleistocene man in Australia: Age and significance of the Mungo skeleton. *Nature,* **240**: 48–50.

BRAIN, C.K., 1981. *The Hunters of the Hunted? An Introduction to African Cave Taphonomy.* University of Chicago Press, Chicago.

BRANAGAN, D.F., ed. 1973. *Rocks, Fossils, Profs: A History of the Geological Sciences Australian Science,* **6** (1): 71–84.

BRANAGAN, D.F. & TOWNLEY, K.A., 1976. The geological sciences in Australi—a brief historical review. *Earth Sci.ence Reviews,* **12**: 323–346.

BRIGGS, J.C., 1989. The historic biogeography of India: Isolation or contact? *Systematic Zoology,* **38**: 322–332.

BRONGNIART, A., 1828. *Prodrome d'une Histoire de Vegetaux Fossiles.* Paris.

BROOM, R., 1896. Report on a bone breccia deposit near the Wombeyan caves, N.S.W. with descriptions of some new species of marsupials. *Proceedings of the Linnean Society of New South Wales,* **21**: 48–61.

BROOME, R. & ROBINSON, J.T., 1948. Some new fossil reptiles from the Karroo Beds of South Africa. *Proceedings of the Zoological Society,* **118**: 392–407.

BROWN, B. & MORGAN, L., 1989. *The Miracle Planet.* Child & Associates, Sydney (Frenchs Forest).

BROWN, D. A, CAMPBELL, K.S.W. & CROOK, D.A.W., 1968. *The Geological Evolution of Australia & New Zealand.* Pergamon Press, Sydney.

BROWN, H.Y.L., 1894. Report of the government geologist for year ended June 30, 1894. *South Australian Parliamentary Papers,* **25**.

BROWN, I.A., 1946. An outline of the history of palaeontology in Australia. *Proceedings of the Linnean Society of New South Wales,* **21**: 48–61.

BROWN, L.R., 1981. *Building a Sustainable Society.* Norton, New York.

BRUNTON, C.H.C., MILES, R.S. & ROLFE, W.D.I., 1969. Gogo expedition 1967. *Proceedings of the Geological Society of London,* **1655**: 79–83.

BUCHSBAUM, R., BUCHSBAUM, M., PEARSE, J. & PEARSE, V., 1987. *Animals without Backbones.* University of Chicago Press, Chicago.

BUCKLAND, W., 1821. Observations on some specimens from the interior of New South Wales, collected during Mr Oxley's Expedition to the River Macquarie, in the year 1818, and transmitted also to Earl Bathurst. *Transactions of the Geological Society of London,* **5**: 480–481.

BURROW, C.J., 1997. Microvertebrate assemblages from the Lower Devonian (*Pesavis/Sulcatus* Zones) of central New South Wales, Australia. *Modern Geology,* **21**: 43–77.

CALLEN, R.A. & TEDFORD, R.H., 1976. New late Cainozoic rock units and depositional environments, Lake Frome area, South Australia. *Transactions of the Royal Society of South Australia,* **100**: 125–167.

CAMP, C.L. & BANKS, M.R., 1978. A proterosuchian reptile from the Early Triassic of Tasmania. *Alcheringa,* **2**: 143–158.

CAMP, C.L., CLEMENS, W.A., GREGORY, J.T. & SAVAGE, D.E., 1967. Ruben Arthur Stirton, 1901–1966. *Journal of Mammalogy,* **48** (2): 298–305.

CAMP, S., ed., 1989. *Population Pressures: Threat to Democracy.* Population Crisis Committee, Washington, D.C.

CAMPBELL, K.S.W. & BARWICK, R.E., 1982. The neurocranium of the primitive dipnoan *Dipnorhynchus sussmilchi. Journal of Vertebrate Paleontology,* **2** (2): 286–327.

CAMPBELL, K.S.W. & BARWICK, R.E., 1987. Palaeozoic lungfishes: A review. *Journal of Morphology Supplement,* **1**: 93–132.

CAMPBELL, K.S.W. & BARWICK, R.E., 1988. Geological and palaeontological information and phylogenetic hypotheses. *Geological Magazine,* **125** (3): 207–227.

CAMPBELL, K.S.W. & BARWICK, R.E., 1998. A new tooth-plated dipnoan from the Upper Devonian Gogo formation and its relationships. *Memoirs of the Queensland Museum,* **42**: 1–42.

CAMPBELL, K.S.W. & BELL, M.W., 1977. A primitive amphibian from the Late Devonian of New South Wales. *Alcheringa,* **1**: 369–381.

CARRIER, D.R., 1987. The evolution of locomotor stamina in tetrapods: Circumventing a mechanical constraint. *Paleobiology,* **13** (3): 326–341.

CARROLL, R.L., 1988. *Vertebrate Paleontology and Evolution.* W. H. Freeman & Co., New York.

CARROLL, R.L., 1989. Developmental aspects of lepospondyl vertebrae in Paleozoic tetrapods. *Historical Biology,* **3**: 1–25.

CAS, R.A.C., 1983. Palaeogeographic and tectonic development of the Lachlan Fold Belt Southeastern Australia. *Special Publication, Geological Society of Australia,* **10**: 1–104.

CASE, J.A., 1985. Differences in prey utilization by Pleistocene marsupial carnivores, *Thylacoleo carnifex* (Thylacoleonidae) and *Thylacinus cynocephalus* (Thylacinidae). *Australian Mammalogy,* **8**: 45–52.

CASE, J.A., 1989. Antarctica: The effect of high latitude heterochroneity on the origin of the Australian marsupials. In: J.A. Crame, ed., *Origins and Evolution of the Antarctic Biota,* Geological Society of London, Special Publication **47**: 217–226.

CASE, J. et al., in press. The first hadrosaur from Antarctica. *Journal of Vertebrate Paleontology.*

CHALCONER, W.G., 1989. Fossil charcoal as an indicator of palaeoatmospheric oxygen level. *Journal of the Geological Society, London,* **146**: 171–174.

CHARIG, A.J., 1971. Faunal provinces on land: evidence based on the distribution of fossil tetrapods, with especial reference to the reptiles of the Permian and Mesozoic. In: F.A. Middlemiss, P.F. Rawson & G. Newall, eds., *Faunal Provinces in Space and Time. Geological Journal,* **Special Issue 4**: 111–128.

CHATTERJEE, S., 1974. A rhynchosaur from the Upper Triassic Maleri Formation of India. *Philosophical Transactions of the Royal Society of London,* B 267 (884): 209–261.

CHIAPPE, L.M., 1996. Early avian evolution in the Southern Hemisphere: The fossil record of birds in the Mesozoic of Gondwana. In: F.A. Novas & R.E. Molnar, eds., *Proceedings of the Gondwanan Dinosaur Symposium, Memoirs of the Queensland Museum,* **39** (3): 533–554.

CHIAPPE, L.M., in press. Cretaceous birds of Latin America. In: Stratigraphic Range of Cretaceous Mega and Microfossil of Latin America. *UNESCO International Geological Correlation Project #242: The Cretaceous of South America.*

CHRISTIDIS, L.& WALTER E. BOLES, 1994. *The Taxonomy and Species of Birds of Australia and Its Territories.* Royal Australian Ornithologists Union, **Monograph 2**: 1–112.

CLARK, I.F. & COOK, B.J., 1986. *Perspectives of the Earth.* Australian Academy of Science, Canberra.

CLARKE, W.B., 1878. *Remarks on the Sedimentary Formations of New South Wales.* 4th ed., Government Printer, Sydney.

CLEMENS, W.A., 1977. Phylogeny of the marsupials. In: B. Stonehouse & D. Gilmore, eds., *The Biology of Marsupials,* Macmillan Press, London: 51–68.

CLIFT, W., 1831. In regard to the fossil bones found in the caves and bone breccia of New Holland. *Edinburgh New Philosophical Journal,* **10**: 394–395.

CLOUD, P., 1988. *Oasis in Space: Earth History from the Beginning.* W. W. Norton & Co., New York.

COGGER, H.C., 1992. *Reptiles and Amphibians of Australia.* Reed Books, Sydney.

COLBERT, E.H., 1967. A new interpretation of *Austropelor,* a supposed Jurassic labyrinthodont amphibian from Queensland. *Memoirs of the Queensland Museum,* **15**: 35–41.

COLBERT, E.H., 1977. Mesozoic tetrapods and the northward migration of India. *Journal of the Palaeontological Society of India,* **20**: 138–145.

COLBERT, E.H., 1984. Triassic tetrapod faunas. In: W.E. Reif & F. Westphal, eds., *Third Symposium on Mesozoic Terrestrial Ecosystems, Short Papers,* Attempto Verlag, Tübingen: 53–59.

COLBERT, E.H., 1985. The petrified forest and its vertebrate fauna in Triassic Pangaea. *Museum of Northern Arizona Bulletin,* **54**: 33–43.

COLBERT, E.H. & COSGRIFF, J.W., 1974. Labyrinthodont amphibians from Antarctica. *American Museum Novitates,* **2552**: 1–30.

COLBERT, E.H. & KITCHING, J.W., 1975. The Triassic *Procolophon* in Antarctica. *American Museum Novitates,* **2566**: 1–23.

COMMONER, B., 1975. How poverty breeds overpopulation and not the other way around. *Ramparts.* Company, London.

CONSTANTINE, A., CHINSAMY, A., VICKERS-RICH, P. & RICH, T.H., 1998. Periglacial environments and polar dinosaurs. *South African Journal of Science,* **94**: 137–141.

COOK, P.J. & SHERGOLD, J.H., 1984. Phosphorous, phosphorites and skeletal evolution at the Precambraian–Cambrian Boundary. *Nature,* **308**: 231–236.

CORIA, R.A., in press. Ornithopod dinosaurs from the Neuquen Group,

Patagonia, Argentina: Phylogeny and biostratigraphy. In: Y. Tomida, T.H. Rich & P. Vickers-Rich, eds., *Proceedings of the 2nd International Gondwana Dinosaur Symposium,* National Science Museum, Tokyo.

COSGRIFF, J.W., 1974. Lower Triassic Temnospondyli of Tasmania. *Geological Society of America, Special Paper* **149**: 1–134.

COSGRIFF, J.W., 1984. The temnospondyl labyrinthodonts of the earliest Triassic. *Journal of Vertebrate Paleontology,* **4** (1): 30–46.

COSGRIFF, J.W. & GARBUTT, N.K., 1972. *Erythrobatrachus noonkanbahensis,* a trematosaurid species from the Blina Shale. *Journal of the Royal Society of Western Australia,* **55**: 5–18.

COSGRIFF, J.W. & HAMMER, W.R., 1983. The labyrinthodont amphibians of the earliest Triassic from Antarctica, Tasmania and South Africa. In: R.L. Oliver, P.R. James & J.B. Jago, eds., *Antarctic Earth Science,* Australian Academy of Science, Canberra: 590–592.

COVACEVICH, J. & ARCHER, M., 1975. The distribution of the Cane Toad, *Bufo marinus,* in Australia and its effects on indigenous vertebrates. *Memoirs of the Queensland Museum,* **17**: 305–310.

CROCHET, J.Y., 1984. *Garatherium mahboubii* nov. gen., nov. sp., marsupial de l'Éocene inférieur d'el Kohol (Sud-Oranais, Algérie). *Annales Páleontologie (Vert.-Invert.),* **70**: 275–294.

CUDMORE, F.A., 1926. Extinct vertebrates from Beaumaris. *Victorian Naturalist,* **43**: 78–82.

CURRIE, P., VICKERS-RICH, P. & RICH, T.H., 1996. Possible oviraptorosaur (Theropoda, Dinosauria) specimens from the Early Cretaceous Otway Group of Dinosaur Cove, Australia. *Alcheringa,* **20**: 73–79.

CUTHBERT, T.P., ed., 1973. *The Classic Maya Collapse.* University of New Mexico Press, Albuquerque.

DARLINGTON, P.J., 1957. *Zoogeography: The Geographical Distribution of Animals.* John Wiley, London.

DARRAGH, T.A., 1986. The Cainozoic Trigoniidae of Australia. *Alcheringa,* **10**: 1–34.

DARRAGH, T.A., 1987. The Geological Survey of Victoria under Alfred Selwyn, 1852–1868. *Historical Records of Australian Science,* **7** (1): 1–25.

DAVID, T.W.E., 1950. *The Geology of the Commonwealth of Australia.* Vol. 1. Edited and much supplemented by W.R. Browne. Edward Arnold, London.

DE BLIJ, H.J., 1988. *Earth '88: Changing Geographic Perspectives.* National Geographic Society, Washington, D.C.

DE POMEROY, A.M., 1995. Biostratigraphy of Devonian microvertebrates from Broken River, North Queensland. *Records of the Western Australian Museum,* **17**: 417–438.

DE VIS, C.W., 1887. On an extinct mammal of a genus apparently new. *Brisbane Courier,* **9224 (**44) Aug. 8, 1887: 6.

DE VIS, C.W., 1888. A glimpse of the post-Tertiary avifauna of Queensland. *Proceedings of the Linnean Society of New South Wales,* **3**: 1275–1292.

DE VIS, C.W., 1900. Bones and diet of *Thylacoleo. Annals of the Queensland Museum,* **5**: 7–11.

DE VIS, C.W., 1907. Fossils from the Gulf Watershed. *Annals of the Queensland Museum,* **7**: 3–7.

DE WIT, M., JEFFERY, M., BERGH, H. & NICOLAYSEN, L., 1988. *Geological Map of Sectors of Gondwana. Reconstructed to Their Disposition ~ 150 Ma.* American Association of Petroleum Geologists and University of Witwatersrand, Tulsa.

DEFAUW, S.L., 1989. Temnospondyl amphibians: A new perspective on the last phases in the evolution of the Labyrinthodontia. *Michigan Academician,* **21**: 7–32.

DENISON, R.H., 1967. Ordovician vertebrates from western United States. *Fieldiana: Geology,* **16**: 131–192.

DENISON, R.H., 1974. The structure and evolution of teeth in lungfishes. *Fieldiana: Geology,* **33**: 31–58.

DENISON, R.H., 1983. Further consideration of placoderm evolution. *Journal of Vertebrate Paleontology,* **3** (2): 69–83.

DENNIS, K.D. & MILES, R.S., 1979. New durophagous arthrodires from Gogo, Western Australia. *Zoological Journal of the Linnean Society,* **69**: 43–85.

DENNIS, K.D. & MILES, R.S., 1982. A eubrachythoracid arthrodire with a snub-nose from Gogo, Western Australia. *Zoological Journal of the Linnean Society,* **90**: 153–166.

DETTMANN, M.E., 1963. Upper Mesozoic microfloras from south-eastern Australia. *Proceedings of the Royal Society of Victoria,* **77**: 1–148.

DETTMANN, M.E., 1986. Early Cretaceous palynoflora of subsurface strata correlative with the Koonwarra fossil bed, Victoria. *Association of Australasian Palaeontologists Memoir,* **3**: 79–110.

DONOVAN, S.K., ed., 1989. *Mass Extinctions.* Columbia University Press, New York.

DRINNAN, A.N. & CHAMBERS, T.C., 1986. Flora of the lower Koonwarra fossil bed (Korumburra Group), South Gippsland, Victoria. *Association of Australasian Palaeontologists Memoir,* **3**: 1–77.

DU TOIT, A.L., 1937. *Our Wandering Continents: An Hypothesis of Continental Drifting.* Oliver & Boyd, Edinburgh.

DUGAN, K.G., (1979) 1980. Darwin and *Diprotodon:* The Wellington Cave fossils and the Law of Succession. *Proceedings of the Linnean Society of New South Wales,* **104** (4): 265–272.

DUN, W.S. & RAINBOW, W.A., 1926. Obituary. Robert Etheridge, Junior. *Records of the Australian Museum,* **25** (1): 1–27.

DUNN, W.J., 1910. Biographical sketch of the founders of the Geological Survey of Victoria. *Bulletin of the Geological Survey,* **23**.

DWYER, T.D., 1978. Rats, pigs and men: Disturbance and diversity in the New Guinea highlands. *Australian Journal of Ecology,* **3**: 213–232.

EHRLICH, P.R. & EHRLICH, A., 1981. *Extinction: The Causes and Consequences of the Disappearance of Species.* Random House, New York.

EHRLICH, P.R. & EHRLICH, A.H., 1990. *The Population Explosion.* Simon & Schuster, New York.

EHRLICH, P.R., EHRLICH, A. & HOLDREN, J., 1977. *Ecoscience: Population, Resources, Environment.* Freeman, San Francisco.

ELLIOT, D.H., COLBERT, E.H., BREED, W.J., JENSEN, J.A. & POWELL, J.S., 1970. Triassic tetrapods from Antarctica: Evidence of continental drift. *Science,* **169**: 1197–1201.

ESTES, R., 1984. Fish, amphibians and reptiles from the Etadunna Formation Miocene of South Australia. *Australian Zoologist,* **21**: 335–343.

ETHERIDGE, R., JR., 1878. *A Catalogue of Australian Fossils (Including Tasmania and the Island of Timor) Stratigraphically and Zoologically Arranged.* Cambridge University Press, Cambridge.

ETHERIDGE, R., JR., 1894. Report on the discovery of bones near Callabonna Station. *Annual Report of the Government Geologist of South Australia,* 7–8.

ETHERIDGE, R., JR. & JACK, R.L., 1882. *Catalogue of works on the geology, mineralogy . . . etc. of the Australian continent and Tasmania.* Government Printer, Sydney.

FALCONER, H., 1863. On the American fossil elephant of the regions bordering the Gulf of Mexico (*E. columbia,* Falc.); with general observations on the living and extinct species. *Nat. Hist. Rev.,* **10**: 43–114.

FISCHER, A.G., 1984. The two Phanerozoic supercycles. In: W. Berggren A. & J.A. van Couvering, eds., *Catastrophies and Earth History: The New Uniformitarianism.* Princeton University Press, Princeton, New Jersey: 129–150.

FISHMAN, J. & KALISH, R., 1990. *Global Alert: The Ozone Pollution Crisis.* Plenum, New York.

FLANNERY, T.F., 1988. Origins of the Australo-Pacific land mammal fauna. *Australian Zoological Review,* **1**: 15–24.

FLANNERY, T.F., 1989a. Phylogeny of the Macropodoidea: A study in convergence. In: G. Griggs, P. Jarman & J. Hume, eds., *Kangaroos, Wallabies and Rat Kangaroos.* Surrey Beatty & Sons Pty Ltd, Chipping North: 1–46.

FLANNERY, T.F., 1989b. Plague in the Pacific. *Australian Natural History,* **23** (1): 20–29.

FLANNERY, T.F., 1990. Pleistocene faunal loss: Implications of the aftershock for Australia's past and future. *Archaeology Oceania,* **25**: 45–67.

FLANNERY, T.F., 1994. *The Future Eaters: An Ecological History of the Australasian Lands and Peoples.* Reed Books, Sydney.

FLANNERY, T.F., in press. the mystery of the Meganesian meat-eaters. *Australian Natural History.*

FLANNERY, T.F. & GOTT, B., 1984. The Spring Creek locality, southwestern Victoria, a late surviving megafaunal assemblage. *Australian Zoology,* **21**: 385–422.

FLANNERY, T.F. & HANN, L., 1984. A new macropodine genus and species (Marsupialia: Macropodidae) from the early Pleistocene of south western Victoria. *Australian Mammalogy,* **7**: 193–204.

FLANNERY, T.F., HOCH, E. & APLIN, K., 1989. Macropodines from the Pliocene Otibanda Formation, Papua New Guinea. *Alcheringa,* **13**: 145–152.

FLANNERY, T.F., MOUNTAIN, M.J. & APLIN, K., 1982. Quaternary kangaroos (Macropodidae: Marsupialia) from Nombe Rock Shelter, Papua New Guinea, with comments on the nature of the megafaunal extinction in the New Guinea highlands. *Proceedings of the Linnean Society of New South Wales,* **107** (2): 75–97.

FLANNERY, T.F. & PLANE, M., 1986. A new late Pleistocene diprotodontid (Marsupialia) from Pureni, Southern Highlands Province, Papua New Guinea. *Bureau of Mineral Resources Journal of Geology and Geophysics of Australia,* **10** (1): 65–76.

FLANNERY, T.F. & RICH, T.H., 1981. Dinosaur digging in Victoria. *Australian Natural History,* **20**: 195–198.

FLANNERY, T.F. & SZALAY, F., 1982. *Bohra paulae,* a new giant fossil tree kangaroo (Marsupialia: Macropodidae) from New South Wales, Australia. *Australian Mammalogy,* **5**: 83–94.

FLETCHER, J.J., 1920. The society's heritage from the Macleays. *Proceedings of the Linnean Society of New South Wales,* **45:** 592–629.

FLINDERS, M., 1814. *A Voyage to Terra Australis.* 2 vols. Nichol, London.

FLOOD, J., 1973. Pleistocene human occupation and extinct fauna in Cloggs Cave, Buchan S.E. Australia. *Nature, 246* (5431): 303.

FLOOD, J., 1983. *Archaeology of the Dreamtime.* Collins, Sydney.

FOOTE, M., HUNTER, J.P., JANIS, C.M. & SEPKOSKI, JR., J.J., 1999. Evolutionary and preservational constraints on the origins of biologic groups: Divergence times of eutherian mammals. *Science,* **283:** 1310–1314.

FORDYCE, R.E., 1983. Rhabdosteid dolphins (Mammalia: Cetacea) from the middle Miocene, Lake Frome area, South Australia. *Alcheringa,* **7:** 27–40.

FORDYCE, R.E., 1994. *Waipatia maerewhenua,* new genus and new species (Waipatiidae, new family), an archaic Late Oligocene dolphin (Cetacea: Odontoceti: Platanistoidea) from New Zealand. In: A. Berta and T. Demere, eds., *Contributions in Marine Mammal Paleontology Honoring Frank C. Whitmore, Jr., Proceedings of the San Diego Museum of Natural History,* **29:** 147–176.

FORDYCE, R.E. & WATSON, A.G., 1998. Vertebral pathology in an Early Oligocene whale (Cetacea: ?Mysticeti) from Wharekuri, North Otago, New Zealand. *Karlheinz Rothausen-Festschrift, Mainzer Naturwissenschaftliches Archi Beihefte,* **21:** 161–176.

FOREY, P.L., 1984. Yet more reflections on agnathan-gnathostome relationships. *Journal of Vertebrate Paleontology,* **4** (3): 330–343.

FOSTER, W.C., 1985. *Sir Thomas Livingston Mitchell and His World 1792–1855.* Inst. Surveyors, New South Wales, Sydney.

FOX, R.C., CAMPBELL, K.S.W., BARWICK, R.E. & LONG, J.A., 1995. A new osteolepiform fish from the Lower Carboniferous Raymond Formation, Drummond Basin, Queensland. *Memoirs of the Queensland Museum,* **38:** 97–221.

FRAKES, L.A., 1979. *Climates throughout Geologic Time.* Elsevier, Oxford.

FRAKES, L.A. & FRANCIS, J.E., 1988. A guide to Phanerozoic cold polar climates from high-latitude ice-rafting in the Cretaceous. *Nature,* **333:** 547–549.

FRAKES, L.A., MCGOWRAN, B. & BOWLER, J.M., 1987. Evolution of Australian environments. In: G.R. Dyne & D.W. Walton, eds., *Fauna of Australia: General Articles.* Australian Government Publishing Service, Canberra.

FRITH, H.J., 1973. *Wildlife Conservation.* Angus and Robertson, Sydney.

FROUDE, D.O., IRELAND, T.R., KINNY, P.D., WILLIAMS, I.S., COMPSTON, W., WILLIAMS, I.R. & MYERS, J.S., 1983. Ion microprobe identification of 4,100–4,200 million year old terrestrial zircons. *Nature,* **304:** 616–618.

GAFFNEY, E.S., 1981. A review of the fossil turtles of Australia. *American Museum Novitates,* **2720:** 1–38.

GAFFNEY, E.S., 1989. Chelid turtles from the Miocene freshwater limestones of Riversleigh Station, northwestern Queensland, Australia. *American Museum Novitates,* **2959:** 1–10

GAFFNEY, E.S. & BARTHOLOMAI, A., 1979. Fossil trionychids of Australia. *Journal of Paleontology,* **53:** 1354–1360.

GAGNIER, P.-Y., 1989. The oldest vertebrate: A 470-million-year-old jawless fish, *Sacabambaspis janvieri,* from the Ordovician of Bolivia. *National Geographic Research,* **5** (2): 250–253.

GAGNIER, P.-Y., BLIECK, A. & RODRIGO, G.S., 1986. First Ordovician vertebrate from South America. *Geobios,* **19** (5): 629–634.

GALTON, P.M., 1986. Hypsilophodontid dinosaurs from Lightning Ridge, New South Wales, Australia. *Geobios,* **19:** 231–239.

GARDINER, B.G., 1984. The relationships of palaeoniscoid fishes: A review based on new specimens of *Mimia* and *Moythomasia* from the Upper Devonian of Western Australia. *Bulletin of the British Museum (Natural History), Geology,* **37:** 173–427.

GASPARINI, Z., PEREDA-SUBERBIOLA, X. & MOLNAR, R.E., 1996. New data on the ankylosaurian dinosaur from the Late Cretaceous of the Antarctic Peninsula. In: F.A. Novas & R.E. Molnar, eds., *Proceedings of the Gondwanan Dinosaur Symposium, Memoirs of the Queensland Museum,* **39** (3): 583–594.

GEHLING, J.G., 1987. Earliest known echinoderm: A new Ediacaran fossil from the Pound Subgroup of South Australia. *Alcheringa,* **11** (4): 337–345.

GEORGE, T. N., 1975. Geologists at the University of Glasgow. *College Courant Journal of the University Graduate Association,* **27:** 23–30.

GERVAIS, P., 1848–1852. *Zoologie et paléontologie françaises (animaux vertèbres) ou nouvelles recherches sur les animaux vivants et fossiles de la France.* Tome I. Tome II. Arthur Bertrand, Paris.

GEYH, M.A. & SCHLEICHER, H., 1990. *Absolute Age Determination: Physical and Chemical Dating Methods and Their Application.* Springer Verlag.

GILKESON, C.F. & LESTER, K.S., 1989. Ultrastructural variation in enamel of Australian marsupials. *Scanning Microscopy,* **3:** 177–191.

GILL, E.D., 1965. *Palaeontology of Victoria.* Vict. Yearbook 1965. Government Printer, Melbourne.

GILL, E.D. & BANKS, M.R., 1956. Cainozoic history of Mowbray Swamp and other areas of North-western Tasmania. *Records of the Queen Victoria Museum,* New Series, **6:** 1–42.

GILLESPIE, R., HORTON, D.R., LADD, P., MACUMBER, P.G., RICH, T.H., THORNE, R. & WRIGHT, R.V.S., 1978. Lancefield Swamp and the extinction of the Australian megafauna. *Science,* **200:** 1044–1048.

GILMORE, D.P., 1977. The success of marsupials as introduced species. In: B. Stonehouse & D. Bilmore, eds., *The Biology of Marsupials,* Macmillan Press, London: 169–178.

GLAESSNER, M.F., 1961. *The Dawn of Animal Life.* Cambridge University Press, Cambridge.

GLANTZ, M., 1987. *The Decline of African Agriculture.* Cambridge University Press, Cambridge.

GLAUERT, L., 1914. The Mammoth Cave. *Records of the Western Australian Museum,* **1** (3): 244–252.

GLEADOW, A.J.W. & DUDDY, I.R., 1980. Early Cretaceous volcanism and the early breakup history of southeastern Australia: Evidence from fission-track dating of volcaniclastic sediments. In: M.M. Creswell & P. Vella, eds., *Proceedings of the 5th International Gondwana Symposium, Wellington, New Zeland.* Balkema, Rotterdam: 295–300.

GLEN, W., 1982. *The Road to Jaramillo.* Stanford University Press, Palo Alto.

GLEN, W., 1990. What killed the dinosaurs? *American Scientist,* **78:** 354–370.

GODTHELP, H., ARCHER, M., CIFELLI, R., HAND, S. & GILKESON, C.F., 1992. Earliest known Australian Tertiary mammal fauna. *Nature,* **356:** 514-516.

GOLDSMITH, E. & HILDYARD, N., 1988. *The Earth Report: The Essential Guide to Global Ecological Issues.* Price Stern Sloan, New York.

GOMANI, E.M., JACOBS, L.L. & WINKLER, D.A., in press. Comparison of the African titanosaurian, *Malawisaurus,* with a North American Early Cretaceous sauropod. In: Y. Tomida, T.H. Rich & P. Vickers-Rich, eds., *Proceedings of the 2nd International Gondwana Dinosaur Symposium,* National Science Museum, Tokyo.

GORECKI, P.P., HORTON, D.R., STERN, N. & WRIGHT, R.V.S., 1984. Coexistence of humans and megafauna in Australia: Improved stratified evidence. *Archaeology in Oceania,* **19:** 117–120.

GOULD, S.J., 1968. Trigonia and the origin of species. *Journal of the History of Biology,* **1:** 41–56.

GRAY, J. & BOUCOT, A.J., 1976. *Historical Biogeography, Plate Tectonics & the Changing Environment.* Oregon State University Press, Corvallis.

GREEN, R.H., 1983. *An Illustrated Key to the Skulls of the Mammals in Tasmania.* Queen Victoria Museum and Art Gallery, Launceston.

GREGORY, J.W., 1906. *The Dead Heart of Australia.* John Murray, London.

GREGORY, R.T., DOUTHITT, C.B., DUDDY, I.R., RICH, P.V. & RICH, T.H., 1989. Oxygen isotopic composition of carbonate concretions from the lower Cretaceous of Victoria: Implications for the evolution of meteoric waters on the Australian continent in a paleopolar environment. *Earth and Planetary Science Letters,* **92:** 27–42.

GRIBBIN, J., 1989. The end of the ice ages? *New Scientist,* 17 June: 22–26.

GUNN, R.C., 1848a. Edmund Charles Hobson, M. D. *Tasmanian Journal of Natural Sciences, Agriculture & Statistics,* **3:** 406–407.

GUNN, R.C., 1848b. A trilobite received from Mrs E. C. Hobson. *Tasmanian Journal of Natural Sciences, Agriculture & Statistics,* **3:** 407.

HALE, H.M., 1956. The first hundred years of the South Australian Museum. 1856–1956. *Records of the South Australian Museum,* **12:** 1–225.

HALL, L.S., 1981. The biogeography of Australian bats. In: A. Keast, ed., *Ecological Biogeography in Australia,* W. Junk, The Hague: 1557–583.

HALL, L.S. & RICHARDS, G.C., 1979. *Bats of Eastern Australia.* Queensland Museum Booklet No. 12, Brisbane.

HALL, R. & HOLLOWAY, J.D., 1998. *Biogeography and Geological Evolution of SE Asia.* Backhuys Publishers, Leiden.

HALL, T.S., 1911. On the systematic position of the species of *Squalodon* and *Zeuglodon* described from Australia and New Zealand. *Proceedings of the Royal Society of Victoria,* **23:** 257–265.

HALL, T.S. & PRITCHARD, G.B., 1897. Note on a tooth of *Palorchestes* from Beaumaris. *Proceedings of the Royal Society of Victoria,* **10:** 57–59.

HALLAM, A., 1973a. *A Revolution in the Earth Sciences: From Continental Drift to Plate Tectonics.* Oxford University Press, London.

HALLAM, A., ed., 1973b. *Atlas of Palaeobiogeography.* Elsevier, Amsterdam.

HAMBLIN, W.K., 1985, 1989. *The Earth's Dynamic Systems.* Macmillan Publishing Co., New York.

HAMMER, W.R., 1995. New therapsids from the Upper Fremouw Formation (Triassic) of Antarctica. *Journal of Vertebrate Paleontology,* **15** (1): 105–112.

HAMMER, W.R., 1997. Jurassic dinosaurs from Antarctica. In: D.L. Wolberg, E. Stump & G.D. Rosenberg, eds., *Dinofest International, Proceedings of a Symposium sponsored by Arizona State University,* A Publication of The Academy of Natural Sciences: 249–251.

HAMMER, W.R., COLLINSON, J.W. & RYAN, W. J. III, 1990. A new Triassic vertebrate fauna from Antarctica and its depositional setting. *Antarctic*

Science, **2** (2): 163–167.

HAMMER, W.R. & HICKERSON, W.J., in press. Gondwana dinosaurs from the Jurassic of Antarctica. In: Y. Tomida, T.H. Rich & P. Vickers-Rich, eds., *Proceedings of the 2nd International Gondwana Dinosaur Symposium*, National Science Museum, Tokyo.

HAND, S.J., 1985. New Miocene megadermatids (Chiroptera: Megadermatidae) from Australia with comments on megadermatids phylogeny. *Australian Mammalogy*, **8**: 5–43.

HAND, S.J., 1990. First Tertiary molossid (Microchiroptera: Molossidae) from Australia: Its phylogenetic and biogeographic implications. *Memoirs of the Queensland Museum*, **28**: 175–192.

HAND, S. & ARCHER, M., 1987. *The Antipodean Ark: Creatures from Prehistoric Australia.* Angus and Robertson, Sydney.

HAND, S.J., MURRAY, P., MEGIRIAN, D., ARCHER, M. & GODTHELP, H., 1998. Mystacinid bats (Microchiroptera) from the Australian Tertiary. *Journal of Paleontology*, **72**: 538–545.

HAND, S.J., NOVACEK, M., GODTHELP, H. & ARCHER, M., 1994. First Eocene bat from Australia. *Journal of Vertebrate Paleontology*, **14**: 375–381.

HAQ, B.U., HARDENBOL, J. & VAIL, P.R., 1987. Chronology of fluctuating sea levels since the Triassic. *Science*, **235**: 1156–1164.

HARDJASASMITA, H.S., 1985. Fosil diprotodontidi *Zygomaturus* Owen 1859 Dari Nimboran, Irian Jaya. *Psekmnan Ilmiah Arkeologi III. Jakartai PPAN:* 999–1004.

HARLAND, W.B., ARMSTRONG, R.L., COX, A.V., CRAIG, L., SMITH, A.G. & SMITH, D.G., 1989. *A Geologic Time Scale 1989.* Cambridge University Press, Cambridge.

HARMAN, M.E., FERRELL, W.K. & FRANKLIN, J.F., 1990. Effects on carbon storage of old-growth forests to young forests. *Science*, **247**: 699–702.

HARPER, F., 1945. *Extinct and Vanishing Mammals of the Old World.* New York, Amer. Comm. Internat. Wildlife 2 Protection; New York Zoological Park, Special Publication **12**.

HECHT, M.K. & ARCHER, M., 1977. Presence of xiphodont crocodilians in the Tertiary and Pleistocene of Australia. *Alcheringa*, **1**: 383–385.

HEDGES, S.B., PARKER, P.H., SIBLEY, C.G. & KUMAR, S., 1996. Continental breakup and the ordinal diversification of birds and mammals. *Nature*, **381**: 226–229.

HENDERSON-SELLERS, B. & HENDERSON-SELLERS, A., 1989. Modelling the ocean climate for the early Archaean. *Global and Planetary Change*, **1** (3): 195–221.

HILLS, E.S., 1958. A brief review of Australian fossil vertebrates. In: T.S. Westoll, ed., *Studies on Fossil Vertebrates:* 86–107.

HOBSON, E.C., 1841a. On the *Callorhynchus Australis*. *Tasmanian Journal of Natural Sciences, Agriculture, Statistics, & etc.*, **1**: 14–20.

HOBSON, E.C., 1845. On the fossil bones from Mount Macedon, Port Phillip. *Tasmanian Journal of Natural Sciences, Agriculture, Statistics & etc.*, **2**: 344–347.

HOBSON, E.C., 1846a. A fossil bone of gigantic dimensions. . . . *Tasmanian Journal of Natural Sciences, Agriculture, Statistics & etc.*, **2**: 460.

HOBSON, E.C., 1846b. Drawings of a fossil skull. . . . *Tasmanian Journal of Natural Scioences, Agriculture, Statistics & etc.*, **2**: 464.

HOBSON, E.C., 1848. On the jaw of the *Diprotodon Australis*, and its dental formula. *Tasmanian Journal of Natural Sciences, Agriculture, Statistics & etc.*, **3**: 387–388.

HOCH, E. & HOLM, P.M., 1986. New K/Ar age determinations of the Awe Fauna Gangue, Papua New Guinea: Consequences for Papuaustralian late Cenozoic biostratigraphy. *Modern Geology*, **10**: 181–195.

HOCHSTETTER, F., 1859. Notizen über einiger fossile Thierreste und deren Lagerstatten in Neu-Holland, gesammelt daselbst wahrend des Aufenthaltes Sr. Majestat Fregatte *Novara* im Monate December 1858. *Sber. K. Akademie Wissenschaften Wien*, **35**: 349–358.

HOFFMAN, A., 1989a. Arguments on Evolution. Oxford University Press, New York.

HOFFMAN, A., 1989b. Mass extinctions: The view of a sceptic. *Journal of the Geological Society, London*, **146**: 21–35.

HOLDREN, J.P. & EHRLICH, P.R., 1974. Human population and the global environment. *American Scientist*, **62**: 282–292.

HOLLAND, H.D., 1984. *The Chemical Evolution of the Atmosphere and Oceans.* Princeton University Press, Princeton, New Jersey.

HOLMES, E.B., 1985. Are lungfishes the sister group of tetrapods? *Biological Journal of the Linnean Society*, **25** (4): 379–397.

HOLMES, R., 1989. Functional interpretations of the vertebral structure in Paleozoic labyrinthodont amphibians. *Historical Biology*, **2** (2): 111–124.

HOOKER, J.J., MILNER, A. C. & SEQUEIRA, S.E.K., 1991. An ornithopod dinosaur from the Late Cretaceous of West Antarctica. *Antarctic Science*, **3** (3): 331–332.

HOPE, J.H., 1973b. Mammals of the Bass Strait Islands. *Proceedings of the Royal Society of Victoria*, **85** (2): 163–196.

HOPE, J.H., 1978. Pleistocene mammal extinctions: The problem of Mungo and Menindee, New South Wales. *Alcheringa*, **2** (1): 65–82.

HOPE, J.H., 1982. Fossil vertebrates from Wombeyan Caves. In: H.J. Dyson, R. Ellis & J.M. James, eds., *Wombeyan Caves, Sydney Speleological Society Occasional Paper*, **8**: 155–164.

HOPE, J.H., DARE-EDWARDS, A. & MCINTYRE, M.L., 1983. Middens and megafauna: Stratigraphy and dating of Lake Tandou Lunette, western New South Wales. *Archaeololgy of Oceania*, **18**: 45–53.

HOPSON, J.A., 1987. The mammal-like reptiles: A study of transitional fossils. *The American Biology Teacher*, **49** (1): 16–26.

HORTON, D.R., 1978. Extinction of the Australian Megafauna. *Australian Institute for Aboriginal Studies Newsletter*, **9**: 72–75.

HORTON, D.R., 1979. The great megafaunal extinction debate 1879–1979. *The Artefact*, **4** (1): 11–25.

HORTON, D.R., 1980. A review of the extinction question: Man, climate and megafauna. *Archaeology and Physical Anthropology of Oceania*, **15**: 86–97.

HORTON, D.R., 1984. Red kangaroos: Last of the Australian megafauna. In: P.S. Martin & R.G. Klein, eds., *Quaternary Extinctions. A Prehistoric Revolution*, University of Arizona Press, Tucson: 639–680.

HORTON, D.R. & WRIGHT, R.V.S., 1981. Cuts on Lancefield bones: Carnivorous *Thylacoleo*, not humans, the cause. *Archaeology and Physical Anthropology of Oceania*, **16**: 73–79.

HOTTON, N. III, MACLEAN, P.D., ROTH, J.J. & ROTH, E.C., 1986. *The Ecology and Biology of Mammal-like Reptiles.* Smithsonian Institution Press, Washington, D.C.

HOUGHTON, R.A. & WOODWELL, G.M., 1989. Global climatic change. *Scientific American*, April: 18–26.

HOUSE, M.R., 1977. *The Origin of Major Invertebrate Groups.* Academic Press, London.

HSÜ, K.J., 1983. *The Mediterranean Was a Desert: A Voyage of the Glomar Challenger.* Princeton University Press, Princeton, New Jersey.

HUDSON, J.D., 1989. Palaeoatmospheres in the Phanerozoic. *Journal of the Geological Society, London*, **146**: 155–160.

HUGHES, N.F., ed., 1973. Organisms and continents through time. *Spec. Papers Palaeontology*, **No. 12**.

HUNT, G. S., 1974. Dr. Robert Broom, Taralga. *Helictite*, **12**: 31–52.

HUTCHINS, B. & SWAINSTON, R., 1986. *Sea Fishes of Southern Australia.* Swainston Publishing.

HUXLEY, T.H., 1859. On some amphibian and reptilian remains from South Africa and Australia. *Quarterly Journal of the Geological Society of London*, **15**: 642–658.

HUXLEY, T.H.,1862. On the premolar teeth of *Diprotodon*, and on a new species of that genus. *Quarterly Journal of the Geological Society of London*, **18**: 422–427.

HYETT, J. & SHAW, N., 1980. *Australian Mammals: A Field Guide for New South Wales, Victoria, South Australia and Tasmania.* Thomas Nelson, Melbourne.

INGRAM, G., 1986. "Thickthorn" and his birds. *The Sunbird*, **16** (2): 25–32.

ISETT, G.A., 1990. The Cretaceous/Tertiary boundary interval, Raton Basin, Colorado and New Mexico, and its content of shock-metamorphosed minerals: Evidence relevant to the K/T boundary impact-extinction theory. *Geological Society of America Special Paper*, **249**: 1–100.

JACK, R.L. & ETHERIDGE, R., JR., 1892. *The geology and palaeontology of Queensland and New Guinea, with sixty-eight plates and a geological map of Queensland.* Government Printer, Brisbane.

JACOBS, L.L., WINKLER, D.A. & GOMANI, E.M., 1996. Cretaceous dinosaurs of Africa: Examples from Cameroon and Malawi. In: F.A. Novas & R.E. Molnar, eds., *Proceedings of the Gondwanan Dinosaur Symposium, Memoirs of the Queensland Museum*, **39** (3): 595–610.

JACOBSEN, J., 1988. *Environmental Refugees: A Yardstick of Habitability.* Worldwatch Paper **86**, Worldwatch Institute, Washington, D.C.

JACOBSEN, T. & ADAMS, R.M., 1958. Salt and silt in ancient Mesopotamian agriculture. *Science*, **128**: 1251–1258.

JAIN, S.L., 1973. New specimens of Lower Jurassic holostean fishes from India. *Palaeontology*, **16** (1): 149–177.

JAIN, S.L., 1983. A review of the genus *Lepidotes* (Actinopterygii Semionotiformes) with special reference to the species from Kota Formation (Lower Jurassic), India. *Journal of the Palaeontological Society of India*, **26**: 7–42.

JAIN, S.L., KUTTY, T.S., ROYCHOWDHURY, T.K. & CHATTERJEE, S., 1975. The sauropod dinosaur from the Lower Jurassic Kota Formation of India. *Proceedings of the Royal Society of London*, A **188**: 221–228.

JAIN, S.L., KUTTY, T.S., ROYCHOWDHURY, T.K. & CHATTERJEE, S., 1979. Some characteristics of *Barapaasaurus tagorei*, a sauropod dinosaur from the Lower Jurassic of Deccan, India. In: B. Laskar and C.S. Rajarao, eds., *Proceedings Fourth International Gondwana Symposium*, **1** (3): 204–216. Hindusthan Publishing Corp., Delhi.

JAMESON, R., 1831a. On the fossil bones found in the bone caves and bone breccia of New Holland. *Edinburgh New Philosophical Journal,* **10:** 393, 395–396.

JAMESON, R., 1831b. Further notices in regard to the fossil bones found by Major Mitchell, Surveyor General of New South Wales, in Wellington County, New South Wales. *Edinburgh New Philosophical Journal,* **10:** 179–180.

JANVIER, P., 1981. The phylogeny of the Craniata, with particular reference to the significance of fossil agnathans. *Journal of Vertebrate Paleontology,* **1:** 121–159.

JEFFERIES, R.P.S., 1975. Fossil evidence concerning the origin of the chordates. *Symposium of the Zoological Society of London,* **36:** 253–318.

JELL, P.A. & DUNCAN, P.M., 1986. Invertebrates, mainly insects, from the freshwater, Lower Cretaceous, Koonwarra fossil bed (Korumburra Group), south Gippsland, Victoria. *Association of Australasian Palaeontologists Memoir,* **3:** 111–205.

JOHANSON, Z. & AHLBERG, P.E., 1997. A complete primitive rhizodontid from Australia. *Nature,* **394:** 569–573.

JONES, J.G., CONAGHAN, P.J. & MCDONNELL, K.L., 1987. Coal measures of an orogenic recess: Late Permian Sydney Basin, Australia. *Palaeogeography, Palaeoclimatology, Palaeoecology,* **58:** 203–219.

JONES, R., 1970. Tasmanian Aborigines and dogs. *Mankind,* **7:** 256–271.

JOYSEY, K.A. & KEMP, T.S., eds., 1972. *Studies in Vertebrate Evolution.* Winchester Press, New York.

JUPP, R. & WARREN, A.A., 1986. The mandibles of the Triassic temnospondyl amphibians. *Alcheringa,* **10:** 99–124.

KATES, R.W., CHEN, R.S., DOWNING, T.E., KASPERSON, J.X., MESSER, E. & MILLMAN, S.R., 1988. *The Hunger Report.* Alan Shawn Feinstein World Hunger Program, Brown University, Providence.

KAUFMANN, W.J. III, 1985. *Universe.* W. H. Freeman, New York.

KÄMÄRI, J., BRAKKE, D.F., JENKINS, A., NORTON, S.A. & WRIGHT, R.F., eds., 1990. *Regional Acidification Models: Geographic Extent and Time Development.* Springer-Verlag, Frankfurt.

KEAST, A., ed., 1981. *Ecological Biogeography in Australia.* W. Junk, The Hague.

KEMP, A., 1983. *Ceratodus nargun,* a new early Cretaceous lungfish from Cape Lewis, Victoria. *Proceedings of the Royal Society of Victoria,* **95:** 23–24.

KEMP, A., 1992. New cranial remains of neoceratodonts (Osteichthyes: Dipnoi) from the Late Oligocene to Middle Miocene of northern Australia, with comments on generic characters for Cenozoic dipnoans. *Journal of Vertebrate Paleontology,* **12:** 284–293.

KEMP, A., 1997. Four species of *Metaceratodus* (Osteichthyes: Dipnoi, family Ceratodontidae) from Australian Mesozoic and Cenozoic deposits. *Journal of Vertebrate Paleontology,* **17:** 26–33.

KEMP, A. & MOLNAR, R.E., 1981. *Neoceratodus forsteri* from the Lower Cretaceous of New South Wales, Australia. *Journal of Paleontology,* **55:** 211–217.

KENNEDY, M., ed., 1990. *Australia's Endangered Species: The Extinction Dilemma.* Simon & Schuster, Brookvale.

KEYFITZ, N., 1989. The growing human population. *Scientific American,* September: 71–77A.

KIELAN-JAWOROWSKA, Z., CIFELLI, R.L. & LUO, Z., 1998. Alleged Cretaceous placental from down under. *Lethaia,* **31:** 267–268.

KING, G.M., 1990. *Life and death in the Permo-Triassic: The fortunes of the dicynodont mammal-like reptiles.* South African Museum, Sidney Houghton Memorial Lecture, Capetown.

KIRK, R.L. & THORNE, A.G., 1976. Introduction. In: R.L. Kirk & A.G. Thorne, eds., *The Origin of the Australians,* Australian Institute of Aboriginal Studies, Canberra: 1–8.

KIRSCH, J.A.W., 1977a. The six-percent solution: Second thoughts on the adaptedness of the Marsupialia. *American Scientist,* **65:** 276–288.

KIRSCH, J.A.W., 1977b. The comparative serology of marsupials. *Australian Journal of Zoology, Supplemental Series,* **52:** 1–152.

KIRSCH, J.A.W. & CALABY, J.H., 1977. The species of living marsupials: An annotated list. In: B. Stonehouse & D. Gilmore, eds., *The Biology of Marsupials,* Macmillan Press, London: 9–26.

KOCH, C.F., 1978. Bias in the published fossil record. *Paleobiology,* **4** (3): 67–372.

KOHLER, R., 1995. A new species of the fossil turtle *Psephophorus* (Order Testudines) from the Eocene of South Island, New Zealand. *Journal of the Royal Society of New Zealand,* **25** (3): 371–384.

KOHLER, R. & FORDYCE, R.E., 1997. An archaeocete whale (Cetacea: Archaeoceti) from the Eocene Waihao Greensand, New Zealand. *Journal of Vertebrate Paleontology,* **17:** 574–583.

KOHLSTEDT, S.G., 1983. Australian museums of natural history: Public priorities and scientific initiatives in the 19th century. *Historical Records of Australian Science,* **5** (4): 1–29.

KONIG, C., 1825. *Icones Fossilium Sectiles.* London.

KRAUSE, D.W., PRASAD, G.V.R., VON KOENIGSWALD, W., SAHNI, A. &

GRINE, F.E., 1997. Cosmopolitanism among Gondwana Late Cretaceous mammals. *Nature,* **390:** 504–507.

KREFFT, G., 1866. On the dentition of *Thylacoleo carnifex* (Ow.). *Annual Magazine of Natural History,* **18** (3): 148–149.

KREFFT, G., 1867. Fossil remains of mammals, birds and reptiles from the caves of Wellington Valley; collected and described by Gerard Krefft. In: *Catalogue of the natural and industrial products of New South Wales, forwarded to the Paris Universal Exhibition of 1867, by the New South Wales Exhibition Commissioners.* Government Printer, Sydney.

KREFFT, G., 1870. *Guide to the Australian fossil remains exhibited by the trustees of the Australian Museum, and arranged and named by Gerard Krefft, F.L.S., Curator and Secretary.* Trustees of the Australian Museum, Sydney.

KREFFT, G., 1873. Natural history. Review of Professor Owen's papers on the fossil mammals of Australia. *Sydney Mail and New South Wales Advertiser,* **16** (686), Aug. 23, 1873: 238.

KUMAR, S. & HEDGES, S.B., 1998. A molecular timescale for vertebrate evolution. *Nature,* **392:** 917–920.

KUROCHKIN, E.N. & MOLNAR, R.E., 1997. New material of enantiornithine birds from the Early Cretaceous of Australia. *Alcheringa,* **21** (4): 291–297.

KURTZBACK, J.E. & GALLIMORE, R.G., 1989. Pangaean climates: Megamonsoons of the megacontinent. *Journal of Geophysical Research,* **94:** 3341–3357.

LANG, J.D., 1831. Account of the discovery on bone caves in Wellington Valley about 210 miles west from Sydney in New Holland. *Edinburgh New Philosophical Journal,* **10:** 364–368.

LANGER, W., 1964. The Black Death. *Scientific American,* February.

LAUDER, G.V. & LEIM, K.F., 1983. The evolution and interrelationships of the actinopterygian fishes. *Bulletin of the Museum of Comparative Zoology,* **150:** 95–197.

LEGRAND, H.E., 1988. *Drifting Continents and Shifting Theories.* Cambridge University Press, Melbourne and New York.

LEICHHARDT, F.W.L., 1855. Beitrage zur Geologie von Australien. *Abhandlungen der naturforschungen Gesellshaft Halle,* **3** (1): 1–62.

LESTER, K.S., ARCHER, M., GILKESON, G.F. & RICH, T., 1988. Enamel of *Yalkaparidon coheni:* Representative of a distinct order of Tertiary zalambdodont marsupials. *Scanning Microscopy,* **2:** 1491–1501.

LEU, M., 1989. A Late Permian freshwater shark from eastern Australia. *Palaeontology,* **32** (2): 265–286.

LONG, J.A., 1985a. A new osteolepidid fish from the Upper Devonian Gogo Formation, Western Australia. *Records of the Western Australian Museum,* **12:** 361–377.

LONG, J.A., 1985b. The structure and relationships of a new osteolepiform fish from the Late Devonian of Victoria, Australia. *Alcheringa,* **9** (1): 1–22.

LONG, J.A., 1988a. Late Devonian fishes from Gogo, Western Australia. *National Geographic Research,* **4** (4): 436–450.

LONG, J.A., 1988b. New palaeoniscoid fishes from the Late Devonian and Early Carboniferous of Victoria. In: P.A. Jell, ed., *Devonian and Carboniferous Fish Studies,* Association of Australasian Palaeontologists, Sydney: 1–64.

LONG, J.A., 1988c. The extraordinary fishes of Gogo. *New Scientist,* **1639:** 40–44.

LONG, J.A., 1989. A new rhizodontiform fish from the Early Carboniferous of Victoria, Australia, with remarks on the phylogenetic position of the group. *Journal of Vertebrate Paleontology,* **9** (1): 1–17.

LONG, J.A., 1990. Heterochrony and the origin of tetrapods. *Lethaia,* **23** (2): 157–166.

LONG, J.A., 1993a. Cranial ribs in Devonian lungfishes and the origin of dipnoan air breathing. *Memoirs of the Australasian Association of Palaeontologists,* **15:** 199–210.

LONG, J.A., ed., 1993b. *Palaeozoic Vertebrate Biostratigraphy and Biogeography.* Belhaven Press, London: 1–369.

LONG, J.A., 1995a. A new plourdosteid arthrodire from the Late Devonian Gogo Formation, Western Australia: Systematics and phylogenetic implications. *Palaeontology,* **38:** 1–24.

LONG, J.A., 1995b. *The Rise of Fishes: 500 Million Years of Evolution.* University of New South Wales Press (Sydney) and Johns Hopkins University Press (USA).

LONG, J.A., 1998. *Dinosaurs of Australia and New Zealand and Other Animals of the Mesozoic Era.* University of New South Wales Press, Sydney.

LONG, J.A. & BURRETT, C.F., 1989. Early Devonian conodonts from the Kuan Tung Formation, Thailand: Systematics and biogeographic considerations. *Records of the Australian Museum,* **41:** 121–133.

LONG, J.A., CAMPBELL, K.S.W., & BARWICK, R.E., 1994. A new dipnoan genus *Ichnomylax,* from the Lower Devonian of Victoria, Australia. *Journal of Vertebrate Paleontology,* **14:** 127–131.

LONG, J.A. & YOUNG, G.C., 1988. Acanthothoracid remains from the Early Devonian of New South Wales, including a complete sclerotic capsule and pelvic girdle. In: P.A. Jell, ed., *Devonian and Carboniferous Fish Studies*, Association of Australasian Palaeontologists, Sydney: 65–85.

LOYAL, R.S., KHOSLA, A. & SAHNI, A., 1996. Gondwanan dinosaurs of India: Affinities and palaeobiogeography. In: F.A. Novas & R.E. Molnar, eds.., *Proceedings of the Gondwanan Dinosaur Symposium, Memoirs of the Queensland Museum*, **39** (3): 627–638.

LUNDELIUS, E.L., JR., 1963. Vertebrate remains from the Nullarbor Caves, Western Australia. *J. R. Soc. W. Aust.*, **46**: 3, 10, 75–80.

LUNDELIUS, E.L., JR., 1966. Marsupial carnivore dens in Australian caves. *Studies in Speleology*, **1**: 174–180.

LUNDELIUS, E. L., JR., 1983. Climatic implications of Late Pleistocene and Holocene faunal associations in Australia. *Alcheringa*, **7** (2): 125–150.

LUNDELIUS, E.L., JR. & TURNBULL, W.D., 1989. The mammalian fauna of Madura Cave, Western Australia, Part VII: Macropodidae: Sthenurinae, Macropodinae, with a Review of the Marsupial Portion of the Fauna. *Fieldiana Geology, New Series*, **17**: 1–71.

LUNDELIUS, E.L., JR. & WARNE, S., 1960. Mosasaur remains from the Upper Cretaceous of Western Australia. *Journal of Paleontology*, **34**: 1215–1217.

LYDEKKER, R., 1887. *Catalogue of the fossil Mammalia in the British Museum (Natural History) Cromwell Road, S.W.* Part 5, British Museum of Natural History, London.

LYDEKKER, R., 1896a. *A Geographical History of Mammals.* Cambridge University Press, Cambridge.

LYDEKKER, R., 1896b. *A Handbook to the Marsupialia and Monotremata.* Edward Lloyd, London.

LYNE, A.G., 1975. The rabbit. *Australian Wildlife*, **7**: 3176–3183.

MACFADDEN, B.J., WHITELAW, M.J., MCFADDEN, P. & RICH, T.H.V., 1987. Magnetic polarity stratigraphy of the Pleistocene section at Portland (Victoria), Australia. *Quaternary Research*, **28**: 364–373.

MACINTOSH, N.W.G., 1971. Analysis of an Aboriginal skeleton and a pierad tooth necklace from Lake Nitchie, Australia. *Anthropology*, **9** (1): 49–62.

MACK, G., 1956. The Queensland Museum, 1855–1955. *Memoirs of the Queensland Museum*, **13** (2): 107–119.

MAHONEY, J.A. & RIDE, W.D.L., 1975. Index to the genera and species of fossil Mammalia described from Australia and New Guinea between 1838 and 1968 (including citations of type species and primary type specimens). *Western Australian Museum Species Publication*, **6**.

MAIN, A.R., 1978. Ecophysiology: Towards an understanding of late Pleistocene marsupial extinction. In: D. Walker & J.C. Guppy, eds., *Biology and Quaternary Environments*, Australian Academy of Science, Canberra.

MAISEY, J.G., 1984. Chondrichthyan phylogeny: A look at the evidence. *Journal of Vertebrate Paleontology*, **4** (3): 359–371.

MAISEY, J.G., 1986. Heads and tails: a chordate phylogeny. *Cladistics*, **2**: 201–256.

MAISEY, J.G., ED., 1991. *Santana Fossils: An Illustrated Atlas.* T. F. H. Publications, Neptune City, New Jersey.

MANN, P., 1987. Fossil find rewrites Australia's past. *Omega Science Digest*, January/February: 14–21.

MARENSSI, S.A. et al., 1994. Eocene land mammals from Seymour Island, Antarctica: Palaeobiogeographical implications. *Antarctic Science*, **6** (1): 3–15.

MARGULIS, L., 1984. *Early Life.* Jones & Bartlett, Boston.

MARIN, H.A., 1989. Evolution of mallee and its environment. In: Noble, J.C. & Bradstock, R.A., *Mediterranean Landscapes in Australia*, Commonwealth Scientific and Industrial Organization, Canberra: 83–92.

MARSHALL, L.G., 1973. Fossil vertebrate faunas from the Lake Victoria region, S.W. New South Wales, Australia. *Memoirs of the National Museum of Victoria*, **34**: 151–172.

MARSHALL, L.G., 1977. Cladistic analysis of borhyaenoid, dasyuroid, didelphoid, and thylacinid (Marsupialia: Mammalia) affinity. *Systematic Zoology*, **26**: 410–425.

MARSHALL, L.G., 1980. Marsupial paleobiogeography. In: L.L. Jacobs, ed., *Aspects of vertebrate history: Essays in honor of Edwin Harris Colbert*, Museum of Northern Arizona Press, Flagstaff: 345–386.

MARSHALL, L.G., 1981. The families and genera of Marsupialia. *Fieldiana Geology, New Series*, **8**: 1–65.

MARSHALL, L.G., 1982. Evolution of South American Marsupialia. In: M.A. Mares & H.H. Genoways, eds., *Mammalian Biology in South America*, Special Publication Series, Pymatuning Laboratory of Ecology, University of Pittsburgh, Linesville, **6**: 251–272.

MARSHALL, L.G., CASE, J.A. & WOODBURNE, M.O., 1989. Phylogenetic relationships of the families of marsupials. *Current Mammalogy* **2**: 433–502.

MARSHALL, L.G. & CORRUCCINI, R.S., 1978. Variability, evolutionary rates, and allometry in dwarfing lineages. *Paleobiology*, **4**: 101–118.

MARSHALL, L.G. & MUIZON, C., 1988. The dawn of the age of mammals in South America. *National Geographic Research*, **4**: 23–55.

MARTIN, H.A., 1978. Evolution of the Australian flora and vegetation through the Tertiary: Evidence from pollen. *Alcheringa*, **2** (3): 181–202.

MARTIN, H.A., 1989. Vegetation and climate of the late Cainozoic in the Murray Basin and their bearing on the salinity problem. *Bureau of Mineral Resources, Journal of Australian Geology and Geophysics*, **11**: 291–299.

MARTIN, P.S. & KLEIN, R.G., eds., 1984. *Quaternary Extinctions: A Prehistoric Revolution.* University of Arizona Press, Tucson.

McNAMARA, G.C., 1990. The Wyandotte Local Fauna: A new, dated, Pleistocene vertebrate fauna from northern Queensland. *Memoirs of the Queensland Museum*, **28**: 285–297.

McALESTER, A.L., 1977. *The History of Life.* Prentice-Hall, Englewood Cliffs.

MCCLEAN, D.M., 1981. Size factor in the late Pleistocene extinctions. *American Journal of Science*, **281**: 1144–1152.

McKENNA, M.C., 1972. Possible biological consequences of plate tectonics. *BioScience*, **22** (9): 519–525.

McMENAMIN, M.A., 1987. The emergence of animals. *Scientific American*, **April**.

MCNEILL, W., 1976. *Plagues and People.* Doubleday, New York.

MERRILEES, D., 1968. Man the destroyer: Late Quaternary changes in the Australian marsupial fauna. *Journal of the Royal Society of Western Australia*, **51** (1): 1–24.

MERRILEES, D., 1979. The prehistoric environment in Western Australia. *Journal of the Royal Society of Western Australia*, **62**: 109–128.

METCALFE, I., 1998. Palaeozoic and Mesozoic geological evolution of the SE Asian region: Multidisciplinary constraints and implications for biogeography. In: R. Hall & D. Holloway, eds., *Biogeography and Geological Evolution of SE Asia*, Backbuys Publications, Leiden: 25–41.

MILES, R.S., 1977. Dipnoan (lungfish) skulls and the relationships of the group: A study based on new species from the Devonian of Australia. *Zoological Journal of the Linnean Society*, **61**: 1–328.

MILLER, A.H., 1963. The fossil flamingos of Australia. *Condor*, **65**: 289–299.

MILNER, A., 1989. Late extinctions of amphibians. *Nature*, **338**: 117.

MITCHELL T.L., 1831. An account of the limestone caves at Wellington Valley, and of the situation, near one of them, where fossil bones have been found. *Proceedings of the Geological Society*, **1**: 321–322.

MITCHELL, T.L., 1838. *Three Expeditions into the Interior of Eastern Australia, with Descriptions of the Recently Explored Region of Australia Felix, and of the Present Colony of New South Wales.* 2 vols. T. & W. Boone, London.

MOLNAR, P., 1986. The geologic history and structure of the Himalayas. *American Scientist*, **74**: 144–154.

MOLNAR, R.E., 1986. An enantiornithine bird from the Lower Cretaceous of Queensland, Australia. *Nature*, **322** (6081): 736–738.

MOLNAR, R.E., 1987. A pterosaur pelvis from western Queensland, Australia. *Alcheringa*, **11**: 87–94.

MOLNAR, R.E., 1989. Terrestrial tetrapods in Cretaceous Antarctica. In: A. Crame, ed., *Origins and Evolution of the Antarctic Biota*, Geological Society Special Publication No. 47: 131–140.

MOLNAR, R.E., ANGRIMAN, A.L. & GASPARINI, Z., 1996. An Antarctic Cretaceous theropod. In: F.A. Novas & R.E. Molnar, eds., *Proceedings of the Gondwanan Dinosaur Symposium, Memoirs of the Queensland Museum*, **39** (3): 669–674.

MOLNAR, R.E., FLANNERY, T.F. & RICH, T.H., 1985. Aussie *Allosaurus* after all. *Journal of Paleontology*, **59**: 1511–1513.

MOLNAR, R.E., FLANNERY, T.F. & RICH, T.H.V., 1981. An allosaurid theropod dinosaur from the early Cretaceous of Victoria, Australia. *Alcheringa*, **5**: 141–146.

MOLNAR, R.E. & PLEDGE, N.S., 1980. A new theropod dinosaur from South Australia. *Alcheringa*, **4**: 281–287.

MOLNAR, R.E. & THULBORN, R.A., 1980. First pterosaur from Australia. *Nature*, **288**: 361–363.

MORRISON, R. & MORRISON, M., 1988. *The Voyage of the Great Southern Ark.* Lansdowne Press, Sydney.

MOSIMANN, J.E. & MARTIN, P.S., 1975. Simulating overkill by paleoindians. *American Scientist*, **63**: 304–313.

MOY-THOMAS, J.A. & MILES, R.S., 1971. *Palaeozoic Fishes.* Chapman and Hall, London.

MOYAL, A.M., 1975. Richard Owen and his influence on Australian zoological and palaeontological science. *Records of the Australian Academy of Science*, **3** (2): 41–56.

MOYAL, A.M., 1976. *Scientists in Nineteenth Century Australia: A Documentary History.* Cassell Australia, Melbourne.

MUIRHEAD, J. & FILAN, S.L., 1995. *Yarala burchfieldi*, a plesiomorphic bandicoot (Marsupialia, Peramelemorphia) from Oligo-Miocene deposits of Riversleigh, northwestern Queensland. *Journal of Paleontology*, **69**: 127–134.

MUKHERJEE, R.N. & SENGUPTA, D.P., 1988. New capitosaurid amphibians from the Triassic Denwa Formation of the Satpura Gondwana Basin, Central India. *Alcheringa*, **22**: 317–327.

MURRAY, P.F., 1984. *Australia's Prehistoric Animals.* Methuen Australia, North Ryde, N.S.W.

MURRAY, P.F., 1985. Ichthyosaurs from Cretaceous Mullaman beds near Darwin, Northern Territory. *The Beagle,* **2:** 39–55.

MURRAY, P.F. & MEGIRIAN, D., 1998. The skull of dromornithid birds: Anatomical evidence for their relationship to Anseriformes. *Records of the South Australian Museum,* **31** (1): 51–97.

MYERS, N., 1979. *The Sinking Ark.* Pergamon Press, New York.

MYERS, N., 1983. *A Wealth of Wild Species: Storehouse for Human Welfare.* Westview, Boulder.

NEWELL, N.D., 1963. Crises in the history of life. *Scientific American,* **208** (2): 76–92.

NEWELL, N.D., 1971. An outline history of tropical organic reefs. *American Museum Novitates,* **2465:** 1–37.

NICOLL, R.S., 1977. Conodont apparatuses in an Upper Devonian palaeoniscoid fish from the Canning Basin, Western Australia. *Bureau of Mineral Resources Journal of Australian Geology and Geophysics,* **2:** 217–228.

NOMA, E. & GLASS, A.L., 1987. Mass extinction pattern: Result of chance. *Geological Magazine,* **124** (4): 319–322.

NORIEGA, J.I. & TAMBUSSI, C. P., 1995. A Late Cretaceous Presbyornithidae (Aves: Anseriformes) from Vega Island, Antarctic Peninsula: Paleobiogeographic implication. *Ameghiniana,* **32** (1): 57–61.

NORTHCUTT, R.G. & GANS, C., 1983. The genesis of neural crest and epidermal placodes: A reinterpretation of vertebrate origins. *Quarterly Review of Biology,* **58** (1): 1–28.

NOVACEK, M.J., 1990. Morphology, paleontology, and the higher clades of mammals. *Current Mammalogy,* **2:** 507–543.

NOVAS, F., 1996. Alvarezsauridae, Cretaceous basal birds from Patagonia and Mongolia. In: F. A. Novas & R. E. Molnar, eds., *Proceedings of the Gondwanan Dinosaur Symposium, Memoirs of the Queensland Museum,* **39** (3): 675–702.

OLIVER, W.R.B., 1949. The moas of New Zealand and Australia. *Dominion Museum Bulletin,* **15:** 1–206.

OLSON, E.C., 1979. Biological and physical factors in the dispersal of Permo-Carboniferous terrestrial vertebrates. In: J. Gray & A.J., Boucot, eds., *Historical Biogeography, Plate Tectonics, & the Changing Environment.* Oregon State University Press: 227–238.

OLSON, S.L., 1986. The fossil record of birds. In: D.S. Farner & J.R. King, *Avian Biology,* **8:** 79–252.

OWEN, R., 1838. Fossil remains from Wellington Valley, Australia. *See* Mitchell, T.L. (1838). Marsupialia.

OWEN, R., 1845. Report on the extinct mammals of Australia with descriptions of certain fossils indicative of the former existence in that continent of large marsupial representatives of the order Pachydermata. *Report of the British Association for the Advancement of Science, York* (1844), **14:** 223–240.

OWEN, R., 1870. On the fossil mammals of Australia—Part III. *Diprotodon australis,* Owen. *Philosophical Transactions of the Royal Society of London,* **160:** 519–578.

OWEN, R., 1872d. The fossil mammals of Australia. *Nature,* **5:** 503–504.

OWEN, R., 1877. *Researches on the fossil remains of the extinct mammals of Australia.* J. Erxleben, London.

OWEN, R., 1879a. *Memoirs on the extinct wingless birds of New Zealand with an appendix on those in England, Australia, Newfoundland, Mauritius and Rodriquez.* J. van Voorst, London.

OWEN, R., 1879b. On *Dinornis* containing a restoration of the skeleton of *Dinornis maximus* (Owen), with an appendix on additional evidence of the genus *Dromornis* in Australia. *Transactions of the Zoological Society of London,* **10** (3): 147–188.

OWEN, R., 1879c. On *Dinornis.* Pars. 19: containing a description of a femur indica-tive of a new genus of large wingless birds (*Dromornis australis*) from a post-terti-ary deposit in Qld. *Transactions of the Zoological Society of London,* **8:** 381–384.

OWEN, R., 1882. Description of portions of a tusk of a proboscidean mammal (*Notelephas australis,* Owen). *Philosopical Transactions of the Royal Society,* **173:** 777–781.

OWEN, R.M., 1843. On the discovery of the remains of a mastodontoid pachyderm in Australia. *Annual Magazine of Natural History,* **11:** 7–12.

OWEN, R.M., 1844. Description of a fossil molar tooth of a Mastodon discovered by Count Strzelecki in Australia. *Annual Magazine of Natural History,* **14:** 268–271.

PALFREYMAN, W.D., 1984. Guide to the geology of Australia. *Bureau of Mineral Resources Bulletin,* **181:** 1–111.

PANT, D.D., 1988. The origin, rise and decline of *Glossopteris* Flora: With notes on its palaeogeographical northern boundary and age. *The Palaeobotanist,* **36:** 106–117.

PARRISH, J.T. & SPICER, R.A., 1988. Late Cretaceous terrestrial vegetation: A near polar temperature curve. *Geology,* **16:** 22–25.

PASCUAL, R. et al., 1992. The first non-Australian monotreme: An early Paleocene South American platypus (Monotremata, Ornithorhynchidae). In: M.L. Augee, ed., *Platypus and Echidnas,* The Royal Zoological Society of New South Wales, Sydney: 2–15.

PASCUAL, R. & CARLINI, A.A., 1987. A new superfamily in the extensive radiation of South American Paleogene marsupials. *Fieldiana Zoology,* New Series, **39:** 99–110.

PATTERSON, B., 1958. Affinities of the Patagonian fossil mammal, *Necrolestes. Breviora,* **94:** 1–14.

PATTERSON, B. & PASCUAL, R., 1972. The fossil mammal fauna of South America. In: A. Keast, F.C. Erk & B. Glass, eds., *Evolution, Mammals, and Southern Continents,* State University of New York Press, Albany: 247–309.

PATTERSON, C. & RICH, P.V., 1987. The fossil history of the emus, *Dromaius* (Aves: Dromaiinae). *Records of the South Australian Museum,* **21** (2): 85–117.

PEABODY, F.E., 1959. Trackways of living and fossil salamanders. *University of California Publication in Zoology,* **63:** 1–72.

PEARCE, F., 1989. *Turning Up the Heat.* Paladin Grafton Books, Collins, Sydney.

PENTLAND, J.B., 1832. On the fossil bones of Wellington Valley, New Holland or New South Wales. *Edinburgh New Philosophical Journal,* **12** (24): 301–308.

PESCOTT, R.T.M., 1954. *Collections of a Century: The History of the First Hundred Years of the National Museum of Victoria.* National Museum of Victoria, Melbourne.

PETTIGREW, J.D. & JAMIESON, B.G.M., 1987. Are flying-foxes (Chiroptera: Pteropodidae) really primates? *Australian Mammalogy,* **10:** 119–124.

PHILIP, G.G., 1979. Carpoids—echinoderms or chordates? *Biological Review,* **54:** 439–471.

PILLING, A.R. & WATERMAN, R.A., eds., 1970. *Diprotodon to Detribalization: Studies of Change among Australian Aborigines.* Michigan State University Press, East Lansing.

PIMM, S., 1990. *The Balance of Nature?* University of Chicago Press, Chicago.

PIZZEY, G. & DOYLE, R., 1980. *A Field Guide to the Birds of Australia.* William Collins Sons & Co. Ltd, Sydney.

PLANE, M.D., 1967. The stratigraphy and vertebrate fauna of the Otibunda Formation, New Guinea. *Bulletin of the Australian Bureau of Mineral Resources,* **86:** 1–64.

PLAYER, A.V., 1983. *The Archer letters.* Argyle Press, Goulburn.

PLEDGE, N.S., 1985. An early Pliocene shark tooth assemblage in South Australia. *Special Publication of the South Australian Department of Mines and Energy,* **5:** 287–299.

PLEDGE, N.S., 1990. The Upper Fossil Fauna of the Henschke Fossil Cave, Naracoorte, South Australia. *Memoirs of the Queensland Museum,* **28:** 247–262.

PLOMLEY, N.J.B., 1969. The Tasmanian Journal of Natural Science. *Pap. Proceedings of the Royal Society of Tasmania,* **103:** 13–15.

POLCYN, M.J., TCHERNOV, E. & JACOBS, L.L., in press. The Cretaceous biogeographic of the eastern Mediterranean with a description of a new basal mosasauroid from Ein Yabrud, Israel. In: Y. Tomida, T.H. Rich & P. Vickers-Rich, eds., *Proceedings of the 2nd International Gondwana Dinosaur Symposium,* National Science Museum, Tokyo.

POWELL, G.V.N., POWELL, A.H. & PAUL, N.K., 1988. "Brother, can you spare a fish?" *Natural History,* February: 34–38.

PRASAD, G.V.R., JAEGER, J.J., SAHNI, A., GHEERBRANT, E. & KHAJURIA, C.K., 1994. Eutherian mammals from the Upper Cretaceous (Maastrichtian) Intertrappean Beds of Naskal, Andhra Pradesh, India. *Journal of Vertebrate Paleontology,* **14:** 260–277.

PRASAD, G.V.R. & KHAJURIA, C.K., 1995. Implication of the infra-and inter-trappean biota from the Deccan, India, for the role of volcanism in Cretaceous–Tertiary boundary extinctions. *Journal of the Geological Society,* London, **152:** 289–296.

PRESS, M.M., 1979. *Julian Tenison Woods.* Catholic Theological Faculty, Sydney.

PRIDMORE, P.A. & BARWICK, R.E., 1993. Post-cranial morphologies of the Late Devonian dipnoans *Griphognathus* and *Chirodipterus* and the locomotor implication. *Memoirs of the Association of Australasian Palaeontologists,* **15:** 161–182.

PRIDMORE, P.A., CAMPBELL, K.S.W. & BARWICK, R.E., 1994. On the holodipteran dipnoans of the Late Devonian Gogo Formation of north-western Australia. *Philosophical Transactions of the Royal Society of London,* B **344:** 105–164.

PRINCE, J.H., 1979. *The First One Hundred Years of the Royal Zoological Society of N.S.W.* Surrey Beatty & Sons, Sydney.

QUIRK, S. & ARCHER, M., 1983. *Prehistoric Animals of Australia.* Australian Museum, Sydney.

RAATH, M.A., 1996. Earliest evidence of dinosaurs from Central Gondwana.

In: F.A. Novas & R.E. Molnar, eds., *Proceedings of the Gondwanan Dinoaur Symposium, Memoirs of the Queensland Museum, 39* (3): 703–709.

RAMANATHAN, V., 1988. The Greenhouse theory of climate change: A test by an inadvertent global experiment. *Science,* **240:** 293–299.

RANKEN, C.G., 1916. *The Rankens of Bathurst.* S.D. Townsend, Sydney.

RAUP, D.M. & BOYAJIAN, G.E., 1988. Patterns of generic extinction in the fossil record. *Paleobiology,* **14** (2): 109–125.

RAUP, D. M. & SEPKOSKI, J.J., JR., 1982. Mass extinctions in the marine fossil record. *Science,* **215** (4539): 1501–1503.

RAVEN, P.H. & AXELROD, D.I., 1972. Plate tectonics and Australasian paleobiogeography. *Science,* **176** (4042): 1379–1386.

RAVEN, P.H. & JOHNSON, G.B., 1986. *Biology.* Times Mirror/Mosby College Publishing, St. Louis.

REPETSKI, E., 1978. A fish from the Upper Cambrian of North America. *Science,* **200:** 529–531.

RETALLACK, G.J., 1977. Reconstructing Triassic vegetation of eastern Australasia: A new approach for the biostratigraphy of Gondwanaland. *Alcheringa,* **1** (3): 245–278.

RETALLACK, G.J., 1996. Early Triassic therapsid footprints from the Sydney Basin, Australia. *Alcheringa,* **20:** 301–314.

RICE, A., 1990. The role of volcanism in K/T extinctions. In: Lockley, M.G. & Rich, A., eds., *Vulcanism and Fossil Biotas, Geological Society of America Special Paper* **244.**

RICH, P.V., 1975a. Antarctic dispersal routes, wandering continents and the origin of Australia's non-passeriform avifauna. *Memoirs of the National Museum of Victoria,* **36:** 63–126.

RICH, P.V., 1975b. Changing continental arrangements and the origin of Australia's non-passeriform continental avifauna. *Emu,* **75:** 97–112.

RICH, P.V., 1979. The Dromornithidae, a family of large, extinct ground birds endemic to Australia. *Bulletin of the Bureau of Mineral Resources, Geology and Geophysics, Australia,* **184.**

RICH, P.V. & BAIRD, R.F., 1986. History of the Australian avifauna. In: R. F. Johnston, ed., *Current Ornithology,* Plenum, New York, **4:** 97–139.

RICH, T.H., FLANNERY, T.F. & ARCHER, M., 1989. A second Cretaceous mammalian specimen from Lightning Ridge, N.S.W. *Alcheringa,* **13:** 85–88.

RICH, T.H. & RICH, P.V., 1989. Polar dinosaurs and biotas of the Early Cretaceous of southeastern Australia. *National Geographic Research,* **5** (1): 15–53.

RICH, P.V., RICH, T.H. & FENTON, M.A., 1989. *The Fossil Book: A Record of Prehistoric Life.* Doubleday, New York.

RICH, P.V., RICH, T.H., WAGSTAFF, B., MCEWAN-MASON, J., DOUTHITT, C.G., GREGORY, R.T. & FELTON, E.A., 1988. Biotic and geochemical evidence for low temperatures and biologic diversity in Cretaceous high latitudes of Australia. *Science,* **242:** 1403–1406.

RICH, P.V. & THOMPSON, E.M., eds., 1982. *The Fossil Vertebrate Record of Australasia.* Monash University Offset Printing Unit, Clayton.

RICH, P.V. & VAN TETS, G. F., eds., 1985. *Kadimakara: Extinct Vertebrates of Australia.* Pioneer Design Studio, Lilydale.

RICH, P.V., VAN TETS, G.F., RICH, T.H.V. & MC EVEY, A.R., 1987. The Pliocene and Quaternary flamingoes of Australia. *Memoirs of the Queensland Museum,* **25:** 207–225.

RICH, T.H. & VICKERS-RICH, P., 1994. Neoceratopsians and ornithomimosaurs: Dinosaurs of Gondwana origin? *National Geographic Research and Exploration,* **10** (1): 132–134.

RICH, T.H., VICKERS-RICH, P., CONSTANTINE, A., FLANNERY, T.F., KOOL, L. & VAN KLAVEREN, N., 1997. A tribosphenic mammal from the Mesozoic of Australia. *Science,* **278:** 1438–1442.

RICH, T.H., VICKERS-RICH, P., CONSTANTINE, A., FLANNERY, T.H., KOOL, L. & VAN KLAVEREN, N., 1999. Early Cretaceous mammals from Flat Rocks, Victoria, Australia. *Records of the Queen Victoria Museum,* **106:** 1–29.

RICH, T.H. et al., in press. The metapodial of a hadrosaur from West Antarctica. In: Y. Tomida, T.H. Rich & P. Vickers-Rich, eds., *Proceedings of the 2nd International Gondwana Dinosaur Synmposium,* National Science Museum, Tokyo.

RIDE, W.D.L., 1959b. On the evolution of Australian marsupials. *Symp. R. Soc. Vict., Evolution of living organisms,* Royal Society of Victoria, Melbourne: 281–306.

RIDE, W.D.L., 1960. The fossil mammalian fauna of the *Burramys parvus* Breccia from the Wombeyan Caves, New South Wales. *Journal of the Royal Society of Western Australia,* **43** (3): 74–80.

RIDE, W.D.L., 1962b. On the evolution of Australian marsupials. In: G.W. Leeper, ed., *The Evolution of Living Organisms,* Melbourne University Press, Melbourne: 281–306.

RIDE, W.D.L., 1964. A review of Australian fossil marsupials. *Journal of the Royal Society of Western Australia,* **47** (4): 97–131.

RIDE, W.D.L., 1980. *A Guide to the Native Mammals of Australia.* Oxford University Press, Melbourne.

RITCHIE, A., 1973. *Wuttagoonaspis* gen. nov., an unusual arthrodire from the Devonian of Western New South Wales, Australia. *Palaeontographica,* **A 143:** 58–72.

RITCHIE, A., 1974. From Greenland's icy mountains, a detective story in stone. *Australian Natural History,* **18:** 28–35.

RITCHIE, A., 1975. *Groenlandaspis* in Antarctica, Australia and Europe. *Nature,* **254:** 569–573.

RITCHIE, A., 1981. The first complete specimen of the dipnoan, *Gosfordia truncata* Woodward from the Triassic of New South Wales. *Records of the Australian Museum,* **33** (11): 606–616.

RITCHIE, A., 1985. *Arandaspis prionotolepis.* The Southern four-eyed fish. In: P.V. Rich & G.F. van Tets, eds., *Kadimakara: Extinct Vertebrates of Australia,* Pioneer Design Studio, Lilydale.

RITCHIE, A., 1987. The great Somersby fossil fish dig. *Australian Natural History,* **22** (4): 146–150.

RITCHIE, A. & GILBERT-TOMLINSON, J., 1977. First Ordovician vertebrates from the southern hemisphere. *Alcheringa,* **1:** 351–368.

ROBIN, L., 1987. Thomas Sergeant Hall (1858–1915): Scholar and enthusiast. *Historical Records of Australian Science,* **6** (4): 485–492.

ROBINSON, J.M., 1989. Phanerozoic O_2 1variation, fire, and terrestrial ecology. *Palaeogeography, Palaeoclimatology, Palaeoecology,* **75:** 223–240.

ROLLS, E.C., 1969. *They All Ran Wild.* Angus and Robertson, Sydney.

ROMER, A.S., 1956. *Osteology of the Reptiles.* University of Chicago Press, Chicago.

ROMER, A.S. & LEWIS, A.D., 1960. A mounted skeleton of the giant plesiosaur *Kronosaurus. Breviora,* **112:** 1–15.

ROMER, A.S. & PARSONS, T.S., 1977. *The Vertebrate Body,* 5th edition. Saunders, Philadelphia.

ROSEN, D.E., FOREY, P.L., GARDINER, B.G. & PATTERSON, C., 1981. Lungfishes, tetrapods, paleontology and plesiomorphy. *Bulletin of the American Museum of Natural History,* **167:** 163–275.

ROSS, C.A., ed., 1974. *Paleogeographic Provinces and Provinciality.* Soc. Econ. Paleontologists and Mineralogists Special Publication **21.** Tulsa.

ROYCHOWDHURY, T.K., 1965. A new metoposaurid amphibian from the Upper Triassic Maleri Formation of Central India. *Philosophical Transactions of the Royal Society of London,* **B 250:** 1–52.

ROYCHOWDHURY, T.K., 1970a. Two new dicynodonts from the Yerrapalli Formation of Central India. *Palaeontology,* **13:** 132–144.

ROYCHOWDHURY, T.K., 1970b. A new capitosaurid amphibian from the Triassic Yerrapalli Formation of the Pranhita-Godavari valley. *Journal Geological Society of India,* **11:** 155–162.

ROZHDESTVENSKY, A.K., 1977. The study of dinosaurs in Asia. *Journal of the Palaeontological Society of India,* **20:** 102–119.

RUNNEGAR, B., 1983. A *Diprotodon* ulna chewed by the marsupial lion, *Thylacoleo carnifex. Alcheringa,* **7** (1): 23–26.

SAGAN, C., 1980. *Cosmos.* Ballantine Books, New York.

SAMPSON, S.D. et al., 1998. Predatory dinosaur remains from Madagascar: Implications for the Cretaceous biogeography of Gondwana. *Science,* **280:** 1048–1051.

SANSON, G., 1978. The evolution and significance of mastication in the Macropodidae. *Australian Mammalogy,* **2:** 23–28.

SATSANGI, P.P., 1988. Vertebrate faunas from the Indian Gondwana Sequence. *The Palaeobotanist,* **36:** 245–253.

SAVAGE, D.E. & RUSSELL, D.E., 1983. *Mammalian Paleofaunas of the World.* Addison-Wesley, Reading, Massachusetts.

SCHAEFFER, B., 1967. Comments on elasmobranch evolution. In: P.W. Gilbert, R.F. Mathewson & D.P. Rall, *Sharks, Skates and Rays,* Johns Hopkins University Press, Baltimore: 3–35.

SCHAEFFER, B. & ROSEN, D., 1961. Major adaptive levels in the evolution of the actinopterygian feeding mechanism. *American Zoologist,* **1** (2): 187–204.

SCHAEFFER, B. & WILLIAMS, M., 1977. Relationships of fossil and living elasmobranchs. *American Zoologist,* **17:** 293–302.

SCHMID, E., 1972. *Atlas of Animal Bones.* Elsevier, Amsterdam.

SCHOPF, J.W. & PACKER, B.M., 1987. Early Archean (3.3-Billion to 3.5-billion-year-old) microfossils from Warrawoona Group, Australia. *Science,* **237:** 70–73.

SCIENTIFIC AMERICAN SPECIAL ISSUE, September 1989. Managing Planet Earth. *Scientific American,* **261** (3): 1–134.

SCOTESE, C.R. & SAGER, W.W., eds., 1989. *Mesozoic and Cenozoic Plate Reconstructions.* Elsevier, Amsterdam.

SEELEY, H.G., 1891. On *Agrosaurus macgillivrayi,* a saurischian reptile from the NE-coast of Australia. *Quarterly Journal of the Geological Society of London,* **47:** 164–165.

SEILACHER, A., BOSE, P.K. & PFLUGER, F., 1998. Triploblastic animals more than 1 billion years ago: Trace fossil evidence from India. *Science,* **282:** 80–83.

SENGUPTA, D.P., 1995. Chigutisaurid temnospondyls from the Late Triassic of India and a review of the Family Chigutisauridae. *Palaeontology*, **38**: 313–339.

SEPKOSKI, J.J., JR., 1983. Precambrian-Cambrian boundary: The spike is driven and the monolith crumbles. *Paleobiology*, **9** (3): 199–206.

SEPKOSKI, J.J., JR., 1989. Periodicity in extinction and the problem of catastrophism in the history of life. *Journal of the Geological Society, London*, **146**: 7–19.

SERENO, P.C., in press. Dinosaurian biogeography: Vicariance, dispersal and regional extinction. In: Y. Tomida, T. Rich & P. Vickers-Rich, eds., *Proceedings of the 2nd International Gondwana Dinosaur Symposium*, National Science Museum, Tokyo.

SERENO, P.C. et al., 1996. Predatory dinosaurs from the Sahara and Late Cretaceous faunal differentiation. *Science*, **272**: 986–991.

SERENO, P.C. et al., 1998. A long-snouted predatory dinosaur from Africa and the evolution of spinosaurids. *Science*, **282**: 1298–1302.

SHISHKIN, M.A., 1993 (Abs.). A first find of Gondwana rhytiodosteid (Temnospondyli) in the lower Triassic of Eastern Europe. *Journal of Vertebrate Paleontology*, **13**, Supplement to Number 3: 57A.

SHOTWELL, J.A., 1955. An approach to the paleoecology of mammals. *Ecology*, **36**: 326–337.

SHOTWELL, J.A., 1958. Inter-community relationshiups in Hemphillian (mid-Pliocene) mammals. *Ecology*, **39**: 271–282.

SIGÉ, B., HAND, S. & ARCHER, M., 1982. An Australian Miocene *Brachipposideros* (Mammalia, Chiroptera) related to Miocene representatives from France. *Palaeovertebrata*, **12**: 149–172.

SIGOGNEAU-RUSSELL, D. & ENSOM, P., 1994. *Thereuodon* (Theria: Symmetrodonta) from the Lower Cretaceous of North Africa and Europe, and a brief review of symmetrodonts. *Cretaceous Research*, **19**: 445–470.

SIGOGNEAU-RUSSELL, D., EVANS, S.E., LEVINE, J.F. & RUSSELL, D.A., 1998. The Early Cretaceous microvertebrate locality of Anoual, Morocco: A glimpse at the small vertebrate assemblages of Africa. In: S.G. Lucas, J.I. Kirkland & J.W. Estep, eds., *Lower and Middle Cretaceous Terrestrial Ecosystems, New Mexico Museum of Natural History and Science Bulletin*, **14**: 177–181.

SIMPSON, A., 1988. The Somersby fossil dig. *Australian Science Magazine*, **1**: 16–21.

SIMPSON, G.G., 1930. Post-Mesozoic Marsupialia. *Fossilium Catalogium*, **47**: 8, 67–68.

SIMPSON, G.G., 1945. The principles of classification and a classification of mammals. *Bulletin of the American Museum of Natural History*, **85**: 1–350.

SIMPSON, G.G., 1970. Miocene penguins from Victoria, Australia and Chubut, Argentina. *Memoirs of the National Museum of Victoria*, **31**: 17–24.

SIMPSON, G.G., 1977. Too many lines: The limits of the Oriental and Australian Zoogeographic Regions. *Proceedings of the American Philosophical Society*, **121** (2): 107–120.

SIMPSON, G.G., 1980. *Splendid Isolation: The Curious History of South American Mammals*. Yale University Press, New Haven, Connecticut.

SIMPSON, K., ed., 1986. *The Birds of Australia: A Book of Identification*. Lloyd O'Neil Pty Ltd, Melbourne.

SKIBA, U. & CRESSER, M., 1989. The ecological significance of increasing carbon dioxide. *Endeavour*, **12** (3): 143–147.

SLUITER, I.R. & KERSHAW, A.P., 1982. The nature of Late Tertiary vegetation in Australia. *Alcheringa*, **6** (3): 211–222.

SMITH, A.G., HURLEY, A.M. & BRIDEN, J.C., 1981. *Phanerozoic Paleocontinental World Maps*. Cambridge University Press, Cambridge.

SMITH, M.J., 1971. Small fossil vertebrates from Victoria Cave, Naracoorte, South Australia. I. (Marsupialia). Potoroinae (Macropodidae) Petauridae and Burramyidae (Marsupialia). *Transactions of the Royal Society of South Australia*, **95** (4): 185–198.

SMITH, M.J., 1972. Small fossil vertebrates from Victoria Cave, Naracoorte, South Australia. II. Peramelidae, Thylacinidae and Dasyuridae (Marsupialia). *Transactions of the Royal Society of South Australia*, **96** (3): 127–137.

SMITH, M.J., 1976. Small fossil vertebrates from Victoria Cave, Naracoorte, South Australia. IV. Reptiles. *Transactions of the Royal Society of South Australia*, **100**: 39–51.

SMITH, M.J. & PLANE, M., 1985. Pythonine snakes (Boidae) from the Miocene of Australia. *Bureau of Mineral Resources, Journal of Australian Geology and Geophysics*, **9**: 191–195.

SOKOLOFF, J., 1988. *The Politics of Food*. Sierra Club Books, San Francisco.

SPENCER, B. & KERSHAW, J.A., 1910. A collection of sub-fossil bird and marsupial remains from King Island, Bass Strait. *Memoirs of the National Museum of Victoria*, **3**: 5–35.

STAINES, H.R.E., 1954. Dinosaur footprints at Mount Morgan. *Queensland Government Mining Journal*, **55**: 483–485.

STANHOPE, M.J., WADDELL, V.G., MADSEN, O., DE JONG, W., HEDGES, S.B., CLEVEN, G.C., KAO, D. & SPRINGER, M.K., 1998. Molecular evidence for multiple origins of Insectivora and for a new order of endemic African insectivore mammals. *Proceedings of the National Academy of Sciences*, **95**: 9967–9972.

STANLEY, S. M., 1989. *Earth and Life through Time*. W. H. Freeman & Co., New York.

STEBBINS, G.L., 1982. *Darwin to DNA, Molecules to Humanity*. Basic Books, New York.

STEHLI, F.G. & WEBB, S.D., eds., 1985. *The Great American Biotic Interchange*. Plenum Press, New York.

STIRLING, E.C., 1894. The recent discovery of fossil remains at Lake Callabonna, South Australia. *Nature*, **50**: 184–188, 206–211.

STIRLING, E.C., 1896. The newly discovered extinct gigantic bird of South Australia. *Ibis*, **7** (11): 593.

STIRLING, E.C., 1907. Reconstruction of *Diprotodon* from the Callabonna deposits, South Australia. *Nature*, **76**: 543–544.

STIRLING, E.C., 1913. Fossil remains of Lake Callabonna. Part IV. 2. *Phascolonus gigas*, Owen, and *Sceparnodon ramsayi*, Owen, with a description of some parts of its skeleton. *Memoirs of the Royal Society of South Australia*, **1** (4): 127–178.

STIRLING, E.C. & ZEITZ, A.H.C., 1896. Preliminary notes on *Genyornis newtoni*: A new genus and species of fossil struthious bird found at Lake Callabonna, South Australia. *Transactions and Proceedings of the Royal Society of South Australia*, **20**: 171–190.

STIRLING, E.C. & ZEITZ, A.H.C., 1899. Preliminary notes on *Phascolonus gigas*, Owen (*Phascolomys* [*Phascolonus*] Owen), and its identity with *Sceparnodon ramsayi*, Owen. *Transactions and Proceedings of the Royal Society of South Australia*, **23**: 123–135.

STIRLING, E.C. & ZEITZ, A.H.C., 1900a. Description of the bones of the manus and pes of *Diprotodon australis*, Owen. *Memoirs of the Royal Society of South Australia*, **1**: 1–40.

STIRLING, E.C. & ZEITZ, A.H.C., 1900b. Fossil remains of Lake Callabonna. I. *Genyornis newtoni*. A new genus and species of fossil struthious bird. *Memoirs of the Royal Society of South Australia*, **1** (2): 41–80.

STIRLING, E.C. & ZEITZ, A.H.C., 1905. Fossil remains of Lake Callabonna. III. Description of the vertebrae of *Genyornis newtoni*. *Memoirs of the Royal Society of South Australia*, **1** (3): 81–110.

STIRLING, E.C. & ZEITZ, A.H.C., 1913. Fossil Remains of Lake Callabonna. IV. Descriptions of some further remains of *Genyornis newtoni*. *Memoirs of the Royal Society of South Australia*, **1** (4): 111–126.

STIRTON, R.A., 1955. Late Tertiary marsupials from South Australia. *Records of the South Australian Museum*, **11**: 247–268.

STIRTON, R.A., 1967. The Diprotodontidae from the Ngapakaldi fauna, South Australia. *Bulletin Australian Bureau of Mineral Resources*, **85**: 1–44.

STIRTON, R.A., TEDFORD, R.H. & MILLER, A.H., 1961. Cenozoic stratigraphy and vertebrate paleontology of the Tirari Desert, South Australia. *Records of the South Australian Museum*, **14** (1): 19–61.

STIRTON, R.A., TEDFORD, R.H. & WOODBURNE, M.O., 1968. Australian Tertiary deposits containing terrestrial mammals. *University of California Publications in Geological Science*, **77**: 1–30.

STIRTON, R.A., WOODBURNE, M.O. & PLANE, M.D., 1967. A phylogeny of the Tertiary Diprotodontidae and its significance in correlation. *Bulletin of the Bureau of Mineral Resources, Geology and Geophysics of Australia*, **85**: 149–160.

STRAHAN, R., 1979. *Rare and Curious Specimens: An Illustrated History of the Australian Museum, 1827–1979*. Australian Museum, Sydney.

SUN, A.L., 1989. *Before Dinosaurs: Land Vertebrates of China 200 Million Years Ago*. China Ocean Press, Beijing.

SZALAY, F.S., 1982. A new appraisal of marsupial phylogeny and classification. In: Archer, M., *Carnivorous Marsupials*, Royal Zoological Society of New South Wales and Surrey Beatty & Son, Chipping North: 621–640.

TALENT, J.A., DUNCAN, P.M. & HANDLEY, P.L., 1966. Early Cretaceous feathers from Victoria. *Emu*, **66**: 81–86.

TANBURY, P., ed., 1975. *100 Years of Australian Scientific Explorations*. Holt, Rinehart & Winston, Sydney.

TARLO, L.B., 1964. The origin of bone. Discussion. In: H. J. J. Blackwood, ed., *Bone and Tooth: Proceedings of the First European Symposium*, Oxford, April 1963, Macmillan, New York: 3–17.

TATE, R., 1893. A century of geological progress. *Report of the Australasian Association for the Advancement of Science*, **5**: 1–69.

TAYLOR, D.W. & HICKEY, L.J., 1990. An Aptian plant with attached leaves and flowers: Implications for angiosperm origin. *Science*, **247**: 702–704.

TEDFORD, R.H., 1955. Report on the extinct mammalian remains at Lake Menindee, N.S.W. *Records of the South Australian Museum*, **11**: 299–305.

TEDFORD, R.H., 1966. A review of the macropodid genus *Sthenurus. University of California, Publications in Geological Science,* **57:** 1–72.

TEDFORD, R.H., 1970. Principles and practices of mammalian geochronology in North America. In E.L. Yochelson, ed., *Proceedings of the North American Paleontological Convention,* Allen Press, Lawrence, Kansas, **1:** 666–703.

TEDFORD, R.H., 1973. The *Diprotodons* of Lake Callabonna. *Australian Natural History,* **17:** 349–354.

TEDFORD, R.H., ARCHER, M., BARTHOLOMAI, A., PLANE, M., PLEDGE, N.S., RICH, T., RICH, P. & WELLS, R.T., 1977. The discovery of Miocene vertebrates, Lake Frome area, South Australia. *Bureau of Mineral Resources Journal of Australian Geology and Geophysics,* **2:** 53–57.

TEDFORD, R.H., BANKS, M.R., KEMP, N.R., MCDOUGALL, I. & SUTHERLAND, F.L., 1975. Recognition of the oldest known fossil marsupials from Australia. *Nature,* **255:** 141–142.

TEDFORD, R.H. & WELLS, R.T., 1990. Pleistocene deposits and fossil vertebrates from the "Dead Heart of Australia." *Memoirs of the Queensland Museum,* **28:** 263–284.

TEDFORD, R.H., WILLIAMS, D. & WELLS, R.T., 1986. Lake Eyre and Birdsville Basins: Late Cainzoic sediments and fossil vertebrates. In R.T. Wells & R.A. Callen, eds., *The Lake Eyre Basin: Cainozoic Sediments, Fossil Vertebrates and Plants, Landforms, Silcretes and Climatic Implications,* Australasian Sedimentologists Group, Field Guide Series No. 4, Geological Society of Australia, Sydney: 42–72.

TEICHERT, D. & MATHESON, R.S., 1944. Upper Cretaceous ichthyosaurian and plesiosaurian remains from Western Australia. *Australian Journal of Science,* **6:** 167–170.

THOMSON, K.S. & CAMPBELL, K.S.W., 1971. The structure and relationships of the primitive Devonian lungfish—*Dipnorhynchus sussmilchi* (Etheridge). *Bulletin of the Peabody Museum of Natural History,* **38:** 1–109.

THORNE, A.G. & MACUMBER, P.G., 1972. Discoveries of late Pleistocene man at Kow Swamp, Australia. *Nature,* **238** (5363): 316–319.

THULBORN, R.A., 1983. A mammal-like reptile from Australia. *Nature,* **303:** 330–331.

THULBORN, R.A., 1986a. Early Triassic tetrapod faunas of southeastern Gondwana. *Alcheringa,* **10:** 297–313.

THULBORN, R.A., 1986b. The Australian Triassic reptile *Tasmaniosaurus triassicus* (Thecodontia: Proterosuchia). *Journal of Vertebrate Paleontology,* **6** (2): 123–142.

THULBORN, R.A. & WADE, M., 1979. Dinosaur stampede in the Cretaceous of Queesland. *Lethaia,* **12:** 275–279.

THULBORN, R.A. & WADE, M., 1984. Dinosaur trackways in the Winton Formation (mid-Cretaceous) of Queensland. *Memoirs of the Queensland Museum,* **21:** 413–517.

THULBORN, R.A. & WARREN, A., 1980. Early Jurassic plesiosaurs from Australia. *Nature,* **285:** 224–225.

THULBORN, T., 1990. *Dinosaur Tracks.* Chapman and Hall, Melbourne.

THULBORN, T., WARREN, A., TURNER, S. & HAMLEY, T., 1996. Early Carboniferous tetrapods in Australia. *Nature,* **381:** 777–780.

TREZISE, P., 1971. *Rock Art of South-east Cape York.* Australian Institute for Aboriginal Studies, Canberra.

TURNBULL, W.D. & LUNDELIUS, E.L. Jr., 1970. The Hamilton fauna: A late Pliocene mammalian fauna from the Grange Burn, Victoria, Australia. *Fieldiana Geology,* **19:** 1–163.

TURNBULL, W.D., LUNDELIUS, E.L., JR. & MCDOUGALL, I., 1965. A Potassium/argon dated Pliocene marsupial fauna from Victoria, Australia. *Nature,* **206**(4986): 816.

TURNER, S., 1982. Middle Palaeozoic elasmobranch remains from Australia. *Journal of Vertebrate Paleontology,* **2** (2): 117–131.

TURNER, S., 1982. *Saurichthys* (Pisces, Actinopterygii) from the Early Triassic of Queensland. *Memoirs of the Queensland Museum,* **20** (3): 545–551.

TURNER, S. & PICKETT, J., 1982. Silurian vertebrates in Australia. *Search,* **13:** 11–12.

TURNER, S. & YOUNG, G.C., 1987. Shark teeth from the Early–Middle Devonian Cravens Peak Beds, Georgina Basin, Queensland. *Alcheringa,* **11:** 233–244.

TYLER, M.J., 1976. Comparative osteology of the pelvic girdle of Australian frogs and description of a new fossil genus. *Transactions of the Royal Society of South Australia,* **100:** 3–14.

TYLER, M.J., 1979. Herpetofaunal relationships of South America with Australia. In: W. F. Duellman, ed., *The South American Herpetofauna: Its Origin, Evolution, and Dispersal,* University of Kansas, Museum of Natural History, Monograph **7:** 73–106.

TYLER, M.J., ed., 1979. *The Status of Endangered Australasian Wildlife.*

TYLER, M.J., 1982. Tertiary frogs from South Australia. *Alcheringa,* **6:** 101–103.

TYLER, M.J., 1989. *Australian Frogs.* Viking O'Neil Penguin Books, Melbourne.

TYNDALE-BISCOE, H., 1973. *Life of Marsupials.* American Elsevier Publishing Co., New York.

VALLANCE, T.G., 1975. Origins of Australian geology. *Proceedings of the Linnean Society of New South Wales,* **100:** 13–43.

VALLANCE, T.G., 1978. Pioneers and leaders: A record of Australian palaeontology in the nineteenth century. *Alcheringa,* **2:** 243–250.

VALLANCE, T.G., 1981. The fuss about coal: Troubled relations between palaeobotany and geology. In D. J. & S. G. M. Carr, eds., *Plants and Man in Australia,* Academic Press, Sydney: 136–176.

VANDERWAL, R. & FULLAGAR, R., 1989. Engraved *Diprotodon* tooth from the Spring Creek locality, Victoria. *Archaeology of Oceania,* **24:** 13–16.

VEEVERS, J.J., 1986. *Phanerozoic Earth History of Australia.* Oxford Science Publications, Clarendon Press, Oxford.

VEEVERS, J.J., 1988. Gondwana facies started when Gondwanaland merged in Pangea. *Geology,* **16:** 732–734.

VICKERS-RICH, P. & MOLNAR, R.E., 1996. The foot of a bird from the Eocene Redbank Plains Formation of Queensland. *Alcheringa,* **20** (1): 21–29.

VICKERS-RICH, P., MONAGHAN, J.M., BAIRD, R.F. & RICH, T.H., eds., 1996. *Vertebrate Palaeontology of Australasia* (revised and reprinted). Pioneer Design Studio in cooperation with the Monash University Publications Committee, Melbourne.

VICKERS-RICH, P., RICH, T.H., MCNAMARA, G. & MILNER, A., 1999. Is *Agrosaurus macgillivrayi* Australia's oldest dinosaur? *Records of the Western Australian Museum Supplement,* **57:** 191–200.

VICKERS-RICH, P., TRUSLER, P., ROWLEY, M.JH., COOPER, A., CHAMBERS, G.K.; BOCK, W.J., MILLENER, P.R., WORTHY, T.H. & YALDWYN, J.C., 1995. Morphology, myology, collagen and DNA of a Mummified Upland Moa, *Megalapteryx didinus* (Aves: Dinornithiformes) from New Zealand. *Tuhinga, Records of the Museum of New Zealand Te Papa Tongarewa,* **4:** 1–26.

VIDAL, G., 1983. The oldest eukaryotic cells. *Scientific American,* September.

VOORHIES, M., 1969. Taphonomy and population dynamics of an early Pliocene vertebrate fauna, Knox County, Nebraska. *Contributions in the Geological Sciences, Special Paper* **1,** University of Wyoming Press, Laramie.

VOORHIES, M., 1970. Sampling difficulties in reconstructing late Tertiary mammalian communities. *Proceedings of the North American Paleontological Congress,* **6:** 545–568.

WADE, M., 1984. *Platypterygius australis,* an Australian Cretaceous ichthyosaur. *Lethaia,* **17:** 99–113.

WADE, R.T., 1931. The fossil fishes of the Australian Mesozoic rocks. *Journal and Proceedings of the Royal Society of New South Wales,* **4:** 115–147.

WADE, R.T., 1949. The Triassic fishes of Gosford, New South Wales. *Journal and Proceedings of the Royal Society of New South Wales,* **73:** 206–217.

WAGSTAFF, B.E. & MASON, J.M., 1989. Palynological dating of Lower Cretaceous coastal vertebrate localities, Victoria, Australia. *National Geographic Research,* **5** (1): 54–63.

WAKEFIELD, N.A., 1967a. Mammal bones in the Buchan district. *Victorian Naturalist,* **84:** 211–214.

WAKEFIELD, N.A., 1972. Palaeoecology of fossil mammal assemblages from some Australian Caves. *Proceedings of the Royal Society of Victoria,* **85** (1): 1–26.

WALDMAN, M., 1970. A third specimen of a lower Cretaceous feather from Victoria, Australia. *Condor,* **72:** 377.

WALDMAN, M., 1971. Fish from the freshwater Lower Cretaceous of Victoria, Australia, with comments on the palaeo-environment. *Special Papers in Palaeontology,* **9:** 1–124.

WALDMAN, M., 1973. A fossil lake fauna of Koonwarra, Victoria. *Australian Natural History,* **17:** 317–321.

WALKER, C.A., 1981. New subclass of birds from the Cretaceous of South America. *Nature,* **292:** 51–53.

WALLACE, A.R., 1869. *The Malay Archipelago.* 2 vols. Macmillan, London.

WALLACE, A.R., 1876. *The Geographical Distribution of Animals.* 2 vols. Macmillan, London.

WALLS, G.L., 1963. *The Vertebrate Eye and Its Adaptive Radiation.* Hafner, New York.

WALTER, M.R., 1977. Interpreting stromatolites. *American Scientist,* **65** (5): 563–571.

WARREN, A.A., 1972. Triassic amphibians and reptiles of Australia in relation to Gondwanaland. *Australian Natural History:* 279–283.

WARREN, A.A., 1980. *Parotosuchus* from the Early Triassic of Queensland and Western Australia. *Alcheringa,* **4** (1): 25–36.

WARREN, A.A., in press. *Karroo tupilakosaurid:* A relict from Gondwana. *Transactions of the Royal Society of Edinburgh,* **89.**

WARREN, A.A. & BLACK, T., 1985. A new rhytidosteid (Amphibia, Labyrinthodontia) from the Early Triassic Arcadia Formation of

Queensland, Australia, and the relationships of Triassic temnospondyls. *Journal of Vertebrate Paleontology,* 5 (4): 303–327.

WARREN, A.A. & HUTCHINSON, M.N., 1983. The last labyrinthodont: A new brachyopid (Amphibia, Temnospondyli) from the Early Jurassic Evergreen Formation of Queensland, Australia. *Philosophical Transactions of the Royal Society of London,* **Ser. B, 303** (1113): 1–24.

WARREN, A.A. & HUTCHINSON, M.N., 1987. The skeleton of a new hornless rhytidosteid (Amphibia, Temnospondyli). *Alcheringa,* **11**: 291–302.

WARREN, A.A., JUPP, R. & BOLTON, B., 1986. Earliest tetrapod trackway. *Alcheringa,* **10**: 183–186.

WARREN, A. et al., 1997. The last last labyrinthodonts? *Palaeontographica,* **A 247**: 1–24.

WARREN, J.W., 1969. A fossil chelonian of probably Lower Cretaceous age from Victoria. *Memoirs of the National Museum of Victoria,* **29**: 23–28.

WARREN, J.W. & WAKEFIELD, N.A., 1972. Trackways of tetrapod vertebrates from the Upper Devonian of Victoria, Australia. *Nature,* **238**: 469–470.

WATSON, D.M.S., 1958. A new labyrinthodont (*Paracyclotosaurus*) from the Upper Trias of New South Wales. *Bulletin of the British Museum of Natural History,* **3**: 233–264.

WATTS, C.H.S. & ASLIN, H.J., 1981. *The Rodents of Australia.* Angus and Robertson Publishers, Melbourne.

WEGENER, A., 1966. *The Origin of Continents and Oceans.* Methuen, London.

WEIGELT, J., 1989. *Recent Vertebrate Carcasses and Their Paleobiological Implications.* University of Chicago Press, Chicago.

WEIJERMARS, R.G., 1989. Global tectonics since the breakup of Pangea 180 million years ago: Evolution maps and lithospheric budget. *Earth-Science Reviews,* **26**: 113–162.

WEINBERG, S., 1977. *The First Three Minutes.* Basic Books, New York.

WELLES, S.P., 1983. *Allosaurus* (Saurischia, Theropoda) not yet in Australia. *Journal of Paleontology,* **57**: 196.

WELLS, R.T., 1975. Reconstructing the past: excavations in fossil caves. *Australian Natural History,* **18** (6): 208–211.

WELLS, R.T., 1978. Fossil mammals in the reconstruction of Quaternary environments with examples from the Australian fauna. In: D. Walker & J.C. Guppy, eds., *Biology and Quaternary Environments,* Australian Academy of Science, Canberra, 103–124.

WELLS, R.T., MORIARTY, K. & WILLIAMS, D.L.G., 1984. The fossil vertebrate deposits of Victoria Fossil Cave Naracoorte: An introduction to the geology and fauna. *Australian Zoologist,* **21**: 305–333.

WELLS, R.T. & MURRAY, P., 1979. A new sthenurine kangaroo (Marsupialia, Macropodidae) from South-eastern Australia. *Transactions of the Royal Society of South Australia,* **103** (8): 213–219.

WELLS, R.T. & NICHOL, B., 1977. On the manus and pes of *Thylacoleo carnifex* Owen (Marsupialia). *Transactions of the Royal Society of South Australia,* **101** (5–6): 139–146.

WELLS, S.P. & ESTES, R., 1969. *Hadrokkosaurus bradyi* from the upper Moenkopi Formation of Arizona. With a review of the brachyopid labyrinthodonts. *University of California Publications in Geological Sciences,* **84**: 1–56.

WHITE, E.I., 1952. Australian arthrodires. *Bulletin of the British Museum of Natural History (Geology),* **1**: 294–304.

WHITE, J.P. & O'CONNELL, J.F., 1979. Australian prehistory; new aspects of antiquity: Recent discoveries make the prehistory of Australia's Aborigines longer and emphasis local developments. *Science,* **203**: 21–28.

WHITE, M.E., 1986. *The Greening of Gondwana.* Reed, Sydney.

WHITELAW, M.J., 1989. Magnetic polarity stratigraphy and mammalian fauna of the Late Pliocene (Early Matuyama) section at Batesford (Victoria) Australia. *Journal of Geology,* **97**: 624–631.

WHITLEY, G.P., 1958–1959. The life and work of Gerard Krefft. *Proceedings of the Royal Zoological Society of New South Wales,* **59**: 21–34.

WHITLEY, G.P., 1967–1968. Gerard Krefft and his bibliography. *Proceedings of the Royal Zoological Society of New South Wales,* **68**: 38.

WHITMORE, T.C., ed., 1987. *Biogeographical Evolution of the Malay Archipelago.* Oxford Science Publications, Oxford University Press, New York.

WIFFEN, J., 1996. Dinosaurian palaeobiology: A New Zealand perspective. In: F.A. Novas & R.E. Molnar, eds., *Proceedings of the Gondwanan Dinosaur Symposium, Memoirs of the Queensland Museum,* **39** (3): 725–731.

WIJKMAN, A. & TIMBERLAKE, L., 1984. Natural Disasters: Acts of God or Acts of Man? *Earthscan,* Washington, D.C.

WILSON, E.O., 1989. Threats to biodiversity. *Scientific American,* September: 60–66.

WILSON, E.O., ed., 1988. *Biodiversity.* National Academy Press, New York.

WOLFF, R.G., 1973. Hydrodynamic sorting and ecology of a Pleistocene mammalian assemblage from California (U.S.A.). *Palaeogeography, Palaeoclimatology, Palaeoecology,* **13**: 19–101.

WOLFF, R.G., 1975. Sampling and sample size in ecological analysis of fossil mammals. *Paleobiology,* **1** (2): 195–204.

WOODBURNE, M.O., 1967. The Alcoota fauna, Central Australia: An integrated palaeontological and geological study. *Bulletin of the Bureau of Mineral Resources, Geology and Geophysics,* **87**: 1–187.

WOODBURNE, M.O., 1969. A lower mandible of *Zygomaturus gilli* from the Sandringham sands, Beaumaris, Victoria, Australia. *Memoirs of the National Museum of Victoria,* **29**: 29–39.

WOODBURNE, M.O. & CLEMENS, W.A., 1986. Introduction. In: M.O. Woodburne & W.A. Clemens, eds., *Revision of the Ektopodontidae (Mammalia: Marsupialia; Phalangeroidea) of the Australian Neogene,* University of California, Publications in Geological Science, **131**: 1–9.

WOODBURNE, M.O. & TEDFORD, R.H., 1975. The first Tertiary monotreme from Australia. *American Museum Novitates,* **2588**: 1–11.

WOODBURNE, M.O. & ZINSMEISTER, W.J., 1984. The first land mammal from Antarctica and its biogeographic implications. *Journal of Paleontology,* **58**: 913–948.

WOODBURNE, M. O. et al., 1993. Land mammal biostratigraphy and magnetostratigraphy of the Etadunna Formation (late Oligocene) of South Australia. *Journal of Vertebrate Paleontology,* **13**: 483–515.

WOODWARD, A.S., 1890. The fossil fishes of the Hawkesbury Series of Gosford, New South Wales. Memoirs of the Geological Survey of New South Wales, *Palaeontology,* **4**: 1–55.

WOODWARD, A.S., 1906. On a tooth of *Ceratodus* and a dinosaurian claw from the Lower Jurassic of Victoria, Australia. *Annals and Magazine of Natural History,* **18**: 1–3.

WORLD COMMISSION ON ENVIRONMENT AND DEVELOPMENT, 1987. *Our Common Future.* Oxford University Press, London.

WORLD DEVELOPMENT REPORT, 1989. The World Bank, Oxford University Press, New York.

WROTE, S., 1998. A new "bone-cracking" dasyurid (Marsupialia) from the Miocene of Riversleigh, northwestern Queensland. *Alcheringa,* **22**: 277–284.

WYLLIE, P. J., 1976. *The Way the Earth Works: An Introduction to the New Global Geology and Its Revolutionary Development.* John Wiley & Sons, Sydney.

YATES, A.M., in press. The Lapillopsidae: A new family of small temnospondyls from the Early Triassic of Australia. *Journal of Vertebrate Paleontology.*

YOUNG, G.C., 1971. New information on the structure and relationships of *Buchanosteus* (Placodermi: Euarthrodira) from the Early Devonian of New South Wales. *Zoological Journal of the Linnean Society,* **66** (4): 309–352.

YOUNG, G.C., 1978. A new Early Devonian petalichthyid fish from the Taemas/Wee Jasper region of New South Wales. Alcheringa, **2**: 103–116.

YOUNG, G.C., 1981a. Biogeography of Devonian vertebrates. *Alcheringa,* **5** (3): 225–243.

YOUNG, G.C., 1981b. New Early Devonian brachythoracids (placoderm fishes) from the Taemas-Wee Jasper region of New South Wales. *Alcheringa,* **5** (4): 245–271.

YOUNG, G.C., 1984. An asterolepidoid antiarch (placoderm fish) from the Early Devonian of the Georgina Basin, central Australia. *Alcheringa,* **8**: 65–80.

YOUNG, G.C., 1986. Early Devonian fish material from the Horlick Formation, Ohio Range, Antarctica. *Alcheringa,* **10** (1): 35–44.

YOUNG, G.C., 1987. Devonian palaeontological data and the Armorica problem. *Palaeogeography, Palaeoclimatology, Palaeoecology,* **60**: 283–304.

YOUNG, G.C., 1988. Palaeontology of the Late Devonian and Early Carboniferous. In: J.G. Douglas & J.A. Ferguson, eds., *Geology of Victoria.* Victorian Division Geological Society of Australia, Melbourne: 191–194.

YOUNG, G.C., 1989. The Aztec fish fauna (Devonian) of Southern Victoria Land: Evolutionary and biogeographic significance. In: J.A. Crame, ed., *Origins and Evolution of the Antarctic Biota. Geological Society Special Publication* No. **47**: 43–62.

YOUNG, G.C., 1997. Ordovician microvertebrate remains from the Amadeus Basin, central Australia. *Journal of Vertebrate Paleontology,* **17**: 1–25.

YOUNG, G.C. & GORTER, J.D., 1981. A new fish fauna of Middle Devonian age from the Taemas/Wee Jasper region of New South Wales. *Bureau of Mineral Resources Bulletin,* **209**: 83–147.

YOUNG, G.C., KARATAJUTE-TALIMAA, V.N. & SMITH, M.M., 1996. A possible Late Cambrian vertebrate from Australia. *Nature,* **383**: 810–812.

YOUNG, G.C. & LAURIE, J.R., 1996. *An Australian Phanerozoic Timescale.* Oxford University Press, Oxford: 1–279.

ZHIMING, D., 1987. *Dinosaurs from China.* China Ocean Press, Beijing.

ZUBAKOV, V.A. & BORZENKOVA, I.I., 1990. *Global Palaeoclimate of the Late Cenozoic.* Elsevier, Amsterdam.

INDEX

aardvarks *211*
Aboriginal art *49*
   peoples 221
   traditions 48
absolute dating 27
Acacia 151
Acanthochelys 239
acanthodians 38,70-72, 93-94, 106
Acanthodii 95
Acanthothoraci 96
acanthothoracids 72, 76, 78
Acerastea wadeae 112
acid preparation of fossils 59, 87
Acre region (Brazil) *176, 177*
Acrobates pygmaeus 216, 217
Acrobatidae 226
Acrolepis tasmanicus 110, 111
Actinistia 97
actinolepid level *72,74*
actinopterygians 88, 93, 95, 108, *110*, 111,*144*,230
Actinopterygii 80, 97
Adamson, Margaret 50
Adanites bifida 134,136, 137
adductor mandibulae 81, 104
Aegothelidae *160*
aetosaurs 119, 228
Africa 82, 94, 118, 119, 206, 228, 230, 238, 239
African Plate 23
agamids 166, 219
Agathis jurrasica 4
"Age Of Mammals" 27
"Age Of Reptiles" 27
Agnatha 66, 67, 70, 93
Agrosaurus macgillivrayi 124
Airley, Qld 105
Alamosaurus 234
Alaska (USA) *132, 133*
albatrosses 206
Alcoota Station, NT *164, 168-170, 172, 173*, 208
Alemania (Argentina) 238
algae 33
algal blooms 32
Allosaurus 45,132, 146, 214
Alps (European) 42
Alwertatherium 170
American Museum Of Natural History (USA) 57
American Plate 23
Ameridelphia 32, *154, 211, 214*
amino acids 32
Amniota 147
amniote egg 40, 105, 116
amphiaspids 94
Amphibolurus pictus 4
Anagalida *209,211*
Anaspida *66, 67*
Andamooka, SA 138,141
Andes Mountains (South America) 26, 58
Angaridium Flora 229
angiosperms 41, 42, 126, 143, *146*
anhingas *210*
ankle joint (CLAPJ) 214
Ankylosauridae 141, *142*
Anomodontia *119*
anoxic basins 32

anoxic conditions 59
Anseriformes *210*
Antarctilamna 93
Antarctic Beech *16*
Antarctic Circle *126, 131*
Antarctic ice sheet 151
Antarctic Plate 23
Antarctica 24, 67, 68, *81, 93*, 112, *116-117*, 118, 119, 123, 202, 228, 239, *240-241*
anteaters *211*
Antechinus stuartii 216, 217
antelopes *211*
anthracosauroids *147*
anthracosaurs 102
Anthropornis 206, 240, 241
A. nordenskjoeldi 206
antiarchs 74, 78, 87, 92
antorbital fenestra 116
Anura 40
ANZAAS 57
Anzaldo Formation (Bolivia) 65
apatite 38
Apatosaurus 124
Aphnelepis australis 126, 127
Apoda 40
Apollo object 33
Aptenodytes forsteri 206
"Araucanian" *244*
Aranda people 65
Arandaspis 36, 65, 66, 67
A. prionotolepis 62, 63, 64, 65
Araripe Plateau (Brazil) 140
Araucariaceae *134,136, 137*, 151
arboreal forms 163
Arcadia Fauna 118
Arcadia Formation, Qld 17, 40, 108, 111, 112, 115, *116-118*
archaeocete 202
Archaeomene 126
A. tenuis 125, 127
Archaeopteris howittii 104
Archaeopteryx 210
Archean Eon 27, 30, *31*, 32
Archer, Michael 56, 57
Archonta *211*
archosaurian reptiles *116, 118*
Archosauromorpha *147*
Arctic Circle *132, 133*
Ardeidae *210*
Argentina 118, 140, 152, 206, 227, 228, 230-232, *234-251*
argyrolagids *211*
arid adaptations 220
   arid regions *214, 215*
aridification 151, 220
aridity 174
armadillo 152, *244*
Armorica 68
Arroyo Chasico Formation (Argentina) *240, 241*
Artamidae 218
arterial system 88
arthrodires *71, 72*, 76, 78, 87, 88-91,92, 94
Arthur River, WA *106*
Artinskian 57
Artiodactyla *211*, 239
Asia 131, 163, 166, 234, 239
aspidorhynchiform *230, 231*
asteroid impact 41

Asterolepidoidei 74,96
astrapotheres *211*
Atlascopcosaurus loadsi 134-135
Atrichornithidae 218
auditory system, evolution of 101, 118
auks *210*
Aurora *123*
Australasian biogeographic realm 45
Australasian Institute Of Mining And Metallurgy 53
Australian Alps 174
Australian And New Zealand Association For The Advancement Of Science 57
Australian Army 57
Australian Institute Of Mining Engineers 53
Australian Museum 50, 51, 55, 57
Australian National University 57
Australian Research Council 57
Australian-Indian Plate 23
Australidelphia *154, 211*, 214
Australobatrachus ilius 166
Australopapuan Fauna *45*
australopithecines 55
Austrobrachyops 114, 115
Austrophyllolepis 92
Austrosaurus 143
A.mckilloppi 141
Aves 126, 140,*147*, 156, *160*, 166, 169, *173, 174, 175, 180, 181, 182, 185, 194*, 206, 210, 215, 221, 224, 238, 239, *240*,
Avon River Group 80, 92
Awe (New Guinea) 208
Aztec Siltstone76-78, *81*, 88, 93

Bacchus Marsh, Vic 17, *182, 183*
Bacon, Francis 21
bacteria 31, 33
Badong Formation 118
Bailly, J.C. 48
Bairstow's Sand Pit 188
Balcombe Bay, Vic 208
baleen whales 202
Bali 45
Baltica 24
banded iron formations (BIFs) 30,31, 32
bandicoots 163, *176-177*, 211, 214, 218
   Eastern Barred 216-217
   Short-nosed *192*
Bangladesh 224
Banks, Sir Joseph 51
Baragwanathia 103
Baragwanathia Flora 103
Barakar Coal Measures (India) *17, 18*
Barameda 97
B. decipiens 106
barbets *210*
Barlow, F.O. 112

Barosaurus 230
Barrancas Del Rio Santa Cruz (Argentina) *250, 251*
barrier reef (Devonian) of Gogo, WA 70, 86, 87, 88-91
Bartholomai, Alan 57
Barwick, Richard 88
basalt *174, 175*
Bass Strait 179, *214,215*
   islands of 196
Bassian biogeographic region 221
Batesford Quarry, Vic 208
bats 43, *163*, 211, 214
   fruit bats 218
   Horseshoe bats 162
Baudin, Nicholas 45, 48, *49*, 196
Beacon Hill Quarry, NSW *106, 108*
Beacon Supergroup *17,18*
Beaufort Series 18
Beaumaris, Vic *170, 171, 176, 177*, 202-205, 206, 208
Beelarongia 97
Belichthys magnidorsalis 106
Belonostomus 141
Bering Land Bridge 150
Bermagui, NSW *70, 71*
Berner, Robert 30
Berry Springs, NT *222, 223*
Bettles (Alaska) *132, 133*
Bettongia moysei 165
BIF's 30,31, 32
Big Bang Hypothesis 30
Big Sink 208
bilbies 218
Billa Kalina Basin, SA 151
Billabong Station Crossing, WA *179*
Billiluna, WA *118*
Bindook Caldera 92
biochemical affinities of vertebrates 37
biogeography 218
bird interelationships *210*
Birdsville Track *222, 223*
Bitter Springs, NT *31*, 32
bivalves *204*, 205
Black Rock Member *170, 171, 176, 177,* 202-206
Black Rock Sandstone *202, 203, 206*
Blackwater, Qld *106-107*
Blackwater Shale 106
Blanche Point Formation 206
Blandford, W.T. *16*, 21
Blandowski, William 51
Blina Shale 108, *111*, 112, 116, 120
Blinasaurus 111
B. townrowi, 112, 113, 120
blue-green algae 31
bluebush 218
Bluff, Qld 111
Bluff Downs, Qld *174*, 176, 208
boids 166, 218
Bokkeveld Series (South Africa) 18
Bolivia 36, 65, *152, 153*
bone, acellular 37, 38

dermal 37, *251*
   endochondral 37
Bone Gulch, NSW 208
Bone Gulch Fauna 176
bony fishes 70, 80, 92, 97, *175*, 176
boobies *210*
Boomerang Range (Antarctica) 76, 88
borhyaenids *211, 242, 243*
Bothriceps australis 105
Bothriolepididae 96
bothriolepidoids 74
Bothriolepis karawaka 76, 77
Bothriolepis 74, 78, 79, 87, 92, 94, 104
B . karawaka 76, 77
bottom dwellers 88
Botucatu Sandstone 18
Boulia, Qld 140
Bow, NSW *176*, 208
Bowler, Jim 220
Boyd Volcanic Complex 104
brachiopods 36, *64*
Brachiosaurus 124, 230
Brachipposideros 163
brachyopid labyrinthodont 105
Brachyopidae 111, 115
Brachyopoidea 111
brachythoracids 72, 73
Braidwood, NSW 78, 79, 118
branchial skeletons 115
branchiostegals 84, 85
Branchiostoma 36, 37, 62, 63
Brazil 104, 116,*176, 177*, 229-231
Breccia Cave, NSW 49
Brindabellaspis stensioi 76
Brisbane River, Qld 166
Brisbane, Qld *160, 161*
British Museum (Natural History) 51, 57, 87
Brittany (France) 94
Broad-toothed Rat 182
Broken River, Qld 71, 86, 118
"Brontosaurus" 124
Brookvale, NSW *106, 108*, 112
Broom, Robert 55
Brown, H.Y.L. 54, 55, 56, *199*
Brown, Robert 49, 51
Bruntonichthys 88
Brushtail possums 156, 174
Bryo Gully, Vic *64*
bryozoans 126, *204, 205*
Buchan, Vic 72, 118
Buchan Basin, 92
Buchan caves, *186*
Buchanosteus 72, 74, 94
Buckeye Tillite (Antarctica) *17, 18*
Buckland, William 49
Buenos Aires Province (Argentina) *240, 241*, 245-*251*
Bufo marinus 224
Bugaldi, NSW *160*
Bulgoo Station, NSW 118
bull ant *190*
Bullerichthys 88
Bullock Creek, NT 57, 59, 162, 163, *164*, 166, 168-170
Bunga Beach, NSW *118*
Bunga Beds 70, *71*, 86, 93

297

Bunga Creek, Vic 208
bunyip 48
Bureau Of Mineral Resources 57
Burgess Shale (Canada) 37
Burhinidae 166, 210
Burnett River, Qld 166
Burramyidae 226
Burramys parvus, 216, 217
Burramys 174
B. parvus 216, 217
Burrinjuck Dam, NSW 78
Burrinjucosteus 74

Cacatua galerita 4
Cacatuinae 214
cadimurka 54
caecilians 40
Cainozoic Era, 27
    record on Gondwana 239-251
calcified cartilage 38
Callistemon 151
Cambridge University (UK) 54
Camfield Beds 164, 169
Camfield Station, NT 162
Campbell, Kenneth S. 56, 88
Campbelltown, Vic 66
camuropiscids 88
Canadian Lead, NSW 208
Canberra, ACT 171, 51
Canberra Magmatic Province 92
Cane Toad 224
Canis familiaris 216
C. familiaris dingoensis 214
Canning Basin, WA 95
Canowindra, NSW 92, 97, 118
Canowindra Fauna 92
Canowindra 92, 97
cantharid beetles 126
Cape Barren Geese 214, 215
Cape Barren Island 214, 215
Cape Range, WA 204, 205
Cape York Peninsula, Qld 48, 49
capitosaurs 111, 108, 115
Capricorn Group, Qld 222
Caprimulgiformes 210
Captains Flat Trough 92
Captorhinomorphia 147
Carbon 14 analysis 27, 182
carbon dioxide 30, 41
carbon isotope ratios 31
carbonate sediments 41
carbonic acid 30
Carcharodon angustidens 204, 205
C. megalodon 204, 205
carettochelyids 218
Cariamae 210
Caribbean impact site 41
Caribbean Plate 23
Carinatae 210
Carnarvon, WA 106
Carnarvon Basin 93
Carnivora 211
carnivorous dinosaurs 132, 144
Carnotaurus sastrei 234, 235
Carter, Frederick 102
Carter, Samuel 102
cartilage 38
cassowaries 45, 166, 210, 214, 218
Casuariidae 218
Casuarina 151, 218

cat, domestic 43
Catamarca Province (Argentina) 244
catastrophic kill 126
Catombal, NSW 104
cats 214
Caviomorpha 175
Ceará (Brazil) 140
Ceduna, SA 143
cell membrane 34
Cemetery Beach (Norfolk Island) 240
Centropus collosus 186, 187
Cephalaspidomorphi 65
cephalaspids 94
cephalopods 88-91
Ceradactylus atrox 140
ceratodontida 82, 166, 176
Ceratodus 111
C. wollastoni 138, 139
ceratopsians 146, 234
Ceratosaurus 230
Cerberean Caldera, Vic 104
Cercatetus lepidus 4
Cereopsis novaehollandiae 214, 215
Cerro Condor (Argentina) 231
Cerro Del Humo (Argentina) 244
cervids 246
Cetacea 202, 204, 211
Chanaresuchus 228
Chapada Do Araripe (Brazil) 230, 231
Chapadmalal (Argentina) 244, 245, 248
Charadriiformes 166, 210
Charters Towers, Qld 174
Chasmatosaurus 120
Chasmatosaurus yuani 116, 117
Chatisgarh (India) 16
cheetah 45
chelids 166, 176, 218, 239
Chelodina insculpta 179
chenopods 218
chigutisaurids 111, 115
Chile 251
chimaeras 72, 85
China 24, 74, 82, 93, 116-119, 124, 152, 209, 230
Chinchilla, Qld 174, 175, 208
Chinchilla Fauna 176
Chiniquodontidae 119
Chiniquodontoidea 119
Chirodipterus 88, 97
C. australis 4, 38, 82, 85
Chiroptera 211, 226
Chondrichthyes 38, 85, 86,95
chondrosteans 80, 97
Chordata 36
    invertebrate chordates 37
Chorobates 244, 245
Christies Beach, SA 206, 208
Chromatobatrachus 112
C. halei 114, 115
chromosome 34
Chrysophrys 168, 169
Chubut Province (Argentina) 230, 231, 234, 235, 242-244
Ciconiiformes 210
Cimolesta 209
Cimoliosaurus leucocephalus 138, 139
Circum-Antarctic Current 43, 150, 202
Cladoselache 96

CLAPJ ankle joint 214
Clarke, Andrew 51
Clarke, William B. 50, 54
claspers 76
clavicle 101
clay galls 136, 137
Cleithrolepis 106-111
cleithrum 101
Clift, William 49
Clifton Bank, Vic 202, 203
climatic indicators 24
climatiids 71
Cloghnan Shale 104
Cluan Formation 111
Club Mosses 103
Clupeomorpha, 141
co-evolution 42
coal seams 17
Coalsack Bluff (Antarctica) 114, 115
Cobar Basin 92
Cobb & Co. 102
Coccolepis 126
C. australis 126, 127
C. woodwardi 128
coccosteomorph level 74, 87
Cochabamba (Bolivia) 65
cockatoos, Sulfur-crested 4
Cocos Plate 23
coelacanths 97
Coimadai, Vic 208
collagen fibres 38
Colombia 26
Colombian Basin 41
colubrids 218
Columbia University (USA) 21
Columbiformes 210
Combienbar River, Vic 118
comet impact 41
comets 30, 33
concretions 87
Conepatus 246
conies 211
conifers 119
Conodon Astolto Formation (Argentina) 230, 231
conodonts 36
continental drift 22
    continental seas 41
    continental shelves 26
continentality 44
convection 23
convection cell 23
convergent evolution 248, 249
Coober Pedy, SA 138
Cooma, NSW 82
Cooper Creek, SA 52, 59, 174, 175
Cooyoo 141
C. australis 144, 145
Cora Lynn Cave, SA 184
Coraciiformes 210
coral islands 222
Cordoba Province (Argentina) 239
Corriguen-Kaik (Peru) 236
cosmine 80
Coturnix pectoralis 218
Cowombat Rift 92
Cowra Trough 92

cows 214
coyotes 45
cranes 210
Cravens Peak Beds 76, 85
Crinia, 166
Crisp And Gunns Quarry, Tas 114, 115
crocodilians 131, 132, 166, 174-176, 197, 214, 234
crocodile, Saltwater 112
Crocodilomorpha 147
Crocodylus, 175
C. (?Botosaurus) selaslophensis, 124
C. porosus 112, 174, 218, 219
crossopterygians 40, 72, 82, 86, 88, 92, 93, 95, 100, 102, 104, 106
Crossopterygii 80
Crust, Earth's 22, 23, 30
crustaceans 204, 205
Cruziana 65
Cryptodira 147
Cryptozoic 27, 30, 31
ctenacanth shark 93
Ctenacanthoidea 96
Ctenurella 96
Cuba 41
Culmacanthus 92, 184, 185
Curie Point 21
Currabubula Tillite 18
Curramulka, SA 184, 185, 208
cuscuses 44
cuticle 220
cutwaters 71
Cuvier, Baron Georges 49, 50
cycads 119, 132, 133
cyclic lake deposits 33
cyclicity 17, 40, 220
cyclothems 40, 41
Cynodontia 119, 211
Cynognathidae 119
Cynognathus 20
C. minor 232, 233
Cynognathus Zone 232, 233
cytroplasm 34

Dacrydium 151
Daly River Basin, NT 67
Damuda Series (India) 18
Dana, J.D. 54
Dani Valley (New Guinea) 207
Dare Plain, NT 118
Darling Downs, Qld 50, 169, 198, 199
Darling River, NSW 179
Darwin, NT 216, 222, 223
Darwin, Charles 45, 50
Dasyuridae 175, 176, 186, 226
Dasyuromorphia 211, 214, 226
dating of fossils 27
"daughter" products 21
de Vis, Charles 55
Deep Sea Drilling Project 42
deer 214, 239
Dempsey's Lake, SA 194
Dendrolagus, 174
dentinal tubercles 37
dentine 102
D'Entrecasteau Channel, Tas 111
depressor mandibulae musculature 104
derived Australian endemics 218
dermal bones 37,
Derwentia warreni 120

Devils Lair, WA 186
Devonian barrier reef of Gogo, WA 70, 86, 87, 88-91
Diabolichthyes, 82
diademodontoids 119
diamonds 26
Diapsida 147
diatomite 160
Dickinsonia costata 34
Dicroidium, 40, 119
D. callipteroidium 108
Dicroidium Flora 40, 119
dicynodonts 17, 111, 116, 118, 119
Didelphimorphia 152, 211
Didymograptus 66
Dingo 43, 214, 216
Dinmore, Qld 124
Dinocephalia 119
Dinosaur Cove, Vic 130, 132, 133, 144, 145
dinosaurs 27, 41, 99, 123, 131-138, 141-146, 147, 230
    carnivorous 132
    duck-billed 141, 238
    juvenile 132, 134
Diodon formosus 186, 187
Diodon 204, 205
Diomedea thyridata 206
Diplorhina 66
Diplorhynchus, 97
dipnoans 72, 88, 89, 90, 91, 93, 100
Dipnoi 80, 97
Dipnorhychus 72, 82, 83
Diprotodon 50, 54, 55, 59, 175, 176-177, 179-184, 186, 194, 197, 221-222
D. optatum 184, 188, 189, 195
Diprotodontia 163, 175, 211, 214, 218
Diprotodontidae 164, 170-172, 174, 182, 226
diprotodontoids 156, 162, 164, 207
dipterids 97
Dipterus 83
disharmonious association 175, 186
diversity, species 213
Diviniidae 119
Divisadero Largo Formation 13
de Vis, Charles 53
DNA, 32, 34
Docodonta 211
Dog Rocks, Vic 208
Dog Rocks Local Fauna, 176
dogs 211, 214
dolichothoracids 72
Dollar Bird 218
dolomites 30
dolphins 41, 58, 149, 156, 208, 211
domestic cat 43
Dorcopsis 172, 174
Dorcopsoides fossilis 172
Dover Cliffs (UK) 41
Dragon, Komodo 197, 224
    Painted 4
Drakensberg dolerites 18
dromaeosaurid 141
dromaines 218
Dromaius 180, 181
    D. ater 48, 196
        D. baudinianus 224
        D. gidju 160, 161, 172, 173
        D. novaehollandiae 166,

175, 196, 214, 215
D. ocypus 175
Dromiciops 211
D. australis 154, 214
Dromornis stirtoni 172, 173
Dromornithidae 49, 66, 148, 172, 184, 185, 218
Dromornithiformes 210
Dryosaurus 230
du Toit, Alexander L. 13, 21, 26
duck-billed dinosaurs 141, 238
ducks 166, 208, 210
Duncanovelia extensa 128
Dulcie Range, NT 118
Dundas Shire Offices, Vic 98, 99
dunes 220, 222, 223
Durham Downs Station, Qld 124
durophagus feeding style 82, 85, 88
Dwarf Emu 196
dwarfing 196
Dwyka Shale 18
Dwyka shales 17
Dwyka Tillite 17
dynamo 21

eagles 210
ear region 143, 204
eardrum 40
Earth history, major events 31
earthquakes 21, 22, 23
Earthwatch 57
East African Rift Valley 22
East Gondwana Province 94
Eastern Barred Bandicoot 216, 217
Eastern Grey Kangaroo 214
Eastmanosteus 68, 69, 72, 87
E. calliaspis 88-91
Ebenaqua 106
E. ritchei 106, 107
Ecca Coal Measures 17
Ecca Series 18
echidnas 45, 211, 216, 218
echinoderms 34, 37
echinoids 204, 205
echolocation 202
ecological stress 41
ecomorphic types 115
ecospace 41
ectoderm 71
ectodermal placodes 37
ectopterygoid 100
Eden, NSW 118
Edentata 209, 211
Ediacara Fauna 27, 34
Edmontosaurus 143
egg 40, 105, 116, 166, 194
Ehrlich, Anne 224
Ehrlich, Paul 224
Eildon Group 92
Ekladelta 197
Ektopodon 209
E. stirtoni 160
Ektopodontidae 160, 226
El Tranquilo Formation 230, 323
Elaphrosaurus 230
elapids 166, 218, 219
elasmobranchs 72, 85, 106
Elburz Mountains (Iran) 94
elephant shrews 211
elephants 50, 211
Elizabeth Bay House, NSW 51

Ellery Creek, NT 32
Elonichthyes 105
Emballonuridae 226
embryological affinities of vertebrates 37
emus 45, 175, 210, 214, 215, 218
Emu, Dwarf 49, 196
mummified 180, 181
Emydura macquarii 155
enamel 102
Enantiornithes 140, 210, 238
endemic 214
Enderby Land (Antarctica) 30
endoplasmic reticulum 34
Enoggera Reservoir, Qld 156, 157
environmental destruction 224
environments of deposition 58
eosuchians 116, 117, 175
Eotitanosuchia 119
Epitheria 211
Eremaean 221
Ericmas Quarry, SA 156, 157, 208
Ernanodon 102
Ernanodon antellios 152
Erskine Range, WA 111, 116
Estancia Canadon Largo (Argentina) 230, 232
Estrada Nova Beds 18
Estuarine Crocodile 218, 219
Etadunna, SA 208
Etadunna Formation 111, 158, 160
Etheridge, Robert Jr. 51, 53, 55
Ettrick Formation 186, 187
euarthrodires 72
Eucalyptus 43, 45, 149, 151, 218
eucaryotic cell 34
Eudyptula minor 206
Eugaleaspida 67
Euramerica 94
Euramerican Province 94
Eurasian Plate 23
Europe 93, 94, 112, 163, 166, 197, 228, 239
European fauna (Devonian) 68
Euryapteryx gravis 184, 185
Eurystomus 218
Euryzygoma 174
Eusthenopteron 100-102
Eutheria 211
evaporite 40
Evergreen Formation 111, 124
Ewing, Maurice 21
exoskeleton 38
exploration map of Australia 49
external gills 40
external naris 100
extinction 115, 105, 123, 146, 182, 221
extinction of dinosaurs 123, 146
Eyrean biogeographic region 221

Fairy Penguin 206
Falconer, Hugh 50
Falconiformes 210
falcons 210
Fallacosteus turnerae 74, 75
Famennian 104
faulting, tensional 22
fauna, modern 207

faunal similarities 94
feathers 128
Feather-tailed Glider 216, 217
Felis atrox 216
ferns, tree 218
Ferrar Dolerite 18
fibulare 101
Field Museum Of Natural History (Chicago) 57
Fig Tree Group 31
filter-feeding 37, 62, 65
firestorm 41
fish-to-tetrapod transition 100
Fisherman's Cliff 208
Fishermans Cliff Fauna 176
fission track dating 27
fission tracks 27
fissure-fills, 186
Fitzroy Basin, WA 57
flagellum 34
flamingoes 58, 156, 166, 175, 180, 208
Flannery, T.F. 56, 57
Flinders Island 201-204
Flinders Ranges 27, 33, 34
Flinders University 57
Flinders, Matthew 48
Flinders expedition 49
flood debris 136, 137
floodplain deposits 130
floodplain faunas 130
floods 21
Floraville, Qld 208
Flores 224
flower (oldest) 128
flowering plants 41
fluoride 38
Fly, HMS 124
flying fish 144
footprints 124, 143
foraminifera 41
Forbes, NSW 104
foreign experts 49, 50
Forest Red Gum 212, 213
formation of the Earth 31
Forsyth's Bank, Vic 208
Fortescue Group 31
fossil assemblage 50
"living fossil" 48, 49
opalized 138, 141, 152
Fossil Bluff, Tas 148, 149, 202, 203
Fossil Bluff Sandstone 148,149, 202, 203
fossil record, bias in the 58, 59
fowl 210
Frasnian 87, 92, 93, 104
Freestone Creek, Vic 118
Fremouw Formation 114-117
freshwater limestones 162
frigate birds 210
frogmouths, 210
frogs 40, 45, 218, 230, 238
tree 166
fruit bats 218
Fulbright scholarship program 57
Fulgurotherium australe 134, 141

Galapagos Penguin 206
galeaspids 76, 94
Galesauridae 119
Galliformes 210
Gambier Limestone 204, 206
ganoine 80
gas rings 30

Gascoyne River, 118
Gates, Dwayne 176, 177
Geelong, Vic 176
geese 210
Cape Barren 214, 215
Geilston Bay, Tas 155, 156, 208
gekkonids 218
geniohyoids 81
Genoa River, Vic 118
Genoa River Beds 104
Genoa River trackway 104
Genyornis 49, 55, 184, 185, 194, 221
G. newtoni 194, 199-201
geochemists 132
Géographe 48, 49
Geologic Time Scale 27
Geological map of Australia 20
Geological Survey (of India) 16
Geological Survey Of New South Wales 55
geological time 27
Georgina Basin 67, 76, 86
Germany 72, 94
Gervais, P. 50
Géographe 48, 49
ghost sharks 96
ghost-fish 72
Giant Coucal 186, 187
Giant Goanna 197
Giant Wombat 195
gibber 58, 222, 223
Gigantopteris Flora 229
Gilberton 118
Giles Creek 47, 49
Gill, E.D. 56
gill arch 38
gill bars 84, 85
gill basket 80
gill slits 36
Ginkgo 119
G. biloba 132, 133
Ginkgoites australis 136, 137, 146
glacial deposits 105
glacial dropstone 17, 67
glaciation 26, 32, 33, 40, 42, 105, 179 179, 220
glaciers 41, 179
Glenidal Formation 111
Glenisla Homestead, Vic 99, 102
Glenisla tetrapod trackway 103
gliders 156
Squirrel Glider 216, 217
Glires 211
Glossopteris, 13, 17, 20, 26, 40, 105, 119, 228, 229
G. browniana 49
Glossopteris Flora 16, 26
Glossotheirum listai 251
glyptodont 248, 249
Gnathorhiza 82
Gnathostomata 67, 70, 71
Gneuda, WA 118
Gneuda Formation 93
goanna 198, 199
Gogo, WA 57, 59, 118
Gogo Formation 4, 38, 39, 39, 74, 75, 82-87
Gogo reef faunas 87
Gogo Station 38, 39, 68, 70, 74, 75, 82-85, 86, 87
Gogonasus andrewsi 88

gold 51
Golgi body 34
Gomphonchus 71
Gond kingdoms 16
Gondwana, 22, 24, 27, 41, 213, 218
Gondwana concept 13, 16
Gondwanaland 13
Gondwanan biota 23
Gondwanan rock sequences 18, 21, 23
Gondwanan supercontinent 27
Gondwanatherium Fauna 152, 153
Gordo Formation 239
gorgonopsians 119
Gosford, NSW 110, 111
Gosfordia, 108 110, 111
Gosse's Lookout, NSW 118
Goulburn, Major 51
Grallinidae 214, 218
Grampians, Vic 99, 102, 118
Grampians Group 92
Grange Burn, Vic 208
graptolites 66, 67
Graveyard Creek Formation 71
Great Australian Bight 143, 150
Great Barrier Reef, 222
Great Buninyong Estate Mine, Vic 208
Great Dividing Range 58
Great Sandy Desert 179
Great Southern Rift Valley 132, 134
Green Waterhole, SA 186-188, 197
greenhouse conditions 34, 40, 41
Greenland, 30, 36, 64, 104
Gregory River, Qld 162
Gregory, John Walter 52, 53, 55, 56, 59, 179
Grenfell, NSW 88
Grey Gum 212, 213
Grey Kangaroo 176, 182, 183
Griphognathus 88, 97
G. whitei 82-85, 88-91
Groberia minopriori 13
Groenlandaspididae 93, 96
Groenlandaspis 78, 92, 94
ground sloths 239, 251
Gruiformes 210
Gulf Of Mexico 41
Gulgong, NSW 126
gulls 210
Gumardee, 160, 161
Gunneraceae 151

Hadean Eon 27, 31
Hadronomas puckridgi 172
hadrosaurs 141, 143, 146, 234, 238
hagfish 38, 65
Haiti 41
Hale River, NT 151
half-life 27
Hall, T.S. 52, 53, 54
Ham, Thomas 50
Hamersley Group 31, 32
Hamilton, Vic 98, 99, 174, 175, 202, 203, 208, 216
Hamilton River, Qld 140
Hand, S. 56
hares 224
Harpagodens ferox 86, 93

Harrytoombsia 87
Harvard University (USA) 57
Hatchery Creek, Vic 86
Hawaiian islands 224
Hawkesbury Sandstone *106-109, 109*, 112
hawks *210*
heat flow 22
heat transfer 23
Heathcote, Vic *64*
hedgehogs *211*
Heezen, Bruce 21
Helicoprion davisi 106
hematite 21
Henschke's Cave, SA *184*
Henschke's Fossil Cave, SA *61, 62, 178,186, 187, 190, 192, 193*
herons *210*
Hervey *104*
Hesperornithiformes *210*
heterocercal tail 71, 72
heterochrony 100
Heterostraci 65, 66, 67, 72, 94
hexanchids 85
Hexanchus agassizi 186, 187
Hill End Trough 92
Hills, E.S. 56
Himalayas 16, 42
Hippidion 246
Hipposideridae *163, 226*
Hobart 108, *112-115*
Hobson, Edmund Charles 50
Hobson, Margaret Adamson 50
Hochstetter 50
Holocephali 85,96
Holodipterus 88, 97
Holonematidae 96
Holoptychius 97
holosteans 80, *81*, 108
Hominidae 226
Homo sapiens 43, 211
Hope, J. 56
Hoplophorus ornatus 248, 249
Horlick Formation *18*
Hornsby Heights, NSW *106-111*
Horse latitudes 44
horses *211, 214, 239, 246, 247*
horse-shoe bats 162
horsetails 103, 119, *132, 133*
hot spot 23
House mice 214
Howittacanthus 92
Howqua River, Vic *80*
Howqualepis 92, 97
H. rostridens 80
Hughenden, Qld 138
humerus *101, 182, 186, 190, 206, 211, 240, 291*
Hunan Province (China) *118*
Hunsrückschiefer sediments 94
Hunter Siltstone 88
Hunter Valley, NSW 17
Hunterian Museum (Glasgow) 49, 87
Hurst, H. 54
Huxley, Thomas Henry 45, 50, 105
Hybodontoidei 96
hydrocarbons 41
    hydrocarbon accumulation 41
hypersaline environments 31
hypobranchial musculature 104
hypothesis testing 132

hypsilophodontids 130-132, *134*, 141, 143
Hyracoidea 211
hyraxes *211*

Ian's Prospect, SA 208
ibises *210*
ice ages 33
icebergs 33
icehouse conditions 41, Chapter 5
Iceland 21
ichthyodectids 141
Ichthyornis 210
ichthyosaurs 115, 116, 138, 143
Ichthyostega 100, 101, 104
Ichthyostegalia 102
ichthyostegaliids 104
Icthyopterygia *147*
Ictidosauria *119*
Idaho (USA) 67
iguanodontid 141
iguanid lizards 43
Iguanodon 141
iguanodontids 143
Ilaria lawsoni 158, 159
Ilbandornis 172, 173
    I. woodburnei 172-173
Illariidae 156, *174*, 226
Illawarra Coal Measures, Qld 17, 18
inarticulate brachiopod 64
index fossils 27, 76
India 94, 116, 118, 119, 126, 228, 239
Indian Geological Survey 13
indicator birds *210*
Indobrachyopidae 115
Indonesia 197
infraorbital 80
Insectivora *211*
insects 42, *128*
intercentra 102, 115
internal naris *100*
internasal *100*
invasion of the land 40
Inverloch, Vic 130, 132
Iran 26, 93, 94
Irati Formation 229
    Irati Shales 17, *18*
Irian Jaya 207
Irian biogeographic area 221
iridium *41*
    iridium-rich boundary clay, 41
iron ore 32
    iron-rich waters 32
Ischnodon 175
    I. australis 176-177
Isle De Pins (New Caledonia) 240
Isodon 192
Isua Group *31*
Isurus hastalis 204, 205
Itarare Series *17, 18*

Jabalpur (India) 16
jacamars *210*
Jack's Lookout, NSW *118*
Jameson, Robert 49
Janjuc, Vic 208
Japan Trench 22, 23
Japan 116
jawed vertebrates 70
jawless fish interrelationships *67*

jawless vertebrates 66
jaws 38
jellyfish 34
Jemalong Gap, NSW *118*
Jemmys Point, Vic 208
Jew Lizard *218, 219*
Jiucaiyuan Formation *116, 117*
Jovian planets 30
Jukes, William 51

Kadimakara, 111
    K. australiensis 116, 117
Kalamurina Waterhole, SA *179*
Kalisuchus rewanensis 116
Kangaroo Fracture Zone (Southern Ocean) 23
Kangaroo Island, SA 196, 220
Kangaroo Well, NT 208
kangaroos 45, 56, 162, *174-176, 211, 214*
    Eastern Grey *214*
    Red *4*, 186
    tree *174*
Kannemeyeria 118
Kansas (USA) 41
Kanunka, SA 208
    Kanunka assemblage *175-177*
Karoo basalts 26
Katipiri Sands *199*
Kazakhstan 24, 94
Kazanian *105*
Kendrickichthys 88
Kentrosaurus 230
Kerang, Vic 220
Keratobrachyops australis 111
kerogen 30
Kimberley region, WA 86
Kimberlite pipes 26
Kimsar (China) *116, 117*
King Island, Tas *48, 49*
    King Island Emu *48*, 224
kingfishers *210*
Kitching Ridge (Antarctica) *116,117*
Knocklofty Formation 108, *111-116, 120*
koalas 45, 156, *174, 176,184, 186, 211-212, 214*, 224
Kockatea Shale 108, 111
Kolopsis, 158, 159, 170
    K. torus, 170, 171
Komodo Dragon, *197, 199*, 224
Koobor 174
Koonwarra, Vic 126, 130, 140, *146*
    Koonwarra Fish Beds, *128*
Krefft, Johann Ludwig Gerard 50, 52, 53, 55
Kritosaurus australis 238
Kronosaurus queenslandicus 138
Krui River, NSW 208
Kudnu, 111
    K. mackinlayi 118
Kuduru kujani 141
Kyancutta Museum, SA 52
Kyarranus 166

La Amagra Formation (Argentina) *152, 153*
La Meseta Formation (Argentina) *240, 241*
La Rioja Province (Argentina) 227, 228, 232, 233
labyrinthine infolding of

dentine *102*
Labyrinthodontia 40, 45, 95, 99, 101, 102, 108, 111, 119, *124, 131, 132, 134*, 146, 214
Lachlan Fold Belt 86, 95, *118*
Lagomorpha *211*
Laguna La Colorada (Argentina) *230, 232*
Laguna Umayo Local Fauna 152
Lake Bungunnia (Central Australia) 176
Lake Callabonna, SA 48, 54, 55, 57, 176, *182, 183, 194, 195*
Lake Eyre, SA 54, *151, 158, 159*, 175, *176, 177*, 220
Lake Eyre Basin, SA 52, 143, 156, *179*
Lake Frome, SA 156, 158, *160, 161*, 166, 208
Lake George, NSW *151*
Lake Kanunka, SA 176
Lake Mungo, NSW *4*, 223, 223
Lake Namba, SA 58
Lake Ngapakaldi, SA *160, 161*, 208
Lake Palankarinna, SA 58, *151*, 156, 158, *159, 160*, 166, *167*, 176, 208
Lake Pinpa, SA 156, *157*
Lake Tarkarooloo, SA *160, 161*
Lake Tyers, Vic 208
Lake Victoria, NSW *178, 197*
Lamarck, Jean Baptiste *48, 49*
Lambie, NSW *104*
Lamont-Doherty Geological Laboratories (USA) 21
lamprey 38, 65, 67
Lancefield, Vic *184, 185*
Lancefield Swamp, Vic *182-184*
lancelet 36
landbridge (Bass Strait) 196
Lapillopsis nana 112
Larapintine Sea (Central Australia) 67
Lark Quarry, Qld 143, *144*
larval forms 37
Las Baja (Argentina) *240*
Las Cutiembres Formation 238
Lasaga, Antonio 30
Lashly Mountains (Antarctica) *81, 84*
lateral line canals 88
    lateral line organs 36
    lateral line system 37, 40, 101, 115
Latocamurus coulhardi 74, 75
Laurasia 26
Laurentia 24
Law Of Succession 50
Law Of Superposition 27
Leadbeater's Possum 156
Leaellynasaura 146
    L. amicagraphica, 131, 134, 136
leaf thickness 220
Leichhardt, Friedrich Wilhelm Ludwig 50
Leigh Creek, SA 111
    Leigh Creek Coal Measures *18*
Leipoa ocellata 214, 215
lemurs, flying *211*

Lepidosauromorpha *147*
lepidosaurs 116
Lepidosiren 82
    L. paradoxa 166
lepidosirenids 82
lepidotrichia *101*
Leptoceratops 234
leptodactylids *166*, 218
Leptolepis 126
    L. koonwarri 128, 129
    L. talbragarensis 4
Lick Hole Limestone 82
Lightning Ridge *124*, 138, *141*, 152, 153
Ligulalepis 72
    L. toombsi 80
Limnodynastes 166
Linnean Society Of New South Wales 54
lions 45, *211*
lithospheric plates 22, 23, 26
litopterns *211*, 246-247
Litoria 166
Little Penguin, *206*
"living" fossil 48, 49
lizards *166*
    Common Scaly-foot *218, 219*
llamas 239
lobe-finned fishes 80
Lomasuchus 236, 237
Lombok 45
London Clay 54
Long, J.A. 56
Loongana, Tas *186, 187*
Lord Howe Island *166-169, 180*, 206
Loriinae *214*
Los Alamitos Formation 152, 154
Los Chañares (Argentina) 228, 232, 233
Los Colorado Formation 227, 228, 232
Lotosaurus adentus 118
Loxton Sands *204, 205*
Lumbrera Formation *243*
Lunaspis 76, 94, 96
Lundelius, E. 56
lungfish 80, 83, 86, 88, 97, 106, 108, 111, 138, *144*, 156-157, 166, 175, 176, 218
    denticle shedding lungfish 97
    lungfish interrelationships 97
    rudimentary tooth-plated 97
Lycaenops 199
Lycopodium 134, 136, 137
lycopods 103
Lycopsid Flora 17
Lydekker, R. 50
    Lydekker's Line 45
lydekkerinids 112, 115
Lyell, Sir Charles 54
lyrebirds *214*, 218
Lystrosaurid Empire 119
Lystrosaurus 17, *18-20, 114-115, 119-120*
    Lystrosaurus Zone 112

Macdonnell Ranges, NT 31, 33, 58, 222, 223
Machaeracanthus 94
Mackunda Formation *141*

Macleay, Alexander 51, 52, 53
Macleay, William Julian 51
Macleay, William Sharp 51
Macraucheniopsis 248, 249
Macroaethes brookvalensis 108, 109
Macropodidae 156, 172,226
Macropodoidea 160, 226
Macropus 174
   M. giganteus 176, 182-184, 186, 214, 221
   M. rufus 4, 186
Macroscelidea 211
Madagascar 170, 182, 228
Madariscus 126
madtsoiids 166
Madura Cave, WA 186
mafic rocks 30
magnetic anomalies 26
magnetic field 21,22, 176
   Earth's 21
magnetic properties of sea floor 22
magnetite 21
Mahadevi Series 18
Mahoney, Jack A. 56, 155
Maiden Hair tree 132, 133
Mair's Cave, SA 192
Malawi 152
Malaysia 26
Malvinokaffric Realm 67
Mallee 4, 222, 223
Mallee Fowl 214, 215
mammal tracks, 234, 235
mammal-like reptiles 99, 105, 111, 118, 120, 211
   interrelationships 119
Mammalodon colliveri 202
manatees 211
Mandasuchus 230
mandible 81, 182
Manna Gum 212, 213
Mansfield Basin, Vic 80, 118
   Mansfield Group 80, 106
Mansfieldiscus sweeti 80, 106
Mantle, Earth's 22,23, 30
   hot spot 23
Manu antiquus 206
Marcus Locality, Vic 179
marine transgression, 115
Marree, SA 222, 223
Mars 30
Marsdenichthys 92
marsupial lions 156, 182, 188-189, 197
marsupial moles 163, 211, 214, 218
Marsupialia 43, 45, 52, 154, 211, 218, 226, 234, 239
marsupicarnivores 218
Mary River, Qld 166
mass extinctions 34, 40
Massetognathus 232, 233
Mastacomys 182
mastodonsaurids 115
mastodont 50
Mathinna Beds 92
Matthew, D.H. 21
Mawsonites spriggi 34
maxilla 81, 100, 135, 240
McCoy, Frederick 50, 52-54
Mcmurdodus 85
Medicott, H.B., 16, 21
Mediterranean Sea 42, 202
Megachiroptera 226
Megadermatidae 226
megafauna 182

Megalania, 54, 216
   M. prisca, 197 198-201
megapodes 214, 215
Megatherm 221
Megatherium 221
Meiolania, 168, 169
   M. oweni 168, 189
   M. platyceps, 166-169
meiolanids 166, 168, 169, 218
Melaleuca 151
Melbourne *Age* 53
Melbourne Trough 92
Melbourne University, Vic 57
Mendoza Province (Argentina) 232, 233
Meniscolophus 175, 176
Menuridae 214, 218
Meridungulata 211
mesoderm 71
mesonychids 202
Mesoplodon 202
Mesosauria 17, 116, 147, 229
Mesosaurus 18, 20
mesosuchians 236, 237
Mesosuchia 236, 237
Mesotherm 221
Mesozoic Era 27
Mesozoic localities, Australia 147
Mesozoic record on Gondwana 230-238
Metaxygnathus denticulus 104
meteorites 41
   meteorite bombardment 31
meteors 30, 31
metoposaurs 115
Mexico 140
Miamia, Vic 88
mice, House 214
   Spinifex Hopping 216-217
microbes 31
Microbiotheria 211, 214
Microbiotheriidae 154
microhylids 218
Micropholis stowi 112
Microtherm 221
microvertebrates 93
Mid-Atlantic Ridge 21, 22
mid-ocean ridge 21, 22,23
Middle East 94
Midway Point, Tas 114, 115
Mihirung paringmal 48
Mihirungs 166, 182, 210, 218
Miles, R. 56
Miller, Stanley 32
millipede 124
Minmi 142
   M. paravertebra 141, 142
Miralinidae 156, 226
Mirrabooka Formation 71
Mitchell, Thomas Livingstone 49, 50, 57
Mithaka Waterhole, Qld 118
mitochondrion 34
moas 184, 185, 210
Mogorafugwa (New Guinea) 208
moles 211
Molnar, Ralph 56, 238
Molossidae 226
Monash University 57
Mongolia 124, 140, 146, 234
monkeys 211
Monitor, Lace 218, 219
Monorhina 66
Monotremata 138, 152, 160, 211, 218, 226
Monte Hermoso (Argentina) 248

Moon 30
Mootwingee, NSW 47, 49, 118
Morenelaphus 246
morganucodonts 211
Morocco 72, 115
Morrison Formation 230
Morwell, Vic 208
mosasaurs 143
Mount Flora Beds 18
Mount *Glossopteris* Formation 18
Mount Narrayer, NT 30
Mountain Pygmy Possum 174, 216, 217
Mouse, Spinifex Hopping 216, 217
Mowbray Swamp, Tas 180,197
Moythomasia 88
Mt Bepcha, Vic 102
Mt Charlotte, NT 62, 63, 118
Mt Crean (Antarctica) 81, 88
Mt Deerina, NSW 118
Mt Everest (Asia) 42
Mt Fuji (Japan) 22
Mt Gambier, SA 204, 206
Mt Grenfell Station, NSW 118
Mt Hotham, Vic 216-217
Mt Howitt, Vic 80, 118
   Mt Howitt Fauna 92
   Mt Howitt Province 104
Mt Jack, NSW 118
   Mt Jack Station, NSW 94
Mt Narrayer, WA 31
Mt Ritchie (Antarctica) 76, 77, 88
Mt Tambo, Vic 118
Mt Watt, NT 64, 65
Mt Wellington sill dolerites, 18
Muddy Creek, Vic 202, 203
Mudgee, NSW 26
mudlarks 214, 218
Mulga Downs Group 94, 104
multicellular animals 31, 33, 34
Multituberculata 211
mummified emu 180, 181
Mungo lacustral phase 220
Murchison River, WA 179
Murchison, Roderick 54
Murgon, Qld 106
Muridae 175, 214, 226
Murndel, Vic 94
Muramura williamsi 158, 159
Murray River 176, 179
Murrindalaspis 76, 96
   M. wallacei 78
Murrumbidgee Group 71-73, 80
Mus musculus 224
muscular hypomere 37
musculature, *depressor mandibulae* 104
   hypobranchial 104
   jaw 81
   locomotor 38
Museum Of Victoria 50, 51, 57
Mussaurus 230, 232
Muttaburrasaurus 143, 146
   M. langdoni 141
Mylodon 251
Myrmecobiidae 226
mysticetes 202
Myxinoidea 67

Nabberu Basin, WA 32
Nagpur (India) 16
Nama System 18

Namba Formation 160, 161
Nambaroo novus 160, 161
   N. tarrinyeri 160, 161
Namilamadeta snideri 158, 159
Nantius eos 140
Naracoorte, SA 61, 62, 186-193
Narrabeen Group 108
Narrien Range, Qld 106
Narrows, Qld 155
nasolacrimal duct 101
National Geographic Society (USA) 57, 87
National Museum Of Victoria 50, 54, 57
native cats 175, 211, 214
native mice 211
nautiloids 36, 88-91
Nazca Plate (Pacific Basin) 23
nebula 30
needles 119
Nemegetosaurus, 234
Neoaetosauroides engaens 227, 228
neoceratodonts, 166, 176
Neoceratodus 83, 97, 175
   N. forsteri 138, 139, 156, 157, 166
   N. gregoryi 156, 157
Neochea (Argentina) 251
Neognathae 210
Neohelos tirariensis 164
Neophoca cineaea 202
Neopterygia 80, 230, 231, 97
neopterygian 126
Neornithes 210
Neoselachii, 96, 106
nerve chord 36, 37, 38
Nesodon 250, 251
Neuquén Province (Argentina) 234, 236, 237
neural arches 100, 112
neural crest 37, 38
Nevada (USA) 67
New Caledonia 240
New Guinea 57, 207
   New Guinea Forest Wallaby 174
New Hebrides Trench 22
New Zealand 182, 184, 185, 202, 206, 228, 240
Newcastle, NSW 118
   Newcastle Coal Measures 17, 18, 105, 106
Ngapakaldia 209
   N. tedfordi 158, 164
Nikolivia 93
Niolamia 168-169
Nix, Henry 221
"Noah's Arks" 150
Nodule Bed 202
Nolan, Jackie 54
Nooraleeba, Qld 208
Norfolk Island 206, 240
North America 93, 112, 163, 166, 182, 228, 234
North American Fauna (Devonian) 68
Northern Territory Museum 52
Nothofagus 149, 151, 218
Notobatrachus degiustai 230
Notochelone 144, 145
notochord 36, 37, 65, 100
Notomys alexis 216, 217
Notopetalichthys 76
Notorhizodon mackelveyi 88
Notoryctemorphia 211, 214
Notoryctidae 163, 218, 226

Notostylops 242, 243
Notosuchus terrestris 234
Nototherium 50
notoungulates 211, 239, 244
Novogonatodus kasantsevae 80
nuclear winter 41
nucleotides 32
nucleus 34
Nullarbor Plain 57, 186

Obdurodon 156, 157
obsidian 27
ocean basins 22
oceanic circulation 42
   oceanic crust 24
   oceanic ridges 23
   oceanic trenches 30
Ocyplanus 175
odontocetes 202
oil 30
oil birds 210
oil shale 105
Old Beach, Tas 112, 113
"Old" Endemic elements 175, 218
Old Settlement Beach, Lord Howe Island 180
Onohippidium compressidens 246, 247
Onverwacht Formation 32
   Onverwacht Group 31
Onychodus 86
opal fields 138, 152, 153
opossums 239
optic lobes 131
Orbost, Vic 208
orchids 42
Oriental Biogeographic Realm 45
Oriental Fauna 45
origin of life 30
Ornithischia 143,147
Ornithocheirus 140
Ornithocheiridae 226
Ornithorhynchus 156, 157, 214
   O. anatinus 216
ornithosuchid 232, 233
Ornithurae 210
Osteichthyes 38, 71, 80, 95, 97
   osteichthyian interrelationships 97
osteoblasts 38
Osteolepiformes 88,97, 100
Osteostraci 66, 67
ostracods 126
ostriches 210
Otariidae 202
otic bulla 204
otic notch 111, 112
otic-occipital portion of braincase 100
Otway Group 27, 126, 132, 133, 144-145
Otway Ranges, Vic 130, 132, 133, 144, 145
Overland Corner, SA 168, 169
overpopulation 224
Owen, Sir Richard 48, 50, 52, 53, 55, 155, 188, 194
owlet-nightjars 160, 210
owls 210
oxygen isotopes 131
   oxygen ratio 132

pachyosteomorph level 72,74
pachyrhizodonts 141
Pachyrhizodus 141, 144, 145

Paddy's Springs, WA 4
Paddy's Valley, WA 38, 70, 84, 85
pademelons 174
Padeotherium 248
paedomorphosis 115
Paenungulata 211
Painted Dragon 4
Pakistan 202
Palaeeudyptes antarcticus 206
Palaelodidae 58, 156, 166, 208, 210
palaeogeographic maps of Australia 28, 29
palaeogeographic reconstruction (Devonian, southeastern Australia) 104
palaeogeographic reconstructions of the Earth's continents 24, 25
Palaeognathae 210
palaeomagnetic measurements 21
palaeomagnetics 70
Palaeonisciformes 80, 86,97, 106, 126
palaeoniscoids 70, 72, 80, 81, 88, 92, 105, 106, 108
palaeopathology 143
Palaeopotorous priscus, 160, 161
palaeotemperature determinations 41
Palaeozoic Era 27
Palaeozoic record on Gondwana 229
Palaeozoic vertebrate-bearing localities 95
palagornithids 210
Palankarinna assemblage 175
palatoquadrate 80
Pallimnarchus 174, 175
Palorchestes 158, 159, 174
P. painei 170
Palorchestidae 226
Pampa Grande (Argentina) 243
Panamanian Isthmus 150
Panchet Series 18
Pangaea 13, 24, 26, 40, 41, 44, 213, 218
pangolins 211
Pantotheria 152
Papua New Guina 26
Paracyclotosaurus 108, 112
P. davidi 112
Paraná basalts (South America) 26
parasphenoid 100
Paratheria 152,211
parietal 100
Paris Basin (France) 54
Parotosuchus 108, 112
P. brookvalensis 112
parrots 43, 45, 210, 214, 218
Parvancorina minchami 34, 35
Pascualgnathus polanskii 232, 233
Paso Del Sapo (Argentina) 234
Passeriformes 210, 218
Patagonia (South America) 152, 153, 202, 230, 238, 248
Paucituberculata 211
Pecopteris Flora 229
pectoral fin structure 101
Pedionomidae 210, 218
peirosaurid 236, 237

Pelecaniformes 210
pelicans 208, 210
Pelodryadidae 45, 166
pelomedusids 239
Peltephilus 248
pelvic fin structure 101
penguins 206, 210, 240-241
Little Penguin 206
peninsular effect 119
Penola, SA 54
Pentland, Joseph Barclay 49, 50
Perameles gunnii 216, 217
Peramelidae 226
Peramelomorphia 211, 214, 226
Percichthys 240
perciform 166, 168, 169
Perikoala palankarinnica 176
Periphragnis 244
Perissodactyla 211, 239
Peroryctidae 226
Persian Gulf 31, 41
Peru 152, 202, 236
petalichthyids 72, 76, 94, 96
Petaurus norfolcensis 216, 217
Petauridae 156, 226
Petauroidea 226
petrels 210
Petromyzontiformes 67
petrosals 204
Petyrogale 174
petroglyph 49
Phalangerida 226
Phalangeridae 156, 160, 226
Phanerozoic Eon 27, 34
Phascolarctidae 226
Phascolarctos cinereus 184, 212, 213
Phascolonus 175, 194
P. gigas 195
Philippine Plate 23
Philosophical Society of Australasia 51
phlyctaenaspid level 74
Phlyctaeniidae 96
Phoberomys burmeisteri 176, 177
Phocidae 202
Phoeniconotius 208
Phoenicopteridae 210
Phoenicopterus 175
P. novaehollandiae 208
Pholidota 211
phororhacoid 240-241
phorusrhacoids 210
phosphate 38
phosphate-enriched bottom waters 34
phosphorus 37
phyllolepid 78
Phyllolepidae 92, 94, 96
Phyllolepis 94, 104
Phyllopteroides dentata 134, 136, 137
Phytosauridae 147
Piciformes 210
pico-passerines 210
pigeons 43, 45, 210, 218
pigs 211
Pikaia 37
pikas 211
Pilbara region, WA 30
Pilkipildridae 156, 226
Pilosisporites notensis 130
Pinpa, SA 208
pipid 238

placental mammals 43, 152, 214, 242-243
Placodermi 38, 66, 70-72, 76, 80, 86, 87-91, 92, 94, 95,104, 228
placoderm interrelationships 96
Placolepis 78
Plagiobatratrachus australis 115
Plagiosauridae 115
Plaisiodon 170
plankton 202
plankton feeders 232, 233
plaster jacketing technique 194
Plataleidae 210
plate tectonics 30, 31, 32
converging plate boundaries 23
diverging plate boundaries 23
platelet tectonics 31
Platypterygius australis 138
platypuses 45, 58,152, 153, 156, 211, 214, 216, 218
playa lakes 58, 156
plesiosauroids 124
plesiosaurs 124, 132, 138, 139, 142, 143
pleuracanths 108
pleurocentra 100, 102, 115
Pleurodira 147
pliosaur 124, 138
Podargus strigoides 214, 215
Podicepididae 210
Podocarpaceae 151
Pogona vitticeps 218, 219
polar wander paths 21, 22, 24
polar wandering 21
polydolopids 211, 240, 241
Polydolops thomasi 240, 241
polytypic fauna 218
porcupines 239
Porolepiformes 97
Porophoraspis 65, 67
Portal Mountain (Antarctica) 78
possums 45, 176, 214
Brushtailed Possum 156, 174
Leadbeater's Possum 156
Pygmy 4
Striped Possum 156
Potassium-Argon method 21, 27
Potoroidae 156, 186, 209, 226
Potoroo, Long-footed 216, 217
Potorus longipes 216, 217
Pratt-Winter, Samuel 54
Precambrian 31
premaxilla 81, 100
preopercular 80, 100
Prestosuchus 230
Primates 211
Prionotemnus 175
P. greyi 224
Priscaleo 156
pristichampsines 175
Proboscidea 211
procaryotes 31,32, 34
procaryotic cell 34
Procellariiformes 210
Procolophonia 111, 116, 118,147
Procoptodon 192, 193
P. goliah 178, 188
Procynosuchidae 119

procyonids 239
Prolacerta 116, 117
prolacertiforms 116
Prolocophon 116, 117, 120
Promecosomina 108
Propalorchestes 170
P. novaculacephalus 164
Propleopus 45, 182, 197, 216
Prosainognathidae 119
prosauropod 124, 230, 232
prospecting techniques 59
Prosqualodon davidis 202, 204
proterochampsids 228
proterosuchian 111
Proterotherium mixtum 246
Proterozoic Eon 27, 30, 31
Protoceratops 234
Protopterus 82
psammosteids 94
Pseudoborhyaena macrodonta 242, 243
Pseudocheiridae 156, 226
Pseudocheirus 184, 185
pseudomyine rodent 192
Pseudoneoreomys mesorhynchus 236
Psilopteridae 58
Psilopterus colzecus 240, 241
Psittaciformes 210
psychrosphere 44, 202, 206
Pteraspidomorphi 65, 66
Pterodaustro 232, 233, 236
Pterodroma solandri 180
Pteropidae 226
Pterosauria 140, 143, 147, 232, 233, 236
ptyctodontids, 76, 88
Ptyctodontomorphia 96
Pucadelphys andinus 152, 153
Puesto Viejo Formation (Argentina) 232, 233
pulp cavity 38, 37
Purtia 158, 159
pygmy elephants 224
Pygmy Possum 4, 174, 216-217
pygopodids 218
Pygopus lepidopodus 218, 219
Pyramios 170
P. alcootense 172
pythonids 166

quadrate 100, 118
quadratojugal 100
quail, Button 210
Stubble 218
Quanbun, WA 208
quartz, shocked 41
Quebrada De Los Jachaleros (Argentina) 232, 233
Queen Victoria Museum 52
Queenscliff, Vic 202
Queensland Museum 52, 55, 57, 238
Quinkan Gallery (Cape York Peninsula, Qld) 49
Quinkana fortirostrum 166, 197
Quipollornis koniberi 160

rabbits 211, 214, 224
Rackham's Roost, Qld 208
racoons 239
radioactive decay 21, 27
radiometric dates 174, 175
radius 101
Raemeotherium yatkolai 158, 159

Raglass, F.B. 54
rainfall, 221, Chapters 5-6
rainforests 174, 207, 222, 223
temperate 220, 221
tropical 163
Rajmahal Dolerite 18
Rajmahal Series 126
Rallidae 210
Rangal Coal Measures, Qld 106, 107
Ranken, George 49, 50, 57
Raptor ornitholestoides 141
rat kangaroo 55
ratites 210
rats 214
Broad-toothed 182
Rattus 175
R. rattus 214
ray-finned fishes 86, 97
rays 85, 204-205
Razorback Beds 124
red beds 31, 32
Red Cliff Mountain 88
Red Kangaroo 4, 186
Red Sea 22
Redbank Plains 160, 161, 208
Redcliff Mountain, NSW 118
reefs 24
reef feeders 88
tropical 42
refuge 45, 48
relative dating 27
relicts 124, 146, 214
Remigolepis 74, 92, 94, 96, 104
reproduction 101
reptile interrelationships 147
restoration 199
Rewan, Qld 40, 110, 111, 116-118
Rewan "Formation" 18
Rewan Group 108, 112
Rewana 112
Rhabdosteidae 208
rheas 210
Rheinische Schiefergeberge (Germany) 94
Rhineland (Germany) 94
Rhinolophidae 162, 226
rhipidistians 40, 101, 102, 104
Rhizodontia 88, 97
Rhoetosaurus, 126
R. brownei, 124, 125
Rhondda Colliery, Qld 124
Rhynchosaur-Diademodontoid Empire 119
Rhynchosauria 147
Rhytiodosteidae 112, 115
Rich, P.V. 56
Ride, W. D. L. 56, 57, 87
Rio Bonito Beds 17, 18
Rio Casa Grande (Argentina) 239
Rio Colorado Formation 140, 238
Rio Cosquin (Argentina) 239
Rio De Janerio (Brazil) 224
Rio Negro Province (Argentina) 240
Riojasuchus tenuiceps 232, 233
Ritchie, Alex 56, 57
River Red Gum 212, 213
Riversleigh, Qld 57, 59, 158, 162, 163, 165, 166, 208
Riversleigh Station 156, 157, 162
rock wallabies 174
Rocky Creek Tillite 18

Rocky Mountains (USA) 42, 58

Rodentia 43, 175, 176, 192, 211, 214

Roma, Qld 124

Royal School Of Mines (London) 54

Royal Society (London) 54

Royal Society Of Victoria 53

rudimentary tooth-plated lungfish 97

rudists 41

Russia 112, 124, 230

Sacabambaspis 36, 65, 66
    S. janvieri, 65

Sagenodus 83

Sakmarian 105

salamanders 40, 131, 218

Saline Series 18

salt deposits 24, 40, 41, 70

Salta Province (Argentina) 238, 243

saltatorial 216

saltbush 218

Saltenia ibanezi 238

Salto (Argentina) 250, 251

Saltwater Crocodile 112

Samfrau Geosyncline 26

San Rafael (Argentina) 232, 233

sand ridge deserts 179

Sandringham Sand 170, 171, 176, 177, 204, 205

Sangzhi (China) 118

Santa Cruz (Argentina) 230
    Santa Cruz Province 230, 232, 234, 235, 250, 251

Santa Maria Formation 18

Santana Formation 140, 230, 231

Sao Bento Dolerite 18

Sao Paulo (Brazil) 229

Saqenodus 83

Sarcophilus 182, 192, 193

Saurichthys 108

Saurischia 147

Sauropodomorpha 141, 143, 147, 234, 236

Sauropterygia 147

savannah 43

Scaldicetus macgeei 202, 203

Scandentia 211

scapulocoracoid 101, 140, 172

scleractinian corals 42

sclerotic capsule 76,78

Scotia Plate (West Antarctica) 23

Screw Palm 222, 223

scrub-birds 218

sea cows 211

sea floor spreading 22, 150

sea levels 41, 179, 196

sea lion, Australian 202

sea pens 34

sea squirts 37

seabirds 210

Seaham Formation 17

Seaham Tillite 18

seals 41, 202, 211

sebecosuchian 175

Sedgwick, Adam 54

seed-bearing plants 105

seed-ferns, 26, 105 119

segmented worms 34

Sehuen (Argentina) 248

seismology 22

Selwyn, A.R.C. 51, 54, 55

semi-circular canals 143

septomaxilla 100

seriamas 210

Serpentes 147

Seton Rock Shelter 196

sexuality 33

Seymour Island 206, 240, 241

Shan-Thai (Asia) 94

Shark Bay, WA 31, 33

sharks 66, 70-72, 76, 85, 93, 94, 96, 204, 206
    ctenacanth 93
    modern 96
    shark interrelationships 96
    White Pointers 204, 205
    xenacanths 93

sheep 214

Sherbonaspis 96

Shearsbyaspis 76

shocked quartz 41

Short-nosed Bandicoot 192

shrews 224
    tree 211

Siberia 24, 76, 94

Siberian Province 94

Siderops, 126
    S. kehli 111, 124, 125

Silcrete Flora 151

siluriforms 166

Silverdale Formation 71

Simosthenurus 188

Simpson Desert 179

Simpson, George Gaylord 45

Sinjiang Province (China) 116, 117

Sinolepididae 96

sinolepidoids 74

siphlonurid mayflies 126

Sirenia 211

"*Skamolepis*" 85

skeletons, 34

skin impression 234-236, 240, 251

skinks 166, 218

skuas 210

skull structure 100

skunk 246

Slippery Rock Locality, Vic 144

sloths 211, 251
    ground sloths 239
    sloth hair 251

Smeaton, Vic 208

Smilodon 216

Smithsonian Institution (USA) 57

Smithton, Tas 180, 181, 197

snakes 147, 166
    Brown 218, 219

solar radiation 33

Solomon Islands island arc 23

Somalian Subplate 23

Somersby, NSW 108, 109

songbirds 210, Chapters 5-6

South Africa 112, 116, 118, 239

South America 94, 111, 116, 118, 154, 166, 175, 176-177, 228, 234, 238, 238

South Australian Museum 52, 54

South China 68, 94
    South China Province 94

South Durras, NSW 17

South Tasman Rise 150

South Victorialand (Antarc-

tica) 88

Southern Rift Valley 130, 131

Soviet Union 234

Spain 94

Sparssodontia 211

species diversity 213

Speonesydrion 83, 94

Sphenacodontidae 119

Spheniscidae 210

Sphenodon 131, 147

sphenopsids 103, 132-133

Sphenopteris warragulensis 134, 136, 137

Speonesydrion 94

spiders 126

"spiny-sharks" 66, 71

Spinifex Hopping Mouse 216, 217

Spitzbergen 36, 64, 76, 116, 126

spores 130

Springsure, Qld 118

Squalodon gambierense 204

Squalodontidae 202

Squamata 147

Squirrel Glider 216, 217

St Peters Quarry, NSW 108, 112

Stairway Sandstone Formation 62-65

stapes 40

Stegosaurus 230

Stereosternum tumidum 229

Steropodon galmani 152-153, 154

Stethacanthus 106

Sthenurus 182, 186, 188, 190, 194
    S. atlas 186, 187
    S. gilli 190, 191

STHP (subtropical high pressure belt) 220

Stigocuscus 175

Stirling, E.C. 54, 55

Stirton, Ruben Arthur 53, 56, 57, 59, 174

storks 210

Stormberg Series 18

Straits Of Gibralter 43

stream channel deposits 130

stream transport of bones 59

streptostyly 147

striated surfaces 17, 33

Strigiformes 210

Striped Possum 156

Struthionidae 210

Strzelecki Group 27, 111, 146
    Strzelecki Ranges, Vic 130

Sturt Pea 222, 223

Stutchbury, Samuel 50

subduction zones 22, 23, 30

subholostean 108

Subsyclotosaurus brookvalensis 108

subtropical high pressure belt 220

Suess, Eduard 16

Sulphur-crested Cockatoos 4

Sunlands, SA 208
    Sunlands Pumping Station 204, 205

Sunset Country 222-223

survival of the fittest 207

Swamp Gum 212, 213

swans 210

swifts 210

Sydney Gazette 49

symmetrodonts 152

Synapsida 119,147, 229

synarcual 72

synyardiids 156

Table Mountain Sandstone 18

tabulate corals 70

Tachyglossidae 226

Tachyglossus 214
    T. aculeatus 216

Tadorna tadornoides 194

Taemas, NSW 76
    Taemas Bridge 78
    Taemas-Wee Jasper fish fauna 82
    Taemas-Wee Jasper, 72, 78, 83, 118

Taeniopteris daintreei 136, 137

Taggerty, Vic 118

Talbragar, NSW 126
    Talbragar Fish Beds 4, 126

Talchir Tillite 17, 18

Talyawalka, SA 208

Tambar Springs, NSW 182, 183, 195

Tambo Series 141

Tanzania 230

taphonomy 58, 59, 126, 132

tapirs 211, 239

Tara Creek, Qld 208

Taralga, NSW 55

Tarkarooloo, SA 208
    Tarkarooloo Basin, 156
    Tarkarooloo Subbasin, 156, 157

Taroona, Tas 155
    Taroona High School, 155

Tarsipedidae 226

Tarsipedoidea 226

tarsometatarsus 172, 184-185

Tasman Sea 26, 150

Tasmanian Devil 182

Tasmanian devils 192, 193

Tasmanian dolerites 18

Tasmanian Museum 52, 224

Tasmanian Society 54

Tasmanian Tiger 45, 186, 187, 197, 224, 225

Tasmaniosaurus 111
    T. triassicus 116

Tate, Ralph 51, 53, 55

Tatong, Vic 86, 118

Tawny Frogmouth 214, 215

Tedford, Richard H. 56, 155, 179

teeth, labyrinthine 101

teleosts 41, 80, 81, 97, 144, 166, 186-187, 204

Telestomi 95

Telemon Station 138

temnospondyls 102, 111, 115

temperate rainforest 220, 221

temperatures, mean annual 132

Tendaguru (Tanzania) 230

Tenison-Woods, Julian 54

tensional faulting 22

tensional features 22

teratornithids 210

termite nests 222, 223

terns 210

Terra Australis collections 48

terrestrial carnivore niche 242, 243

    terrestrial planets 30

terrestrial skeleton 100

tethering (to waterholes) 182, 221

Tethys Sea 42, 43

tetracorals 70

Tetragraptus (Pendeograptus) fruticosus 66
    T. approximatus 66

tetrapod trackway 98, 99

Tetrapoda, 97, 99, 100

Texas (USA) 41

Thailand 93

"*Tharrias*" *ferugloi* 230, 231

Tharp, Marie 21

The Brisbane Courier 55

thecodonts 66, 67, 71, 72, 76, 93, 111, 119, 147, 227, 228, 232, 233

thelodonts 67

Theory of Organic Evolution By Natural Selection 45

Theory of Plate Tectonics 13, 18, 22, 218

Therapsida 119

Theriodontia 119

Therocephalia 119

Theropoda 141, 147, 210, 234, 235

thick-knees 166, 210

"Thickthorn" 55

"*Thingodon*" 162

Thompson River, Qld 141

Thompson, Alexander M. 53

Thrinacodus (Harpagodens) ferox 86, 93

Thrinaxodon 20, 119

Thrinaxodontida 119

Thulborn, Anthony 56, 119

Thylacinidae 226

Thylacinus, 45, 197, 216
    T. cynocephalus, 186, 187, 192, 224, 225

Thylacoleo 55, 182, 197, 216
    T. carnifex 188, 189, 192, 193, 197

Thylacoleonidae 156, 164, 226

Thylacomyidae 175, 218, 226

Thylacosmilus lentis 244

Thylogale 174

Tibet 26

tibia 101

tibiotarsus 172, 180, 196

Ticinosuchus 230

tillites 17, 26, 33

Timor 224

Tinamidae 210

Tinderbox Bay, Tas 111

Tingamarra, Qld 155, 208
    Tingamarra Fauna, 155

Titanosauridae 234, 236, 238

Tiupampa Local Fauna 152, 153

Tonga Trench 23

Toolache Wallaby 224

Toolebuc Formation 138, 140, 142, 145

Toomba Range, Qld 118

Toombs, Harry 87

tooth plates 97

toothed whales 202

Torquay, Vic 202, 208

Torresian biogeographic region 221

tortoise 166-167

toucans 210

Town Well, SA 208

Toxodon 250, 251

toxodont 250,251
trackways 99, 102, 104, 144, 234, 235
Transantarctic Mountains 18, 23, 114-117
transform faults 22, 23, 24
traversodonts 232, 233
tree frogs 166
tree kangaroos 174
tree shrews 211
tree sloths 152
tree-ferns 218
Trematosaurid-Rhytiodosteid Empire 119
Trematosauridae 111, 112, 115
trenches 22, 23, 24
Tribrachidium heraldicum 34
Triceratops 146, 234
Trichosurus 156, 174, 175
Triconodonta 152, 211
Trigonia 45, 48, 214
trilobites 27, 36, 65, 70
trionychids 218
Tritylodontoidea 119
trogons 210
tropic birds 210
tropical rainforest 163
tropical reefs 42
Troposodon kenti 176, 177
true tooth-plated lungfish 97
Tuatara 131, 147
Tubulidentata 211
Tumbunan biogeographic region 221
Tumut Trough 92
Tupe Tillite 17, 18
Turanoceratops 234
Turinia 93
    T. australiensis 76
Turkey 26
Turnbull, W. 56
Turner, S. 56, 57
Turnicidae 210
turtles 132, 134, 135, 147, 166, 167-169, 179, 218, 239
Twenty Mile Quarry, Vic 103
tympanum 40
typhlopids 166
Typotherium pseudopachygnathus 248

Tyrannosaurus 197
Uabryichthys 126
ulnare 101
Ultima Esperanza Cave (Chile) 251
ultraviolet radiation 32
Umberatana Group 33
underwater sieving 59
Ungulata 211,248, 250, 251
University Of California at Berkeley 57
University Of California at Riverside 57
University Of Edinburgh 49
University Of Glasgow 53
University Of Melbourne 52, 53
University Of Sydney 51, 53
University Of Tasmania 57
University Of Texas 57
Uranium 238 27
Uranolophus 97
Urey, Harold 32
Urodela 40
Urosthenes 105
Utting Calcarenite 106
Valle Hermoso (Argentina) 242, 243
van Tets, G.F. 56
varanids 45, 166, 174, 216
Varanus komodensis 199
    V. varius 218, 219
varves 33
Venus 30
vertebral construction 100
vertebrate bone 37
Vertebrate Evolution, Palaeontology And Systematics, Conference On 56
    vertebrate localities (Cainozoic) 208
    vertebrate origins 64
    vertebrates 34, 36
Vespertilionidae 226
vibratools 50
Victoria Cave, SA 186, 188-193
Victorialand (Antarctica) 93
Victorian Chamber of Mines 53

Victorian Colonial Survey 51
Victorian Geological Survey 54
"Viking Funeral Ship" 43
Vincelestes neuquenianus 152, 153
Vinctifer comptoni 230, 231
Vine, Fred 21
Vivero Member 240, 241
volcanic activity 22
volcanic glass 27
volcanic islands 22
volcanoes 41
volcanogenic sediments 132
Vombatidae 226
Vombatiformes 156, 174, 226
Vombatomorphia 226
Vombatus 175, 182, 188, 189
    V. ursinus 44
von Buch, Leopold 48
vulcanism 22
vulturids 210
Wadeichthys 126
    W. oxyops 128, 129
waders 210
Wadikali, SA 208
Waikerie, SA 204, 205
Waite Formation 170, 172, 173
Wakaleo 156
    W. oldfieldi 158, 159
    W. vanderleueri 164
Walgettosuchus woodwardi 141
wallaby, Toolache 224
Wallace, A.R. 45
    Wallace's Line (Indonesia) 45
    Wallacea 45
Walleramany, NSW 26
Walloon Coal Measures, Qld 124
Walpole Island (South Pacific) 168, 169
Warburton River, SA 179, 199
warm-bloodedness 131
Warra Station, NSW 140
Warrawoona, WA 31
Warren, James 57
Warrumbungle Mountains, NSW 160, 208
Waterford, Qld 198, 199

waterholes 221
    waterhole tethering 221
Watson, D.M.S. 112
Wau (New Guinea) 57
Waurn Ponds, Vic 208
Weber's Line (Indonesia) 45
Wee Jasper, NSW 71-73, 78, 95
Weejasperaspis 76, 96
Wegener, Alfred 13
well core 155
Wellington, SA 186, 187
Wellington Valley, NSW 54, 57
    Wellington caves 50, 186
    Wellington Valley caves 49
Wells, Rod 56, 179, 190, 191
West Antarctica 206, 240, 241
West Irian 45
Western Australian Museum 52, 57, 87
Western District of Victoria 102
Westralichthyes 74
wet sclerophyll forests 4
whales 41, 202, 204, 211
    baleen 202
White Cliffs, NSW 138
White Pointers 204, 205
Wianamatta Group 108, 112
    Wianamatta Series 112
Wijdeaspis 76, 94, 96
Wilcannia, NSW 94
wildfire 43
Williamsaspidae 96
Williamsaspis 74
windblown soils 30
Winteraceae 151
Winton, Qld 143, 146
    Winton Fauna 143
    Winton Formation 143, 146
Wipijiri Formation 160, 161, 164
Witteberg Series 18
wolves 45
"Wombaroo" 158, 159
wombats 156, 182, 195, 211
    Common Wombat 44
Wonambi 218
    W. naracoortensis 190
Woodburne, M.O. 56

woodland community 163
woodpeckers 210
woodswallows 218
worms, acorn 37
Wuttagoona Station, NSW 86,118
Wuttagoonaspid-Phyllolepid Province, 94
Wuttagoonaspidae, 94, 96
Wuttagoonaspis 74
    W. fletcheri 94
Wynyard, Tas 148, 149, 156, 158, 202, 203, 208
Wynyardia 156
    W. bassiana 148, 149, 158, 202
Wynyardiidae 158, 226
xenacanth shark 93, 106, 108
Xenacanthiformes 86, 96
Xenarthra 211
Xenobrachyops 110, 111
Xenorhynchopsis 175
Xinjiang Province, China 116
Yalkaparidon coheni 162, 162
Yalkaparidontidae 226
Yalwal-Conerong Rift 104
Yanda, NSW 94
Yarra Track, Vic 103
Yea, Vic 103
Yingabalanaridae 226
Young, G. 56
Yunnanolepidae 94,96
yunnanolepidoids 74
Zaglossus 214
    Z. ramsayi 61, 63
    Z. robusta 190
Zebra Stripes 21, 22, 24
Zietz, A.H.C. 54, 55
ziphodont 166, 197, 214, 236, 237
zircon crystals 27, 30
zygomaturine 207
Zygomaturus 158, 159, 174-176, 182, 197
    Z. gilli 176, 177
    Z. tasmanicus 197
    Z. trilobus 184, 197

PATRICIA VICKERS-RICH holds a Chair in Palaeontology at Monash University, where she lectures in the Earth Sciences Department. She holds a Ph.D. from Columbia University for work on the origin and evolution of the Australasian avifauna. Her main research interests are in Late Mesozoic vertebrates from Gondwana. She and her students are currently investigating the past geographic and environmental changes that have shaped the modern Australasian fauna and flora. Vickers-Rich is interested in making frontline science research accessible to young students, and she is currently Director of the Monash Science Centre, a science communication centre which she founded at Monash University.

THOMAS HEWITT RICH is Curator of Vertebrate Palaeontology at Museum Victoria (formerly the Museum of Victoria) in Melbourne. He received his Ph.D. from Columbia University for work on fossil hedgehogs of the Northern Hemisphere. His main research interests lie in Mesozoic faunas, especially dinosaurs and primitive mammals. His main efforts over the past ten years have been in elucidating the Early Cretaceous vertebrate faunas of Austral-ia's Southeast and of Patagonia in Argentina. He is also interested in the philosophy of science and mathematical modelling in palaeontology.

FRANCESCO COFFA is a freelance photographer based in Melbourne. Born in Italy, he came to Australia at the age of ten, and later graduated from the Royal Melbourne Institute of Technology. He began his photographic career in 1962 and

has illustrated and contributed to many books and scientific papers. He has also produced videos, short films, and audio-visuals on diverse subjects such as dinosaurs, Egyptology, Aboriginal culture, and the environment. Until 1997, Coffa was the manager of photographic and audio-visual services at the Museum of Victoria.

STEVEN MORTON is the head of the scientific photographic services for the Faculty of Science, Monash University. He is a graduate of the Photographic Department at the Royal Melbourne Institute of Technology and holds a Master's of Applied Science in Photography. Morton developed a 360° panoramic camera of unique design. One of his first panoramas taken with this camera appears in this volume along with many other of his photographs. His lab at Monash has now migrated into computer graphics that enhance his conventional photography; some of those new applications have been used in this revised edition of *Wildlife of Gondwana*.

PETER TRUSLER is a Melbourne artist who has received international acclaim for his natural history paintings following the publication of *Birds of Australian Gardens* and his spectacular 1993 stamp series on Australian dinosaurs for Australia Post. He is a science graduate of Monash University and has contributed illustrations to many scientific and popular publications. The blending of his biological and artistic interests has proven particularly successful in presenting informative reconstructions of the past, and he is working currently with Tom and Pat Rich on a number of prehistoric projects.